MITCH SHORT

Fundamentals of Engineering Mechanics

FUNDAMENTALS OF ENGINEERING MECHANICS

S S BHAVIKATTI
Emeritus Fellow
Ex-Professor and Dean
NITK Surathkal, Karnataka, India

New Academic Science Limited
The Control Centre, 11 A Little Mount Sion
Tunbridge Wells, Kent TN1 1YS, UK
www.newacademicscience.co.uk
e-mail: info@newacademicscience.co.uk

Copyright © 2011 by New Academic Science Limited
The Control Centre, 11 A Little Mount Sion, Tunbridge Wells, Kent TN1 1YS, UK
www.newacademicscience.co.uk • email: info@newacademicscience.co.uk
Tel: +44(0) 1892 55 7767, Fax: +44(0) 1892 53 0358

ISBN : 978 1 906574 80 2

All rights reserved. No part of this book may be reproduced in any form, by photostat, microfilm, xerography, or any other means, or incorporated into any information retrieval system, electronic or mechanical, without the written permission of the copyright owner.

British Library Cataloguing in Publication Data
A Catalogue record for this book is available from the British Library

Every effort has been made to make the book error free. However, the author and publisher have no warranty of any kind, expressed or implied, with regard to the documentation contained in this book.

Printed and bound in India by Replika Press Pvt. Ltd.

Preface

Engineering Mechanics is a course taught in the very first year of under graduate curriculum and in this the student develops capability to model an actual problem into an engineering problem and finds the solution using laws of mechanics. An engineering mechanics problem involving forces, is solved to a great extent, if the free body diagram is drawn correctly. Hence throughout the text, authors have given emphasis on drawing neat free body diagrams.

Experience of the authors is that a number of students fail to take note of many important points and commit mistakes. Hence in this book important points are emphasised.

Finding centre of gravity and moment of inertia by tabular method, takes more time and in this process concepts of the terms used is lost. The student studying this method, find it difficult to cope up with the subjects like strength of materials, structural analysis, structural design, machine design, which are taught in higher classes. Hence in this book emphasis is on the method which gives proper concept of the method and is faster also.

Clearcut distinction has been brought out between resultant motion and relative motion. Various methods of solving kinematic problems are clearly separated by devoting different chapters and in each of them, the situations in which they are advantageous is pointed out.

In part III, a chapter on Bending moment and Shear Force Diagrams is added to complete the curriculum requirement of many universities.

With their emphasis on formulating practical problem into an engineering problem and applying equations of mechanics systematically, the authors have tried to prepare students to solve any problem in Engineering Mechanics.

S.S. BHAVIKATTI

Contents

Preface *(v)*

PART I

1. INTRODUCTION TO ENGINEERING MECHANICS — 1

- 1.1. Classification of Engineering Mechanics 1
- 1.2. Basic Terminologies in Mechanics 2
- 1.3. Laws of Mechanics 4
- 1.4. Derived Laws 7
- 1.5. Units 8
- 1.6. Characteristics of a Force 10
- 1.7. System of Forces 11
- 1.8. Vectors 12
- 1.9. Dimensional Homogeneity 12
- 1.10. Idealisations in Mechanics 14
- 1.11. Accuracy of Calculations 14
- 1.12. Approaches to Solution 15
  - *Important Definitions and Concepts* 15

2. RESULTANT AND EQUILIBRIUM OF SYSTEM OF COPLANAR CONCURRENT FORCES — 17

- 2.1. Composition of Two Force System 17
- 2.2. Resolution of Forces 19
- 2.3. General Method of Composition of Forces 21
- 2.4. Equilibrium of Bodies 25
- 2.5. Equilibrium of Connected Bodies 33
  - *Important Definitions and Formulae* 41
  - *Problems for Exercise* 42

3. RESULTANT AND EQUILIBRIUM OF SYSTEM OF COPLANAR NON-CONCURRENT FORCES — 44

- 3.1. Moment of a Force 44
- 3.2. Varignon's Theorem 45
- 3.3. Couple 47
- 3.4. Resolution of a Force into a Force and a Couple 48

3.5.	Resultant of Non-Concurrent Force System	49
3.6.	*x* and *y* Intercepts of Resultant	50
3.7.	Equilibrium of Non-Concurrent System of Forces	56
3.8.	Applications to Beam Problems	64
	Important Definitions and Formulae	78
	Problems for Exercise	79

4. ANALYSIS OF PIN-JOINTED PLANE FRAMES — 83

4.1.	Perfect, Deficient and Redundant Frames	83
4.2.	Assumptions	85
4.3.	Nature of Forces in Members	85
4.4.	Methods of Analysis	86
4.5.	Method of Joint	86
4.6.	Method of Section	97
4.7.	Method of Tension Coefficient	103
	Important Definitions and Formulae	111
	Problems for Exercise	111

5. FRICTION — 117

5.1.	Frictional Force	117
5.2.	Laws of Friction	118
5.3.	Angle of Friction, Angle of Repose and Cone of Friction	118
5.4.	Wedges	129
5.5.	Problems Involving Non-Concurrent Force Systems	132
5.6.	Rope Friction	135
	Important Definitions and Formulae	139
	Problems for Exercise	140

6. LIFTING MACHINES — 144

6.1.	Definitions	144
6.2.	Practical Machines	145
6.3.	Law of Machine	147
6.4.	Variation of Mechanical Advantage	147
6.5.	Variation of Efficiency	148
6.6.	Reversibility of a Machine	151
6.7.	Pulleys	152
6.8.	Wheel and Axle	155
6.9.	Wheel and Differential Axle	156
6.10.	Weston Differential Pulley Block	157
6.11.	Inclined Plane	158
6.12.	Screw Jack	163
6.13.	Differential Screw Jack	166
6.14.	Winch Crabs	167
	Important Definitions and Formulae	171
	Problems for Exercise	172

7. TRANSMISSION OF POWER — 175

- 7.1. Belt Drive 175
- 7.2. Transmission of Power in Belt Drive 180
- 7.3. Centrifugal Tension 184
- 7.4 Rope Drive 187
- 7.5 Gear Drive 189
  - *Important Formulae* 196
  - *Theory Questions* 197
  - *Problems for Exercise* 197

8. VIRTUAL WORK METHOD — 199

- 8.1. Work done by forces and moments 199
- 8.2. The Method 201
- 8.3. Sign Convention 201
  - *Important Definitions and Equations* 223
  - *Problems for Exercise* 223

9. CENTROID AND MOMENT OF INERTIA — 226

- 9.1. Centre of Gravity 226
- 9.2. Centre of Gravity of a Flat Plate 227
- 9.3. Centroid 227
- 9.4. Difference between Centre of Gravity and Centroid 228
- 9.5. Use of Axis of Symmetry 228
- 9.6. Determination of Centroid of Simple Figures from First Principle 229
- 9.7. Centroid of Composite Sections 233
- 9.8. Moment of Inertia 240
- 9.9. Polar Moment of Inertia 241
- 9.10. Radius of Gyration 241
- 9.11. Theorems of Moment of Inertia 242
- 9.12. Moment of Inertia from First Principles 243
- 9.13. Moment of Inertia of Standard Sections 245
- 9.14. Moment of Inertia of Composite Sections 249
  - *Important Definitions and Formulae* 260
  - *Problems for Exercise* 261

10. CENTRE OF GRAVITY AND MASS MOMENT OF INERTIA — 266

- 10.1. Centre of Gravity 266
- 10.2. Use of Symmetry 267
- 10.3. Centre of Gravity from First Principles 267
- 10.4. Centre of Gravity of Composite Bodies 270
- 10.5. Theorems of Pappus-Guldinus 273
- 10.6. Mass Moment of Inertia 276
- 10.7. Radius of Gyration 277
- 10.8. Determination of Mass Moment of Inertia from First Principles 277

10.9. Parallel Axis Theorem/Transfer Formula 285
10.10. Moment of Inertia of Composite Bodies 286
 Important Definitions and Formulae 290
 Problems for Exercise 291

PART II

11. INTRODUCTION TO DYNAMICS   295

11.1. Basic Terms 295
11.2. General Principles in Dynamics 296
11.3. Types of Motion 298
11.4. Outline of the Section 299
 Important Definitions and Formulae 299

12. LINEAR MOTION   300

12.1. Motion Curves 300
12.2. Motion with Uniform Velocity 301
12.3. Motion with Uniform Acceleration 301
12.4. Acceleration Due to Gravity 304
12.5. Motion with Varying Acceleration 315
 Important Definitions and Formulae 321
 Problems for Exercise 321

13. PROJECTILES   323

13.1. Definitions 323
13.2. Motion of Body Projected Horizontally 323
13.3. Inclined Projection on Level Ground 326
13.4. Inclined Projection with Point of Projection and Point of
  Strike at Different Levels 331
13.5. Projection on Inclined Plane 340
 Important Definitions and Formulae 343
 Problems for Exercise 345

14. RELATIVE VELOCITY   347

14.1. Motion on Parallel Paths in Like Directions 347
14.2. Motion on Parallel Path in Opposite Direction 347
14.3. Motion in a Plane in any Direction 347
14.4. Relative Distance 349
14.5. Relative Velocity and Resultant Velocity 349
 Important Definitions and Formulae 358
 Problems for Exercise 358

15. D' ALEMBERT'S PRINCIPLE — 360

- 15.1. Newton's Second Law of Motion — 360
- 15.2. D' Alembert's Principle — 360
  - *Important Definitions* — 374
  - *Problems for Exercise* — 374

16. WORK ENERGY METHOD — 376

- 16.1. Work — 376
- 16.2. Work Done by a Varying Force — 377
- 16.3. Energy — 377
- 16.4. Power — 378
- 16.5. Work Energy Equation for Translation — 379
- 16.6. Motion of Connected Bodies — 385
- 16.7. Work Done by a Spring — 388
  - *Important Definitions and Formulae* — 392
  - *Problems for Exercise* — 392

17. IMPULSE MOMENTUM — 395

- 17.1. Linear Impulse and Momentum — 395
- 17.2. Connected Bodies — 399
- 17.3. Force of Jet on a Vane — 404
- 17.4. Conservation of Momentum — 406
- 17.5. Pile and Pile Hammer — 410
  - *Important Definitions and Equations* — 414
  - *Problems for Exercise* — 414

18. IMPACT OF ELASTIC BODIES — 417

- 18.1. Definitions — 417
- 18.2. Coefficient of Restitution — 417
- 18.3. Direct Central Impact — 420
- 18.4. Oblique Impact — 424
- 18.5. Loss of Kinetic Energy — 430
  - *Important Definitions and Formulae* — 432
  - *Problems for Exercise* — 432

19. CIRCULAR MOTION OF RIGID BODIES — 434

- 19.1. Acceleration During Circular Motion — 434
- 19.2. Motion on Level Road — 435
- 19.3. Need for Banking of Roads and Super Elevation of Rails — 437
- 19.4. Designed Speed — 437
- 19.5. Skidding and Overturning on Banked Roads — 438
  - *Important Formulae* — 445
  - *Problems for Exercise* — 445

20. ROTATION OF RIGID BODIES — 447

- 20.1. Angular Motion 447
- 20.2. Relationship Between Angular Motion and Linear Motion 448
- 20.3. Uniform Angular Velocity 448
- 20.4. Uniformly Accelerated Rotation 448
- 20.5. Kinetics of Rigid Body Rotation 453
- 20.6. Kinetic Energy of Rotating Bodies 453
  - *Important Formulae* 459
  - *Problems for Exercise* 459

21. MECHANICAL VIBRATION — 461

- 21.1. Simple Harmonic Motion 461
- 21.2. Simple Harmonic Motion as a Sine Wave 465
- 21.3. Simple Pendulum 466
- 21.4. Compound Pendulum 467
  - *Important Definition and Formulae* 471
  - *Problems for Exercise* 472

22. GENERAL PLANE MOTION OF RIGID BODIES — 473

- 22.1. Meaning of General Plane Motion 473
- 22.2. Kinematics of General Plane Motion 474
- 22.3. Instantaneous Axis of Rotation 479
- 22.4. Kinetics of Rolling Bodies 480
- 22.5. Kinetics of General Plane Motion 484
  - *Important Formulae* 489
  - *Problems for Exercise* 489

INDEX — 491

PART—I

1
Introduction to Engineering Mechanics

The state of rest and state of motion of the bodies under the action of different forces has engaged the attention of philosophers, mathematicians and scientists for many centuries. The branch of physical science that deals with the state of rest or the state of motion is termed as **Mechanics.** Starting from the analysis of rigid bodies under gravitational force and simple applied forces the mechanics has grown to the analysis of robotics, aircrafts, spacecrafts under dynamic forces, atmospheric forces, temperature forces etc.

Archimedes (287–212 BC), Galileo (1564–1642), Sir Issac Newton (1642–1727) and Einstein (1878–1955) have contributed a lot to the development of mechanics. Contributions by Varignon, Euler, D. Alembert are also substantial. The mechanics developed by these researchers may be grouped as

(*i*) Classical mechanics/Newtonian mechanics

(*ii*) Relativistic mechanics

(*iii*) Quantum mechanics/Wave mechanics.

Sir Issac Newton, the principal architect of mechanics, consolidated the philosophy and experimental findings developed around the state of rest and state of motion of the bodies and put forth them in the form of three laws of motion as well as the law of gravitation. The mechanics based on these laws is called **Classical mechanics** or **Newtonian Mechanics.**

Albert Einstein proved that Newtonian mechanics fails to explain the behaviour of high speed (speed of light) bodies. He put forth the theory of **Relativistic Mechanics.**

Schrödinger (1887–1961) and Broglie (1892–1965) showed that Newtonian mechanics fails to explain the behaviour of particles when atomic distances are concerned. They put forth the theory of **Quantum Mechanics.**

Engineers are keen to use the laws of mechanics to actual field problems. Application of laws of mechanics to field problem is termed as **Engineering Mechanics.** For all the problems between atomic distances to high speed distances Classical/Newtonian mechanics has stood the test of time and hence that is the mechanics used by engineers. Hence in this text classical mechanics is used for the analysis of engineering problems.

1.1. CLASSIFICATION OF ENGINEERING MECHANICS

Depending upon the body to which the mechanics is applied, the engineering mechanics is classified as

(*a*) Mechanics of Solids, and

(*b*) Mechanics of Fluids.

The solid mechanics is further classified as mechanics of rigid bodies and mechanics of deformable bodies. The body which will not deform or the body in which deformation can be neglected in the analysis, are called as **Rigid Bodies**. The mechanics of the rigid bodies dealing with the bodies at rest is termed as **Statics** and that dealing with bodies in motion is called

Dynamics. The dynamics dealing with the problems without referring to the forces causing the motion of the body is termed as **Kinematics** and if it deals with the forces causing motion also, is called **Kinetics**.

If the internal stresses developed in a body are to be studied, the deformation of the body should be considered. This field of mechanics is called **Mechanics of Deformable Bodies/Strength of Materials/Solid Mechanics**. This field may be further divided into **Theory of Elasticity** and **Theory of Plasticity**.

Liquid and gases deform continuously with application of very small shear forces. Such materials are called **Fluids**. The mechanics dealing with behaviour of such materials is called **Fluid Mechanics**. Mechanics of ideal fluids, mechanics of viscous fluid and mechanics of incompressible fluids are further classification in this area. The classification of mechanics is summarised below in flow chart.

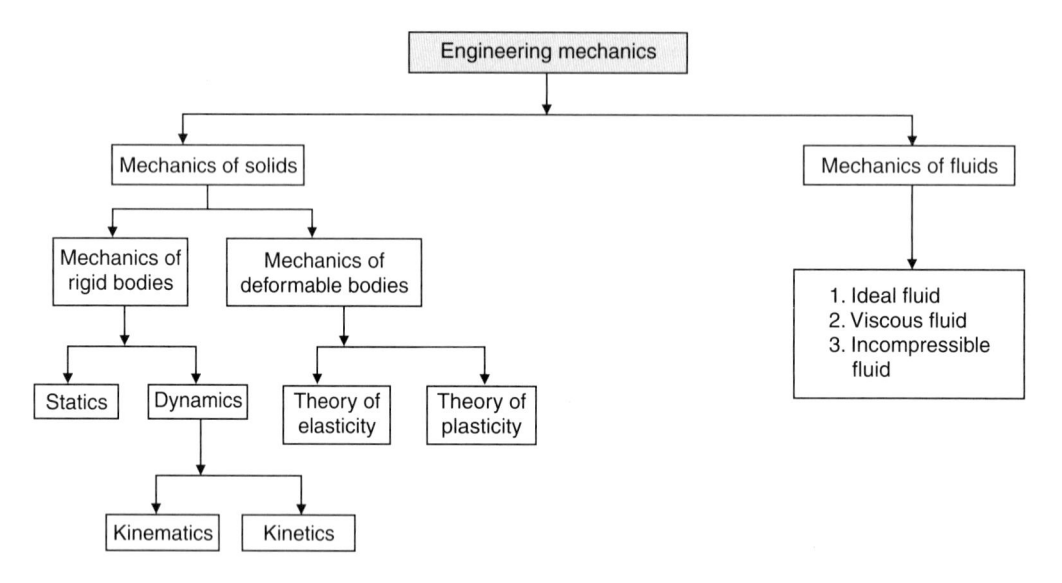

```
                    Engineering mechanics
                    /                    \
          Mechanics of solids      Mechanics of fluids
          /           \                    |
  Mechanics of    Mechanics of       1. Ideal fluid
  rigid bodies    deformable bodies  2. Viscous fluid
    /     \         /      \         3. Incompressible
 Statics Dynamics  Theory  Theory       fluid
            |     of       of
            |    elasticity plasticity
     Kinematics Kinetics
```

1.2. BASIC TERMINOLOGIES IN MECHANICS

The following are the terms basic to study mechanics, which should be understood clearly:

Mass

The quantity of the matter possessed by a body is called mass. The mass of a body will not change unless the body is damaged and part of it is physically separated. When a body is taken out in a spacecraft, the mass will not change but its weight may change due to change in gravitational force. Even the body may become weightless when gravitational force vanishes but the mass remains the same.

Time

Time is the measure of succession of events. The successive event selected is the rotation of earth about its own axis and this is called a day. To have convenient units for various activities, a day is divided into 24 hours, an hour into 60 minutes and a minute into 60 seconds. Clocks are the instruments developed to measure time. To overcome difficulties due to irregularities in the earth's rotation, the unit of time is taken as second which is defined as the duration of 9192631770 period of radiation of the cesium-133 atom.

INTRODUCTION TO ENGINEERING MECHANICS

Space

The geometric region in which study of body is involved is called **space**. A point in the space may be referred with respect to a predetermined point by a set of linear and angular measurements. The reference point is called the origin and set of measurements as '*coordinates*'. If coordinates involve only in mutually perpendicular directions they are known as *Cartesian coordinates*. If the coordinates involve angle and distances, it is termed as *polar coordinate system*.

Length

It is a concept to measure linear distances. The diameter of a cylinder may be 300 mm, the height of a building may be 15 m. Actually metre is the unit of length. However depending upon the sizes involved micro, milli or kilo metre units are used for measurement. A metre is defined as length of the standard bar of platinum-iridium kept at the International Bureau of Weights and Measures. To overcome difficulties of accessibility and reproduction, now meter is defined as 1690763.73 wavelength of krypton-86 atom.

Displacement

Displacement is defined as the distance moved by a body/particle in the specified direction. Referring to Fig. 1.1, if a body moves from position A to position B in the x-y plane shown, its displacement in x-direction is AB' and its displacement in y-direction is $B'B$.

Velocity

The rate of change of displacement with respect to time is defined as velocity.

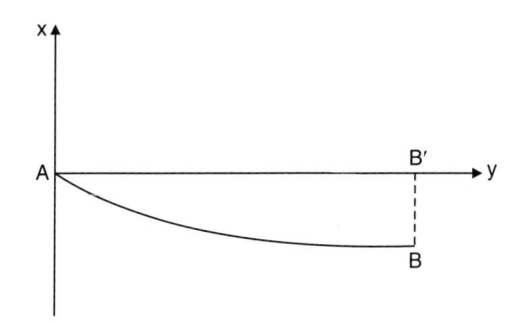

Fig. 1.1

Acceleration

Acceleration is the rate of change of velocity with respect to time. Thus

$$a = \frac{dv}{dt}, \text{ where } v \text{ is velocity} \qquad \ldots(1.1)$$

Momentum

The product of mass and velocity is called momentum. Thus

Momentum = Mass × Velocity ...(1.2)

Continuum

A body consists of several matters. It is a well known fact that each particle can be subdivided into molecules, atoms and electrons. It is not possible to solve any engineering problem by treating a body as a conglomeration of such discrete particles. The body is assumed to consist of a continuous distribution of matter. In other words, the body is treated as **continuum**.

Rigid Body

A body is said to be rigid, if the relative positions of any two particles in it do not change under the action of the forces. In Fig. 1.2 (*a*) points A and B are the original position in a body.

After application of a system of forces F_1, F_2, F_3, the body takes the position as shown in Fig. 1.2 (b). A' and B' are the new positions of A and B. If the body is treated as rigid, the relative position of $A'B'$ and AB are the same i.e.,
$$A'B' = AB.$$

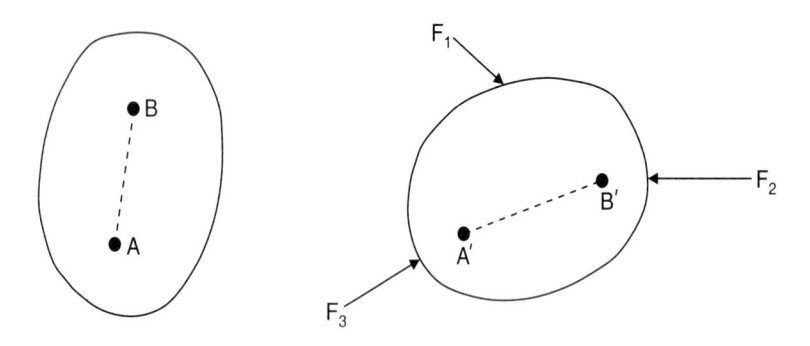

Fig. 1.2

Many engineering problems can be solved satisfactorily by assuming bodies rigid.

Particle

A particle may be defined as an object which has only mass and no size. Such a body cannot exist theoretically. However in dealing with problems involving distances considerably larger compared to the size of the body, the body may be treated as particle, without sacrificing accuracy. Examples of such situations are

— A bomber aeroplane is a particle for a gunner operating from the ground.
— A ship in mid sea is a particle in the study of its relative motion from a control tower.
— In the study of movement of the earth in celestial sphere, earth is treated as a particle.

1.3. LAWS OF MECHANICS

The following are the fundamental laws of mechanics:

Newton's first law
Newton's second law
Newton's third law
Newton's law of gravitation
Law of transmissibility of forces, and
Parallelogram law of forces.

Newton's First Law

It states that every body continues in its state of rest or of uniform motion in a straight line unless it is compelled by an external agency acting on it. This leads to the definition of force as the external agency which changes or tends to change the state of rest or uniform linear motion of the body.

Newton's Second Law

It states that the rate of change of momentum of a body is directly proportional to the impressed force and it takes place in the direction of the force acting on it. Thus according to this law,

INTRODUCTION TO ENGINEERING MECHANICS

Force ∝ rate of change of momentum. But momentum = mass × velocity

As mass do not change,

Force ∝ mass × rate of change of velocity

i.e., Force ∝ mass × acceleration

$$F \propto m \times a \qquad \text{...(1.3)}$$

Newton's Third Law

It states that for every action there is an equal and opposite reaction. Consider the two bodies in contact with each other. Let one body apply a force F on another. According to this law the second body develops a reactive force R which is equal in magnitude to force F and acts in the line same as F but in the opposite direction. Figure. 1.3 shows the action of the ball and the reaction from the floor. In Fig. 1.4 the action of the ladder on the wall and the floor and the reactions from the wall and floor are shown.

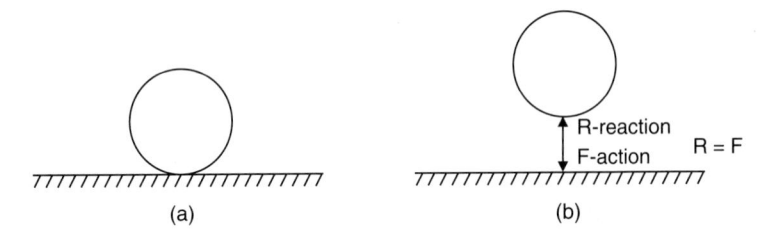

Fig. 1.3

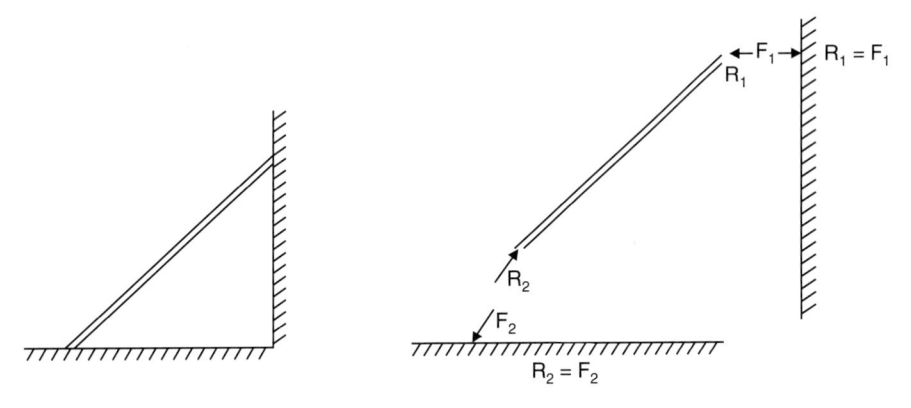

Fig. 1.4

Newton's Law of Gravitation

Everybody attracts the other body. The force of attraction between any two bodies is directly proportional to their masses and inversely proportional to the square of the distance between them. According to this law the force of attraction between the bodies of mass m_1 and mass m_2 at a distance d as shown in Fig. 1.5 is

$$F = G \frac{m_1 m_2}{d^2} \qquad \text{...(1.4)}$$

where G is the constant of proportionality and is known as constant of gravitation.

Fig. 1.5

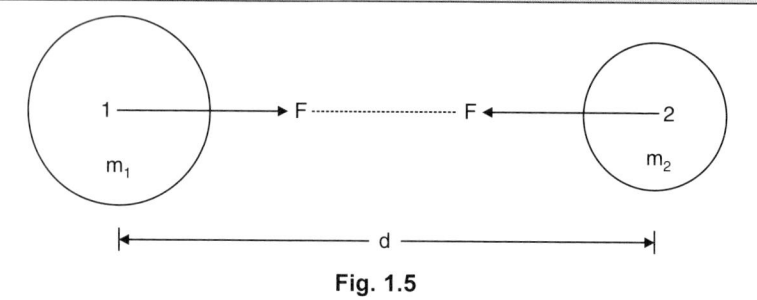

Law of Transmissibility of Force

According to this law *the state of rest or motion of the rigid body is unaltered if a force acting on the body is replaced by another force of the same magnitude and direction but acting anywhere on the body along the line of action of the replaced force.*

Let F be the force acting on a rigid body at point A as shown in Fig. 1.6. According to the law of transmissibility of force, this force has the same effect on the state of body as the force F applied at point B.

In using law of transmissibility of forces it should be carefully noted that it is applicable only if the body can be treated as rigid. In this text, the engineering mechanics is restricted to study of state of rigid bodies and hence this law is frequently used. Same thing cannot be done in the subject 'solid mechanics' where the bodies are treated as deformable and internal forces in the body are studied.

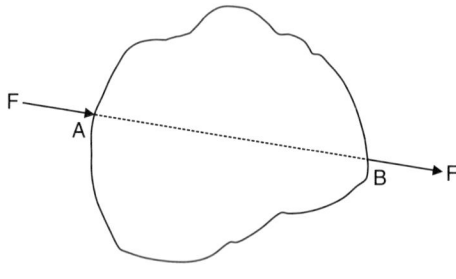

Fig. 1.6

The law of transmissibility of forces can be proved using *the law of superposition,* which can be stated as *the action of a given system of forces on a rigid body is not changed by adding or subtracting another system of forces in equilibrium.*

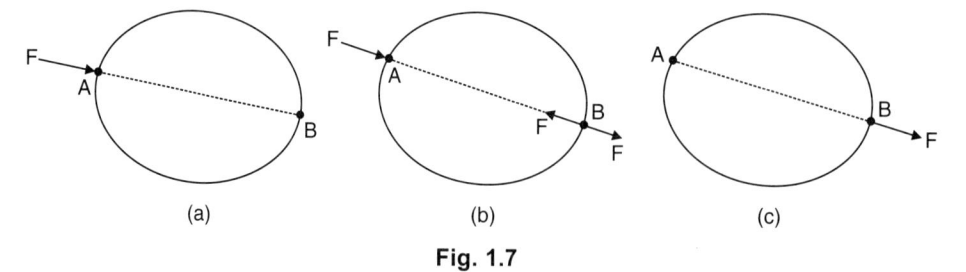

(a)   (b)   (c)

Fig. 1.7

Consider the rigid body shown in Fig. 1.7 (*a*). It is subjected to a force F at A. B is another point on the line of action of the force. From the law of superposition it is obvious that if two equal and opposite forces of magnitude F are applied at B along the line of action of given force F, [Ref. Fig. 1.7 (*b*)] the effect of given force on the body is not altered. Force F at A and opposite force F at B form a system of forces in equilibrium. If these two forces are subtracted from the system, the resulting system is as shown in Fig. 1.7 (*c*). Looking at the system of forces in Figs. 1.7 (*a*) and 1.7 (*c*), we can conclude the law of transmissibility of forces is proved.

INTRODUCTION TO ENGINEERING MECHANICS

Parallelogram Law of Forces

The parallelogram law of forces enables us to determine the single force called resultant which can replace the two forces acting at a point with the same effect as that of the two forces. This law was formulated based on experimental results. Though Stevinces employed it in 1586, the credit of presenting it as a law goes to Varignon and Newton (1687). This law states that *if two forces acting simultaneously on a body at a point are represented in magnitude and direction by the two adjacent sides of a parallelogram, their resultant is represented in magnitude and direction by the diagonal of the parallelogram which passes through the point of intersection of the two sides representing the forces.*

In Fig. 1.8 the force F_1 = 4 units and force F_2 = 3 units are acting on a body at point A. Then to get resultant of these forces parallelogram $ABCD$ is constructed such that AB is equal to 4 units to linear scale and AC is equal to 3 units. Then according to this law, the diagonal AD represents the resultant in the direction and magnitude. Thus the resultant of the forces F_1 and F_2 on the body is equal to units corresponding to AD in the direction α to F_1.

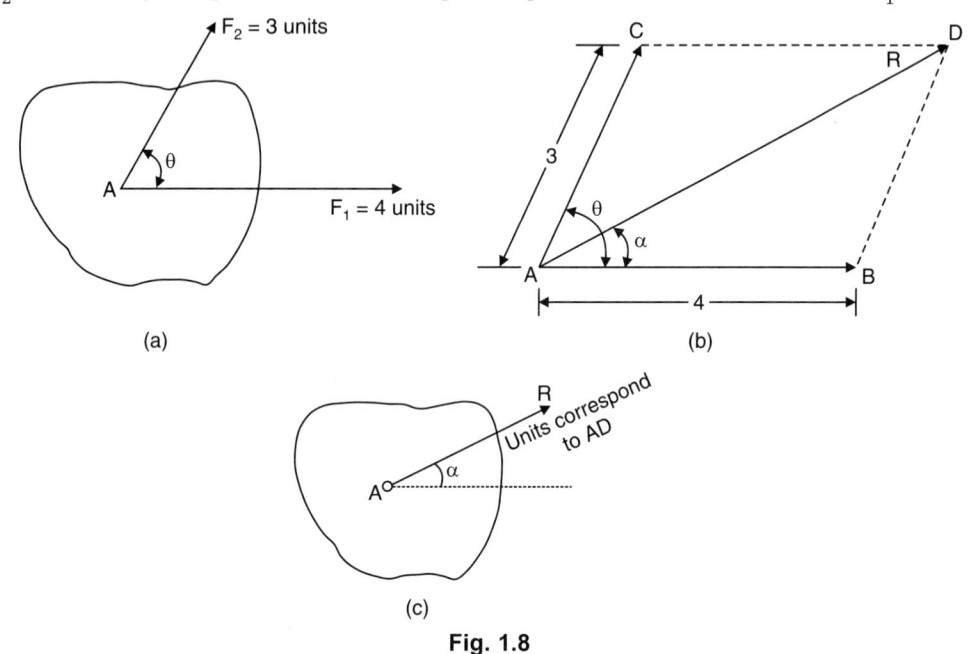

Fig. 1.8

1.4. DERIVED LAWS

Referring to Fig. 1.8 (b), we can get the resultant AD by constructing triangle ABD. Line AB is drawn to represent F_1 and BD to represent F_2. Then AD should represent the resultant of F_1 and F_2. Thus, we have derived triangle law of forces from fundamental law parallelogram law of forces. The **Triangle Law of Forces** may be stated as *If two forces acting on a body are represented one after another by the sides of a triangle, their resultant is represented by the closing side of the triangle taken from first point to the last point.*

If more than two concurrent forces are acting on a body, two forces at a time can be combined by triangle law of forces and finally resultant of all the forces acting on the body may be obtained.

A system of 4 concurrent forces acting on a body are shown in Fig. 1.9. AB represents F_1 and BC represents F_2. Hence according to triangle law of forces AC represents the resultant of F_1 and F_2, say, R_1.

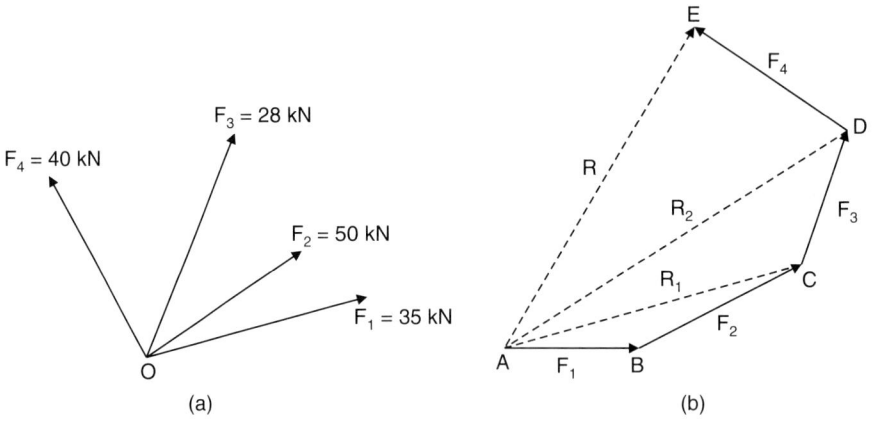

Fig. 1.9

If CD is drawn to represent F_3, then from triangle law of forces AD represents, the resultant of R_1 and F_3. In other words AD represents the resultant of F_1, F_2 and F_3. Let it be called as R_2.

On the same line, logic can be extended to say that AE represents the resultant of F_1, F_2, F_3 and F_4 if DE represents F_4. Thus resultant R is represented by the closing line of the polygon $ABCDE$ in the direction AE. Thus we have derived **polygon of law of forces** and it may be stated as '*If a number of concurrent forces acting simultaneously on a body are represented in magnitude and direction by the sides of a polygon, taken in a order, then the resultant is represented in magnitude and direction by the closing side of the polygon, taken from first point to last point.*

1.5. UNITS

Length (L), Mass (M) and Time (S) are the fundamental units in mechanics. The units of all other quantities may be expressed in terms of these basic units. The three commonly used systems in engineering are

— Metre-Kilogramme—Second (MKS) system
— Centimetre—Gramme—Second (CGS) system, and
— Foot—Pound—Second (FPS) system.

The units of length, mass and time used in the system are used to name the systems. Using these basic units, the units for other quantities can be found. For example, in MKS the units for the various quantities are as shown below:

Quantity	Unit	Notation
Area	Square metre	m^2
Volume	Cubic metre	m^3
Velocity	Metre per second	m/sec
Acceleration	Metre per second per second	m/sec^2

Unit of Forces

Presently the whole world is in the process of switching over to SI *system of units.* SI stands for System Internationale d' units or International System of units. As in MKS system, in SI system also the fundamental units are metre for length, kilogramme for mass and second

INTRODUCTION TO ENGINEERING MECHANICS

for time. The difference between MKS and SI system arise mainly in selecting the unit of force. From eqn. (1.3), we have

$$\text{Force} \propto \text{Mass} \times \text{Acceleration} = k \times \text{Mass} \times \text{Acceleration} \qquad \ldots(1.5)$$

In SI system unit of force is defined as that force which causes 1 kg mass to move with an acceleration of 1 m/sec² and is termed as 1 Newton. Hence the constant of proportionality k becomes unity. Unit of force can be derived from eqn. (1.5) as

$$\text{Unit of Force} = \text{kg} \times \text{m/sec}^2 = \text{kg} - \text{m/sec}^2$$

In MKS, the unit of force is defined as that force which makes a mass of 1 kg to move with gravitational acceleration 'g' m/sec². This unit of force is called kilogramme weight or kg/wt. Gravitational acceleration is 9.81 m/sec² near the earth surface. In all the problems encountered in engineering mechanics the variation in gravitational acceleration is negligible and may be taken as 9.81 m/sec². Hence the constant of proportionality in eqn. (1.5) is 9.81, which means

$$1 \text{ kg-wt} = 9.81 \text{ newton} \qquad \ldots(1.6)$$

It may be noted that in public usage, kg-wt force is called as kg only.

Unit of Constant of Gravitation

From eqn. (1.4),

$$F = G \frac{m_1 m_2}{d^2} \quad \text{or} \quad G = \frac{F d^2}{m_1 m_2}$$

$$\therefore \quad \text{Unit of } G = \frac{\text{N} \times \text{m}^2}{\text{kg} \times \text{kg}} = \text{Nm}^2/\text{kg}^2$$

It has been proved by experimental results that the value of $G = 6.673 \times 10^{-11}$ Nm²/kg². Thus if two bodies one of mass 10 kg and the other of 5 kg are at a distance of 1 m, they exert a force

$$F = \frac{6.673 \times 10^{-11} \times 10 \times 5}{1^2} = 33.365 \times 10^{-10} \text{ N}$$

on each other.

Now let us find the force acting between 1 kg-mass near earth surface and the earth. Earth has a radius of 6371×10^3 m and has a mass 5.96506×10^{24} kg. Hence the force between the two bodies is

$$= \frac{6.673 \times 10^{-11} \times 1 \times 5.96504 \times 10^{24}}{(6371 \times 10^3)^2} = 9.80665 \text{ N}.$$

In common usage we call the force exerted by earth on a body as *weight of the body*. Thus weight of 1 kg mass on earth surface is 9.80665 N, which is approximated as 9.81 N for all practical problems. Compared to this force the force exerted by two bodies near earth surface is negligible as may be seen from the example of 10 kg and 5 kg mass bodies.

Denoting the weight of the body by W, from eqn. (1.4), we get

$$W = \frac{G m M_e}{r^2}$$

where m is the mass of body

M_e is the mass of the earth, and

r is the radius of the earth

Denoting $\dfrac{GM_e}{r^2}$ by g, we get

$$W = mg = 9.81\, m \qquad \ldots(1.7)$$

Unit of g can be obtained as follows:

$$g = \dfrac{GM_e}{r^2}$$

$$\text{Unit of } g = \dfrac{\text{Nm}^2}{(\text{kg})^2} \times \dfrac{\text{kg}}{\text{m}^2} = \dfrac{\text{N}}{\text{kg}}$$

as unit of Newton force is kg-m/sec², we get

$$\text{Unit of } g = \dfrac{\text{kgm/sec}^2}{\text{kg}} = \text{m/sec}^2$$

Hence g may be called as acceleration due to gravity. Any body falling freely near earth surface experiences this acceleration. The value of g is 9.81 m/sec² near the earth surface as can be seen from eqn. (1.7).

The prefixes used in SI system when quantities are too big or too small are shown in Table 1.1.

Table 1.1: Prefixes and Symbols of Multiplying Factors in SI

Multiplying Factor	Prefix	Symbol
10^{12}	tera	T
10^{9}	giga	G
10^{6}	mega	M
10^{3}	kilo	k
10^{0}	—	—
10^{-3}	milli	m
10^{-6}	micro	μ
10^{-9}	nano	n
10^{-12}	pico	p
10^{-15}	femto	f
10^{-18}	atto	a

1.6. CHARACTERISTICS OF A FORCE

From Newton's first law, we defined the force as the agency which tries to change state of stress or state of uniform motion of the body. From Newton's second law of motion we arrived at practical definition of unit force as the force required to produce unit acceleration in a body of unit mass. Thus 1 newton is the force required to produce an acceleration of 1 m/sec² in a body of 1 kg mass. It may be noted that a force is completely specified only when the following four characteristics are specified:

— Magnitude
— Point of application
— Line of action, and
— Direction

INTRODUCTION TO ENGINEERING MECHANICS

In Fig. 1.10 *AB* is a ladder kept against a wall. At point *C*, a person weighing 600 N is standing. The force applied by the person on the ladder has the following characters:
— magnitude is 600 N
— the point of application is at *C* which is 2 m from *A* along the ladder.
— the line of action is vertical, and
— the direction is downward.

Note that the magnitude of the force is written near the arrow. The line of the arrow shows the line of application and the arrow head represents the point of application and the direction of the force.

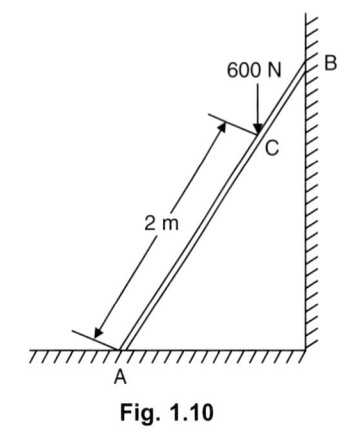

Fig. 1.10

1.7. SYSTEM OF FORCES

When several forces act simultaneously on a body, they constitute a *system of forces*. If all the forces in a system do not lie in a single plane they constitute the *system of forces in space*. If all the forces in a system lie in a single plane, it is called a *coplanar force system*. If the line of action of all the forces in a system pass through a single point, it is called a *concurrent force system*. In a *system of parallel forces* all the forces are parallel to each other. If the line of action of all the forces lie along a single line then it is called a *collinear force system*. Various system of forces, their characteristics and examples are given in Table 1.2 and shown in Fig. 1.11.

Table 1.2: System of Forces

Force System	Characteristics	Examples
Collinear forces	Line of action of all the forces act along the same line.	Forces on a rope in a tug of war.
Coplanar parallel forces	All forces are parallel to each other and lie in a single plane.	System of forces acting on a beam subjected to vertical loads (including reactions).
Coplanar like parallel forces	All forces are parallel to each other, lie in a single plane and are acting in the same direction.	Weight of a stationary train on a rail when the track is straight.
Coplanar concurrent forces	Line of action of all forces pass through a single point and forces lie in the same plane.	Forces on a rod resting against a wall.
Coplanar non-concurrent forces	All forces do not meet at a point, but lie in a single plane.	Forces on a ladder resting against a wall when a person stands on a rung which is not at its centre of gravity.
Non-coplanar parallel forces	All the forces are parallel to each other, but not in same plane.	The weight of benches in a classroom.
Non-coplanar concurrent forces	All forces do not lie in the same plane, but their lines of action pass through a single point.	A tripod carrying a camera.
Non-coplanar non-concurrent forces	All forces do not lie in the same plane and their lines of action do not pass through a single point.	Forces acting on a moving bus.

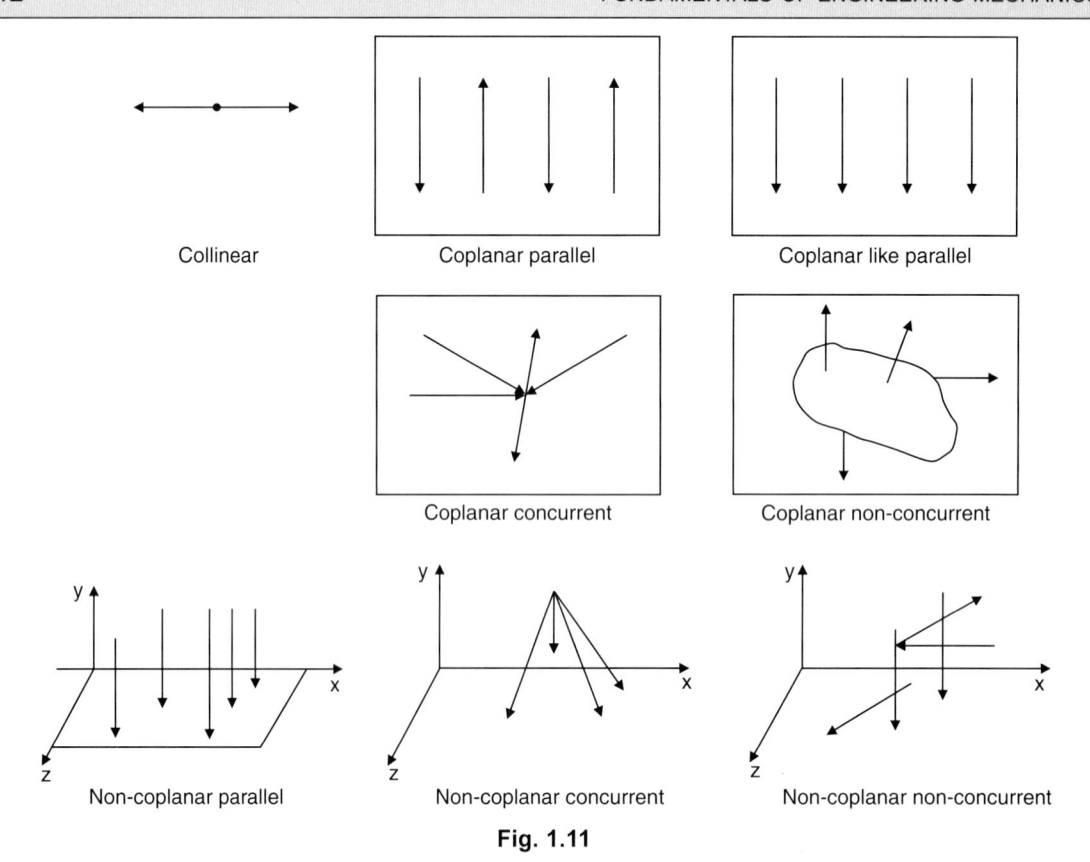

Collinear   Coplanar parallel   Coplanar like parallel

Coplanar concurrent   Coplanar non-concurrent

Non-coplanar parallel   Non-coplanar concurrent   Non-coplanar non-concurrent

Fig. 1.11

1.8. VECTORS

Various quantities used in engineering mechanics may be grouped into scalars and vectors. A quantity is said to be *scalar* if it is completely defined by its magnitude alone. Examples of scalars are length, area, time and mass.

A quantity is said to be *vector* if it is completely defined only when its magnitude as well as direction are specified. Hence force is a vector. The other examples of vector are velocity, acceleration, momentum etc.

1.9. DIMENSIONAL HOMOGENEITY

The qualitative description of physical variable is known as dimension while the quantitative description is known as unit. We come across several relations among the physical quantities. Some of the terms may be having dimensions and some others may be dimensionless. However in any equation dimensions of the terms on both sides must be the same. This is called dimensional homogenity. The branch of mathematics dealing with dimensions of quantities is called dimensional analysis.

There are two systems of dimensional analysis *viz. absolute system* and *gravitational system*. In absolute system the basic quantities selected are Mass, Length and Time. Hence it is known as *MLT-system*. In gravitational system the basic quantities are Force, Length and Time. Hence it is termed as *FLT*-system.

INTRODUCTION TO ENGINEERING MECHANICS

The dimension of acceleration is $\dfrac{L}{T^2} = LT^{-2}$ since its unit is m/sec². From Newton's law we have physical relation

$$\text{Force} = \text{Mass} \times \text{Acceleration}$$

Hence the dimensional relation is,

$$F = \dfrac{ML}{T^2} \qquad \qquad ...(1.8a)$$

or

$$M = \dfrac{FT^2}{L} \qquad \qquad ...(1.8b)$$

Eqn. (1.8) helps in converting dimensions from one system to another. The dimensions of some of the physical quantities are listed in Table 1.3.

Table 1.3: Dimensions of Quantities

Sr. No.	Quantity	MLT-system	FLT-system
1.	Velocity	LT^{-1}	LT^{-1}
2.	Acceleration	LT^{-2}	LT^{-2}
3.	Momentum	MLT^{-1}	FT
4.	Area	L^2	L^2
5.	Volume	L^3	L^3
6.	Force	MLT^{-2}	F
7.	Gravitational Constant	$M^{-1}L^3T^{-2}$	$F^{-1}L^4T^{-4}$

Checking Dimensional Homogenity

As stated earlier all the terms in an equation to the left and right side should have the same dimensions. In other words if,

$$X = Y + Z$$

the terms, X, Y and Z should have same dimension. If,

$$X = bY$$

and if X and Y do not have same dimension, b is not dimensionless constant. The value of this constant will be different in different system of units.

Example 1.1. *Verify whether the following equation has dimensional homogenity:*

$$v^2 - u^2 = 2as$$

where v is final velocity, u is initial velocity, a is acceleration and s is the distance moved.

Solution. Dimensions of velocity $= LT^{-1}$
Dimensions of acceleration $= LT^{-2}$
and dimension of distance $= L$

Substituting these dimensions in the given equation, we find
Dimensions of left hand side terms:
  Dimensions of $v^2 - (LT^{-1})^2$
  $u^2 - (LT^{-1})^2$
Dimension of right hand side term $- LT^{-2}L - (LT^{-1})^2$
Hence it is dimensionally homogeneous equation.

Example 1.2. *In the following equation verify, whether 9.81 is dimensionless constant. If it is not so, what should be its dimension?*

$$s = ut + \frac{1}{2} \, 9.81 \, t^2$$

where s = *distance*
u = *initial velocity*
t = *time*

Solution. Dimensions of various terms are

$$s = L$$
$$u = LT^{-1}$$
$$t = T$$

Substituting these in the given equation, we get,

$$L = LT^{-1} \, T + \frac{1}{2} \times 9.81 \, T^2$$

$$L = L + \frac{1}{2} \times 9.81 \, T^2$$

Hence, 9.81 cannot be dimensionless constant. Its dimension is given by

$$L \equiv \frac{1}{2} \, 9.81 \, T^2$$

9.81 should have dimensions LT^{-2}, same as that of acceleration. We know this is gravitational acceleration term in SI unit *i.e.*, it is in m/sec^2 term. Hence the given equation cannot be straightway used is FPS system or CGS system.

1.10. IDEALISATIONS IN MECHANICS

A number of ideal conditions are assumed to exist while applying the principles of mechanics to practical problems. In fact without such assumptions it is not possible to arrive at practical solutions. The following idealisations are usually made in engineering mechanics.

1. The body is rigid.
2. The body can be treated as continuum.
3. If the size of the body is small compared to other distances involved in the problem, it may be treated as a particle.
4. If the area over which force is acting on a body is small compared to the size of the body, it may be treated as a point force. For example, in Fig. 1.9, 600 N force is the weight of a man. Actually the man cannot apply his weight through a single point. There is certain area of contact, which is, however, small compared to the other dimensions in the problem. Hence, the weight of the man is treated as a point load.
5. Support conditions are idealised (which will be discussed later) as simple, hinged, fixed etc.

1.11. ACCURACY OF CALCULATIONS

As explained in Art. 1.9, a number of ideal conditions are assumed to exist while applying the principles of mechanics to engineering problems. These assumptions have some effect on the accuracy of final results. Further in engineering analysis, many loads are estimated

INTRODUCTION TO ENGINEERING MECHANICS

loads. There are some imperfection in construction and fabrications of structures. To take care all these uncertainities engineers multiply the load by a number (1.5 to 2.25), which is called factor of safety. In view of all these, there is no need in noting the calculations beyond four digits. It is found that 0.2% accuracy is more than sufficient, in the calculations.

1.12. APPROACHES TO SOLUTION

There are two approaches for the solution of engineering mechanics problems *i.e.* classical approach and vector approach. **Classical approach** gives physical feel of the problem. It gives confidence to engineers in accepting the results presented by others and quickly take decisions on site. Development of such feel of the problem is very much essential for engineers. However it becomes difficult to solve three dimensional problems by classical approach. Vector approach is ideally suited for the analysis of three dimensional problems. But the disadvantage of vector approach is physical feel of the problem is lost and the ability of site engineers in quick decision is not developed. Hence there are two school of academicians, one advocating for classical approach and the other advocating for vector approach. In this book author has used classical approach for the solution of engineering mechanics problems.

Important Definitions and Concepts

1. *Displacement* is defined as the distance moved by a body or particle in the specified direction.
2. The rate of change of displacement with time is called *velocity*.
3. *Acceleration* is the rate of change of velocity with respect to time.
4. The product of mass and velocity is called *momentum*.
5. A body is said to be treated as *continuum*, if it is assumed to consist of continuous distribution of matter.
6. A body is said to be *rigid*, if the relative position of any two particles in it do not change under the action of the forces.
7. *Newton's first law* states that everybody continues in its state of rest or of uniform motion in a straight line unless it is compelled by an external agency acting on it.
8. *Newton's second law* states that the rate of change of momentum of a body is directly proportional to the impressed force and it takes place in the direction of the force acting on it.
9. *Newton's third law* states for every action there is an equal and opposite reaction.
10. *Newton's law of gravitation* states everybody attracts the other body, the force of attraction between any two bodies is directly proportional to their mass and inversely proportional to the square of the distance between them.
11. According to the *law of transmissibility of force,* the state of rest or motion of a rigid body is unaltered, if a force acting on a body is replaced by another force of the same magnitude and direction but acting anywhere on the body along the line of action of the replaced force.
12. The *parallelogram law of forces* states that if two forces acting simultaneously on a body at a point are represented by the two adjacent sides of a parallelogram, their resultant is represented in magnitude and direction by the diagonal of the parallelogram which passes through the point of intersection of the two sides representing the forces.
13. The qualitative description of physical variable is known as *dimension* while the quantitative description is known as *unit*.

14. A quantity is said to be *scalar*, if it is completely defined by its magnitude alone.
15. A quantity is said to be *vector* if it is completely defined only when it's magnitude as well as direction are specified.

QUESTIONS

1. Explain the following terms as used in Engineering Mechanics:
   (*i*) Continuum   (*ii*) Rigid Body   (*iii*) Particle.
2. State and explain Newton's three laws of motion.
3. State and explain Newton's law of gravitation.
4. State and explain Law of transmissibility of forces.
5. State and explain parallelogram law of forces. From this derive triangle and polygonal laws of forces.
6. Explain the term 'Force' and list its characteristics.
7. Explain the terms—concurrent and non-concurrent force system; planar and non-planar system of forces.

2

Resultant and Equilibrium of System of Coplanar Concurrent Forces

If all the forces in a system lie in a single plane and pass through a single point, they constitute a coplanar concurrent force system. It is possible to find a single force which will have the same effect as that of a number of forces acting. Such single force is called Resultant force and the process of finding the resultant force is called composition of forces.

Parallelogram law, triangle law and polygonal law may be used to find the resultant graphically. The graphical method of composition gives a clear picture of the work being carried out. However the main disadvantage is that it needs drawing aids like pencil, scale, drawing sheet, etc. Hence there is need for analytical method. In this chapter analytical method of composition is explained for the system of two forces first and then the general method is explained.

A body may be in equilibrium under the action of a concurrent force system, and the analyst is interested in finding the reactions from the other bodies in contact. Since it is concurrent force system, size of the body will not come into analysis. Hence the body may be referred as a particle and the analysis of equilibrium condition as statics of particle. This process is also explained and illustrated with application to many engineering problems.

2.1. COMPOSITION OF TWO FORCE SYSTEM

Consider the two forces F_1 and F_2 acting on a particle as shown in Fig. 2.1 (a). Let the angle between the two forces be θ. If parallelogram $ABCD$ is drawn as shown in Fig. 2.1 (b), with AB representing F_1 and AD representing F_2 to same scale, according to parallelogram law of forces AC represents the resultant R. Drop perpendicular CE to AB.

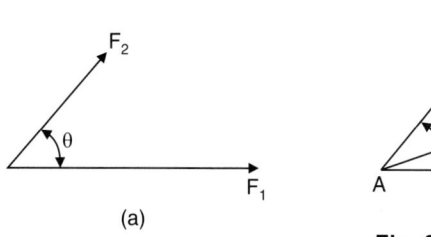

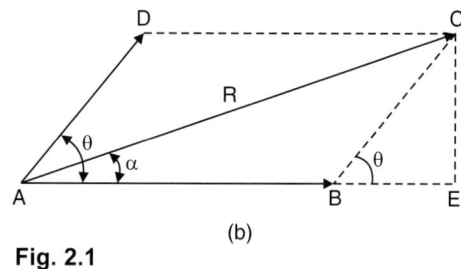

Fig. 2.1

The resultant R of F_1 and F_2 is given by:

$$R = AC = \sqrt{AE^2 + CE^2} = \sqrt{(AB + BE)^2 + CE^2}$$

But $\qquad AB = F_1$

$$BE = BC \cos\theta = F_2 \cos\theta$$
and
$$CE = BC \sin\theta = F_2 \sin\theta$$

$$\therefore \quad R = \sqrt{(F_1 + F_2 \cos\theta)^2 + (F_2 \sin\theta)^2}$$

$$= \sqrt{(F_1^2 + 2F_1F_2 \cos\theta + F_2^2 \cos^2\theta + F_2^2 \sin^2\theta)}$$

$$= \sqrt{F_1^2 + 2F_1F_2 \cos\theta + F_2^2} \qquad \ldots(2.1)$$

Since, $\sin^2\theta + \cos^2\theta = 1$.

The inclination of the resultant to the direction of the force F_1 is given by α, where

$$\tan\alpha = \frac{CE}{AE} = \frac{CE}{AB + BE} = \frac{F_2 \sin\theta}{F_1 + F_2 \cos\theta}$$

Hence
$$\alpha = \tan^{-1} \frac{F_2 \sin\theta}{F_1 + F_2 \cos\theta} \qquad \ldots(2.2)$$

Particular Cases:

1. When $\theta = 90°$ [Ref. Fig. 2.2 a], $\quad R = \sqrt{F_1^2 + F_2^2}$

2. When $\theta = 0°$ [Ref. Fig. 2.2 b], $\quad R = \sqrt{F_1^2 + 2F_1F_2 + F_2^2} = F_1 + F_2$

3. When $\theta = 180°$ [Ref. Fig. 2.2 c], $\quad R = \sqrt{F_1^2 - 2F_1F_2 + F_2^2} = F_1 - F_2$

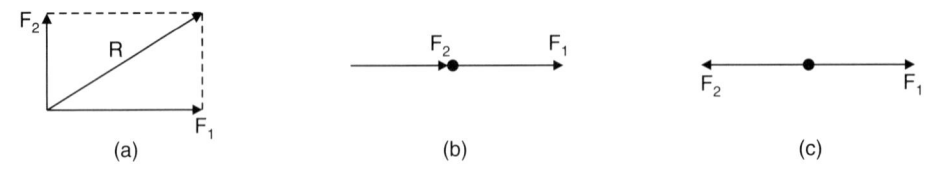

Fig. 2.2

From the particular cases 2 and 3, it is clear that *when the forces acting on a body are collinear, their resultant is equal to the algebraic sum of the forces.*

Example 2.1. *The resultant of two forces, one of which is double the other is 260 N. If the direction of the larger force is reversed and the other remains unaltered, the resultant reduces to 180 N. Determine the magnitude of the forces and the angle between the forces.*

Solution. Let the magnitude of smaller force be F. Hence magnitude of larger force is $2F$.

Thus $\quad F_1 = F$ and $F_2 = 2F$

Let θ be the angle between the two forces

$\therefore$ From the condition 1, we get

$$R = \sqrt{F_1^2 + 2F_1F_2 \cos\theta + F_2^2} = 260.$$

$\therefore \quad F^2 + 2F(2F)\cos\theta + (2F)^2 = 260^2$

i.e., $\quad 5F^2 + 4F^2 \cos\theta = 67600 \qquad \ldots(i)$

RESULTANT AND EQUILIBRIUM OF SYSTEM OF COPLANAR CONCURRENT FORCES

From condition 2, we have

$$\sqrt{F_1^2 + 2F_1F_2 \cos(180° + \theta) + F_2^2} = 180^2$$

i.e., $\quad F^2 - 2F(2F)\cos\theta + (-2F)^2 = 32400$

i.e., $\quad 5F^2 - 4F^2 \cos\theta = 32400 \qquad ...(ii)$

Adding equations (i) and (ii), we get

$$10F^2 = 100000$$

$\therefore \quad F = 100$ N.

Hence, $\quad F_1 = 100$ N $\quad$ and $\quad F_2 = 2F = 200$ N. **Ans.**

Substituting the values of F_1 and F_2 in eqn. (i),

we get, $5(100)^2 + 4(100)^2 \cos\theta = 67600$

$\therefore \quad \cos\theta = 0.44 \quad \therefore \quad \theta = 63.9°$ **Ans.**

2.2. RESOLUTION OF FORCES

Before taking up the general method of finding resultant of any number of concurrent forces, it is necessary to study resolution of a force into its rectangular components. The **resolution of forces** is exactly the opposite process of composition of forces. It is the process of finding a number of component forces which will have the same effect as the given single force. Exactly the opposite process of composition can be employed to get the resolved component of a given force.

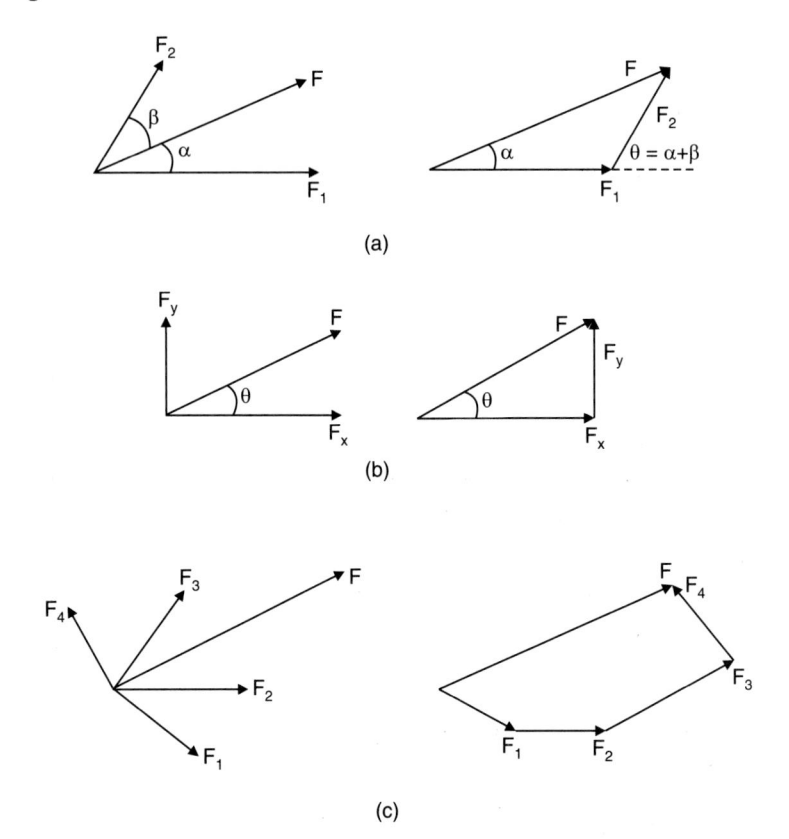

Fig. 2.3

In Fig. 2.3 (a), the given force F is resolved into two components making angles α and β with F.

In Fig. 2.3 (b) the force F is resolved into its rectangular components F_x and F_y.

In Fig. 2.3 (c) the force F is resolved into its four components F_1, F_2, F_3 and F_4.

It should be noted that all component forces act at the same point as the given force. Resolution of forces into its rectangular components is more useful for the analysis. In this case if the force F makes angle θ with x-axis, from Fig. 2.3 (b), it is clear that

$$F_x = F \cos \theta$$
and
$$F_y = F \sin \theta \qquad \qquad ...(2.3)$$

Example 2.2. *The guy wire of the electric pole shown in Fig. 2.4 (a) makes 60° to the horizontal and is subjected to 20 kN force. Find horizontal and vertical components of the force.*

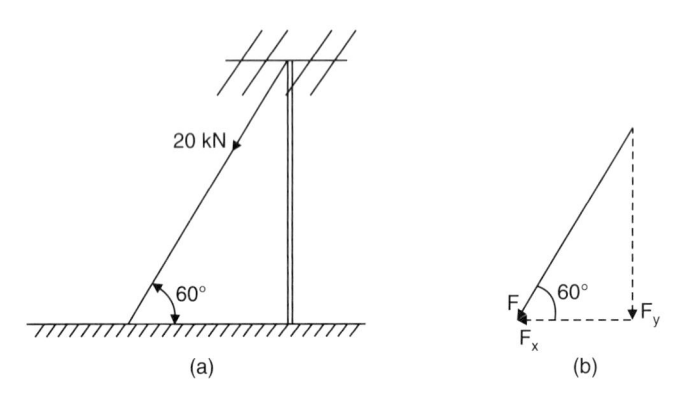

Fig. 2.4

Solution. Figure 2.4 (b) shows the resolution of force F = 20 kN into its components in horizontal and vertical components. From this figure it is clear that

$$F_x = F \cos 60° = 20 \cos 60° = 10 \text{ kN (to left)} \quad \textbf{Ans.}$$
and
$$F_y = F \sin 60° = 20 \sin 60° = 17.32 \text{ kN (downward)} \quad \textbf{Ans.}$$

Example 2.3. *A block weighing W = 10 kN is resting on an inclined plane as shown in Fig. 2.5 (a). Determine its components normal to and parallel to the inclined plane.*

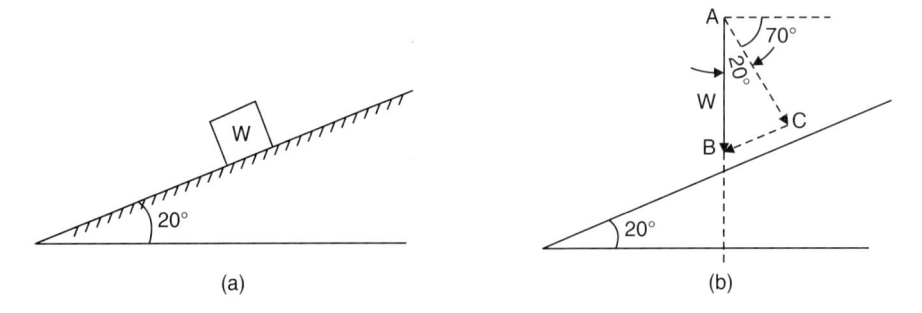

Fig. 2.5

Solution. The plane makes an angle of 20° with the horizontal. Hence the normal to the plane makes an angle of 70° to horizontal *i.e.* 20° to the vertical [Ref. Fig. 2.5 (b)]. In Fig. 2.5 (b), if AB represents the given force W to some scale, AC represents the normal component and CB represents component parallel to the plane.

RESULTANT AND EQUILIBRIUM OF SYSTEM OF COPLANAR CONCURRENT FORCES

From $\triangle ABC$,

Normal component $= AC$
$= W \cos 20°$
$= 10 \cos 20°$
$= 9.4$ kN (Thrust on plane) **Ans.**

Parallel component $= W \sin 20° = 10 \sin 20°$
$= 3.42$ kN (down the plane) **Ans.**

From the above two examples, the following points may be noted:

1. Imagine that the arrow drawn represents the given force to some scale.
2. Travel from tail to arrow head of the force in the direction of the coordinates selected.
3. Then the direction of travel gives the direction of component forces.
4. From the triangle of forces, the magnitude of the components can be calculated.

2.3. GENERAL METHOD OF COMPOSITION OF FORCES

The method explained below can be used to determine the resultant of any number of concurrent forces acting at a point in a plane:

Step 1. Determine the component of each force in two mutually perpendicular coordinate directions.

Step 2. Add algebraically components of all the forces in each coordinate direction to get two component forces.

Step 3. The two component forces which are mutually perpendicular can be combined to get the resultant.

Let F_1, F_2, F_3 and F_4 shown in Fig. 2.6 be the system of four forces, the resultant of which is to be determined.

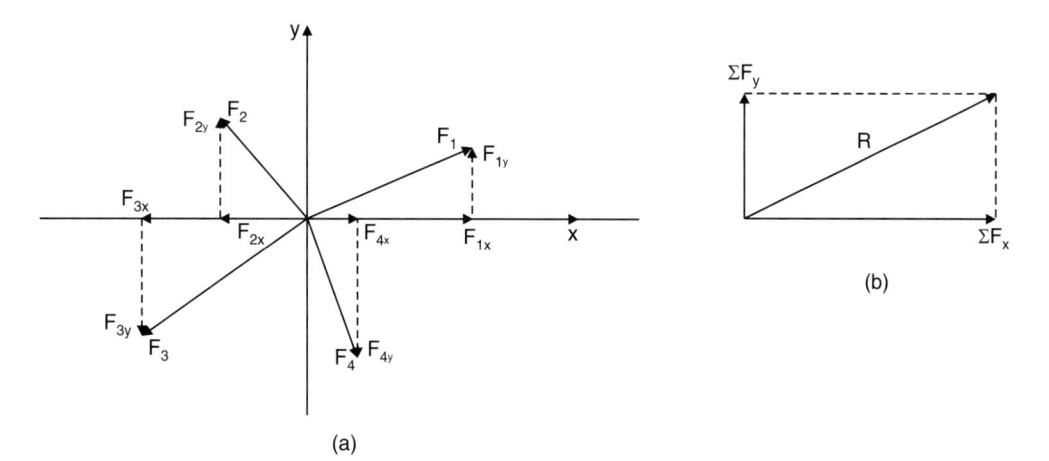

Fig. 2.6

The procedure to be followed is as given below:

Step 1. Find the component of all the forces in x and y directions. Thus $F_{1x}, F_{2x}, F_{3x}, F_{4x}, F_{1y}, F_{2y}, F_{3y}$, and F_{4y}, are obtained.

Step 2. Find the algebraic sum of the component forces in x and y directions.

$$\Sigma F_x = F_{1x} + F_{2x} + F_{3x} + F_{4x}$$

$$\Sigma F_y = F_{1y} + F_{2y} + F_{3y} + F_{4y}$$

[**Note:** In the above example F_{2x}, F_{3x}, F_{3y} and F_{4y} are having negative values.]

Step 3. Now the system of forces is equal to two mutually perpendicular forces, namely, ΣF_x and ΣF_y as shown in Fig. 2.6 (b). Since these two forces are at right angles to each other, the parallelogram of forces becomes a rectangle.

Hence the resultant R is given by:

$$R = \sqrt{(\Sigma F_x)^2 + (\Sigma F_y)^2} \qquad \ldots(2.4)$$

and its inclination to x-axis is given by

$$\alpha = \tan^{-1}\left(\frac{\Sigma F_y}{\Sigma F_x}\right) \qquad \ldots(2.5)$$

Note that $\quad R \cos \alpha = \Sigma F_x \qquad \ldots(2.6)$
and $\quad R \sin \alpha = \Sigma F_y \qquad \ldots(2.7)$

i.e., ΣF_x and ΣF_y are the x and y components of the resultant R.

Example 2.4. *Determine the resultant of the three forces acting on a hook as shown in Fig. 2.7 (a).*

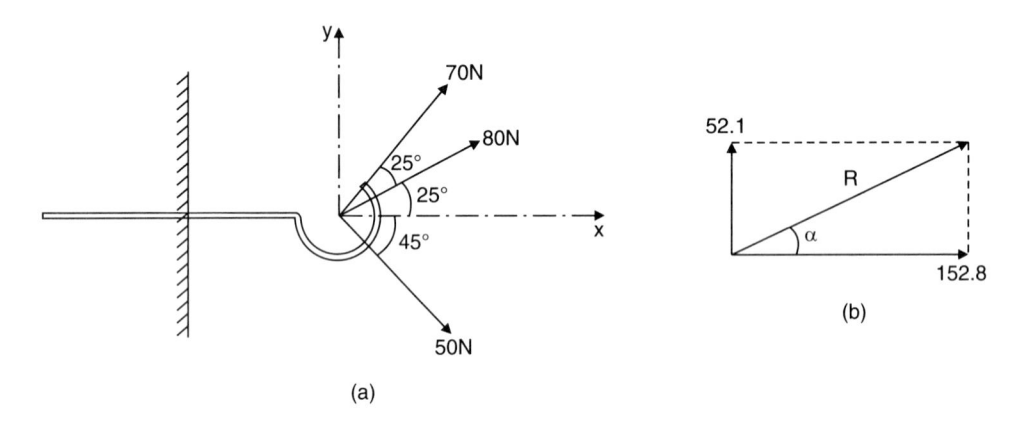

Fig. 2.7

Solution.

Force	x-component	y-component
$F_1 = 70$ N	$70 \cos 50° = 45.0$	$70 \sin 50° = 53.6$
$F_2 = 80$ N	$80 \cos 25° = 72.5$	$80 \sin 25° = 33.8$
$F_3 = 50$ N	$50 \cos 45° = 35.4$	$-50 \sin 45° = -35.4$
	$\Sigma F_x = 152.8$	$\Sigma F_y = 52.1$

∴ $\quad R = \sqrt{152.8^2 + 52.1^2} = 161.5$ N **Ans.**

$$\alpha = \tan^{-1}\frac{52.1}{152.8} = 18.83° \text{ as shown in Fig. 2.7 (b)} \quad \textbf{Ans.}$$

RESULTANT AND EQUILIBRIUM OF SYSTEM OF COPLANAR CONCURRENT FORCES

Example 2.5. *A system of four forces acting at a point on a body is as shown in Fig. 2.8 (a). Determine the resultant.*

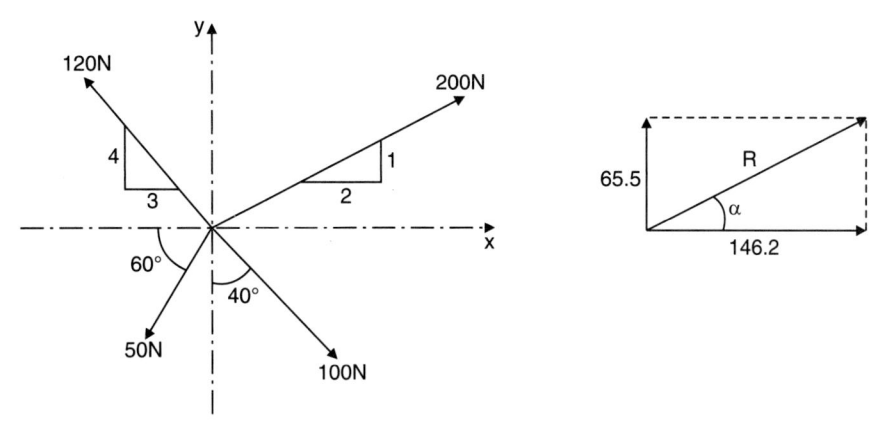

Fig. 2.8

Solution. If θ_1 is the inclination of 200 N force to x-axis.

$$\tan \theta_1 = 1/2 \quad \therefore \quad \theta_1 = 26.565°$$

Similarly, inclination of 120 N force to x-axis is given by

$$\tan \theta_2 = \frac{4}{3} \quad i.e., \quad \theta_2 = 53.13°$$

$\therefore \qquad \Sigma F_x = 200 \cos 26.565° - 120 \cos 53.13° - 50 \cos 60° + 100 \sin 40°$
$\qquad\qquad\quad = 146.2$ N

$\qquad \Sigma F_y = 200 \sin 26.565° + 120 \sin 53.13° - 50 \sin 60° - 100 \cos 40°$
$\qquad\qquad\quad = 65.5$ N

$\therefore \qquad R = \sqrt{146.2^2 + 65.5^2} = 160.2$ N **Ans.**

$$\alpha = \tan^{-1} \frac{65.5}{146.2} = 24.1° \text{ as shown in Fig. 2.8 (b)} \quad \textbf{Ans.}$$

Example 2.6. *A system of forces acting on a body resting on an inclined plane is as shown in Fig. 2.9. Determine the resultant force if $\theta = 60°$, $W = 1000$ N, vertically downward, $N = 500$ newton acting normal to the plane, $F = 100$ N, acting down the plane and $T = 1200$ N, acting parallel to the plane.*

Solution. In this problem, note that coordinate system parallel to and perpendicular to the plane is convenient. Noting that W makes angle θ with y-axis.

$\qquad \Sigma F_x = T - F - W \sin \theta$
$\qquad\qquad = 1200 - 100 - 100 \sin 60°$
$\qquad\qquad = 234.0$ N

$\qquad \Sigma F_y = N - W \cos 60°$
$\qquad\qquad = 500 - 1000 \cos 60°$
$\qquad\qquad = 0$

$\therefore$ Resultant is a force of magnitude 234 N directed up the plane.

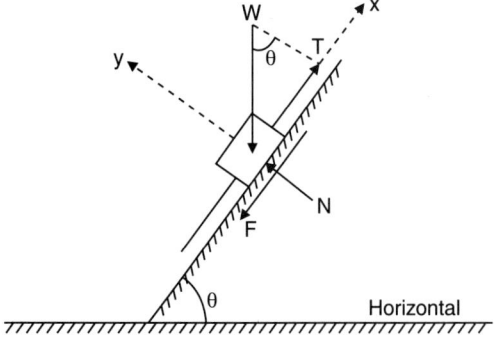

Fig. 2.9

Example 2.7. *Two forces acting on a body are 500 N and 1000 N as shown in Fig. 2.10(a). Determine the third force F such that the resultant of all the three forces is 1000 N directed at 45° to x-axis.*

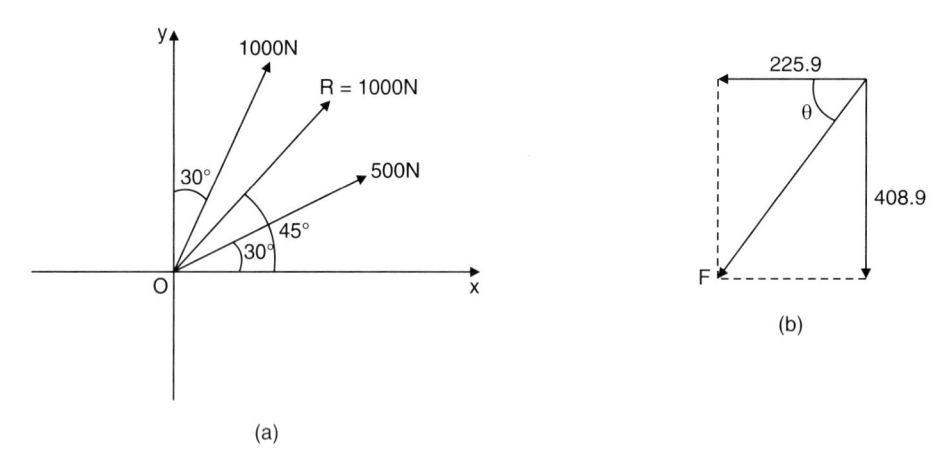

Fig. 2.10

Solution: Let the third force F makes an angle θ with x-axis. Then,
$R \cos \alpha = \Sigma F_x$, gives
$1000 \cos 45° = 500 \cos 30° + 1000 \sin 30° + F \cos \theta$
∴ $\qquad F \cos \theta = -225.9$ N
Similarly, $R \sin 45° = \Sigma F_y$ gives,
$1000 \sin 45° = 500 \sin 30° + 1000 \cos 30° + F \sin \theta$
i.e., $\qquad F \sin \theta = -408.9$
∴ $\qquad F = \sqrt{225.9^2 + 408.9^2}$
i.e., $\qquad F = 467.2$ N **Ans.**

$$\theta = \tan^{-1}\left(\frac{408.9}{225.9}\right) = 61.08° \text{ as shown in Fig. 2.8 }(b) \quad \textbf{Ans.}$$

Example 2.8. *Three forces acting at centre of gravity of a block are shown in Fig. 2.11. The direction of 300 N forces may vary, but the angle between them is always 40°. Determine the value of θ for which the resultant of the three forces is directed parallel to the plane.*

Solution: Let x and y axes be selected as shown in Fig. 2.11. If the resultant is directed along x-axis, the component of reaction in y-direction should be zero.
i.e. $\qquad \Sigma F_y = 0 \rightarrow$
$300 \sin \theta + 300 \sin (40° + \theta) - 500 \sin 30°$
$\qquad = 0$
$\sin \theta + \sin (40° + \theta) = 0.833$
i.e., $\quad 2 \sin \dfrac{40° + \theta + \theta}{2} \times \cos \dfrac{40° + \theta - \theta}{2}$
$\qquad = 0.833$

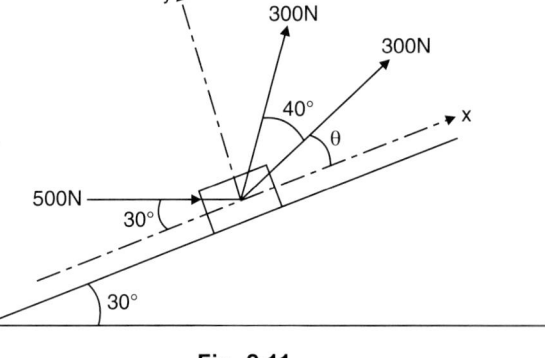

Fig. 2.11

RESULTANT AND EQUILIBRIUM OF SYSTEM OF COPLANAR CONCURRENT FORCES

i.e., $\quad 2 \sin(20° + \theta) \cos 20° = 0.833$

$\therefore \quad \theta = 6.31°$ **Ans.**

Note: $\sin A + \sin B = 2 \sin \dfrac{A+B}{2} \cos \dfrac{A-B}{2}$

2.4. EQUILIBRIUM OF BODIES

A body is said to be in equilibrium if its state of rest or of uniform motion in a straight line is not altered. According to Newton's first law, it means no resultant force acts on the body. In case of coplanar concurrent forces, it means,

$$\Sigma F_x = 0$$

and $\quad\quad\quad\quad\quad\quad \Sigma F_y = 0 \quad\quad\quad\quad\quad\quad\quad\quad\quad\quad\quad$ ...(2.8)

It may be observed that $\Sigma F_x = 0$ means, $R \cos \alpha = 0$. It means the resultant do not have any component in x-direction. This condition ensures that there is no resultant in any direction, except in y-direction ($\alpha = 90°$). Hence the condition $\Sigma F_y = 0$ also should be satisfied to ensure that the resultant R does not exist in any direction. Hence, if a body is in equilibrium under the action of coplanar concurrent system, eqn. (2.8) should be satisfied.

2.4.1. Types of Forces Acting on a Body

While applying equilibrium equations to a body, it is necessary that all forces acting on a body should be considered. The various forces acting on a body may be grouped as:

— Applied forces
— Non-applied forces.

Applied Forces

Applied forces are the forces applied externally to a body. Each force has got point of contact with the body. If a person stands on a ladder, his weight is an applied force to the ladder. If a temple car is pulled, the force in the rope is an applied force for the car.

Non-applied Forces

There are two types of non-applied forces (*a*) self weight, and (*b*) reactions.

(*a*) **Self weight.** Everybody is subjected to gravitational attraction and hence has got self-weight given by the expression,

$$W = mg \quad\quad\quad\quad\quad\quad\quad\quad\quad\quad ...(2.9)$$

where m is the mass of the body and g is the gravitational acceleration (9.81 m/sec^2 near the earth surface). Self weight always acts in vertically downward direction. Self weight is treated as a vertically downward force acting through the centre of gravity of the body. If self weight is very small, it may be neglected.

Reactions

Reactions are self adjusting forces developed by other bodies which are in contact with the body under consideration. According to Newton's third law of motion, reactions are equal and opposite to the actions. The reactions adjust themselves to keep the body in equilibrium.

If the surface of contact is smooth, the direction of reaction is normal to the surface of contact. If the surface of contact is rough, apart from normal reaction, there will be frictional reaction also. Hence the resultant reaction will not be normal to the surface of contact.

2.4.2. Free Body Diagram

For the analysis of equilibrium condition it is necessary to isolate the body under consideration from the other bodies in contact and draw all forces acting on the body. For this, first the body is drawn and then all applied forces, self weight and reactions from the other bodies in contact are drawn. Such *diagram of the body in which the body under consideration is freed from all contact surfaces and is shown with all the forces on it (including self weight and reactions from other contact surfaces) is called the Free Body Diagram* (FBD). Free body diagrams (FBD) are shown for few typical cases in Table 2.1.

Table 2.1: Free Body Diagrams (FBD) for Few Typical Cases

Reacting bodies	*FBD required for*	*FBD*
(ball resting on smooth ground)	Ball	Ball with W acting downward and R acting upward
(ball suspended by string against smooth wall)	Ball	Ball with T, R, W
(ladder 600 N leaning against smooth wall on smooth ground at P)	Ladder	Ladder with $W = 600\,N$ at G, R_1 at top, P and R_2 at bottom
(400 N block on pulley, 600 N block hanging)	Block weighing 600 N	Block with T upward, $600\,N$ downward, R upward

2.4.3. Lami's Theorem

If a body is in equilibrium, it may be analysed using equations of equilibrium 2.3. However, if the body is in equilibrium under the action of only three forces, Lami's theorem can be used conveniently.

Lami's theorem states that *if a body is in equilibrium under the action of only three forces, each force is proportional to the sine of the angle between the other two forces.* Thus for the system of forces shown in Fig. 2.12. (*a*),

$$\frac{F_1}{\sin \alpha} = \frac{F_2}{\sin \beta} = \frac{F_3}{\sin \gamma} \qquad ...(2.10)$$

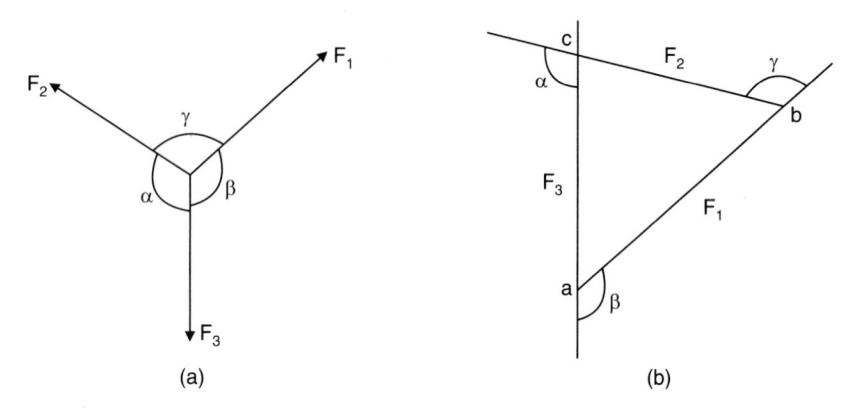

Fig. 2.12

Proof: Draw the three forces F_1, F_2 and F_3 one after the other in direction and magnitude starting from point *a*. Since the body is in equilibrium the resultant should be zero, which means the last point of force diagram should coincide with *a*. Thus, it results in a triangle of forces *abc* as shown in Fig. 2.12 (*b*). Now the external angles at *a*, *b* and *c* are equal to β, γ and α, since *ab*, *bc* and *ca* are parallel to F_1, F_2 and F_3 respectively. In the triangle of forces *abc*,

$$ab = F_1$$
$$bc = F_2 \quad \text{and}$$
$$ca = F_3$$

Applying sine rule for the triangle *abc*,

$$\frac{ab}{\sin(180° - \alpha)} = \frac{bc}{\sin(180° - \beta)} = \frac{ca}{\sin(180° - \gamma)}$$

i.e.,
$$\frac{F_1}{\sin \alpha} = \frac{F_2}{\sin \beta} = \frac{F_3}{\sin \gamma}$$

Note: When a body is in equilibrium under the action of only three forces, these three forces must be concurrent. Proof of this statement is given in the next chapter. However this concept has been used in this article also.

Example 2.9. *A sphere weighing 100 N is tied to a smooth wall by a string as shown in Fig. 2.13 (a). Find the tension T in the string and the reaction R from the wall.*

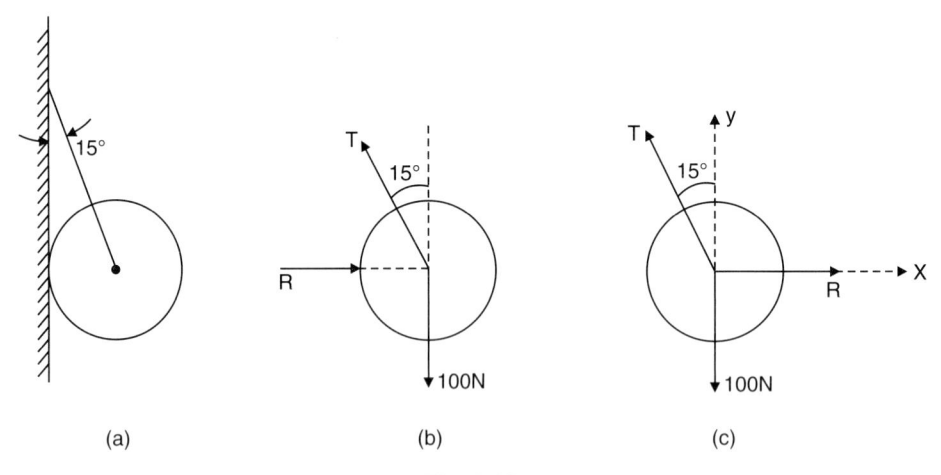

Fig. 2.13

Solution. Free body diagram of the sphere is as shown in Fig 2.13 (b). Figure 2.13 (c) shows all the forces acting away from the centre of the ball, which is permissible as per the law of transmissibility of forces. Applying Lami's theorem to the system of forces, we get

$$\frac{T}{\sin 90°} = \frac{R}{\sin (180° - 15°)} = \frac{100}{\sin (90° + 15°)}$$

$$T = 103.5 \text{ N} \quad \textbf{Ans.}$$

and $\qquad R = 26.8 \text{ N} \quad \textbf{Ans.}$

The above problem may be solved by using equations of equilibrium also. Taking horizontal axis as x-axis and vertical direction as y-axis as shown in Fig. 2.13 (c), we get

$$\Sigma F_y = 0 \quad \rightarrow$$
$$T \cos 15° - 100 = 0$$

∴ $\qquad T = 103.5 \text{ N} \quad \textbf{Ans.}$

$$\Sigma F_x = 0 \quad \rightarrow$$
$$R - T \sin 15° = 0$$

∴ $\qquad R = 103.5 \sin 15° = 26.8 \text{ N} \quad \textbf{Ans.}$

Note the following points:

1. The string can have only tension in it (it can pull a body), but cannot have compression in it (cannot push a body).

2. The wall reaction is a push, but not a pull on the body.

3. The line of action of reactions should be determined accurately, but the direction can be assumed. If assumed direction is correct the value comes out to be positive. If that is exactly opposite, the value comes out to be negetive. Hence in such case, the result may be reported with the reversed direction.

Example 2.10. *Determine the horizontal force F to be applied to the block weighing 1500 N to hold it in position on a smooth inclined plane AB which makes an angle of 30° with the horizontal (Ref. Fig. 2.14 (a)).*

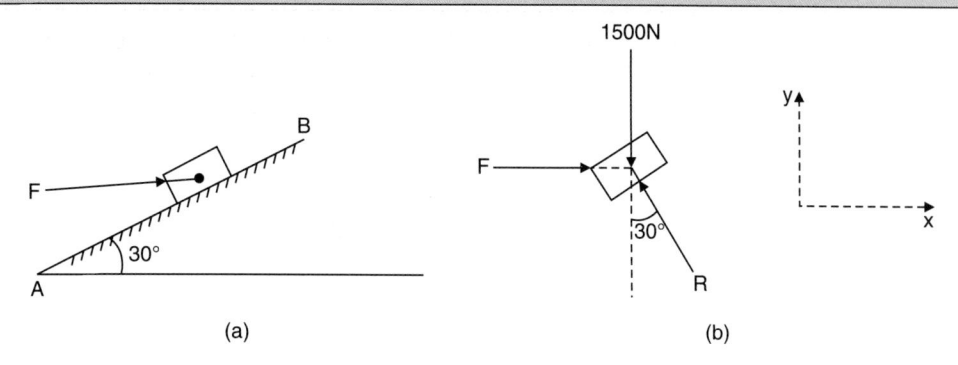

RESULTANT AND EQUILIBRIUM OF SYSTEM OF COPLANAR CONCURRENT FORCES

Fig. 2.14

Solution. The body is in equilibrium under the action of applied force F, self weight 1500 N and reaction R from the plane. Applied force is horizontal and self weight is vertically downward. Reaction is normal to the plane AB, since the plane AB is smooth. Since plane makes 30° to horizontal, normal to it makes 60° to horizontal *i.e.*, 30° to vertical (Ref. Fig. 2.14 (b)).

$$\Sigma F_y = 0 \rightarrow$$
$$R \cos 30° - 1500 = 0$$

∴
$$R = \frac{1500}{\cos 30°} = 1732.0 \text{ N} \quad \textbf{Ans.}$$

$$\Sigma F_x = 0 \rightarrow$$
$$F - R \sin 30° = 0$$

∴
$$F = R \sin 30° = 1732 \sin 30°$$

i.e.,
$$F = 866 \text{ N} \quad \textbf{Ans.}$$

Note: Since the body is in equilibrium under the action of only three forces, the above problem can be solved using Lami's theorem also, as shown below:

$$\frac{R}{\sin 90°} = \frac{F}{\sin (180° - 30°)} = \frac{1500}{\sin (90° + 30°)}$$

∴
$$R = 1732 \text{ N} \quad \text{and} \quad F = 866 \text{N} \quad \textbf{Ans.}$$

Example 2.11. *Find the forces developed in the wires, supporting an electric fixture as shown in Fig. 2.15 (a).*

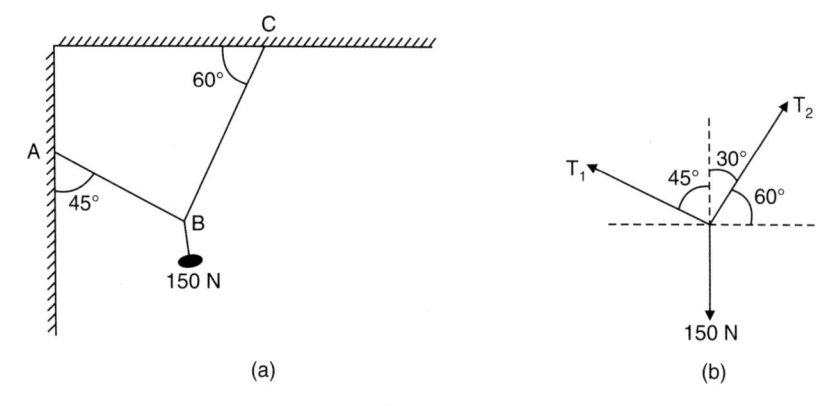

Fig. 2.15

Solution. Let the forces developed in the wires BA and BC be T_1 and T_2 as shown in Fig. 2.15 (b). Applying Lami's theorem to the system of forces, we get

$$\frac{T_1}{\sin(90° + 60°)} = \frac{T_2}{\sin(180° - 45°)} = \frac{150}{\sin(45° + 30°)}$$

∴ $T_1 = 77.6$ N and $T_2 = 109.8$ N **Ans.**

Example 2.12. *A 200 N sphere is resting in a trough as shown in Fig 2.16 (a). Determine the reactions developed at contact surfaces. Assume all contact surfaces are smooth.*

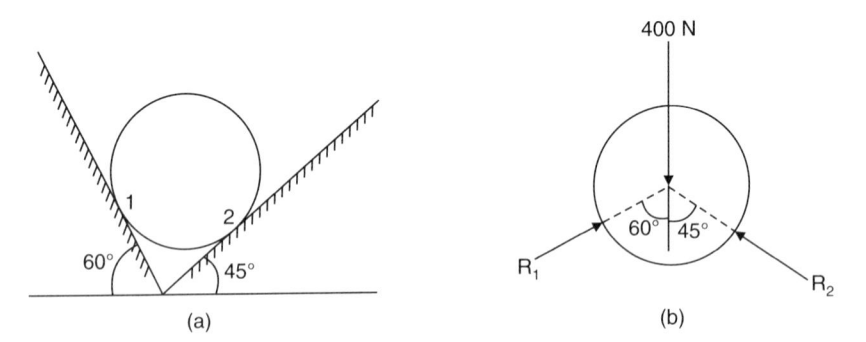

Fig. 2.16

At contact point 1, the surface of contact is making 60° to horizontal. Hence the reaction R_1 which is normal to it, makes 60° with the vertical. Similarly the reaction R_2 at contact point 2, makes 45° to the vertical. FBD is as shown in Fig. 2.16 (b). Applying Lami's theorem to the system of forces, we get

$$\frac{R_1}{\sin(180° - 45°)} = \frac{R_2}{\sin(180° - 60°)} = \frac{400}{\sin(60° + 45°)}$$

∴ $R_1 = 292.8$ N and $R_2 = 358.6$ N **Ans.**

Example 2.13. *A roller weighing 10 kN rests on a smooth horizontal floor and is connected to the floor by the bar AC as shown in Fig. 2.17 (a). Determine the force in the bar AC and reaction from the floor, if the roller is subjected to a horizontal force of 5 kN and an inclined force of 7 kN as shown in the figure.*

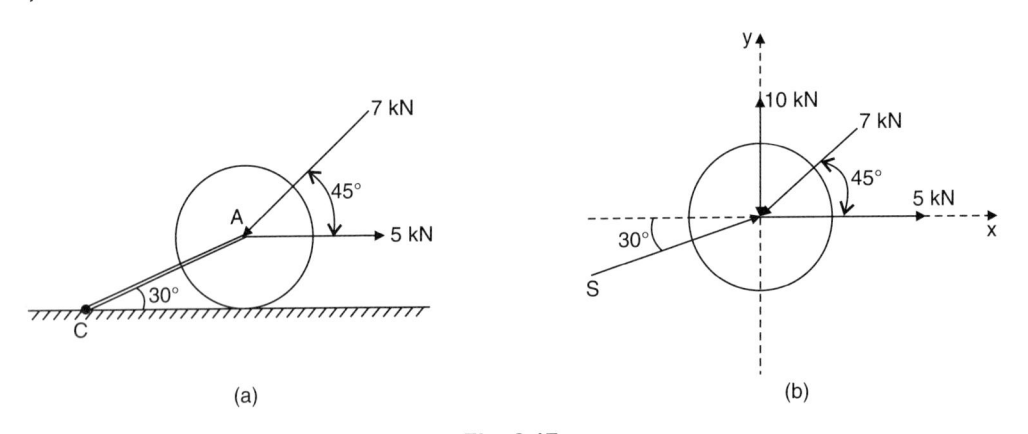

Fig. 2.17

Solution: A bar can develop tensile force or compressive force. Let the force developed be a compressive force S (push on the cylinder). Free body diagram of the cylinder is as shown in Fig. 2.17 (b).

Since there are more than three forces in the system, Lami's equations cannot be used. Consider the equilibrium equations.

$$\Sigma F_H = 0 \rightarrow$$
$$S \cos 30° + 5 - 7 \cos 45° = 0$$
$$S = \frac{7 \cos 45° - 5}{\cos 30°} = -0.058 \text{ kN}$$

Since the value is negative the reaction from the bar is not push, but it is a pull (tensile force in the bar) of magnitude 0.058 kN. **Ans.**

Referring to Fig. 2.17 (b)

$$\Sigma F_y = 0 \rightarrow$$
$$R - 10 - 7 \sin 45° + S \sin 30° = 0$$
$$R = 10 + 7 \sin 45° - S \sin 30°$$
$$= 10 + 7 \sin 45° - (-0.058) \sin 30°$$

i.e., $R = 14.980$ kN **Ans.**

Example 2.14. *A roller of radius r = 300 mm and weighing 2000 N is to be pulled over a curb of height 150 mm, as shown in Fig. 2.18 (a) by applying a horizontal force F applied to the end of a string wound around the circumference of the roller. Find the magnitude of force F required to start the roller move over the curb. What is the least pull F through the centre of the wheel to just turn the roller over the curb?*

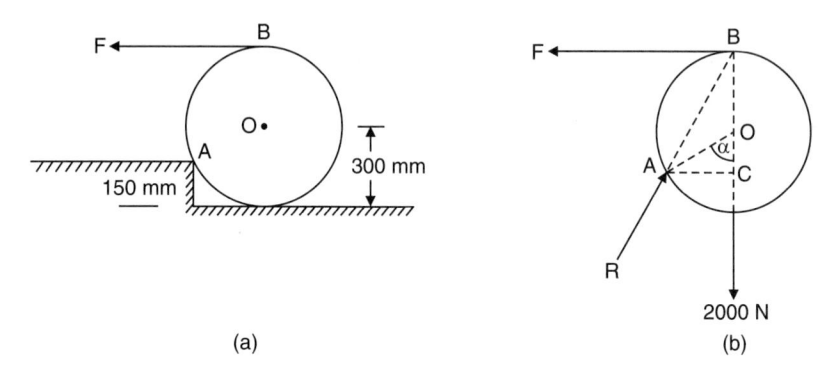

Fig. 2.18

Solution. When the roller is about to turn over the curb, the contact with the floor is lost and hence, there is no reaction from the floor. The body is in equilibrium under the action of three forces, namely,

   (i) Applied force F, which is horizontal
   (ii) Self weight, which is vertically downward, acting through the centre of roller, and
   (iii) Reaction R from the edge of the curb. Since the body is in equilibrium under the action of only three forces, they must be concurrent. It means the reaction at edge A of curb passes through the point B as shown in the Figure 2.18 (b).

Referring to Fig. 2.18 (b),

$$\cos \alpha = \frac{OC}{AO} = \frac{300-150}{300} = \frac{1}{2}$$

∴ $\alpha = 60°$

Now, in ΔAOB, $\angle OAB = \angle OBA$ Since $OA = OB$ = radius of roller.

but $\angle OAB + \angle OBA = \alpha$

∴ $2 \angle OBA = 60°$ or $\angle OBA = 30°$

i.e., reaction makes 30° with the vertical.

$$\Sigma V = 0 \to$$
$$R \cos 30° - 2000 = 0$$

∴ $R = 2309.4$ N.

$$\Sigma H = 0 \to$$
$$F - R \sin 30° = 0 \quad \text{or} \quad F = 2309.4 \sin 30° = 1154.7 \text{ N} \quad \textbf{Ans.}$$

Least force through the centre of Roller:

In this case the reaction from the curb must pass through the centre of the roller since the other two forces pass through that point. Its inclination to vertical is $\theta = 60°$.

(a) (b)

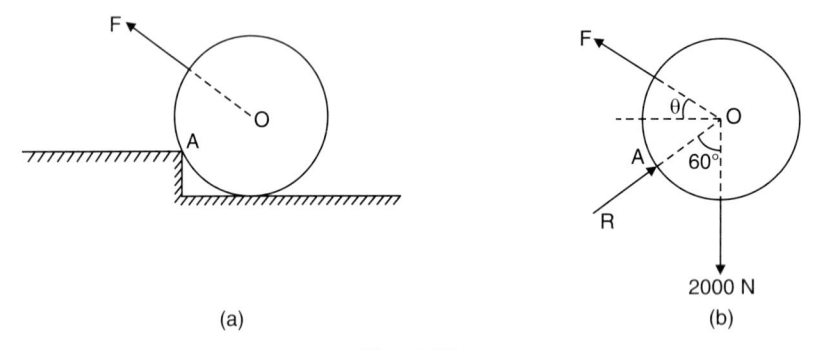

Fig. 2.19

Let force F make angle θ with the horizontal as shown in Fig. 2.19 (b).

$$\Sigma F_H = 0 \to F \cos \theta = R \sin 60°$$
$$\Sigma F_V = 0 \to F \sin \theta + R \cos 60° - W = 0$$

i.e., $$F \sin \theta + \frac{F \cos \theta}{\sin 60°} \cos 60° = W$$

i.e., $$F [\sin \theta + \cot 60° \cos \theta] = W$$

∴ $$\sin \theta + \cot 60° \cos \theta = \frac{W}{F}$$

For $\frac{W}{F}$ to be maximum i.e., F to be least,

$$\frac{d}{d\theta}\left[\frac{W}{F}\right] = 0$$

i.e., $\cos \theta + \cot 60° (-\sin \theta) = 0$

$\cos \theta = \cot 60° \sin \theta$

or $\cot \theta = \cot 60°$

RESULTANT AND EQUILIBRIUM OF SYSTEM OF COPLANAR CONCURRENT FORCES

i.e., $\theta = 60°$,

i.e., F is least when it is at right angles to the reaction R.

$$\therefore \quad F_{min} = \frac{W}{\sin 60° + \cot 60° \cos 60°} = \frac{2000 \sin 60°}{\sin^2 60° + \cos^2 60°}$$

$$= 1732 \text{ N} \quad \textbf{Ans.}$$

Example 2.15. *The frictionless pulley A shown in Fig. 2.20 (a), is supported by two bars AB and AC which are hinged at B and C to a vertical wall. The flexible cable hinged at D, goes over the pulley and supports a load of 20kN at G. The angle made by various members of the system are as shown in the figure. Determine the forces in the bars AB and AC. Neglect the size of the pulley.*

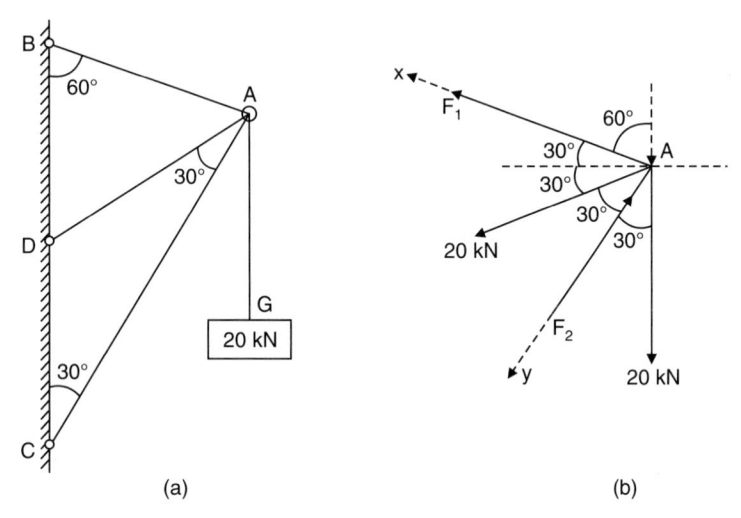

Fig. 2.20

Solution. Since pulley is frictionless, same force exists throughout in the flexible cable. Hence the force in AD is also 20 kN as shown in Fig. 2.20 (b). From the figure it may be observed that AC is perpendicular to AB. Selecting AB and AC as cartesian x and y axes,

$$\Sigma F_x = 0 \rightarrow$$

$$F_1 + 20 \sin 30° - 20 \sin 30° = 0$$

$$\therefore \quad F_1 = 0$$

$$\Sigma F_y = 0 \rightarrow$$

$$-F_2 + 20 \cos 30° + 20 \cos 30° = 0$$

$$\therefore \quad F_2 = 40 \cos 30° = 34.6 \text{ N} \quad \textbf{Ans.}$$

2.5. EQUILIBRIUM OF CONNECTED BODIES

When two or more bodies are in contact with one another, the system of forces appear as though it is a non-concurrent force system. However, when each body is considered separately, in many situations it turns out to be a set of concurrent force system. In such cases, first, the body subjected to only two unknown forces is analysed and then it is followed by the analysis of other connected bodies. This type of examples are illustrated below :

Example 2.16. *A system of connected flexible cables shown in Fig. 2.21 (a) is supporting two vertical forces 200 N and 250 N at points B and D. Determine the forces in various segments of the cable.*

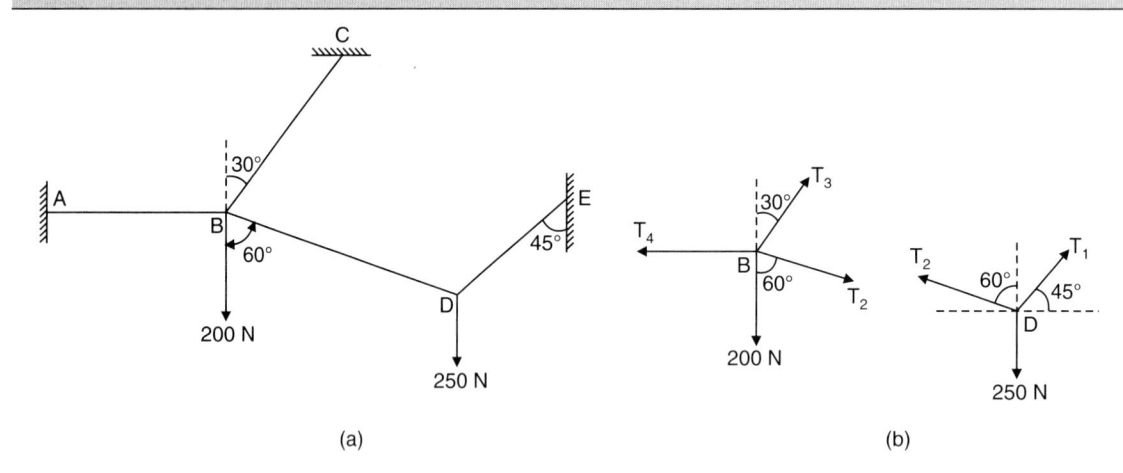

Fig. 2.21

Solution. Free body diagrams of points B and D are as shown in Fig. 2.21 (b). Let the forces in the members be as shown in the figure.

Applying Lami's theorem to the system of forces at point D, we get

$$\frac{T_1}{\sin(180° - 60°)} = \frac{T_2}{\sin(90° + 45°)} = \frac{250}{\sin(60° + 45°)}$$

∴ $T_1 = 224.1$ N and $T_2 = 183$ N **Ans.**

Now, consider the system of forces acting at B.

$$\Sigma F_V = 0 \rightarrow$$

$$T_3 \cos 30° - T_2 \cos 60° - 200 = 0$$

$$T_3 \cos 30° = T_2 \cos 60° + 200 = 183 \cos 60° + 200 = 291.5$$

∴ $T_3 = 336.6$ N **Ans.**

$$\Sigma F_H = 0 \rightarrow$$

$$-T_4 + T_3 \sin 30° + T_2 \sin 60° = 0$$

∴ $T_4 = 336.6 \sin 30° + 183 \sin 60° = 326.8$ N **Ans.**

Example 2.17. *Rope AB shown in Fig. 2.22 (a) is 4.5 m long and is connected at two points A and B at the same level 4 m apart. A load of 1500 N is suspended from a point C on the rope at 1.5 m from A. What load connected at point D on the rope, 1 m from B will be necessary to keep the position CD level?*

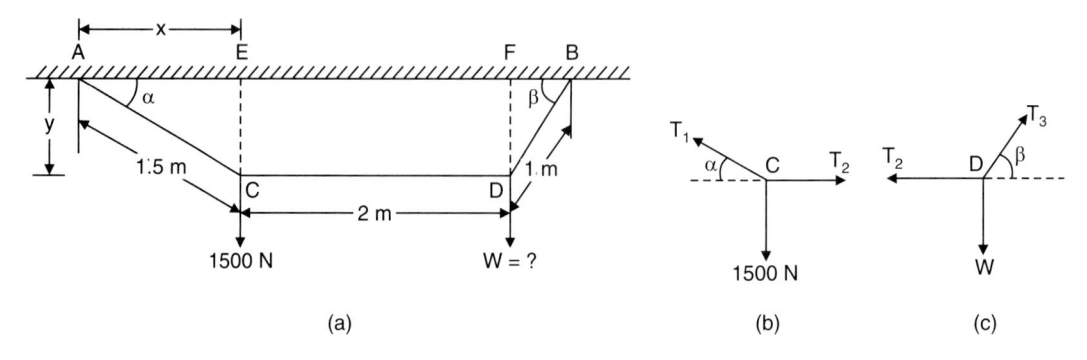

Fig. 2.22

Solution. Drop perpendiculars CE and DF on AB.
Let $\quad CE = y \quad$ and $\quad AE = x$
From $\triangle AEC$, $\quad x^2 + y^2 = 1.5^2 = 2.25$ ...(i)
Now, $\quad AB = 4$ m and $AC + CD + DB = 4.5$ m
$\therefore \quad CD = 4.5 - 1.5 - 1.0 = 2.0$ m
Since $\quad AC = 1.5$ m and $DB = 1$ m
$\therefore \quad EF = 2.0$ m
Now $\quad BF = AB - (AE + EF)$
$\quad\quad\quad = 4 - (x + 2) = 2 - x$ ...(ii)
From $\triangle BFD$,
$$BF^2 + DF^2 = 1^2$$
$$(2 - x)^2 + y^2 = 1 \quad \text{...(iii)}$$
Subtracting eqn. (iii) from eqn. (i), we get
$$x^2 - (2 - x)^2 = 1.25$$
i.e. $\quad x^2 - 4 + 4x - x^2 = 1.25$
$\quad\quad\quad x = 1.3125$ m.

$\therefore \quad \alpha = \cos^{-1} \dfrac{1.3125}{1.5} = 28.955°$

and $\quad \beta = \cos^{-1} \dfrac{2 - 1.3125}{1} = 46.567°$

Applying Lami's theorem to the system of forces acting at point C [Ref. Fig. 2.22 (b)],
$$\frac{T_1}{\sin 90°} = \frac{T_2}{\sin(90° + 28.955°)} = \frac{1500}{\sin(180° - 28.955°)}$$
$\therefore \quad T_1 = 3098.4$ N
$\quad\quad T_2 = 2711.1$ N

Now, applying Lami's theorem to the system of forces at B [Fig. 2.22 (c)],
$$\frac{T_3}{\sin 90°} = \frac{W}{\sin(180° - 46.567°)} = \frac{T_2}{\sin(90° + 46.567°)}$$
$\therefore \quad T_3 = 3943.4$ N and $W = 2863.6$ N **Ans.**

Example 2.18. *A wire rope is fixed at two points A and D as shown in Fig. 2.23 (a). Weights 20 kN and 30 kN are attached to it at B and C, respectively. The weights rest with portions AB and BC inclined at 30° and 50° respectively, to the vertical as shown in the figure. Find the tension in segments AB, BC and CD of the wire. Determine the inclination of the segment CD to vertical.*

Solution. Figures 2.23 (b) and (c) show free body diagrams of points B and C. Applying Lami's theorem for the system of forces B,
$$\frac{T_1}{\sin 50°} = \frac{T_2}{\sin(180° - 30°)} = \frac{20}{\sin(180° + 30° - 50°)}$$
$\therefore \quad T_1 = 44.80$ kN and $T_2 = 29.2$ kN **Ans.**
Consider the equilibrium of forces at C.
$$\Sigma F_H = 0 \rightarrow$$

$$T_3 \sin \theta = T_2 \sin 50° = 22.4 \qquad \text{...(i)}$$
$$\Sigma F_V = 0 \rightarrow$$
$$T_3 \cos \theta + T_2 \cos 50° - 30 = 0$$
i.e., $$T_3 \cos \theta = 30 - 29.2 \cos 50° = 11.20 \qquad \text{...(ii)}$$
From eqns. (i) and (ii), we get
$$\tan \theta = 2$$
∴ $$\theta = 63.43° \quad \textbf{Ans.}$$

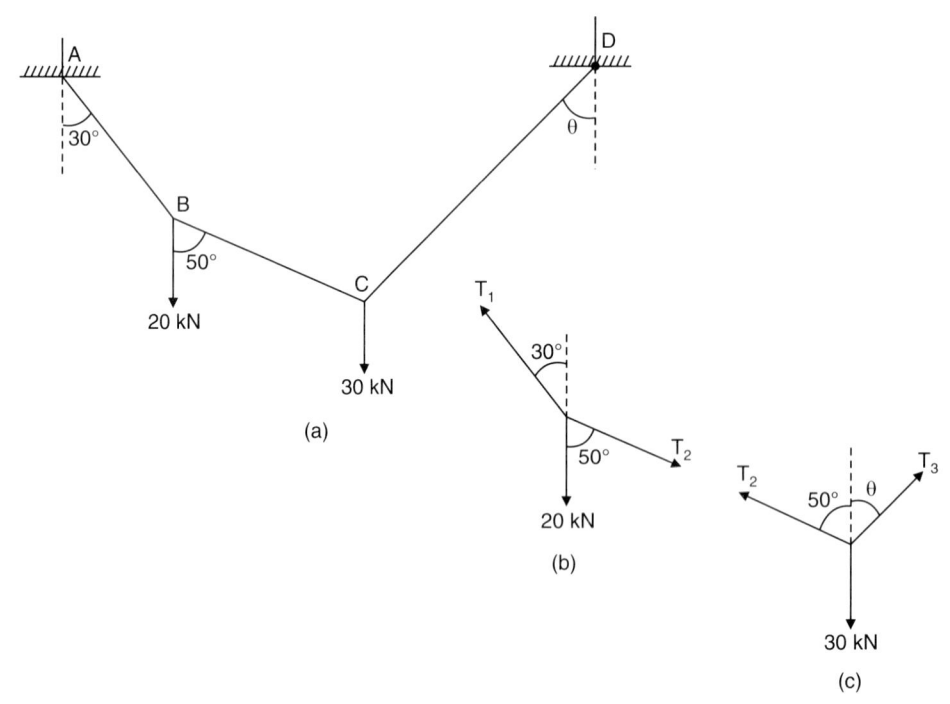

Fig. 2.23

Substituting this value in eqn. (i), we get
$$T_3 = 25.04 \text{ kN} \quad \textbf{Ans.}$$

Example 2.19. *A wire is fixed at A and D as shown in Fig. 2.24 (a). Weight 20 kN and 25 kN are supported at B and C respectively. When equilibrium is reached it is found that inclination of AB is 30° and that of CD is 60° to the vertical. Determine the tension in the segments AB, BC and CD of the rope and also the inclination of BC to the vertical.*

Solution. Free body diagrams of points B and C are shown in Figs. 2.24 (b) and (c) respectively. Considering equilibrium of point B, we get
$$\Sigma F_H = 0 \rightarrow T_2 \sin \theta - T_1 \sin 30° = 0$$
i.e., $$T_2 \sin \theta = T_1 \sin 30° \qquad \text{...(i)}$$
$$\Sigma F_V = 0 \rightarrow -T_2 \cos \theta + T_1 \sin 30° - 20 = 0$$
i.e., $$T_2 \cos \theta = T_1 \sin 30° - 20 \qquad \text{...(ii)}$$
Considering the equilibrium of point C,
$$\Sigma F_H = 0 \rightarrow T_3 \sin 60° - T_2 \sin \theta = 0$$
i.e., $$T_2 \sin \theta = T_3 \sin 60° \qquad \text{...(iii)}$$

$$\Sigma F_V = 0 \rightarrow T_3 \cos 60° + T_2 \cos \theta - 25 = 0$$

i.e., $\qquad T_2 \cos \theta = -T_3 \cos 60° + 25$ ...(iv)

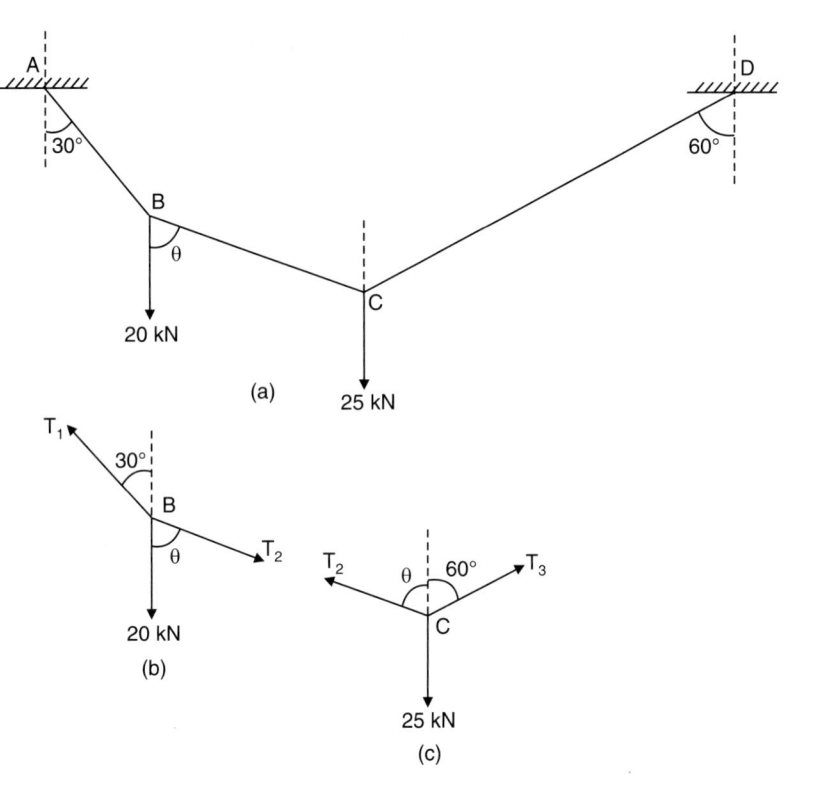

Fig. 2.24

From eqns. (i) and (iii), we get
$$T_1 \sin 30° = T_3 \sin 60°$$

i.e., $\qquad T_1 = \sqrt{3}\, T_3$ ...(v)

From eqns. (ii) and (iv), we get,
$$T_1 \cos 30° - 20 = 25 - T_3 \cos 60°$$

$$\sqrt{3} \times T_3 \times \frac{\sqrt{3}}{2} - 20 = 25 - T_3 \times 0.5$$

i.e., $\qquad 2T_3 = 45 \quad \text{or} \quad T_3 = 22.5$ kN

From eqn. (v) $\quad T_1 = 38.97$ kN
From eqn. (i) $\quad T_2 \sin \theta = 19.48$
and from eqn. (ii) $\quad T_2 \cos \theta = 13.75$
$\therefore \qquad \tan \theta = 1.416$
$\therefore \qquad \theta = 54.78°$ **Ans.**
and $\qquad T_2 = 23.84$ kN **Ans.**

Example 2.20. *Two identical cylinders, each weighing 500 N are placed in a trough as shown in Fig. 2.25 (a). Determine the reactions developed at contact points A, B, C and D. Assume all points of contact are smooth.*

Solution. Free body diagrams of two cylinders are as shown in Figs. 2.25 (b) and (c).

Consider the equilibrium of cylinder 1. Since R_A is at right angles to the plane, it makes 60° to the horizontal i.e., 30° to vertical. R_B is parallel to plane since the cylinders are identical. Thus R_A and R_B are at right angles to each other.

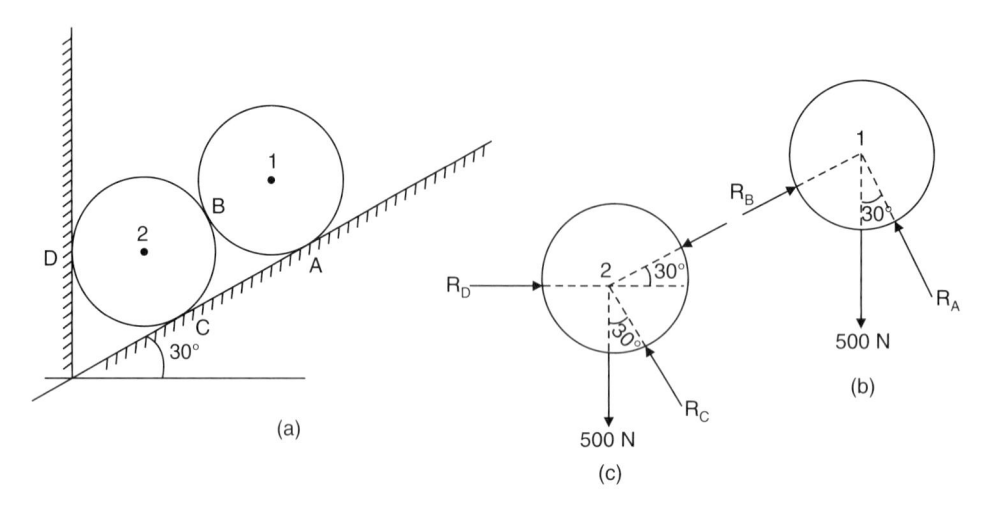

Fig. 2.25

Σ Forces normal to plane = 0, gives

$R_A - 500 \cos 30° = 0$ or $R_A = 433$ N **Ans.**

Σ Forces parallel to plane = 0, gives

$R_B - 500 \sin 30° = 0$ or $R_B = 250$ N **Ans.**

Now, consider equilibrium of cylinder 2. R_D is horizontal since the line of contact is vertical. R_C is normal to the plane and R_B is parallel to the plane.

$\Sigma F_V = 0 \rightarrow$
$- 500 + R_C \cos 30° - R_B \sin 30° = 0$

$\therefore \qquad R_C = \dfrac{500 + 250 \sin 30°}{\cos 30°} = 721.7$ N **Ans.**

$\Sigma F_H = 0 \rightarrow$
$R_D - R_C \sin 30° - R_B \cos 30° = 0$
$R_D = 721.7 \sin 30° + 250 \cos 30°$
$\therefore \qquad R_D = 577.4$ N **Ans.**

Example 2.21. *Cylinder 1 of diameter 200 mm and cylinder 2 of diameter 300 mm are placed in a trough as shown in Fig. 2.26 (a). If cylinder 1 weighs 800 N and cylinder 2 weighs 1200 N, determine the reactions developed at contact surfaces A, B, C and D. Assume all contact surfaces are smooth.*

Solution. Since all contact surfaces reactions are at right angles to the contact surfaces i.e., R_A and R_C are horizontal and R_D makes 45° to horizontal/vertical. R_B should be in the direction of O_1O_2. Let O_1O_2 makes an angle θ with horizontal. Let O_1P and O_2P be vertical and horizontal. Then

RESULTANT AND EQUILIBRIUM OF SYSTEM OF COPLANAR CONCURRENT FORCES

$$\cos \theta = \frac{O_2 P}{O_1 O_2} = \frac{450 - 100 - 150}{100 + 150} = 0.8$$

∴ $\theta = 36.87°$

Consider the equilibrium of cylinder 1,

$$\Sigma F_V = 0 \rightarrow R_B \sin \theta - 800 = 0$$

∴ $R_B = 1333.3$ N **Ans.**

$$\Sigma F_H = 0 \rightarrow R_A - R_B \cos \theta = 0$$

∴ $R_A = 1333.3 \cos 36.87° = 1066.7$ N **Ans.**

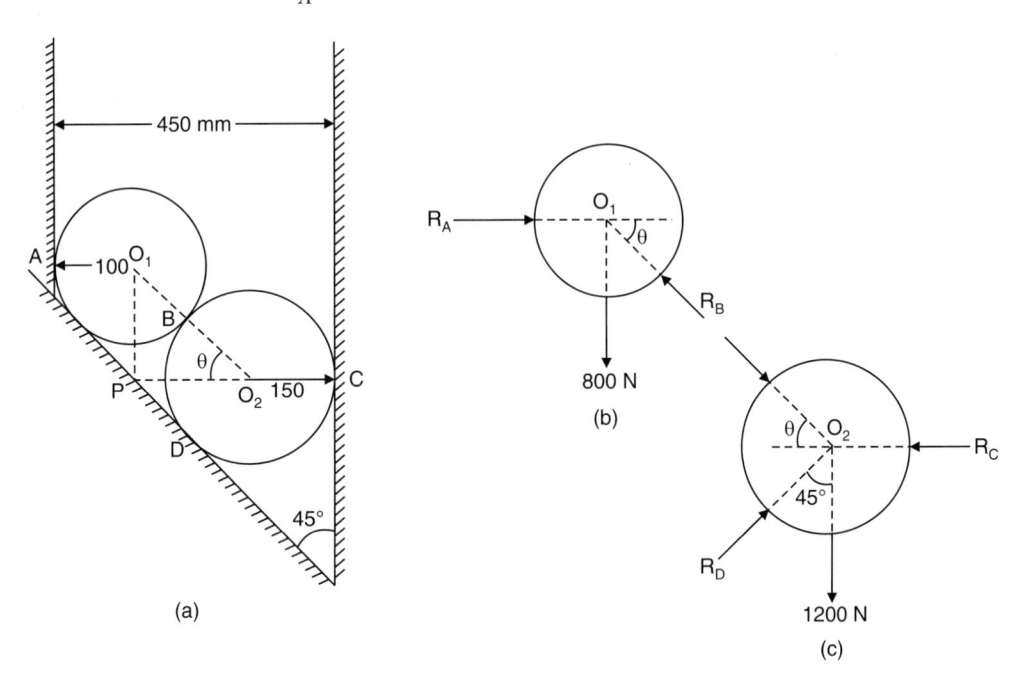

Fig. 2.26

Now consider the equilibrium of cylinder 2,

$$\Sigma F_V = 0 \rightarrow R_D \cos 45° - R_B \sin \theta - 1200 = 0$$

∴ $$R_D = \frac{1333.3 \sin 36.87° + 1200}{\cos 45°} = 2828.4 \text{ N} \quad \textbf{Ans.}$$

$$\Sigma F_H = 0 \rightarrow R_D \sin 45° + R_B \cos \theta - R_C = 0$$
$$2828.4 \sin 45° + 1333.3 \cos 36.87° - R_C = 0.$$

∴ $R_C = 3066.7$ N **Ans.**

Example 2.22. *A 600 N cylinder is supported by the frame ABCD as shown in Fig 2.27 (a). The frame is hinged at D. Determine the reactions developed at contact points A, B, C and D. Neglect the weight of frame and assume all contact surfaces are smooth.*

Solution. Free body diagram of the cylinder and the frame are shown in Figs. 2.27 (b) and (c) respectively. Considering the equilibrium conditions of cylinder, we get,

$$\Sigma F_V = 0 \rightarrow R_D = 600 \text{ N} \qquad \qquad ...(i)$$
$$\Sigma F_H = 0 \rightarrow R_A = R_C \qquad \qquad ...(ii)$$

Consider the equilibrium of the frame. As it is in equilibrium under the action of three forces only, they must be concurrent forces. In other words, the reaction at D has line of action along OD. Hence its inclination to horizontal is given by

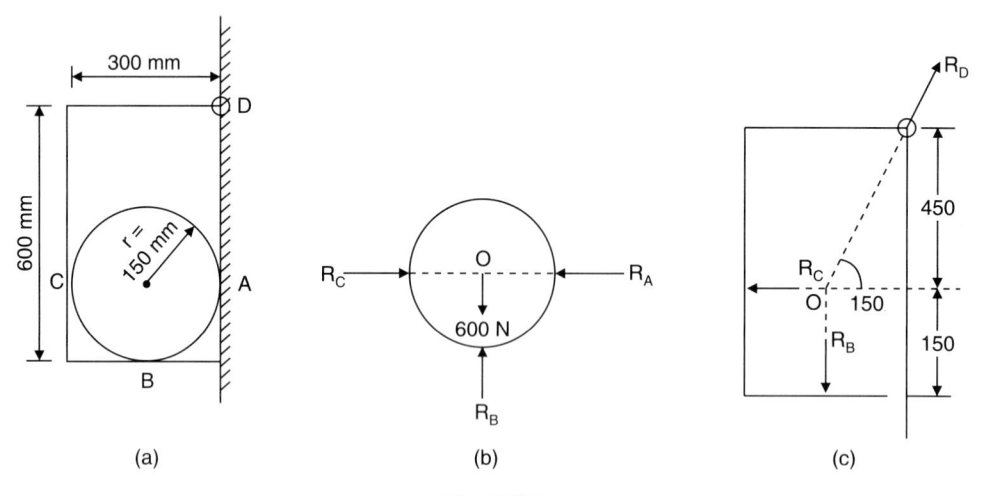

Fig. 2.27

$$\alpha = \tan^{-1} \frac{450}{150} = 71.56°$$

∴ $\Sigma F_V = 0 \to R_D \sin \alpha = R_B = 600$

∴ $R_D = \dfrac{600}{\sin 71.56°} = 632.4$ N **Ans.**

$\Sigma F_H = 0 \to R_D \cos \alpha - R_C = 0$

∴ $R_C = 632.4 \cos 71.56° = 200$ N **Ans.**

Hence from eqn. (*ii*) $R_A = 200$ N **Ans.**

Example 2.23. *Cylinder A weighing 4000 N and cylinder B weighing 2000 N rest on smooth inclines as shown in Fig. 2.28 (a). They are connected by a bar of negligible weight hinged to geometric centres of the cylinders by smooth pins. Find the force P to be applied as shown in the figure such that it will hold the system in the given position.*

Solution. Figures 2.28 (*b*) and (*c*) show the free body diagrams of two cylinders. Applying Lami's theorem to the system of forces on cylinder A, we get

$$\frac{C}{\sin(180° - 60°)} = \frac{4000}{\sin(60° + 90° - 15°)}$$

∴ $C = 4899$ N

Now consider equilibrium of cylinder B. Summation of forces parallel to the inclined plane = 0, gives

$P \cos 15° + 2000 \cos 45° - C \cos(45° + 15°) = 0$

∴ $P = \dfrac{-2000 \cos 45° + 4894 \cos 60°}{\cos 15°} = 1071.8$ N **Ans.**

RESULTANT AND EQUILIBRIUM OF SYSTEM OF COPLANAR CONCURRENT FORCES 41

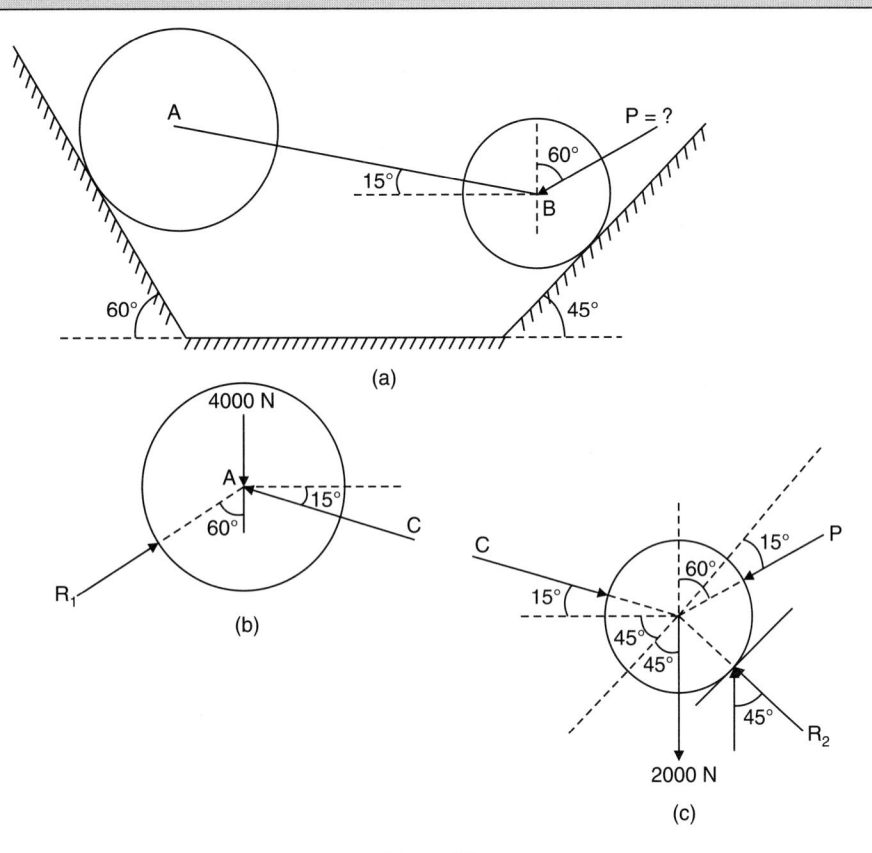

Fig. 2.28

Important Definitions

Resultant: The single force which will have the same effect as the system of forces.

Freebody Diagram: The diagram showing the body freed from all other bodies in contact and showing all the forces acting on it, including self weight and the reactions from the other bodies removed.

Lami's Theorem: If a body is in equilibrium under the action of only three forces, each force is proportional to the sine of the angle between the other two forces.

Important Formulae

1. *Resultant* of any number of concurrent forces may be found using the formulae,

$$R = \sqrt{(\Sigma F_x)^2 + (\Sigma F_y)^2}$$

and $\quad \alpha = \tan^{-1} \dfrac{\Sigma F_y}{\Sigma F_x}$, where α is angle made by resultant with x-axis.

2. *Equilibriant* is same as the resultant in magnitude but its direction is opposite to that of resultant.

3. According to Lami's theorem, $\dfrac{F_1}{\sin \alpha} = \dfrac{F_2}{\sin \beta} = \dfrac{F_3}{\sin \gamma}$

where α, β and γ are angles between $F_2 - F_3$, $F_3 - F_1$ and $F_1 - F_2$ respectively. This formula is to be used, if and only if the body is in equilibrium under the action of three forces only.

4. If a body is in equilibrium under the action of only three forces, they should be concurrent forces.

PROBLEMS FOR EXERCISE

2.1 The resultant of two forces one of which is 3 times the other is 300 N. When the direction of smaller force is reversed, the resultant is 200 N. Determine the two forces and the angle between them. [**Ans.** $F_1 = 80.6$ N, $F_2 = 241.8$ N, $\theta = 50.13°$]

2.2 A body is subjected to the three forces as shown in Fig. 2.29. If possible determine the direction of the force F so that the resultant is in x-direction.
When (a) $F = 5000$ N, (b) $F = 3000$ N

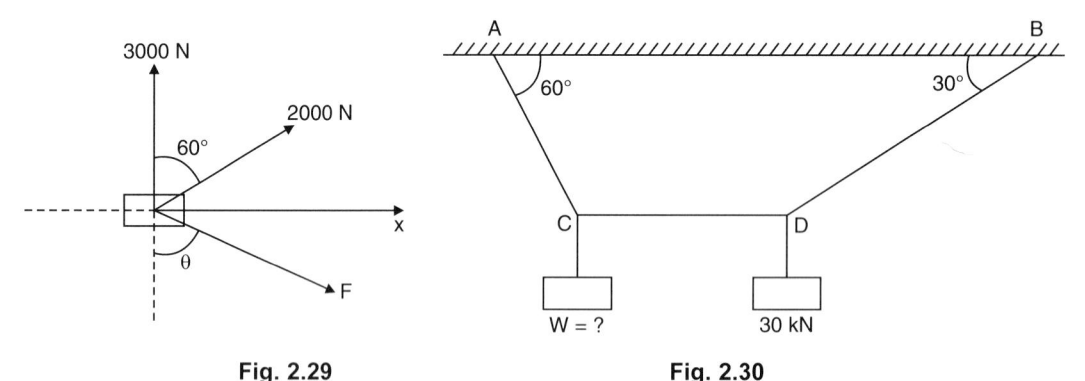

Fig. 2.29

Fig. 2.30

[**Ans.** (a) 36.87° ; (b) Not possible]

2.3 A chord supported at A and B carries a load of 10 kN at D and a load of W at C as shown in Fig. 2.30. Find the value of W so that CD remains horizontal. [**Ans.** $W = 30$ kN]

2.4 Three bars, hinged at A and D and pinned at B and C as shown in Fig. 2.31 form a four-linked mechanism. Determine the value of P that will prevent movement of bars. [**Ans.** 3047.2 N]

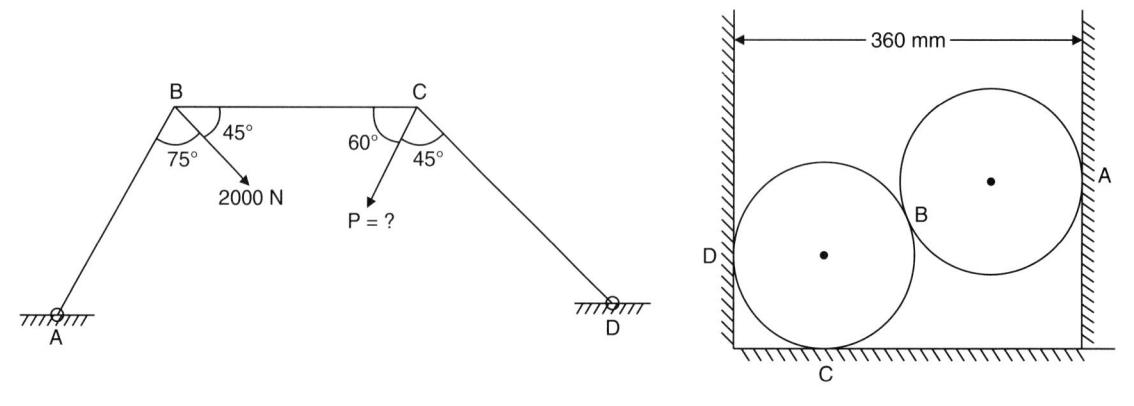

Fig. 2.31

Fig. 2.32

2.5 Two smooth spheres each of radius 100 mm and weighing 100 N, rest in a horizontal channel having vertical walls, the distance between which is 360 mm. Find the reactions at the points of contact A, B, C and D as shown in Fig. 2.32.

[**Ans.** $R_A = 133.33$ N, $R_B = 166.67$ N, $R_C = 200$ N, $R_D = 133.33$ N]

2.6 Three cylinders are placed in a rectangular ditch as shown in Fig. 2.33. Neglecting friction, determine the reaction between cylinder A and the vertical wall. Weights and radii of the cylinders are as given below:

Cylinder	Weight	Radius
A	75 N	100 mm
B	200 N	150 mm
C	100 N	125 mm

[**Ans.** $R = 400$ N]

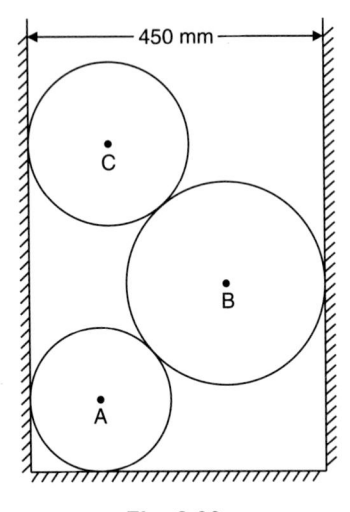

Fig. 2.33

Fig. 2.34

2.7 Three spheres A, B and C having their diameters 500 mm, 500 mm and 800 mm, respectively are placed in a trench with smooth side walls and floor as shown in Fig. 2.34. The centre-to-centre distance of spheres A and B is 600 mm. The cylinders A, B and C weigh 4 kN, 4 kN and 8 kN respectively. Determine the reactions developed at contact points P, Q, R and S.

[**Ans.** $R_P = 2.15$ kN, $R_Q = 7.44$ kN, $R_S = 7.03$ kN and $R_S = 2.29$ kN]

2.8 Three smooth spheres A, B and C weighing 300 N, 600 N and 300 N respectively and having diameters 800 mm, 1200 mm and 800 mm respectively are placed in a trench as shown in Fig. 2.35. Determine the reactions developed at contact points P, Q, R and S.

[**Ans.** $R_P = 61.24$ N, $R_Q = 631.6$ N, $R_R = 1095.2$ N and $R_S = 290.4$ N]

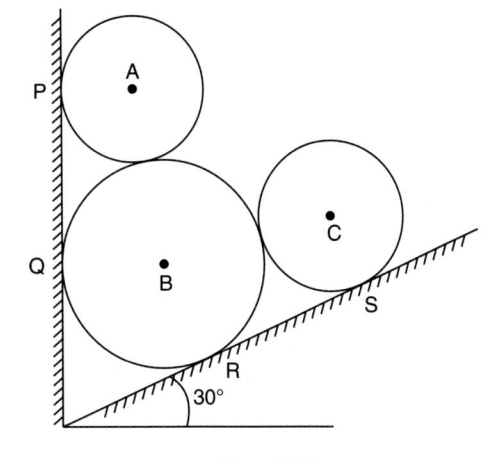

Fig. 2.35

3

Resultant and Equilibrium of System of Coplanar Non-Concurrent Forces

If all the forces in a system lie in a single plane and the lines of action of all the forces do not pass through a single point, the system is said to be **coplanar non-concurrent force system.** Three such systems are shown in Fig. 3.1. The parallel force systems shown in Figs. 3.1(b) and (c) are special cases of non-concurrent force system. In this chapter few technical terms are explained first, and then preliminaries are treated. Then the method of finding resultant and analysing equilibrium conditions are illustrated. The method of determining the reactions in statical beams is explained in detail, which is a common case of system of coplanar non-concurrent system of forces.

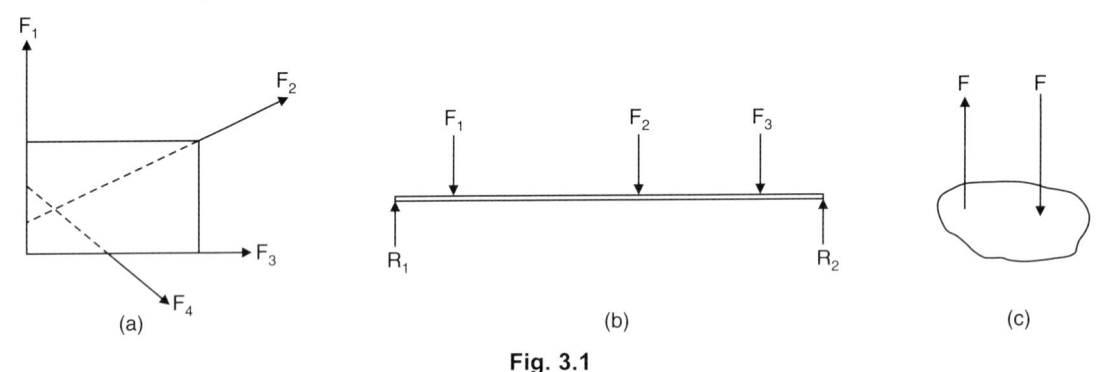

Fig. 3.1

3.1. MOMENT OF A FORCE

Moment of a force about a point is the measure of rotational effect of the force. Moment of a force about a point is defined as the product of the magnitude of the force and the perpendicular distance of the point from the line of action of the force. The point about which the moment is considered is called **moment centre** and the perpendicular distance of the point from the line of action of the force is called **moment arm**. Referring to Fig. 3.2, if d_1 is the perpendicular distance of point 1 from the line of action of force F, the moment of F about point 1 is given by

$$M_1 = Fd_1 \qquad \ldots(3.1a)$$

Similarly moment of F about point 2 is given by

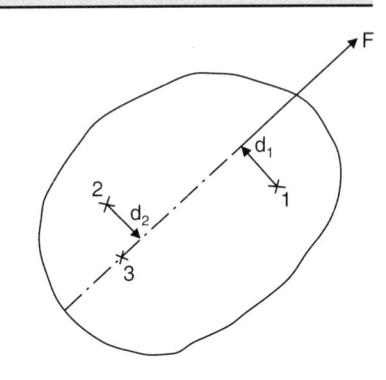

$$M_2 = Fd_2 \qquad \ldots(3.1b)$$

Fig. 3.2

44

RESULTANT AND EQUILIBRIUM OF SYSTEM OF COPLANAR NON-CONCURRENT FORCES

If the moment centre 3 lies on the line of action of the force F, the moment arm is zero and hence

$$M_3 = F \times 0 = 0 \qquad \ldots(3.1c)$$

Thus it may be noted that, **if a point lies on the line of action of a force, the moment of the force about that point is zero.**

The moment of a force has got direction also. In Fig. 3.2 it may be noted that M_1 is **clockwise** and M_2 is **anticlockwise**. To find the direction of moment, imagine that the line of action of the force is connected to the point by a rigid rod pinned at the moment centre and is free to move around the point by the force. The direction of the rotation of the rod indicate the direction of the moment by the force.

If the force is in newton unit and the distance is in millimeter, the unit of moment will be N-mm. Commonly used units of moment in engineering are kN-m, N-m, kN-mm and N-mm.

3.2. VARIGNON'S THEOREM

French mathematician Varignon (1654–1722) gave the following theorem which is also known as **principles of moments:**

The algebraic sum of moments of a system of coplanar forces about a moment centre is equal to the moment of their resultant force about the same moment centre.

Proof: Referring to Fig. 3.3, let R be the resultant of forces F_1 and F_2 and O be the moment centre. Let d, d_1 and d_2 be the moment arms of the forces R, F_1 and F_2 respectively. Then in this case we have to prove that

$$Rd = F_1 d_1 + F_2 d_2 \qquad \ldots(3.2)$$

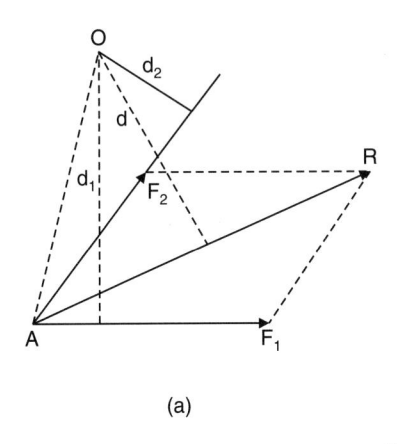

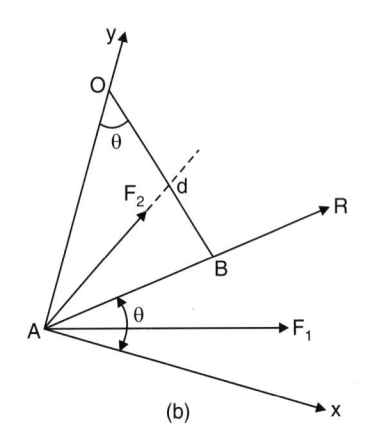

Fig. 3.3

Join OA and consider it as y-axis. Draw x-axis to it with A as origin. [Ref. Fig. 3.3(b)]. Let the resultant R makes an angle θ with x-axis. Noting that angle AOB is also θ, we can write

$$\begin{aligned} Rd &= R \times AO \cos \theta \\ &= AO \times (R \cos \theta) \\ &= AO \times R_x \qquad \ldots(i) \end{aligned}$$

where R_x denotes the component of R in x-direction. Similarly, if F_{1x} and F_{2x} are the components of F_1 and F_2 in x-direction respectively, then

$$F_1 d_1 = AO\, F_{1x} \qquad ...(ii)$$
and
$$F_2 d_2 = AO\, F_{2x} \qquad ...(iii)$$

∴ From eqns. (*ii*) and (*iii*) we get,
$$F_1 d_1 + F_2 d_2 = AO\,(F_{1x} + F_{2x})$$
$$= AO\, R_x \qquad ...(iv)$$

From eqns. (*i*) and (*iv*) we observe
$$F_1 d_1 + F_2 d_2 = Rd$$

Thus we find sum of the moment of forces about a moment centre is same as moment of their resultant about the same moment centre.

If a system of forces consists of more than two forces, the above result can be extended as given below:

Let F_1, F_2, F_3 and F_4 be four concurrent forces and R be their resultant. Referring to Fig. 3.4, d_1, d_2, d_3, d_4 and a be moment arms of F_1, F_2, F_3, F_4 and R about moment centre O.

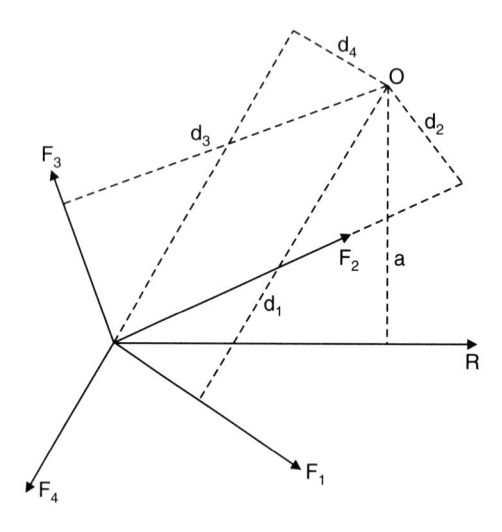

Fig. 3.4

If R_1 is the resultant of forces F_1 and F_2 and its moment arm is a_1, then from the above proof for two force system, we get,
$$R_1 a_1 = F_1 d_1 + F_2 d_2$$

If R_2 is the resultant of R_1 and F_3 and its moment arm from O is a_2, we can say
$$R_2 a_2 = R_1 a_1 + F_3 d_3$$
$$= F_1 d_1 + F_2 d_2 + F_3 d_3$$

Now considering R_2 and F_4, we can write
$$Ra = R_2 a_2 + F_4 d_4$$
$$= F_1 d_1 + F_2 d_2 + F_3 d_3 + F_4 d_4 \qquad ...(3.3)$$

Thus, the moment of the resultant of a number of forces about a moment centre is equal to the sum of the moments of its component forces about the same moment centre.

RESULTANT AND EQUILIBRIUM OF SYSTEM OF COPLANAR NON-CONCURRENT FORCES

Example 3.1. *Determine the moment of 100 N force acting at B about moment centre A as shown in Fig. 3.5.*

Solution. 100 N force may be resolved into its horizontal components 100 cos 60° and vertical component 100 sin 60°. From Varignon's theorem, moment of 100 N force about moment centre A is equal to sum of moment of its components about A. Taking clockwise moment as positive,

$$M_A = 100 \cos 60° \times 300 - 100 \sin 60° \times 500$$
$$= -28301 \text{ N-mm}$$

i.e., $M_A = 28301$ N-mm, anticlockwise. **Ans.**

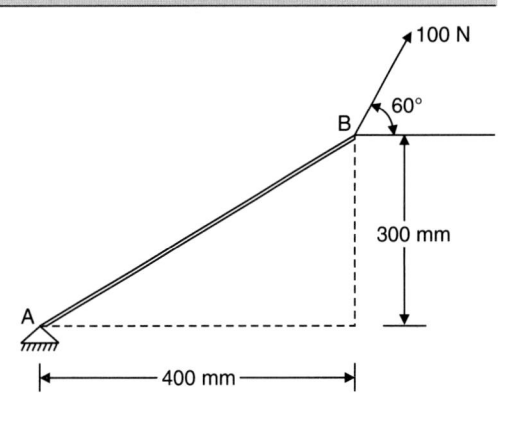

Fig. 3.5

Example 3.2. *What will be y-intercept of 5000 N force shown in Fig. 3.6, if its moment about A is 8000 N-m ?*

Solution. 5000 N force is shifted to point B along the line of action (law of transmissibility) and then it is resolved into its x and y components F_x and F_y as shown in the figure.

Noting that $\cos \theta = 4/5$ and $\sin \theta = 3/5$,

$$F_x = 5000 \cos \theta = 5000 \times 4/5$$
$$= 4000 \text{ N}$$

and
$$F_y = 5000 \sin \theta = 5000 \times 3/5$$
$$= 3000 \text{ N}$$

By Varignon's theorem, moment of 5000 N force about A is equal to the moment of its component forces about the same point.

∴ $\quad 8000 = 4000 \times y + 3000 \times 0$

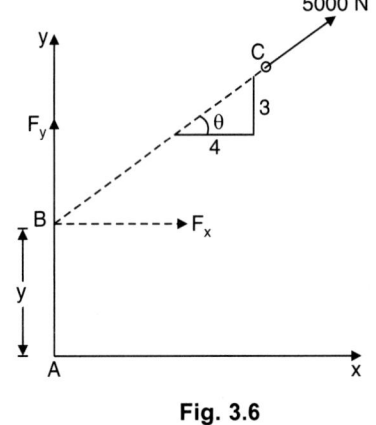

∴ $\quad y = 2$ m **Ans.**

Fig. 3.6

3.3. COUPLE

Two parallel forces equal in magnitude and opposite in direction and separated by a definite distance are said to form a couple. The sum of the forces forming a couple in any direction is zero, which means the translatory effect of the couple is zero. An interesting property of the couple can be observed, if we consider the rotational effect. Let the magnitude of the forces forming the couple be F and the perpendicular distance between the two forces be d. Consider the moment of the two forces constituting a couple about point 1 as shown in Fig. 3.7(a).

Let the moment be M_1. Then
$$M_1 = Fd_1 + Fd_2 = F(d_1 + d_2)$$
i.e., $\quad M_1 = Fd$

Now consider the moment of the forces about point 2, which is outside the two forces as shown in Fig. 3.7 (b). Let M_2 be the moment. Then,
$$M_2 = Fd_3 - Fd_4 = F(d_3 - d_4)$$
i.e., $\quad M_2 = Fd$

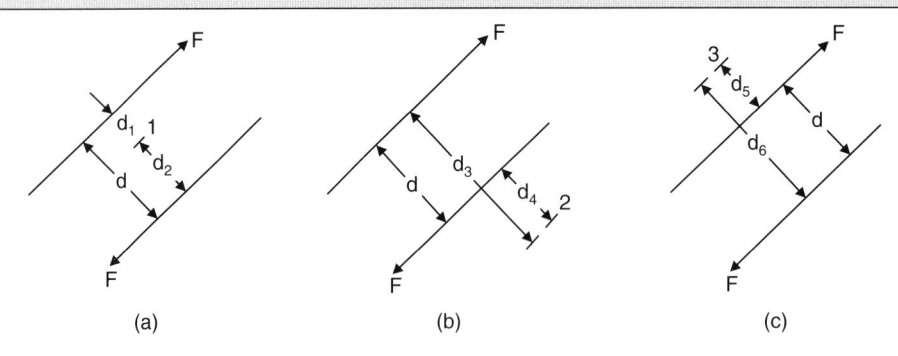

Fig. 3.7

Similarly, referring to Fig. 3.6 (c), we get
$$M_3 = Fd_6 - Fd_5 = F(d_6 - d_5)$$
i.e.,
$$M_3 = Fd.$$

Thus it can be observed that moment of a couple about any point is the same. Now we can list the following properties of a couple:

— A couple consists of a pair of equal and opposite parallel forces which are separated by a definite distance.
— The translatory effect of a couple on the body is zero.
— The rotational effect (moment) of a couple about any point is a constant and it is equal to the product of the magnitude of the forces and the perpendicular distance between the two forces.

Since only effect of a couple is the moment, which is the same at all points, the effect of a couple is unchanged if,

— The couple is rotated through any angle.
— The couple is shifted to any other position.
— The couple is replaced by another pair of forces whose rotational effect is the same.

3.4. RESOLUTION OF A FORCE INTO A FORCE AND A COUPLE

It will be advantageous to resolve a force acting at a point on a body into a force acting at some other point on the body and a couple moment. In Fig. 3.8 (a), F is a force acting on a body at point A.

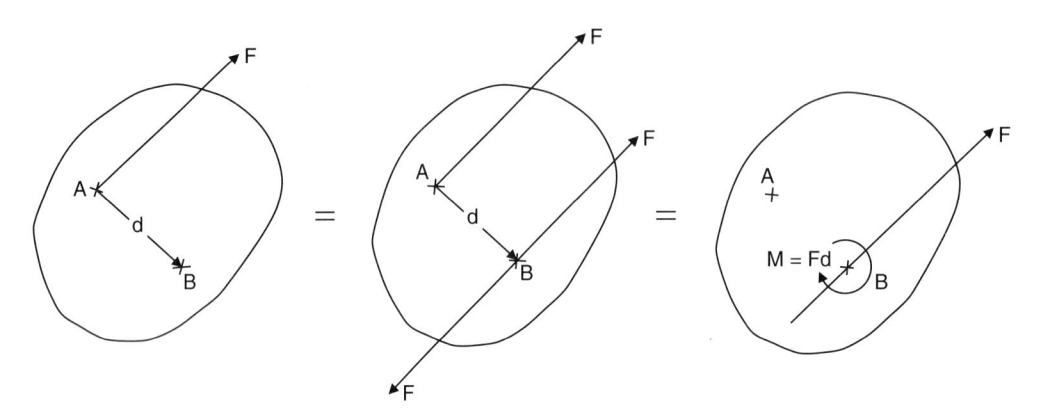

Fig. 3.8

RESULTANT AND EQUILIBRIUM OF SYSTEM OF COPLANAR NON-CONCURRENT FORCES 49

By applying equal and opposite forces of magnitude F at B, the effect of system of forces is unaltered (Law of superposition). Hence the system of forces in Fig. 3.8 (b) is the same as the system given in Fig. 3.8 (a). Now the original force F at A and the opposite force F at B form a couple moment of magnitude $F \times d$. Hence the system in Fig. 3.8 (b) is equal to the system of forces in Fig. 3.8 (c). Thus the given force F at A is equal to a force F at B and a moment Fd.

3.5. RESULTANT OF NON-CONCURRENT FORCE SYSTEM

Resultant of a force system is the one which will have the same rotational and translatory effect as the given system of forces. It may be a single force, a pure moment or a force and a moment.

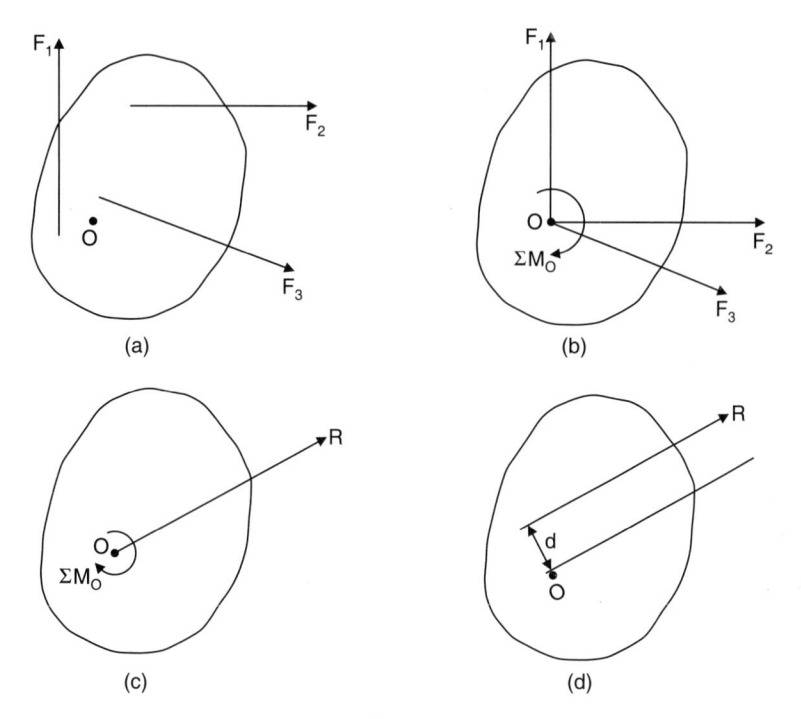

Fig. 3.9

Let F_1, F_2 and F_3, shown in Fig. 3.9 (a) constitute a system of forces acting on a body. Each force can be replaced by a force of same magnitude, acting in the same direction at point O, and a moment about O. Thus the given system in Fig. 3.9 (a) is equal to the system shown in Fig. 3.9 (b), where ΣM_O is the algebraic sum of the moments of the given forces about O.

At O, the concurrent forces F_1, F_2 and F_3 can be combined as to get the resultant force R. Now the resultant of the given system is equal to a force R at O and a moment ΣM_O as shown in Fig. 3.9 (c).

The force R and ΣM_O shown in Fig. 3.9 (c), can be replaced by a single force R acting at a distance d from O such that the moment produced by this force R about O is equal to ΣM_O [Ref. Fig. 3.9 (d)].

Thus we get a single force R acting at a distance d from the point O which gives the same effect as the constituent forces of the system. Mathematically,

$$R = \sqrt{(\Sigma F_x)^2 + (\Sigma F_y)^2}$$

$$\alpha = \tan^{-1}\frac{\Sigma F_y}{\Sigma F_x} \qquad \ldots(3.4)$$

and
$$d = \frac{\Sigma M_O}{R}$$

where ΣF_x – algebraic sum of components of all forces in x-direction
ΣF_y – algebraic sum of component of all forces in y-direction
α – inclination of the resultant R to x-direction.
ΣM_O – algebraic sum of all the forces about a point O.
and d – is the distance of the line of action of resultant from point O.

Note:
1. R is marked at distance d from point O such that it produces the same direction of moment as ΣM_O.
2. Sometimes the values of ΣF_x and ΣF_y may become zero, but ΣM_O may exist. This means that the resultant of the system gets reduced to a pure moment.

3.6. x AND y INTERCEPTS OF RESULTANT

In the previous article, we have seen how to find a point of application on the resultant, by finding the perpendicular distance d from the reference point 'O'. Many times it is convenient to locate the point of intersection of the line of action on the x or y axis through the reference point O.

Let d be the distance of the resultant from O and α be its inclination to x-axis (Ref. Fig. 3.10). Then x and y intercepts of the resultant are given by:

$$x = \frac{d}{\sin \alpha} \quad \text{and} \quad y = \frac{d}{\cos \alpha}$$

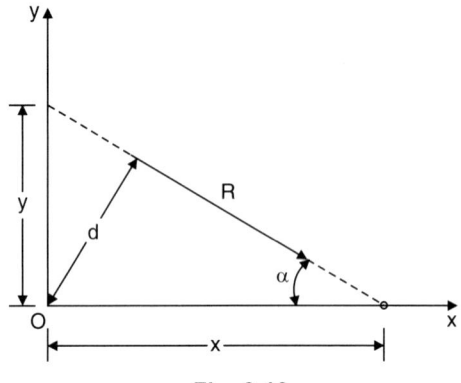

$\ldots(3.5)$

Fig. 3.10

Another method of finding the intercepts is given below:

Let $R_x = \Sigma F_x$ and $R_y = \Sigma F_y$ be the components of the resultant R in x and y directions. Considering the moment of R about O as the sum of its moments of its components at A about O (Varignon's theorem), we get (ref. Fig. 3.11)

$$Rd = \Sigma M_O$$

i.e., $\qquad R_x \times 0 + R_y x = \Sigma M_O$

$$\therefore \quad x = \frac{\Sigma M_O}{R_y} = \frac{\Sigma M_O}{\Sigma F_y} \qquad \ldots(3.6)$$

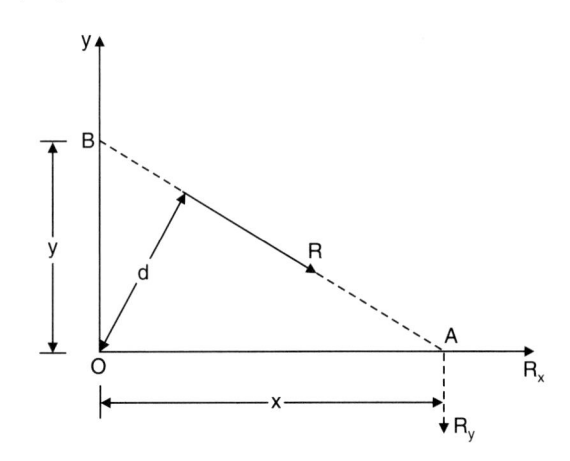

Fig. 3.11

RESULTANT AND EQUILIBRIUM OF SYSTEM OF COPLANAR NON-CONCURRENT FORCES

Similarly, resolving the resultant into its components at B (Ref. Fig. 3.11) and then applying Varignon's theorem for moments about O, it can be shown that

$$y = \frac{\Sigma M_O}{R_x} = \frac{\Sigma M_O}{\Sigma F_x} \qquad \ldots(3.7)$$

Note: A resultant in non-concurrent force system is completely determined only when its magnitude (R), direction (finding α) and a point on its line of action (represented by d or x or y intercepts) are determined.

Example 3.3. *Determine the resultant of the system of forces acting on a beam as shown in Fig. 3.12 (a).*

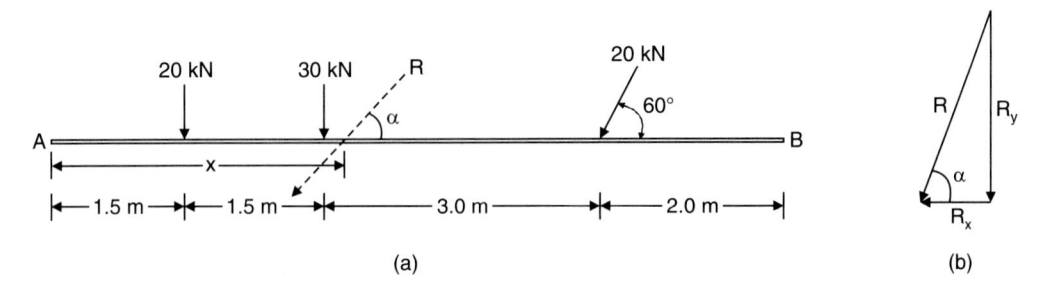

Fig. 3.12

Solution. Taking horizontal direction as x-axis and vertical upward direction as y-axis.

$$R_x = \Sigma F_x = -20 \cos 60° = -10 \text{ kN} = \overleftarrow{10 \text{ kN}}$$
$$R_y = \Sigma F_y = -20 - 30 - 20 \sin 60° = -67.32 \text{ kN} = 67.32 \text{ kN}\downarrow$$

∴
$$R = \sqrt{(\Sigma F_x)^2 + (\Sigma F_y)^2} = \sqrt{10^2 + 67.32^2} = 68.06 \text{ kN}$$

$$\alpha = \tan^{-1} \frac{\Sigma F_y}{\Sigma F_x} = \frac{67.32}{10.0} = 81.55°, \text{ as shown in Fig. 3.12}(b).$$

Taking moment about A,
$$\Sigma M_A = 20 \times 1.5 + 30 \times 3.0 + 20 \sin 60° \times 6 = 223.92 \text{ kN-m}.$$
x-intercept of the resultant is given by

$$x = \frac{\Sigma M_A}{R_y} = \frac{223.92}{67.32} = 3.326 \text{ m, as shown in the figure.}$$

Note: While finding x, signs of ΣM_A and R_y need not be considered, but while showing its location in the figure, see that x is such that the resultant R produces moment of the same nature (clockwise in this case) as ΣM_A.

Example 3.4. *Find the resultant of the force system shown in Fig. 3.13(a) acting on a lamina of equilateral triangular shape.*

Solution.
$$R_x = \Sigma F_x = 60 - 100 \cos 60° - 120 \cos 30°$$
$$= -93.9 \text{ N} = \overleftarrow{93.9 \text{ N}}$$
$$R_y = \Sigma F_y = -80 + 100 \sin 60° - 120 \sin 30°$$
$$= -53.40 = 53.40 \text{ N}\downarrow$$

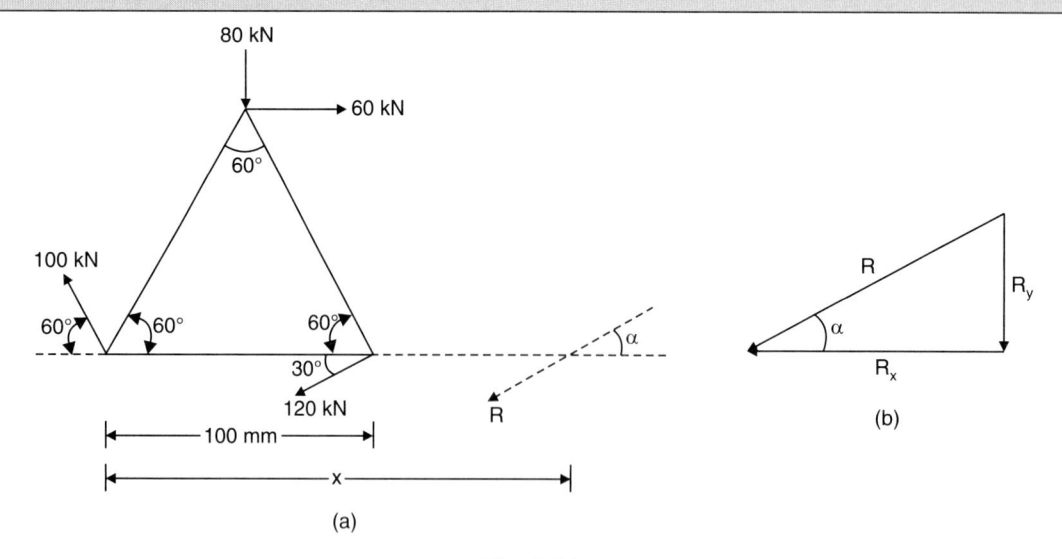

Fig. 3.13

$$R = \sqrt{(93.9)^2 + (53.4)^2}$$

i.e., $\quad R = 108.0 \text{ N}$ **Ans.**

$$\alpha = \tan^{-1}\frac{R_y}{R_x} = \tan^{-1}\frac{53.40}{93.9} = 29.62°, \text{ as shown in Fig. 3.13}(b) \quad \textbf{Ans.}$$

Let x be the intercept on x-axis from A. Then taking moment about A

$$x = \frac{\Sigma M_A}{R_y} = \frac{80 \times 100 \cos 60° + 60 \times 100 \sin 60° + 120 \sin 30° \times 100}{53.40}$$

i.e., $\quad x = 284.6$ mm, as shown in Fig. 3.13(a). **Ans.**

Example 3.5. *Find the resultant of the system of coplanar forces acting on a lamina as shown in Fig. 3.14(a). Each square has a side of 10 mm.*

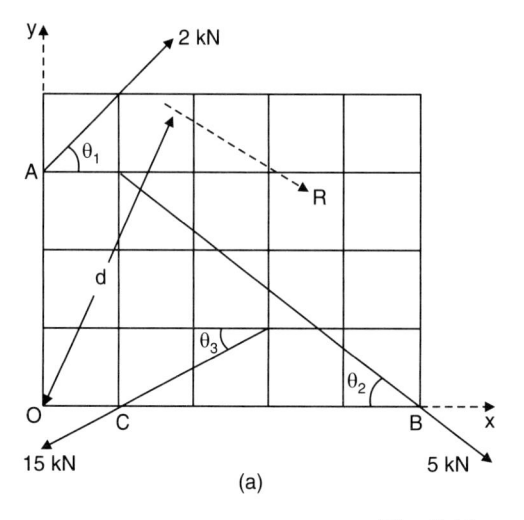

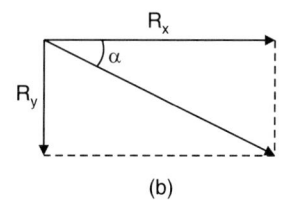

Fig. 3.14

RESULTANT AND EQUILIBRIUM OF SYSTEM OF COPLANAR NON-CONCURRENT FORCES

Solution. If θ_1, θ_2 and θ_3 are the inclinations of forces 2 kN, 5 kN and 1.5 kN with respect to x-axis, then

$$\tan \theta_1 = \frac{10}{10} = 1 \qquad \therefore \theta_1 = 45°$$

$$\tan \theta_2 = \frac{30}{40} \qquad \therefore \theta_2 = 36.87°$$

and

$$\tan \theta_3 = \frac{10}{20} \qquad \therefore \theta_3 = 26.565°$$

$\therefore$
$$R_x = \Sigma F_x = 2 \cos 45° + 5 \cos 36.87° - 1.5 \cos 26.565°$$
$$= 4.072 \text{ kN}$$
$$R_y = \Sigma F_y = 2 \sin 45° - 5 \sin 36.87° - 1.5 \sin 26.565°$$
$$= -2.257 \text{ kN}$$

$\therefore$
$$R = \sqrt{(4.072)^2 + (-2.257)^2} = 4.655 \text{ kN} \quad \textbf{Ans.}$$

$$\alpha = \tan^{-1} \frac{2.257}{4.072} = 29°, \text{ as shown in Fig. 3.14}(b), \quad \textbf{Ans.}$$

Resolving the forces 2 kN, 5 kN and 1.5 kN into their x, y components at A, B and C respectively as shown in Fig. 3.14a and then finding moment of the forces about O, we get

$$\Sigma M_O = 2 \cos 45° \times 30 + 5 \sin 36.87° \times 50 + 1.5 \sin 26.565° \times 10$$
$$= 199.13 \text{ kN-mm}$$

$\therefore$ Distance d of the resultant from O is given by,

$$d = \frac{199.13}{R} = \frac{199.13}{4.655} = 42.8 \text{ mm as shown in Fig. 3.14}(b) \quad \textbf{Ans.}$$

Example 3.6. *The system of forces acting on a bell crank is shown in Fig. 3.15(a). Determine the magnitude, direction and the point of application of the resultant.*

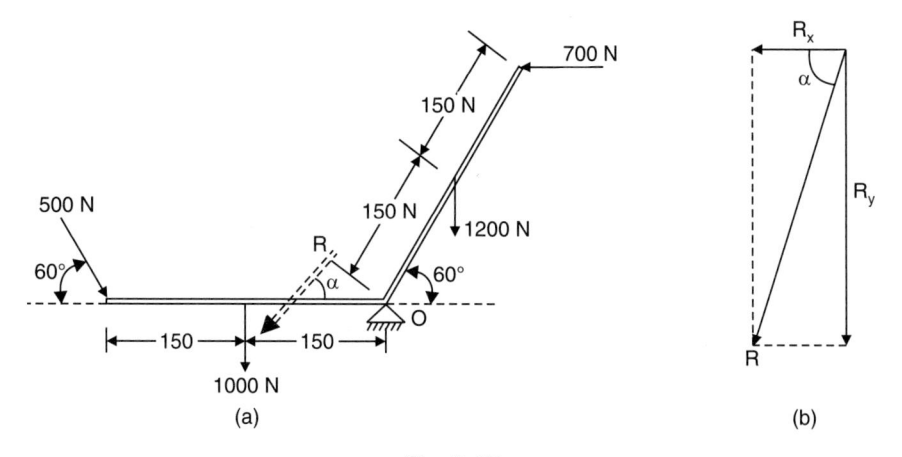

Fig. 3.15

Solution.
$$R_x = \Sigma F_x = 500 \cos 60° - 700 = -450 \text{ N} = \overleftarrow{450 \text{ N}}$$
$$R_y = \Sigma F_y = -500 \sin 60° - 1000 - 1200 = -2633 \text{ N} = 2633 \downarrow$$

$\therefore \qquad R = \sqrt{(450)^2 + (2633)^2} = 2671.2 \text{ N}$ **Ans.**

$$\alpha = \tan^{-1} \frac{2633}{450} = 80.30°, \text{ as shown in Fig. 3.15}(b) \quad \textbf{Ans.}$$

Let the point of application of the resultant be at a distance x from O along the horizontal arm. Then

$$x = \frac{\Sigma M_O}{R_y} = \frac{-500 \sin 60° \times 300 - 1000 \times 150 + 1200 \times 150 \cos 60° - 700 \times 300 \sin 60°}{2633}$$

$\therefore \qquad x = 141.2 \text{ mm}$ **Ans.**

Example 3.7. *Various forces to be considered for the stability analysis of a dam are shown in Fig. 3.16. The dam is safe if the resultant force passes through middle third of the base. Verify whether the dam is safe.*

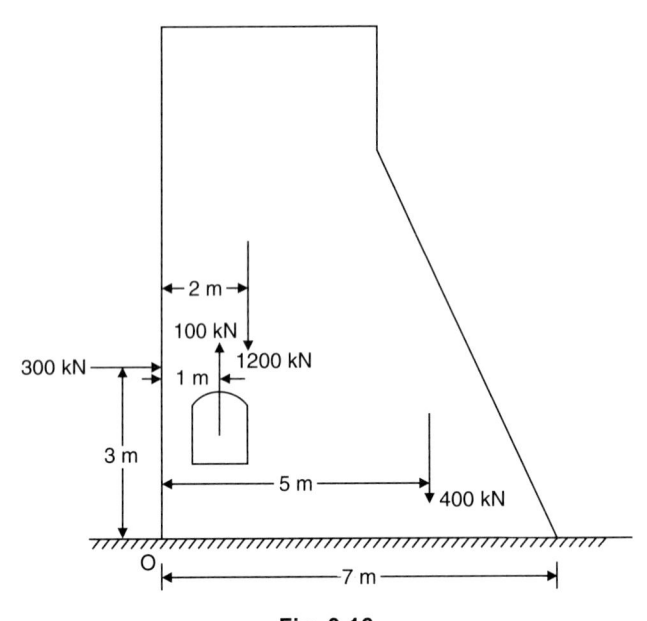

Fig. 3.16

Solution.
$$R_x = \Sigma F_x = \overrightarrow{300} \text{ kN}$$
$$R_y = \Sigma F_y = 100 - 1200 - 400 = -1500 = 1500 \text{ kN} \downarrow$$

Let resultant cut the base at a distance x from 'O'.

Then,
$$x = \frac{\Sigma M_O}{R_y} = \frac{300 \times 3 - 100 \times 1 + 1200 \times 2 + 400 \times 5}{1500}$$
$$= 3.467 \text{ m}.$$

RESULTANT AND EQUILIBRIUM OF SYSTEM OF COPLANAR NON-CONCURRENT FORCES

The resultant lies in the middle third of the base i.e., x is between 7/3 and 2 × 7/3. Hence the dam is safe. **Ans.**

Example 3.8. *A bracket is subjected to three forces and a couple as shown in Fig 3.17(a). Determine magnitude, direction and the line of action of the resultant.*

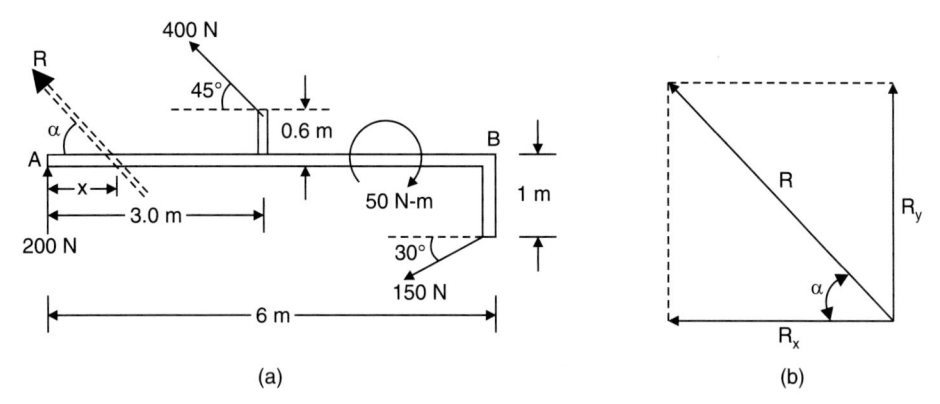

Fig. 3.17

Solution.

$$R_x = \Sigma F_x = -400 \cos 45° - 150 \cos 30°$$
$$= -412.7 \text{ N} = \overleftarrow{412.7}$$
$$R_y = \Sigma F_y = 200 + 400 \sin 45° - 150 \sin 30°$$
$$= 407.8 \text{ N}$$

∴ $$R = \sqrt{(412.7)^2 + (407.8)^2} = \mathbf{580.2 \text{ N}}$$

$$\alpha = \tan^{-1}\left(\frac{R_y}{R_x}\right) = \tan^{-1}\frac{407.8}{412.7} = \mathbf{44.76°}, \text{ as shown in Fig. 3.17(b)}$$

Let the resultant intersect arm AB at a distance x from A.
$$M_A = -400 \sin 45° \times 3 - 400 \cos 45° \times 0.6 + 50 + 150 \sin 30° \times 6 + 150 \times \cos 30° \times 1$$
$$= -438.3 \text{ N-m}$$
$$= 438.3 \text{ N-m (anticlockwise)}$$

∴ $$x = \frac{M_A}{R_y} = \frac{438.3}{407.8} = \mathbf{1.074 \text{ m}, \text{ as shown in Fig. 3.17(a)}}.$$

Example 3.9. *Determine the magnitude, direction and line of action of the equilibriant of the given set of coplanar forces acting on a planar structure shown in Fig. 3.18(a).*

Solution. The two 40 kN forces acting on the smooth pulley may be replaced by a pair of 40 kN forces acting at centre of pulley C and parallel to the given forces, since the sum of moments of the two given forces about C is zero.

∴ $$R_x = \Sigma F_x = 20 \cos 45° - 30 \cos 60° - 50 \cos 30° + 40 \cos 20° - 40 \sin 30°$$
$$= -26.57 \text{ kN}$$
$$= \overleftarrow{26.57} \text{ kN}$$
$$R_y = \Sigma F_y = -20 \sin 45° - 20 + 20 - 30 \sin 60° - 50 \sin 30°$$
$$- 40 \sin 20° - 40 \cos 30°$$

$$= -113.44 \text{ kN}$$
$$= 113.44 \text{ kN} \downarrow$$

$$\therefore \quad R = \sqrt{(26.57)^2 + (113.44)^2} = 116.51 \quad \textbf{Ans.}$$

$$\alpha = \tan^{-1} \frac{113.44}{26.57} = 76.82° \quad \textbf{Ans.}$$

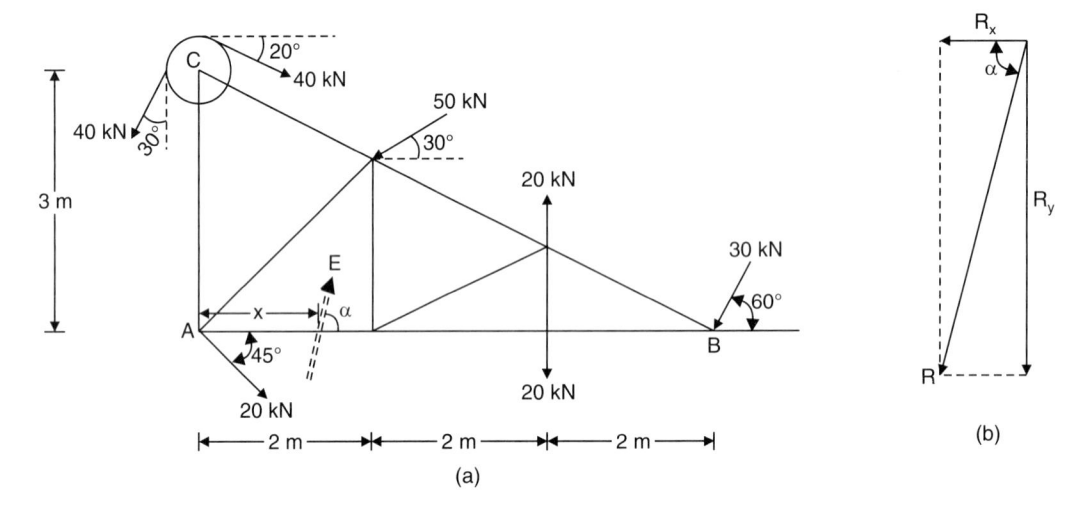

Fig. 3.18

Let the resultant intersect AB at a distance x from A. Then

$$R_y x = \Sigma M_A = 20 \times 4 - 20 \times 4 + 30 \sin 60° \times 6 + 50 \sin 30° \times 2$$
$$- 50 \cos 30° \times 2 + 40 \cos 20° \times 3 - 40 \sin 30° \times 3$$
$$= 172.04$$

$$\therefore \quad x = \frac{172.04}{113.44} = 1.516 \text{ m, as shown in Fig 3.18}(a) \quad \textbf{Ans.}$$

The equilibriant is equal and opposite to the resultant. Hence it is as shown in Fig. 3.18(a) in which

$$R = 116.51 \text{ kN}, \quad \alpha = 76.82° \quad \text{and} \quad x = 1.516 \text{ m.} \quad \textbf{Ans.}$$

3.7. EQUILIBRIUM OF NON-CONCURRENT SYSTEM OF FORCES

A body is said to be in equilibrium when it does not have any translatory or rotational movement. This means when the body is in equilibrium under the action of coplanar forces, the following conditions are satisfied:

(i) The algebraic sum of the components of forces along any two mutually perpendicular directions, is zero.

(ii) The algebraic sum of the moments of forces about any point in the plane, is zero.

Mathematically,

$$R_x = \Sigma F_x = 0 \, ; \quad R_y = \Sigma F_y = 0 \quad \text{and} \quad \Sigma M_A = 0 \qquad ...(3.8)$$

RESULTANT AND EQUILIBRIUM OF SYSTEM OF COPLANAR NON-CONCURRENT FORCES

Referring to Fig. 3.19, A, B, C are three points which are not collinear. Let R be the resultant of the system of forces on the body. Then,

$\Sigma M_A = 0$ means moment of all the forces about point A is zero. In other words, resultant passes through point A.

If $\Sigma M_B = 0$, then B is also the point through which resultant passes. In other words AB is the line of action of the resultant. If point C is not collinear with AB, $\Sigma M_C = 0$ means, R itself should be zero, since R cannot pass through C. Thus, if A, B, C are not collinear the following conditions are necessary and sufficient conditions of equilibrium

$$\Sigma M_A = 0, \qquad \Sigma M_B = 0 \quad \text{and} \quad \Sigma M_C = 0 \qquad \ldots(3.9)$$

The equilibrium conditions (3.9) are not independent of conditions (3.8). Two of them are common to the two sets. Referring to Fig. 3.19(b), R can be resolved into its components perpendicular to and parallel to AC.

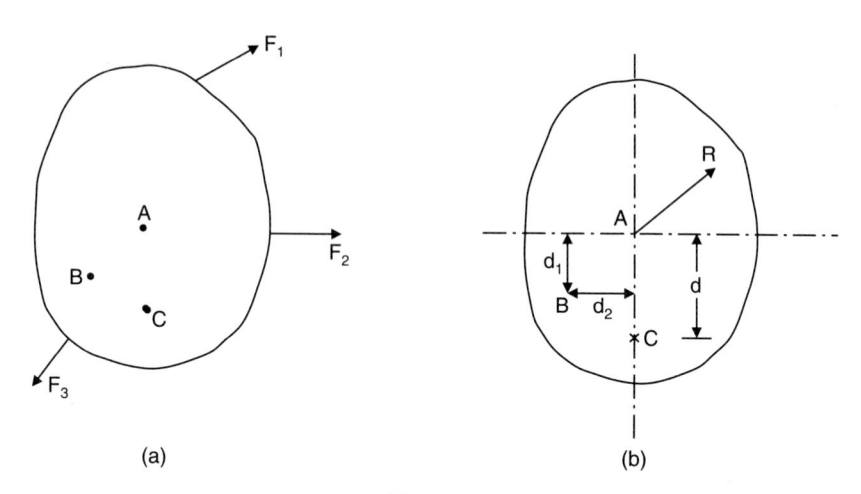

Fig. 3.19

Then, $\Sigma M_C = 0$
$\Rightarrow \qquad R_x d = 0$
i.e. $\qquad R_x = 0$
Now $\qquad \Sigma M_B = 0$
$\Rightarrow \qquad R_x d_1 + R_y d_2 = 0$
i.e., $\qquad R_y d_2 = 0$, since $d_2 \neq 0$.
$\therefore \qquad R_y = 0$

Thus $\Sigma M_C = 0$ is equivalent to $R_x = 0$ and
$\Sigma M_B = 0$ is equivalent to $R_y = 0$. Thus eqn. (10.9) is same as eqn. (10.8).

In fact any of the following set of equilibrium equations can be used:
1. $R_x = \Sigma F_x = 0$, $R_y = \Sigma F_y = 0$ and $M_A = 0$
2. If line AB is not in y-direction,
$$R_y = \Sigma F_y = 0, \Sigma M_A = 0 \text{ and } \Sigma M_B = 0 \qquad \ldots(3.10)$$
3. If line AB is not in x-direction,
$$R_x = \Sigma F_x = 0, \Sigma M_A = 0 \text{ and } \Sigma M_B = 0$$

4. If A, B and C are non-collinear
$$\Sigma M_A = 0, \Sigma M_B = 0 \text{ and } \Sigma M_C = 0$$

Now we are in a position to prove that *if a system is in equilibrium under the action of only three forces, they must be concurrent.*

Let F_1, F_2 and F_3 be the three forces acting on a body and the system is in equilibrium. Then, if A is the point of intersection of forces F_1 and F_2, the equilibrium condition gives
$$\Sigma M_A = 0$$
i.e.,
$$F_3 d = 0$$
where d is the distance of line of action of F_3 from A. Since F_3 is not zero,
$$d = 0$$

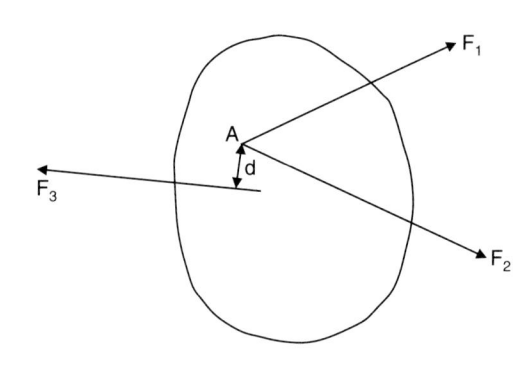

Fig. 3.20

In other word the third force F_3 also should pass through A. Hence the proposition is proved.

Example 3.10. *The 12m boom AB weighs 10kN, the distance of the centre of gravity G being 6m from hinge A. For the position shown, determine the tension T in the cable BC and the reaction at hinge A. [Ref. Fig. 3.21(a)].*

Solution. The free body diagram of the boom is shown in Fig. 3.21(b). Since A is a hinged end,

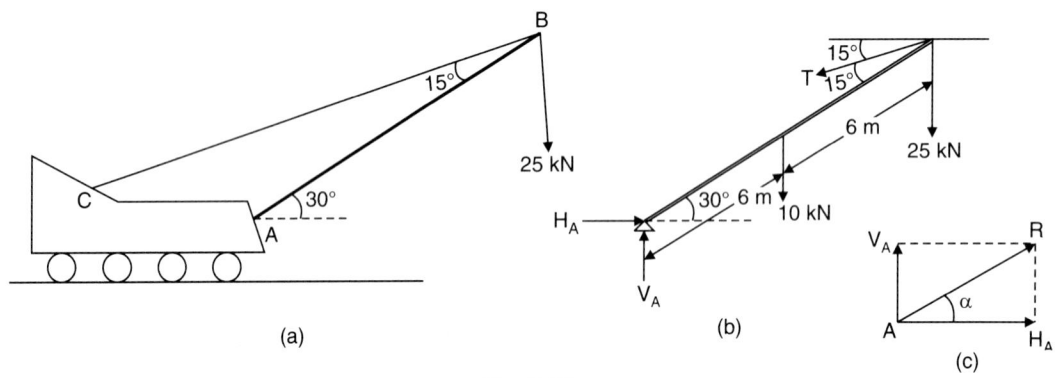

Fig. 3.21

$$\Sigma M_A = 0 \rightarrow$$
$$- T \times 12 \sin 15° + 25 \times 12 \cos 30° + 10 \times 6 \cos 30° = 0$$
$$\therefore \quad T = 100.38 \text{ kN} \quad \textbf{Ans.}$$
$$\Sigma F_H = 0 \rightarrow$$
$$H_A - T \cos 15° = 0$$
$$\therefore \quad H_A = 100.38 \cos 15° = 96.992 \text{ kN.}$$
$$\Sigma F_V = 0 \rightarrow$$
$$V_A - 10 - 25 - T \sin 15° = 0$$
$$V_A = 60.980 \text{ kN}$$
$$R_A = \sqrt{96.992^2 + 60.980^2} = 114.569 \text{ kN}$$

RESULTANT AND EQUILIBRIUM OF SYSTEM OF COPLANAR NON-CONCURRENT FORCES

$$\alpha = \tan^{-1} \frac{60.980}{96.992} = 32.16° \text{ as shown in Fig. 3.21 (c)}.$$

Example 3.11. *A ladder weighing 100 N is to be kept in the position shown in Fig. 3.22 (a), resting on a smooth floor and leaning on a smooth wall. Determine the horizontal force required at floor level to prevent it from slipping when a man weighing 700 N is at 2m above floor level.*

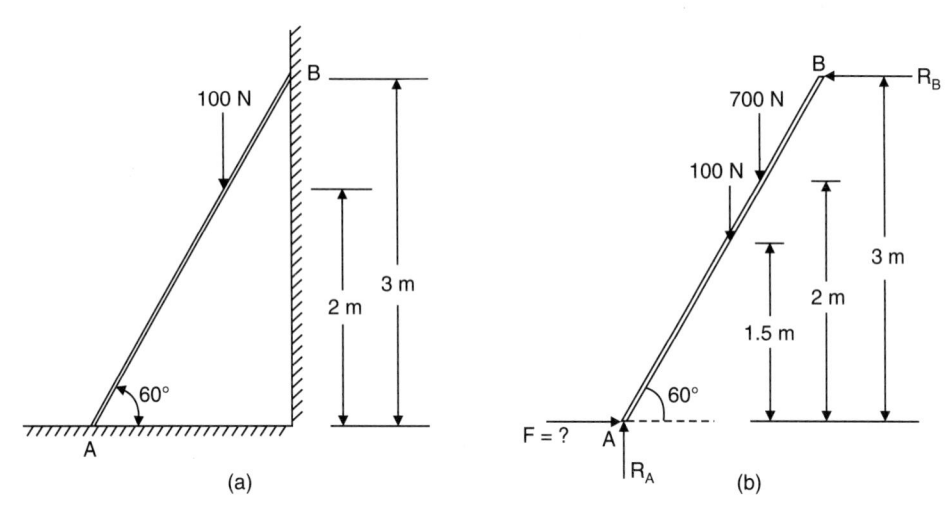

Fig. 3.22

Solution. Free body diagram of the ladder is as shown in Fig. 3.22 (b). R_A is vertical and R_B is horizontal because the surfaces of contact are smooth. Self weight of 100 N acts through centre point of ladder in vertical direction. Let F be the horizontal force required to be applied to prevent slipping.

Then, $\Sigma M_A = 0 \rightarrow -R_B \times 3 + 700 \times 2 \cot 60° + 100 \times 1.5 \cot 60° = 0$

∴ $R_B = 298.3$ N

$\Sigma F_H = 0 \rightarrow$

$F - R_B = 0$

∴ $F = R_B = 298.3$ N **Ans.**

Example 3.12. *In the above ladder problem, if the horizontal force F is to be applied at a height of 1 m above the ground level, how much should F be?*

Solution. Figure 3.23 shows the free body diagram of the ladder for this case.

$\Sigma F_H = 0 \rightarrow F - R_B = 0$ or $F = R_B$ ...(i)

$\Sigma M_A = 0 \rightarrow -R_B \times 3 + 700 \times 2 \cot 60° + 100 \times 1.5 \cot 60° + F \times 1 = 0$ ...(ii)

From (i) and (ii), we get

$3F - F = (700 \times 2 + 100 \times 1.5) \cot 60°$.

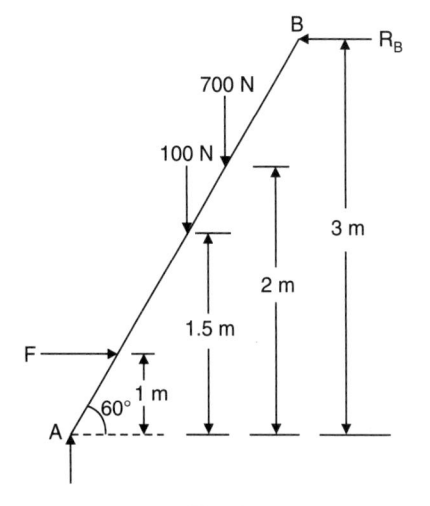

∴ $F = 447.45$ N **Ans.**

Fig. 3.23

Example 3.13. *A roller weighing 2000 N rests on an inclined bar weighing 800 N as shown in Fig. 3.24(a). Assuming weight of bar AB is negligible, determine the reactions developed at supports C and D.*

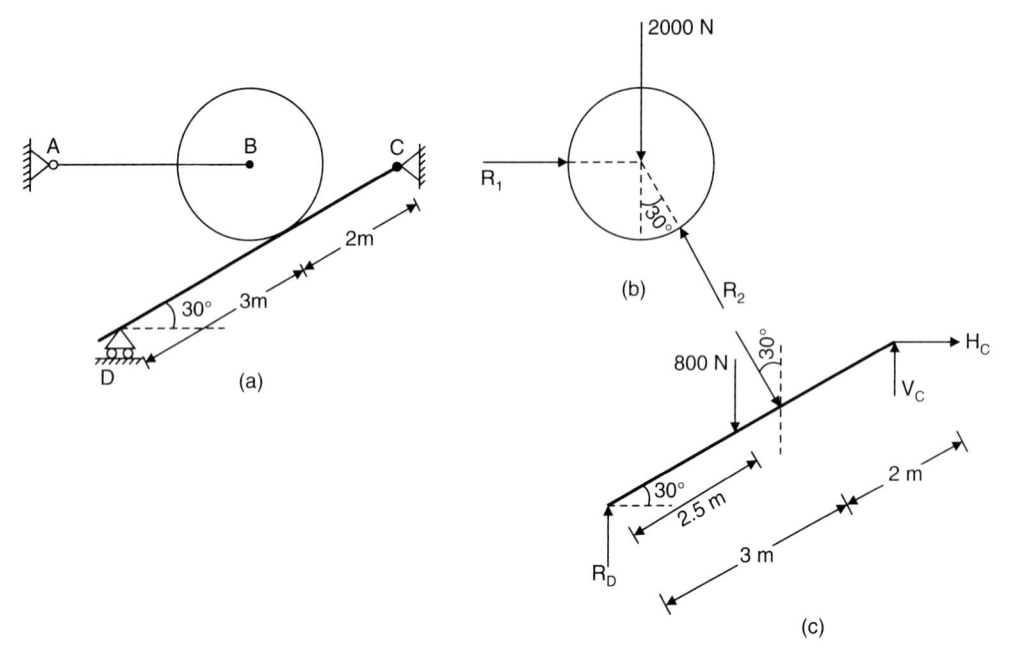

Fig. 3.24

Solution. Free body diagrams of roller and bar *CD* are as shown in Figs. 3.24(b) and (c).

[**Note:** R_D is vertical since at *D* roller support is horizontal. At *C* reaction can be in any direction, since it is hinged support. Hence components of reaction at *C* are taken as H_C and V_C].

For roller,
$$\Sigma F_V = 0 \quad \rightarrow \quad R_2 \cos 30° = 2000$$
∴ $\quad R_2 = 2309.4$ N

Considering the equilibrium of the bar *CD*,
$$\Sigma M_C = 0 \quad \rightarrow \quad R_D \, 5 \cos 30° - 800 \times 2.5 \cos 30° - R_2 \times 2 = 0$$

$$R_D = \frac{800 \times 2.5 \cos 30° + 2309.4 \times 2}{5 \cos 30°}$$

i.e., $\quad R_D = 1466.7$ N **Ans.**

$\Sigma F_V = 0 \quad \rightarrow$
$$R_D - 800 - 2309.4 \cos 30° + V_C = 0$$
∴ $\quad V_C = 800 + 2309.4 \cos 30° - 1466.7 = 1333.3$ N **Ans.**

$\Sigma F_H = 0 \quad \rightarrow$
$$R_2 \sin 30° - H_C = 0.$$
∴ $\quad H_C = 2309.4 \sin 30° = 1154.7$ N **Ans.**

Example 3.14. *A cable car used for carrying materials in a hydroelectric project is at rest on a track formed at an angle of 60° to horizontal. The gross weight of the car and its load is 60 kN and its centroid is at a point 800 mm from the track half way between the axles. The car is held by a cable as shown in Fig. 3.25. The axles of the car are at a distance 1.2 m. Find the tension in the cables and reaction at each of the axles neglecting friction of the track.*

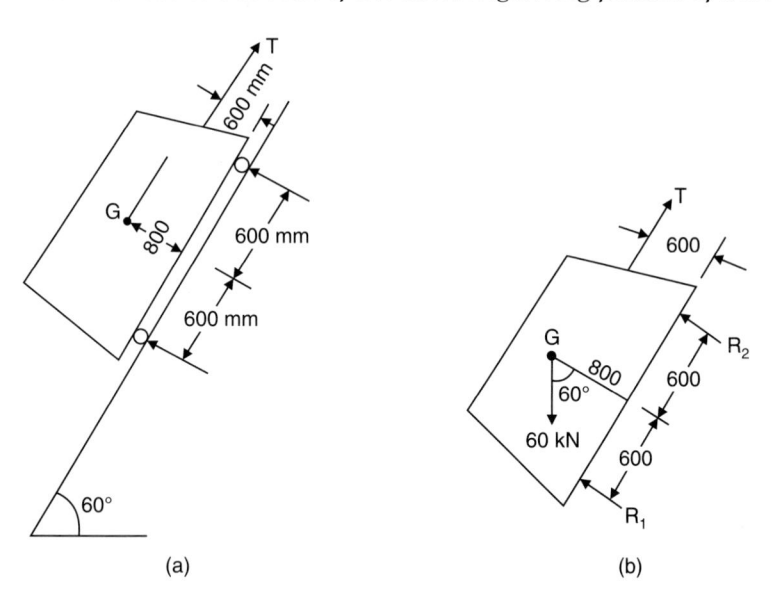

Fig. 3.25

Solution. The free body diagram of the car is shown in Fig. 3.25(b). The self weight is vertical and hence makes 60° to the normal to the inclined plane. Let T be the tension in the rope which is parallel to track.

Now, ΣForces parallel to track = 0, gives,
$$T - 60 \sin 60° = 0 \quad \text{or} \quad T = 51.961 \text{ kN} \quad \textbf{Ans.}$$

Applying moment equilibrium condition about upper axle point on track, we get
$$R_1 \times 1200 + T \times 600 - 60 \sin 60° \times 800 - 60 \cos 60° \times 600 = 0$$
∴ $\quad R_1 = 23.660$ kN. **Ans.**

ΣForces normal to the plane = 0, gives,
$$R_1 + R_2 - 60 \cos 60° = 0$$
∴ $\quad R_2 = 60 \cos 60° - R_1 = 60 \cos 60° - 23.660$
i.e., $\quad R_2 = 6.34$ kN **Ans.**

Example 3.15. *A hollow right circular cylinder of diameter 1600 mm is open at both ends and rests on a smooth horizontal plane as shown in Fig. 3.26(a). A sphere of radius 600 mm weighing 3 kN is first put in the cylinder and then a sphere weighing 1 kN and of radius 400 mm was placed. Neglecting the friction, find the minimum weight of hollow cylinder so that it will not tip over.*

Solution. Join the centres of spheres O_1 and O_2 and drop O_1D perpendicular to horizontal through O_2.

Now, $\quad O_1O_2 = 400 + 600 = 1000$ mm
$\quad O_2D = 1600 - 400 - 600 = 600$ mm

If α is the inclination of O_2O_1 to horizontal,

$$\alpha = \cos^{-1}\frac{600}{1000} = 53.13°$$

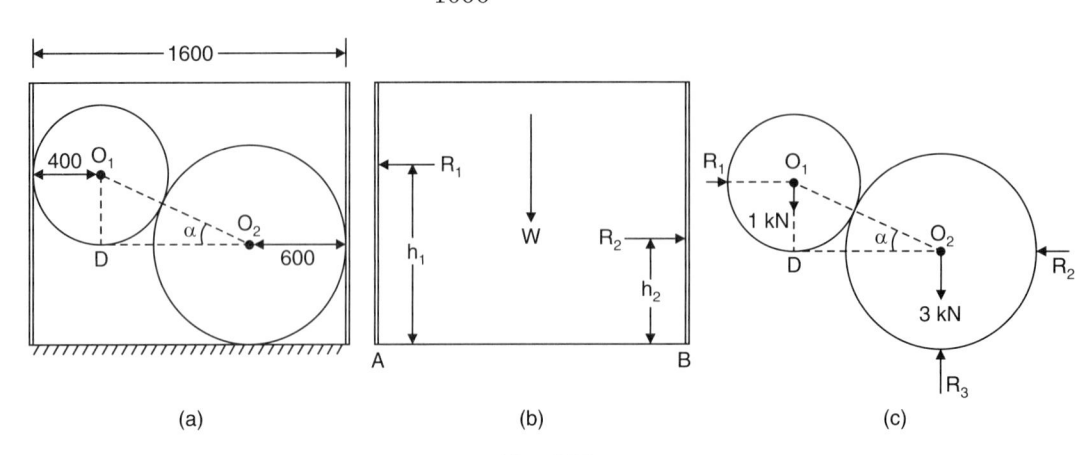

Fig. 3.26

Free body diagrams of cylinders and spheres are shown in Figs. 3.26 (b) and (c). Considering the equilibrium of the spheres,

$$\Sigma M_{O_2} = 0, \text{ gives}$$

$$R_1 O_1O_2 \sin \alpha - 1 \times O_2 D = 0$$

i.e., $\quad R_1 \times 1000 \sin 53.13° = 1 \times 600$

∴ $\quad\quad\quad R_1 = 0.75$ kN.

$\Sigma F_H = 0$, gives

$\quad\quad R_2 = R_1 = 0.75$ kN.

$\Sigma F_V = 0$, gives

$\quad\quad R_3 = 1 + 3 = 4$ kN.

Now, consider the equilibrium of the cylinder. Let the minimum weight be W. Tipping will take place over point A. Hence at this stage there will not be any reaction at point B. Hence,

$$\Sigma M_A = 0, \text{ gives,}$$

$$R_1 h_1 - R_2 h_2 - W \times 800 = 0$$

$$R_1(h_1 - h_2) = W \times 800, \quad \text{since} \quad R_1 = R_2.$$

i.e., $\quad\quad R_1 \times O_1 D = W \times 800$

∴ $\quad\quad W = \dfrac{0.75 \times 1000 \sin 53.13°}{800} = 0.75$ kN **Ans.**

Example 3.16. *A 500 N cylinder of 1 m diameter is loaded between the cross pieces which makes an angle of 60° with each other and are pinned at C. Determine the tension in the horizontal rope DE, assuming a smooth floor [Ref. Fig. 3.27(a)].*

RESULTANT AND EQUILIBRIUM OF SYSTEM OF COPLANAR NON-CONCURRENT FORCES

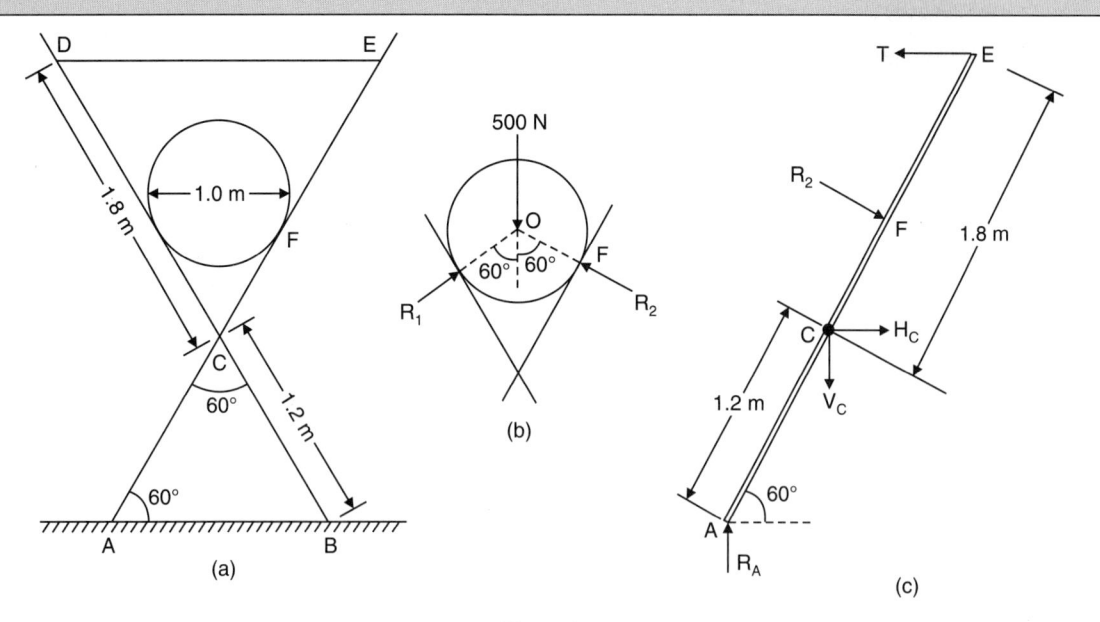

Fig. 3.27

Solution. Consider the equilibrium of the entire system.

Due to symmetry, $R_A = R_B$
$$\Sigma F_V = 0 \rightarrow R_A + R_B = 500$$
$$\therefore \quad R_A = R_B = 250 \text{ N}$$

Now, consider the equilibrium of the cylinder [Ref. Fig. 3.27(b)]. The members AE and BD make 60° with the horizontal. Hence, the reaction R_2 and R_1 make 60° with the vertical.

$\Sigma F_H = 0$, gives
$$R_1 \sin 60° - R_2 \sin 60° = 0$$
$$\therefore \quad R_1 = R_2$$
$\Sigma F_V = 0$, gives
$$R_1 \cos 60° + R_2 \cos 60° - 500 = 0.$$
Since, $R_1 = R_2$, we get
$$2R_1 \cos 60° = 500 \quad i.e. \quad R_1 = 500 \text{ N}$$
Hence $R_2 = 500 \text{ N}$

Now, consider the equilibrium of the member AE. It's free body diagram is shown in Fig. 3.27 (c).
$$\Sigma M_C = 0 \rightarrow$$
$$T \sin 60° \times 1.8 - R_2 \times CF - R_A \times 1.2 \cos 60° = 0$$
Now $\quad CF = OF \tan 60° = 0.5 \tan 60°$
$$= 0.866 \text{ m}.$$
$$\therefore \quad T = \frac{500 \times 0.866 + 250 \times 1.2 \times 0.5}{1.8 \sin 60°} = 374.0 \text{ N} \quad \textbf{Ans.}$$

Note: If the reactions on the pin at C are required, the remaining two equations, namely, $\Sigma F_x = 0$ and ΣF_y, are to be considered.

3.8. APPLICATIONS TO BEAM PROBLEMS

A beam is a structural element which has one dimension (length) considerably larger than the other two dimensions in the cross-section and is supported at few points. It is subjected to lateral loads. Due to applied loads, reactions develop at supports and the system of forces consisting of applied loads, self weight (many times neglected) and reactions keep the beam in equilibrium. The forces constitute a system of coplanar non-concurrent system in equilibrium. If the support reactions can be determined using the equations of equilibrium only, then the beam is said to be **statically determinate.** In this article, types of supports are types of beams and types of loading are explained first and then the method of finding reactions developed at the supports of determinate beams is illustrated.

3.8.1. Types of Supports

Various types of supports and reactions developed at these supports are listed below:

Simple Support. The end of the beam rests simply on a rigid support as shown in Fig. 3.28. In the idealised simple support there is no resistance to the force in the direction of the support. Hence the reaction is always normal to the support. There will not be any restraint from the support for the rotation of the end of the beam. In other words there is no moment resistance at support. [Ref. Fig. 3.28].

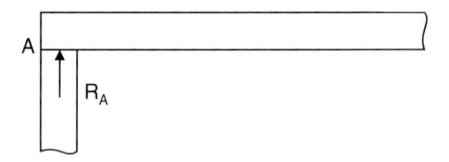

Fig. 3.28. Simple support

Roller Support. In this case, beam end is supported on rollers. In such cases, reaction is always normal to the support, since rollers are free to roll along the supports. In idealised condition rolling friction is neglected and hence there is no resistance in the line of support. The ends are free to rotate also. Hence there is no resistance to moment. Many mechanical components are provided with roller support which roll between guides. In such cases reactions can be normal to guides in either directions (Ref. Fig. 3.29).

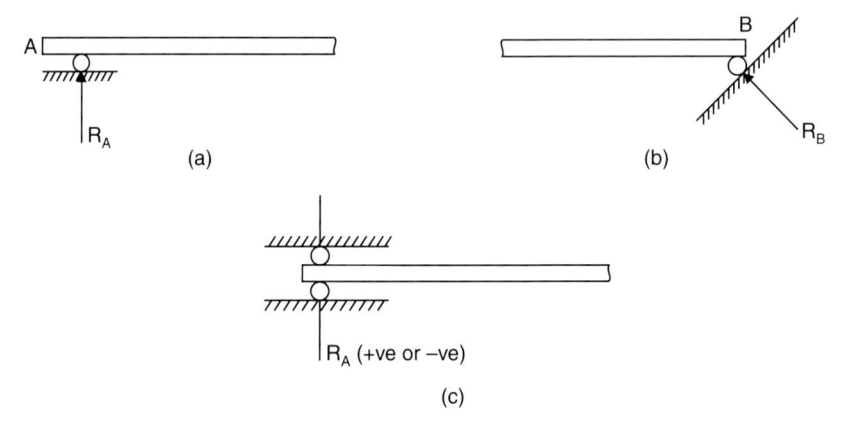

Fig. 3.29. Roller supports

Hinged or Pinned Support. In such cases, the position of the end of the beam is fixed but the end is free to rotate. This idealised conditions can be achieved by using mechanical devices. A typical hinged end condition is shown in Fig. 3.30. At such supports the reaction can be in any direction which is usually represented by its components in two mutually perpendicular directions. This type of support do not provide any resistance to moment, in other words it permits rotation freely at the end.

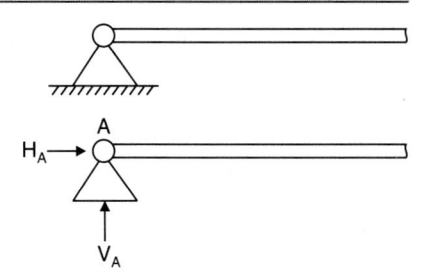

Fig. 3.30. Hinged or Pinned support

Fixed Support. At fixed support the end of the beam is neither permitted to move in any direction nor allowed to rotate. Hence support reactions are a force in any direction and the resisting moment (Ref. Fig. 3.31). Reacting force in any direction is conveniently represented by its components in two mutually perpendicular directions. This end condition may be achieved by taking the end of the beam considerably inside the support or by specially designing the brackets to resist the movement and rotation.

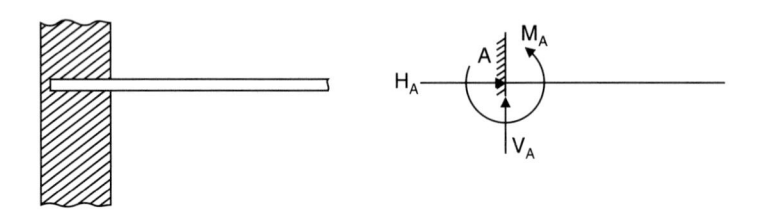

Fig. 3.31. Fixed supports

3.8.2. Types of Beams

Depending upon the types of supports, beams may be classified as following:

(i) Cantilever

(ii) Simply supported

(iii) One end hinged and other on roller

(iv) Overhanging

(v) Both ends hinged

(vi) Propped cantilever, and

(vii) Continuous.

(i) **Cantilever.** If a beam is fixed at one end and free at other end, it is called a cantilever beam. In this there are three reaction components at fixed end [V_A, H_A, M_A as shown in Fig. 3.31] and no reaction component at free end.

(ii) **Simply Supported Beam.** In this type of beam both ends are simply supported as shown in Fig. 3.32. There is one reaction component at each end (R_A and R_B). They act at right angles to the support. This type of beam can resist forces normal to the beam axis. In other words, the equilibrium condition that summation of forces parallel to axis equal to zero is to be satisfied automatically by the loading condition. Hence two equations of equilibrium are available.

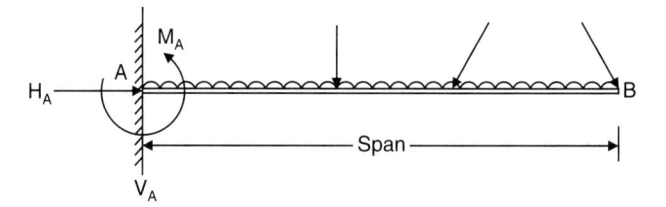

Fig. 3.32. Cantilever beam

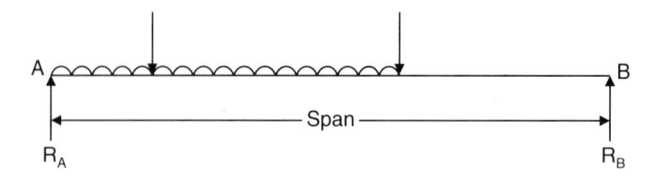

Fig. 3.33. Simple supported beam

(*iii*) **One End Hinged and the Other on Roller.** As the name suggests one end of the beam is hinged and the other end is on roller [Ref. Fig. 3.34]. At hinge, reaction can be in any direction and at roller end it is at right angles to the roller support. The hinged end reaction at any direction can be represented by its two components perpendicular to each other. Thus the reaction components for such beam shown in Fig. 3.34 are V_A, H_A and R_B.

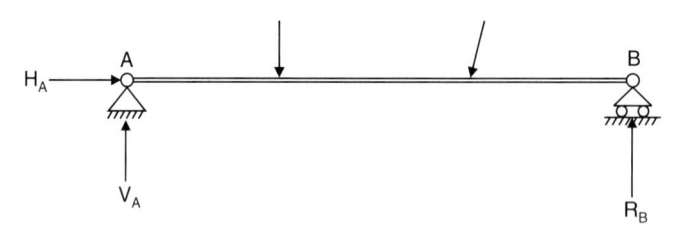

Fig. 3.34. One end hinged other on roller

(*iv*) **Overhanging Beam.** If a beam is projecting beyond the support/supports, it is called an overhanging beam [Fig. 3.35]. The overhang may be on only one side or may be on both sides.

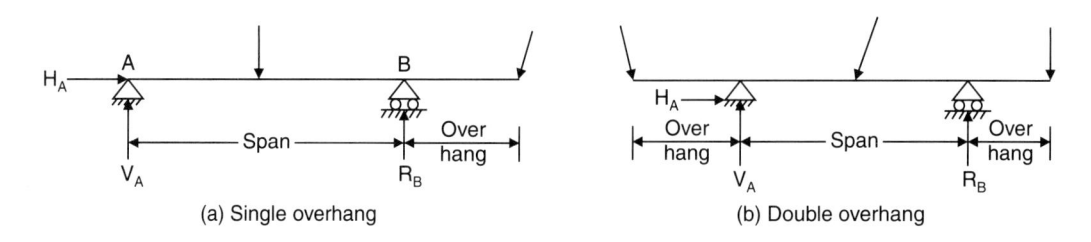

(a) Single overhang   (b) Double overhang

Fig. 3.35. Overhanging beam

(*v*) **Both Ends Hinged.** As the name suggests both ends of the beam are hinged. There are two reaction components at each end, hence total reaction components are four.

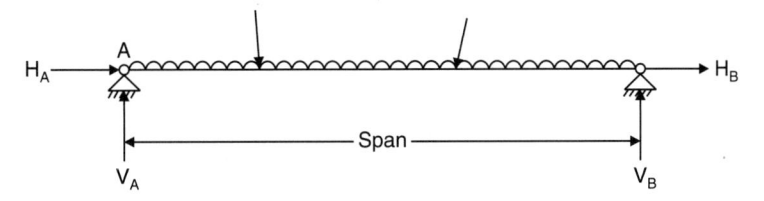

RESULTANT AND EQUILIBRIUM OF SYSTEM OF COPLANAR NON-CONCURRENT FORCES

Fig. 3.36. Both ends hinged beam

(*vi*) **Propped Cantilever.** In this type of beam one end of the beam is fixed and the other end is simply supported or is on rollers. It has got four reaction components (V_A, H_A, M_A, and R_B) as shown in Fig. 3.37.

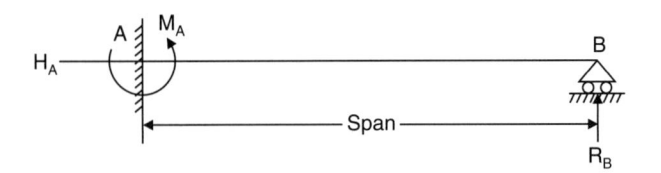

Fig. 3.37. Propped cantilever beam

(*vii*) **Continuous Beam.** A beam having three or more supports is called continuous beam. In such beams three or more reaction components exist. [Ref. Fig. 3.38].

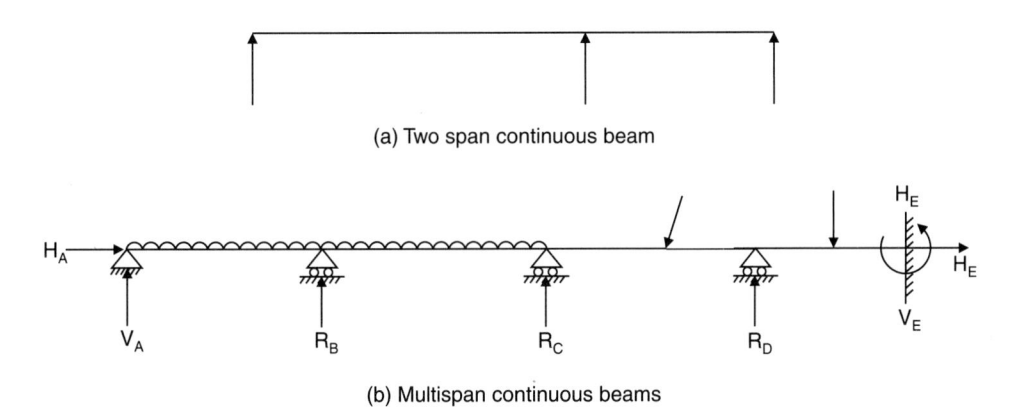

(a) Two span continuous beam

(b) Multispan continuous beams

Fig. 3.38. Continuous beams

In cantilever beams, simply supported beams, one end hinged and the other on roller and overhanging beams, number of unknown reactions are equal to the number of independent equations of equilibrium. Hence for any given loading, reactions can be determined by using equations of equilibrium only. Such beams are known as statically determinate beams. Thus we can define **statically determinate beams as these beams in which all reaction components can be found using the equations of equilibrium only.**

In case of beams with both ends hinged, propped cantilever and continuous beams, number of unknown reaction components are more than the number of available equations of equilibrium. This type of beams are known as statically indeterminate beams.

3.8.3. Types of Loading

Usual types of loadings on the beams are discussed below:

(i) **Concentrated Loads.** If a load is acting over a very small length compared to span of the beam, it is approximated as a point load. It is represented by an arrow as shown in the Fig. 3.39.

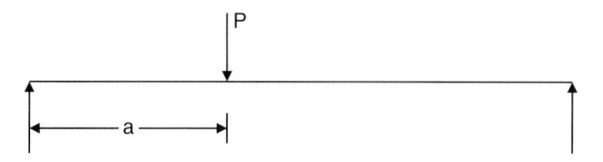

Fig. 3.39. Concentrated load

(ii) **Uniformly Distributed Load (UDL).** A load which has got same intensity over a considerable length is called uniformly distributed load. It is shown as indicated in Fig. 3.40 (a) or as in Fig. 3.40(b).

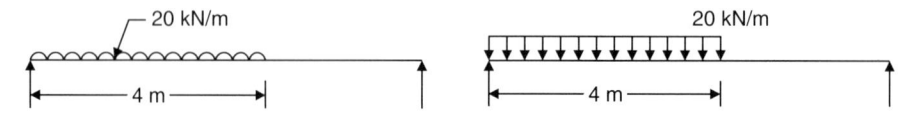

Fig. 3.40. Uniformly distributed load (UDL)

When equilibrium of entire beam is to be considered, this load may be imagined as total load (intensity × length) acting at the middle of length.

(iii) **Uniformly Varying Load.** If the intensity of the load increases linearly along the length, it is called linearly varying load (Fig. 3.41). In the Fig. 3.41(a) load varies from C to D with zero intensity at C and 20 kN/m intensity at D. In the load diagram shown, the ordinate represents the intensity of load and the abscissa represents the position of the load on the beam. Hence if intensity at one end is zero, load diagram takes the shape of a triangle. Therefore it may be called as triangular load. It may be noted that the area of the triangle represents the total load and the centroid of the triangle represents the centre of gravity of the load. Thus total load in Fig. 3.41 (a) is $\frac{1}{2} \times 3 \times 20 = 30$ kN and the centre of gravity of this loading is at $\frac{1}{3} \times 3 = 1$ m from D. For finding the reaction we can imagine that the given load is equivalent to 30 kN acting at $1 + 3 - 1 = 3$ m from end A.

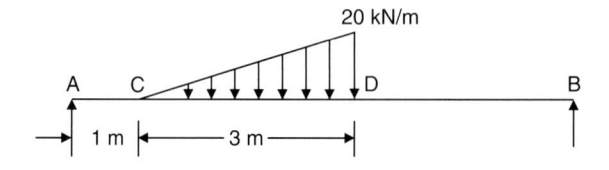

Fig. 3.41. Uniformly varying load

(iv) **General Loading.** Figure 3.42 shows the case of general loading. Here also ordinate represents the intensity of loading and abscissa represents the position of the loading. For simplicity, in the analysis such loadings are replaced by a set of equivalent concentrated loads.

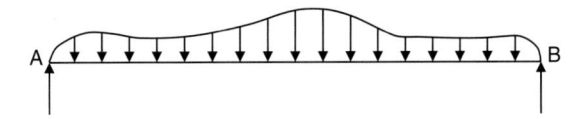

RESULTANT AND EQUILIBRIUM OF SYSTEM OF COPLANAR NON-CONCURRENT FORCES

Fig. 3.42. General loading

(*v*) **External Moment.** A beam may be subjected to external moments at certain points. In Fig. 3.43 the beam is subjected to a clockwise moment of 30 kN-m at a distance 2 m from the left support.

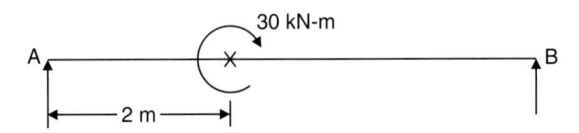

Fig. 3.43. External moment

3.8.4. Finding Reactions at Supports

To resist the applied loads reactions develop at supports of the beam. Reactions are self adjusting forces which will keep the beams in equilibrium. Hence the equations of equilibrium may be written for the system of forces consisting of reactions and the loads. Solutions of these equations give the unknown reactions. The procedure for finding the reactions in statically determinate beams is illustrated with the examples below:

Example 3.17. *Determine the reactions developed in the cantilever beam shown in Fig. 3.44(a).*

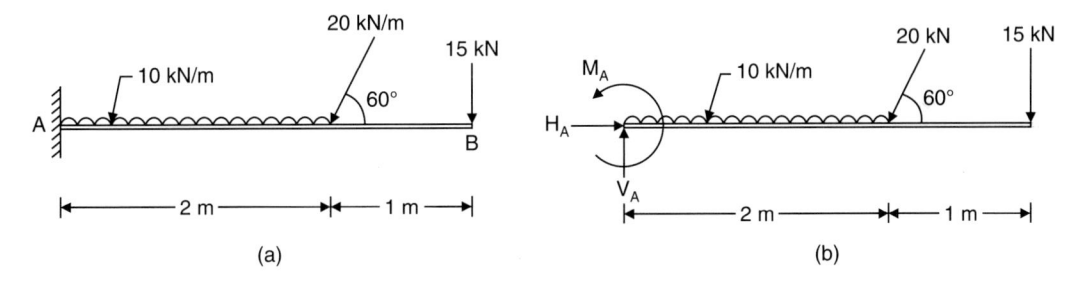

Fig. 3.44

Solution. Let the reactions developed at fixed support A be V_A, H_A and M_A as shown in Fig. 3.44 (*b*).

ΣForces in vertical direction = 0 $\rightarrow$
$$V_A - 10 \times 2 - 20 \sin 60° - 15 = 0.$$
$\therefore$ $\qquad V_A = \mathbf{52.32 \ kN}$ **Ans.**

ΣForces in horizontal direction = 0 $\rightarrow$
$$H_A - 20 \cos 60° = 0 \quad \therefore \ H_A = \mathbf{10 \ kN} \quad \textbf{Ans.}$$

ΣMoments about A = 0 $\rightarrow$
$$-M_A + 10 \times 2 \times 1 + 20 \sin 60° \times 2 + 15 \times 3 = 0.$$
$\therefore$ $\qquad M_A = \mathbf{99.64 \ kN\text{-}m}$ **Ans.**

[**Note:** UDL is treated as a load 10×2 kN acting at its centre of gravity which is at 1m from A].

Example 3.18. *Determine the reactions developed in the cantilever beam shown in Fig. 3.45(a).*

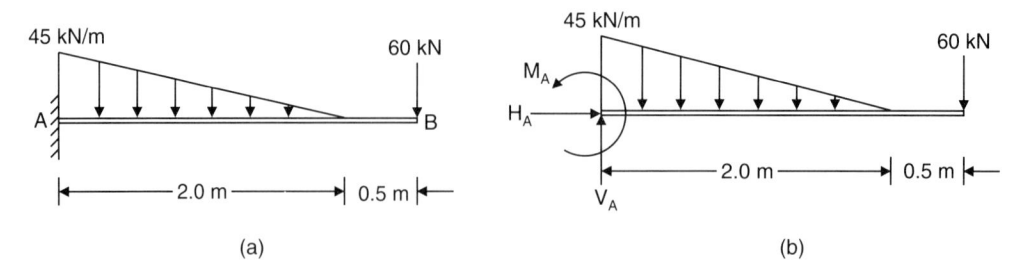

Fig. 3.45

Solution. Let the reactions developed at the fixed end A be V_A, H_A and M_A as shown in Fig. 3.45(b).

$$\Sigma V_A = 0 \rightarrow$$

$$V_A - \frac{1}{2} \times 45 \times 2 - 60 = 0$$

$$V_A = 105 \text{ kN} \quad \text{Ans.}$$

$$\Sigma H = 0 \rightarrow$$

$$H_A - 0 = 0 \quad \therefore \quad H_A = 0 \quad \text{Ans.}$$

$$\Sigma M_A = 0 \rightarrow$$

$$- M_A + \frac{1}{2} \times 45 \times \frac{2}{3} \times 2 + 60 \times 2.5 = 0$$

$$\therefore \quad M_A = 180 \text{ kN-m} \quad \text{Ans.}$$

[**Note:** Total triangular load = $\frac{1}{2} \times 45 \times 2 = 45$ kN and its centroid is at $\frac{2}{3} \times 2$ m from end A].

Example 3.19. *Determine the reaction developed in the simply supported beam shown in Fig. 3.46.*

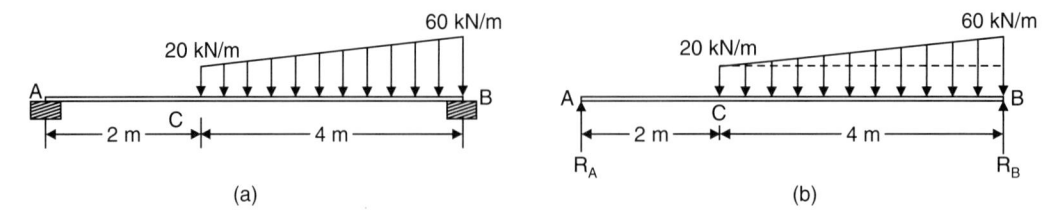

Fig. 3.46

Solution. Let R_A and R_B be the reactions developed at the simply supported ends A and B. The uniformly varying load may be split into a uniformly distributed load of 20kN/m intensity and a triangular load of intensity zero at point C and intensity $60 - 20 = 40$ kN/m at B. Then

$$\Sigma M_B = 0 \rightarrow$$

$$R_A \times 6 - 20 \times 4 \times 2 - \frac{1}{2} \times 4 \times 40 \times \frac{4}{3} = 0$$

$$\therefore \quad R_A = 44.44 \text{ kN} \quad \text{Ans.}$$

$$\Sigma V = 0 \rightarrow$$

$$R_A + R_B - 20 \times 4 - \frac{1}{2} \times 4 \times 40 = 0$$

$$\therefore \qquad R_B = 80 + 80 - 44.44, \qquad \text{since} \quad R_A = 44.44 \text{ kN}$$

$$\therefore \qquad R_B = 115.56 \text{ kN} \quad \text{Ans.}$$

Example 3.20. *The beam AB of span 12 m shown in Fig. 3.47 is hinged at A and is on rollers at B. Determine the reactions developed at A and B due to the loading shown in the figure.*

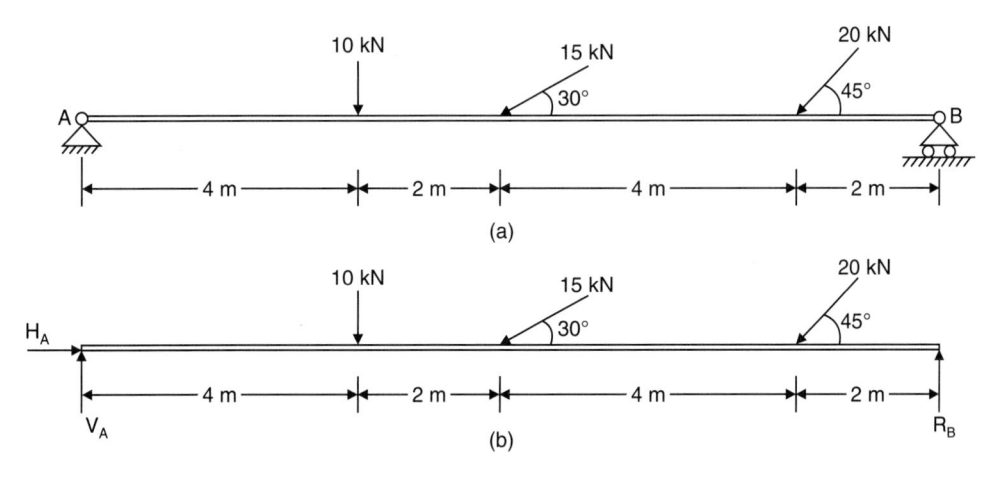

Fig. 3.47

Solution. At A the reaction can be in any direction. Let it be represented by its components V_A and H_A as shown in Fig. 3.47(b). At B the reaction is in vertical direction only. Let its magnitude be R_B. The beam is in equilibrium under the action of system of forces shown in Fig. 3.47(b).

Now, $\qquad \Sigma H = 0 \rightarrow$

$$H_A - 15 \cos 30° - 20 \cos 45° = 0$$

$$\therefore \qquad H_A = 27.13 \text{ kN} \quad \text{Ans.}$$

$$\Sigma M_A = 0 \rightarrow$$

$$-R_B \times 12 + 10 \times 4 + 15 \sin 30° \times 6 + 20 \sin 45° \times 10 = 0$$

$$\therefore \qquad R_B = 18.87 \text{ kN} \quad \text{Ans.}$$

$$\Sigma V = 0 \rightarrow$$

$$V_A + R_B - 10 - 15 \sin 30° - 20 \sin 45° = 0$$

$$V_A + 18.87 - 10 - 15 \sin 30° - 20 \sin 45° = 0$$

$$V_A = 12.77 \text{ kN} \quad \text{Ans.}$$

Example 3.21. *Find the magnitude and direction of reactions at supports A and B in the beam AB shown in Fig. 3.48.*

Solution. The reaction R_B will be at right angles to the inclined support *i.e.* at $90° - 30° = 60°$ to horizontal as shown in the Fig. 3.48 (b). Let the components of reactions at A be V_A and H_A. Then

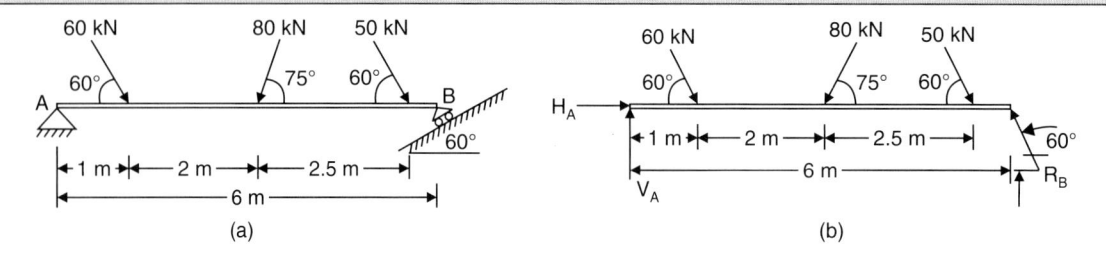

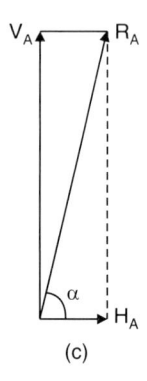

Fig. 3.48

$$\Sigma M_A = 0 \rightarrow$$

$$60 \sin 60° \times 1 + 80 \sin 75° \times 3 + 50 \sin 60° \times 5.5 - R_B \sin 60° \times 6 = 0$$

∴ $R_B = 100.45$ kN, at 60° to horizontal as shown in figure. **Ans.**

$$\Sigma H = 0 \rightarrow$$

$$H_A + 60 \cos 60° - 80 \cos 75° + 50 \cos 60° - R_B \cos 60° = 0$$

$$H_A = -60 \cos 60° + 80 \cos 75° - 50 \cos 60° + 100.45 \cos 60°$$

∴ $H_A = 15.93$ kN

$$\Sigma V = 0 \rightarrow$$

$$V_A + R_B \sin 60° - 60 \sin 60° - 80 \sin 75° - 50 \sin 60° = 0$$

∴ $V_A = -100.45 \sin 60° + 60 \sin 60° + 80 \sin 75° + 50 \sin 60°$

i.e., $V_A = 85.54$ kN

∴ $R_A = \sqrt{H_A^2 + V_A^2} = \sqrt{15.93^2 + 85.54^2}$

i.e., **$R_A = 87.02$ kN**

$$\alpha_A = \tan^{-1} \frac{85.54}{15.93} = 79.45°, \text{ as shown in Fig. 3.48 } c.\ \textbf{Ans.}$$

Example 3.22. *Find the reactions developed at supports A and B of the loaded beam shown in Fig. 3.49.*

Solution. Reaction at A is vertical, since roller support is horizontal. Let H_B and V_B be the components of the reaction at B.

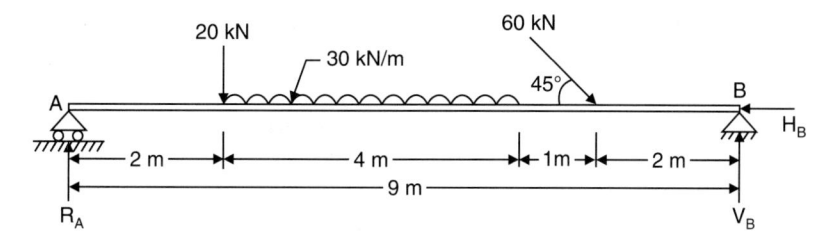

RESULTANT AND EQUILIBRIUM OF SYSTEM OF COPLANAR NON-CONCURRENT FORCES

Fig. 3.49

$$\Sigma M_B = 0 \rightarrow$$

$$R_A \times 9 - 20 \times 7 - 30 \times 4 \times 5 - 60 \sin 45° \times 2 = 0.$$

∴ $R_A = 91.65$ kN **Ans.**

$$\Sigma H_A = 0 \rightarrow$$

$$60 \cos 45° - H_B = 0.$$

∴ $H_B = 42.43$ kN **Ans.**

$$\Sigma V = 0$$

$$R_A + V_B - 20 - 30 \times 4 - 60 \sin 45° = 0.$$

∴ $V_B = 20 + 30 \times 4 + 60 \sin 45° - 91.65$, since $R_A = 91.65$ kN

i.e., $V_B = 90.78$ kN **Ans.**

Example 3.23. *Determine the reactions at supports A and B of the overhanging beam shown in Fig. 3.50.*

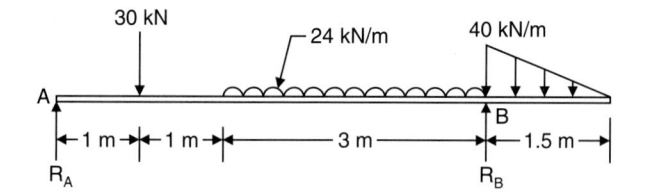

Fig. 3.50

Solution. $\Sigma M_A = 0 \rightarrow$

$$30 \times 1 + 24 \times 3 \times (2 + 1.5) + \frac{1}{2} \times 1.5 \times 40 \times (5 + \frac{1}{3} \times 1.5) - R_B \times 5 = 0.$$

∴ $R_B = 89.4$ kN **Ans.**

$$\Sigma V = 0 \rightarrow$$

$$R_A + R_B - 30 - 24 \times 3 - \frac{1}{2} \times 1.5 \times 40 = 0.$$

∴ $R_A = 30 + 72 + 30 - 85.4$, since $R_B = 85.4$ kN

$$R_A = 42.6 \text{ kN} \quad \textbf{Ans.}$$

Example 3.24. *Determine the reactions developed at supports A and B of overhanging beam shown in Fig. 3.51.*

Solution. Let the reaction components developed at support A be V_A and H_A and that at B be R_B.

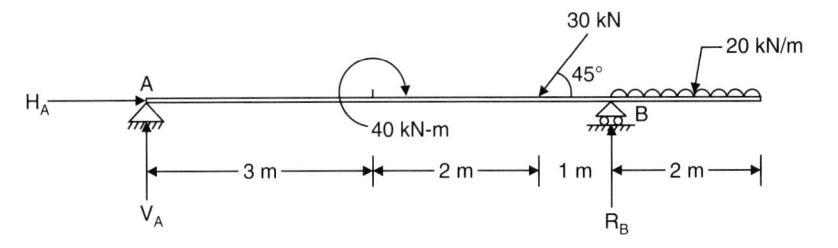

$$\Sigma M_A = 0 \rightarrow$$
$$40 + 30 \sin 45° \times 5 + 20 \times 2 \times 7 - R_B \times 6 = 0$$
∴ $\quad R_B = \textbf{71.01 kN} \quad \textbf{Ans.}$
$$\Sigma F_H = 0 \rightarrow$$
$H_A - 30 \cos 45° = 0 \quad$ ∴ $\quad H_A = \textbf{21.21 kN} \quad \textbf{Ans.}$
$$\Sigma F_V = 0 \rightarrow$$
$V_A - 30 \sin 45° + R_B - 20 \times 2 = 0.$
∴ $\quad V_A = 30 \sin 45° - 71.01 + 40, \quad$ since $R_B = 71.01.$
i.e., $\quad V_A = -9.79$ kN
$\quad V_A = \textbf{9.79}\downarrow \textbf{ kN} \quad \textbf{Ans.}$

Example 3.25. *Find the reactions developed at supports A and B of the loaded beam shown in Fig. 3.52.*

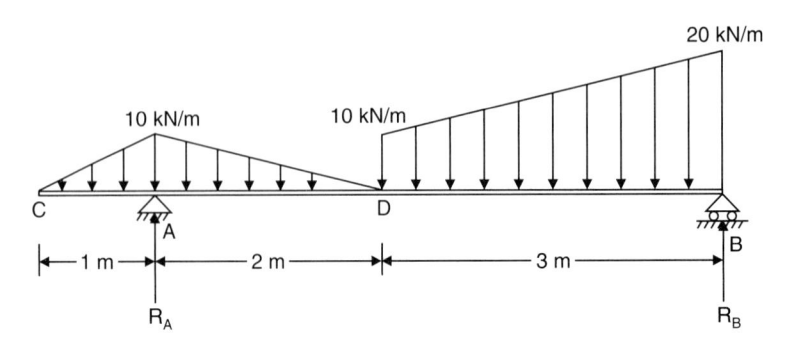

Fig. 3.52

Solution. The load is divided into a triangular load between C to A, another triangular load between A to D, a uniformly distributed load of 10 kN/m over portion DB and a triangular load of intensity zero at D and 10 kN/m [20 – 10 = 10 kN/m] at B.

$$\Sigma M_B = 0 \rightarrow$$

$$-\frac{1}{2} \times 1 \times 10 \times \left(\frac{1}{3}+5\right) - \frac{1}{2} \times 2 \times 10 \times \left(5-\frac{2}{3}\right) - 10 \times 3 \times 1.5$$

$$-\frac{1}{2} \times 3 \times 10 \times 1 + R_A \times 5 = 0.$$

∴ $\quad R_A = \textbf{26 kN} \quad \textbf{Ans.}$
$$\Sigma F_V = 0 \rightarrow$$

RESULTANT AND EQUILIBRIUM OF SYSTEM OF COPLANAR NON-CONCURRENT FORCES

$$R_A + R_B - \frac{1}{2} \times 1 \times 10 - \frac{1}{2} \times 2 \times 10 - 10 \times 3 - \frac{1}{2} \times 3 \times 10 = 0$$

$$26 + R_B - 5 - 10 - 30 - 15 = 0.$$

or $\qquad R_B = 34$ kN   Ans.

Example 3.26. *A beam 20 m long supported on two intermediate supports, 12 m apart, carries a udl of 6 kN/m and two concentrated loads of 30 kN at left end A and 50 kN at the right end B as shown in Fig. 3.53. How far away should the first support C be located from the end A so that the reactions at both the supports are equal.*

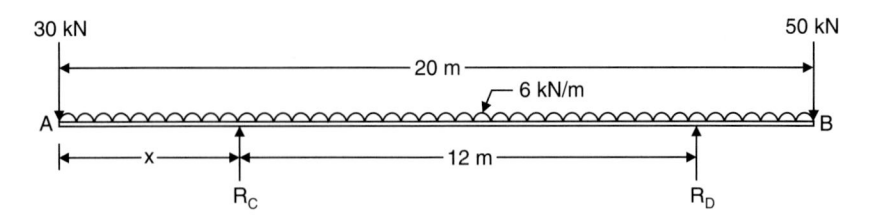

Fig. 3.53

Solution. Let the support C be at a distance x metres from end A.

Now, it is given that $R_C = R_D$

$$\Sigma F_V = 0 \;\rightarrow$$

$$R_C + R_D - 30 - 6 \times 20 - 50 = 0$$

i.e., $\qquad 2R_C = 30 + 120 + 50,$ since $R_D = R_C$

∴ $\qquad R_C = 100$ kN

Hence $\qquad R_D = 100$ kN

$\Sigma M_A = 0$, gives.

$$R_C x + R_D(12 + x) - 6 \times 20 \times 10 - 50 \times 20 = 0$$

$$100x + 100(12 + x) - 1200 - 1000 = 0, \quad \text{since } R_C = R_D = 100 \text{ kN}$$

∴ $\qquad x = 5$ m   Ans.

Example 3.27. *A beam AB supports uniformly distributed load of intensity w_1 and rests on soil which exerts a uniformly varying upward reaction as shown in Fig. 3.54. Determine w_2 and w_3 corresponding to equilibrium. Note that soil can exert only an upward reaction at any point on beam. Hence, state for what range of values of $\frac{a}{L}$ the results obtained are valid.*

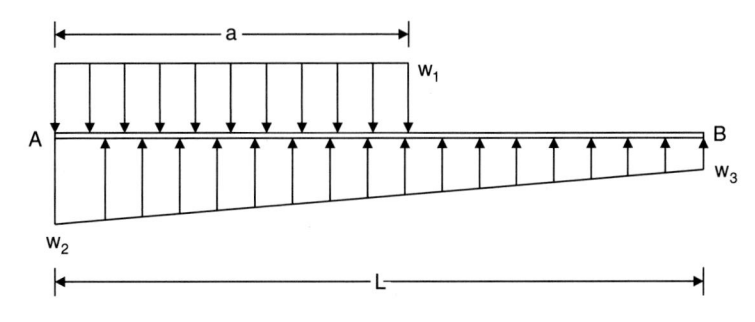

Fig. 3.54

Solution.
$$\Sigma F_V = 0 \rightarrow aw_1 = \frac{w_2 + w_3}{2} L \qquad \ldots(i)$$

$$\Sigma M_A = 0 \rightarrow aw_1 \frac{a}{2} = w_3 L \times \frac{L}{2} + \frac{1}{2}(w_2 - w_3) L \times \frac{L}{3}$$

i.e.,
$$w_1 a^2 = w_3 L^2 + (w_2 - w_3)\frac{L^2}{3} \qquad \ldots(ii)$$

From (i),
$$w_2 = \frac{2aw_1}{L} - w_3 \qquad \ldots(iii)$$

From (ii) and (iii), we get

$$w_1 a^2 = w_3 L^2 + \left[\frac{2aw_1}{L} - w_3 - w_3\right]\frac{L^2}{3}$$

$$= \frac{2aL}{3} w_1 + w_3 \left(1 - \frac{2}{3}\right) L^2$$

∴
$$w_1 \left(a^2 - \frac{2aL}{3}\right) = w_3 \frac{L^2}{3}$$

∴
$$w_3 = \frac{3w_1}{L^2}\left(a^2 - \frac{2aL}{3}\right) \quad \text{Ans.}$$

Substituting this value of w_3 in (iii), we get

$$w_2 = \frac{2aw_1}{L} - \frac{3w_1}{L^2}\left(a^2 - \frac{2aL}{3}\right)$$

∴
$$w_2 = \left(\frac{4a}{L} - \frac{3a^2}{L^2}\right) w_1 \quad \text{Ans.}$$

For the above expressions to be valid W_2 and W_3 should not be negative. The limiting cases are when they are zero.

∴
$$0 = \frac{3w_1}{L^2}\left(a^2 - \frac{2aL}{3}\right)$$

i.e.,
$$a = \frac{2L}{3} \quad \text{or} \quad \frac{a}{L} = \frac{2}{3}$$

and
$$0 = \frac{4a}{L} - \frac{3a^2}{L^2}$$

i.e.,
$$\frac{a}{L} = \frac{4}{3}$$

Hence the results obtained are valid for the range $\frac{a}{L} = \frac{2}{3}$ to $\frac{4}{3}$ only. However, there is no possibility of loading beyond $\frac{a}{L} = 1$. Hence, **the valid range of** $\frac{a}{L}$ **is from** $\frac{2}{3}$ **to 1. Ans.**

Example 3.28. *Determine the reactions at A, B and D of the compound beam shown in Fig. 3.55(a). Neglect the self weight of the members.*

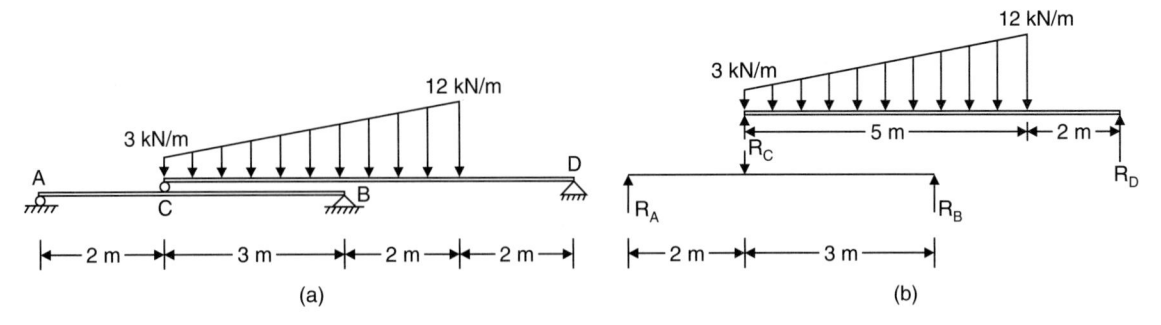

Fig. 3.55

Solution. Free body diagrams of beams CD and AB are as shown in Fig. 3.55(b). The load may be divided into a *udl* of 3 kN/m and a triangular load of intensity zero at C and 9 kN/m at 2m from D. Consider beam CD.

$$\Sigma M_C = 0 \rightarrow$$

$$- R_D \times 7 + 3 \times 5 \times 2.5 + \frac{1}{2} \times 5 \times 9 \times \frac{2}{3} \times 5 = 0$$

∴ $\quad R_D = \textbf{16.07 kN}\quad$**Ans.**

$$\Sigma F_V = 0 \rightarrow$$

$$R_C + R_D - 3 \times 5 - \frac{1}{2} \times 5 \times 9 = 0$$

$$R_C = 15 + 22.5 - 16.07, \quad \text{since} \quad R_D = 16.07 \text{ kN}$$

i.e., $\quad R_C = 21.43$ kN

Consider beam AB.

$$\Sigma M_A = 0 \rightarrow$$

$$R_C \times 2 - R_B \times 5 = 0.$$

$$R_B = \frac{2 \times 21.43}{5}, \quad \text{since} \quad R_C = 21.43 \text{ kN}$$

∴ $\quad R_B = \textbf{8.57 kN}\quad$**Ans.**

$$\Sigma F_V = 0 \rightarrow$$

$$R_A + R_B - R_C = 0.$$

$$R_A + 8.57 - 21.43 = 0.$$

∴ $\quad R_A = \textbf{12.86 kN}\quad$**Ans.**

Example 3.29. *The beam AB and CF are arranged as shown in Fig. 3.56(a). Determine the reactions at A, C and D due to the loads acting on the beam as shown in the figure.*

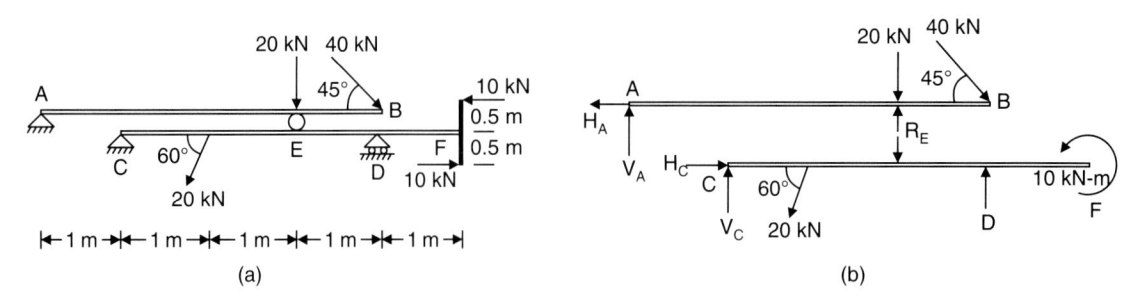

Fig. 3.56

Solution. The free body diagrams of beams AB and CF are as shown in Fig. 3.56(b). The two 10 kN forces constitute a couple moment of $10 \times 1 = 10$ kN-m as shown in Fig. 3.56(b). Now consider beam AB.

$$\Sigma M_A = 0 \rightarrow$$
$$-R_E \times 3 + 20 \times 3 + 40 \sin 45° \times 4 = 0$$
$$\therefore \quad R_E = 57.71 \text{ kN}$$
$$\Sigma F_H = 0 \rightarrow$$
$$-H_A + 40 \cos 45° = 0.$$
or $\quad\quad H_A = \mathbf{28.28 \text{ kN}}$ **Ans.**
$$\Sigma F_V = 0 \rightarrow$$
$$V_A - 20 - 40 \sin 45° + R_E = 0$$
$$V_A = 20 + 40 \sin 45° - 57.71, \quad \text{since} \quad R_E = 57.71 \text{ kN}$$
$$\therefore \quad V_A = -9.43 \text{ kN}$$
i.e., $\quad\quad V_A = \mathbf{9.43 \text{ kN downward}}$ **Ans.**

Consider beam CF.
$$\Sigma M_C = 0 \rightarrow$$
$$-R_D \times 3 + 20 \sin 60° \times 1 + R_E \times 2 - 10 = 0$$
$$\therefore \quad R_D = 20 \sin 60° - 10 + 57.71 \times 2, \quad \text{since,} \quad R_E = 57.71 \text{ kN}$$
$$\therefore \quad R_D = \mathbf{40.91 \text{ kN}} \text{ **Ans.**}$$
$$\Sigma F_H = 0 \rightarrow$$
$$H_C - 20 \cos 60° = 0$$
$$\therefore \quad H_C = \mathbf{10 \text{ kN}} \text{ **Ans.**}$$
$$\Sigma F_V = 0 \rightarrow$$
$$V_C - 20 \sin 60° - R_E + R_D = 0$$
$$V_C - 20 \sin 60° - 57.71 + 40.91 = 0$$
$$\therefore \quad V_C = \mathbf{34.12 \text{ kN}} \text{ **Ans.**}$$

Important Definitions

1. **Moment** of a force about a point is the measure of rotational effect of the force and is measured as the product of the magnitude of the force and the perpendicular distance of the point from the line of action of the force.

RESULTANT AND EQUILIBRIUM OF SYSTEM OF COPLANAR NON-CONCURRENT FORCES 79

2. **Varignon's theorem** states that the sum of moments of a system of coplanar forces about a moment centre is equal to the moment of their resultant about the same moment centre.
3. Two parallel forces equal in magnitude but opposite in direction and separated by a definite distance are said to form a **couple**.
4. **Statically determinate beams** may be defined as beams in which all reaction components can be found using the equation of equilibrium only.

Important Formulae

1. Moment produced by a couple about any point is equal to the product of magnitude of equal and opposite parallel forces and perpendicular distance between them.
$$M = Fd.$$
2. Any one set of following equations of equilibrium may be used for the analysis of non-concurrent coplanar system of forces.
   (i) $\Sigma F_x = 0$, $\Sigma F_y = 0$, $\Sigma M_A = 0$.
   (ii) If AB is not in x-direction,
   $\Sigma F_x = 0$, $\Sigma M_A = 0$, $\Sigma M_B = 0$.
   (iii) If AB is not in y-direction,
   $\Sigma F_y = 0$, $\Sigma M_A = 0$, $\Sigma M_B = 0$.
   (iv) If A, B and C are not collinear,
   $\Sigma M_A = 0$, $\Sigma M_B = 0$, $\Sigma M_C = 0$.

PROBLEMS FOR EXERCISE

3.1 Determine the resultant of the parallel coplanar force system shown in Fig. 3.57.

[**Ans.** $R = 800$ N (towards left), $d = 627.5$ mm]

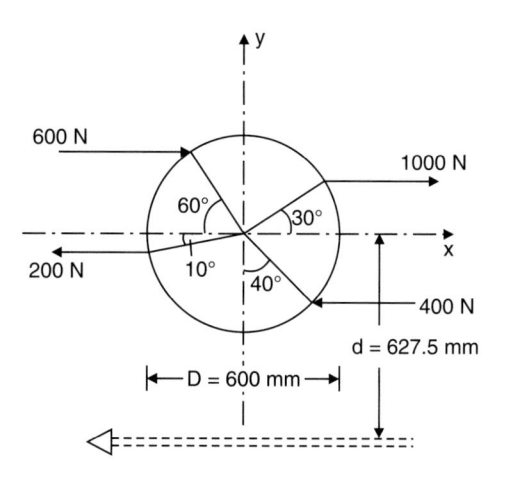

Fig. 3.57

3.2 Determine the magnitude, direction and the line of application of the equilibriant of the set of forces shown in Fig. 3.58, which will keep the plane body $ABCDEFGH$ in equilibrium.

[**Ans.** $E = 23.65$ kN, $\alpha = 24.37°$, x from $A = 1.04$ m]

3.3 An equilateral triangular plate of sides 200 mm is acted upon by four forces as shown in Fig. 3.59. Determine the magnitude and direction of the resultant of this system of forces and its position.

[**Ans.** $R = 57.35$ kN ; $\alpha = 6.70°$, $d = 11.51$ mm]

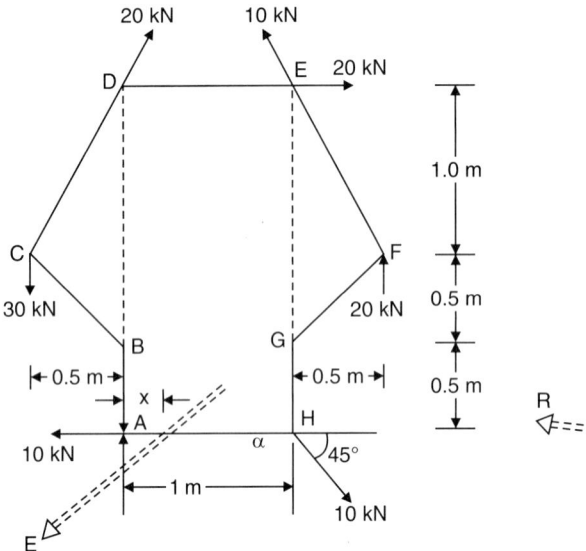

Fig. 3.58

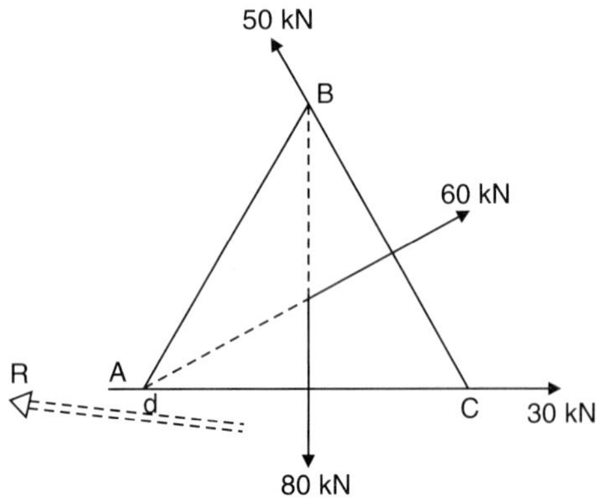

Fig. 3.59

3.4 Determine the resultant of the four forces acting on a body as shown in Fig. 3.60.

[**Ans.** $R = 200$ N, $\alpha = 60°$, at a distance $y = 8.77$ m below O]

3.5 A bracket is subjected to the system of forces and couples as shown in Fig. 3.61. Find the resultant of the system and the point of intersection of its line of action with (a) line AB, (b) line BC and line CD.

[**Ans.** $R = 485.4$ N, $\alpha = 34.50°$, $y_{CD} = 93.75$ mm, $y_{BA} = 112.5$ mm, $x_{BC} = 163.6$ mm]

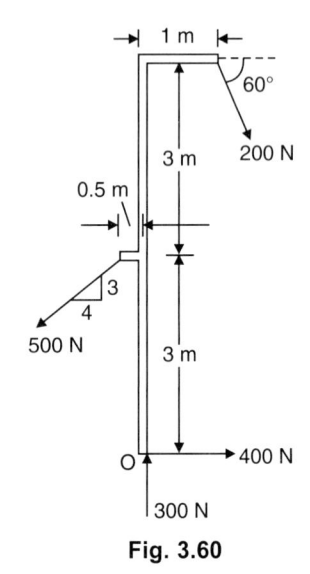

3.6 Determine the resultant of the three forces acting on the dam section shown in Fig. 3.62 and locate its intersection with the base AB. For a safe design this intersection should be within the middle third. Is it a safe design ?

Fig. 3.60

[**Ans.** Resultant intersects AB at 3.33 m from A. It is a safe design]

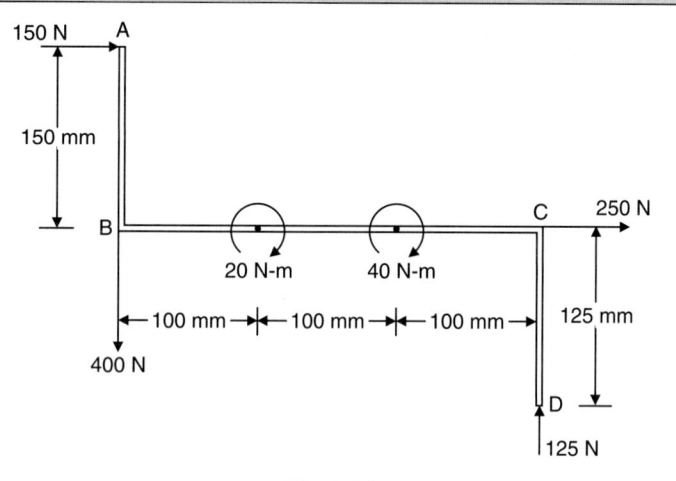

RESULTANT AND EQUILIBRIUM OF SYSTEM OF COPLANAR NON-CONCURRENT FORCES 81

Fig. 3.61

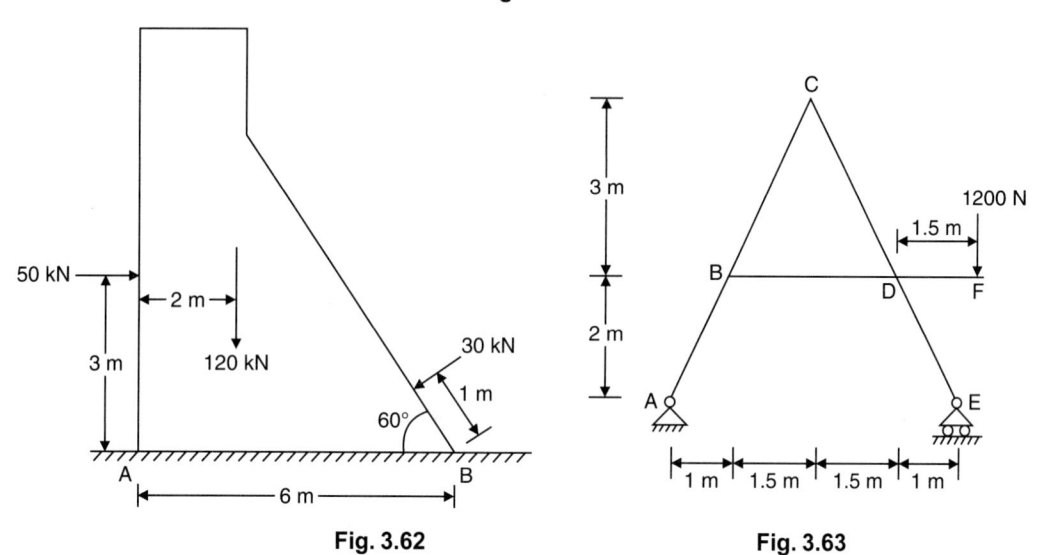

Fig. 3.62

Fig. 3.63

3.7 The frame shown in Fig. 3.63 is supported by a hinge at A and a roller at E. Compute the horizontal and vertical components of the reactions at hinge B and C as they act upon member AC.

[**Ans.** $H_B = 200$ N (towards right). $V_B = 600$ N downward
$H_C = 200$ N (towards left). $V_C = 480$ N (downward)]

3.8 The frame shown in Fig. 3.64 is supported by a hinge at E and by a roller at D. Determine the horizontal and vertical components of the reaction at hinge C as it acts upon member BD.

[**Ans.** $H_C = 140$ N towards right, $V_C = 35$ N, upward]

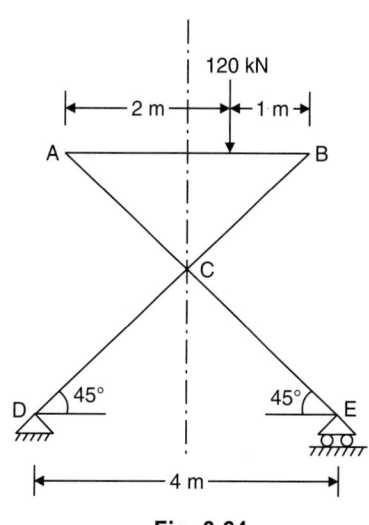

Fig. 3.64

3.9 Beam *AB*, 8.5 m long, is hinged at *A* and has a roller support at *B*. The roller support is inclined at 45° to the horizontal. Find the reactions at *A* and *B*, if the loads acting are as shown in Fig. 3.65. [**Ans.** $H_A = 51.67$ kN, $V_A = 73.04$ kN, $R_B = 101.25$ kN]

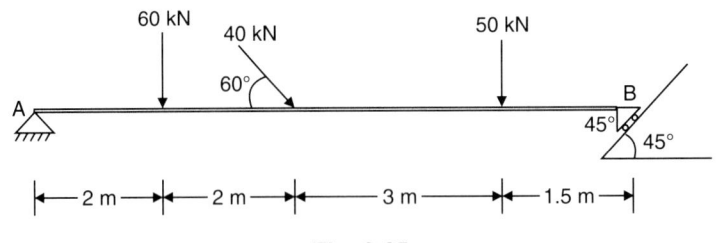

Fig. 3.65

3.10 Determine reactions developed at supports in the beam shown in Fig. 3.66.
[**Ans.** $H_A = 21.21$ kN, $V_A = 9.80$ kN, downward, $R_B = 71.01$ kN]

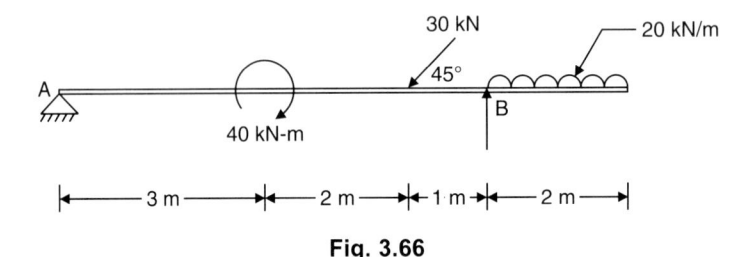

Fig. 3.66

3.11 Overhanging beam shown in Fig. 3.67 is on roller at end *A* and is hinged at the end *B*. Determine the reactions developed for the loadings shown in figure. [**Ans.** $R_A = 160$ kN, $R_B = 90$ kN]

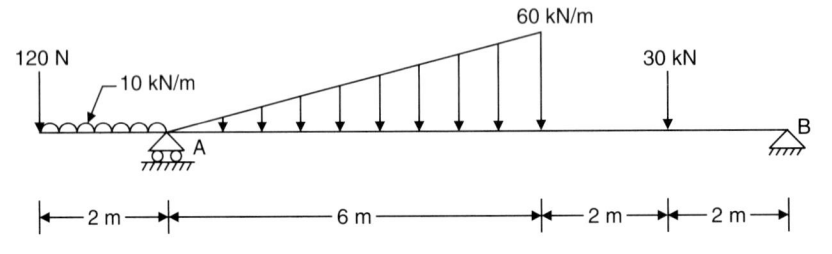

Fig. 3.67

3.12 Determine the reactions at supports *A*, *C* and *D* in the compound beam shown in Fig. 3.68.
[**Ans.** $R_A = 20$ kN, $R_C = 67.5$ kN, $R_D = 7.5$ kN, downward]

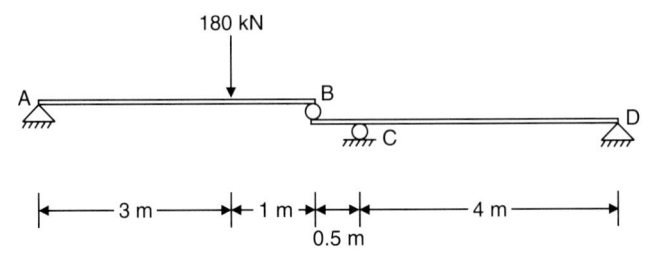

Fig. 3.68

4

Analysis of Pin-Jointed Plane Frames

A pin jointed frame is a structure made of slender (cross-sectional dimensions quite small compared to length) members pin connected at ends and capable of taking loads at joints. Such frames are used as roof trusses to support sloping roofs and as bridge trusses to support deck. Frames are used as arms of many machines. Transmission towers, observation towers are also examples of frames. In case of wooden frames ends of members are connected by nailing or by bolting whereas in steel frames end joints are made by riveting or welding to gusset plates.

A frame, in which all members lie in a single plane, is called **plane frame**. They are designed to resist the forces acting in the plane of frame. Roof trusses and bridge trusses are the examples of plane frames. If all the members of frame do not lie in a single plane, they are called as **space frame**. Tripod, transmission and observation towers are the examples of space frames. In this chapter analysis of only plane frame is dealt. The analysis involves application of equilibrium conditions to system of concurrent and non concurrent forces.

4.1. PERFECT, DEFICIENT AND REDUNDANT FRAMES

A pin jointed frame which has got just sufficient number of members to resist the loads without undergoing appreciable deformation in shape is called a **perfect frame**. Triangular frame is the simplest perfect frame and it has three joints and three members (Ref. Fig. 4.1). Perfect frames with four and five joints are shown in Figs. 4.2 and 4.3 respectively.

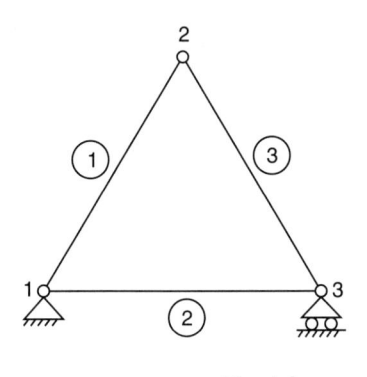

Fig. 4.1

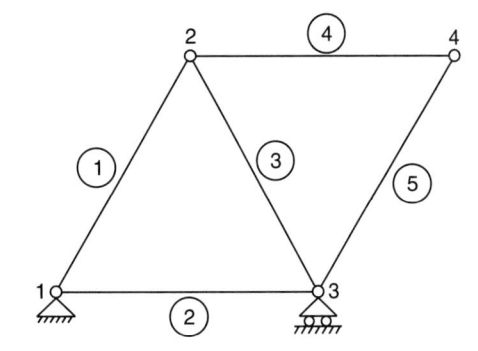

Fig. 4.2

It may be observed that to increase one joint in a perfect frame, two more members are required. Hence the following expression may be written as the relationship between number of joint j, and the number of members m in a perfect frame.

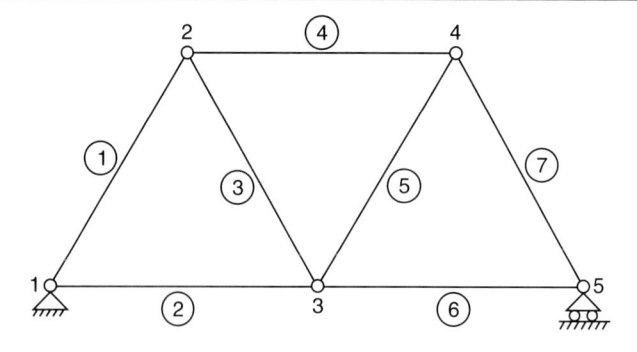

Fig. 4.3

$$m = 2j - 3 \qquad \ldots(4.1)$$

However, the above equation gives only a necessary, but not a sufficient condition of a perfect frame. For example, the two frames, shown in Figs. 4.4(a) and (b) have the same number of members and joints. The frame shown in Fig. 4.4(a) is a perfect frame, since it can retain its shape for loading at any joint. The frame shown in Fig. 4.4(b), is not capable of retaining its shape, if loaded at joint 6. Hence the necessary and sufficient condition for a perfect frame is that it should retain its shape when load is applied in the plane of frame at any joint.

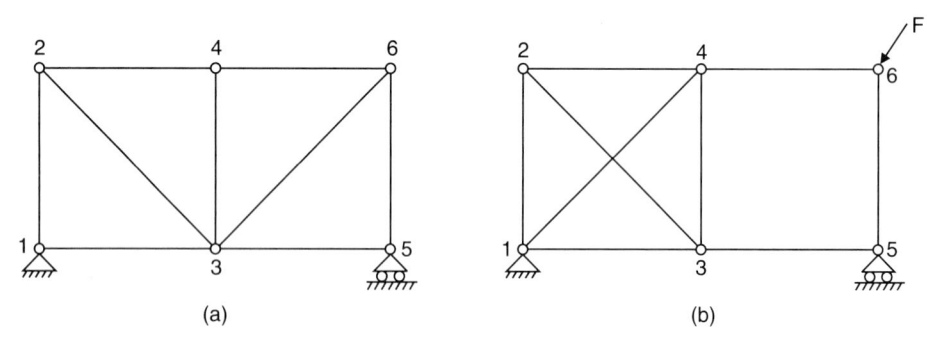

Fig. 4.4

A frame is said to be **deficient** if the number of members in it are less than that required for a perfect frame. These frames cannot retain their shape when loaded. A deficient frame is shown in Fig. 4.5.

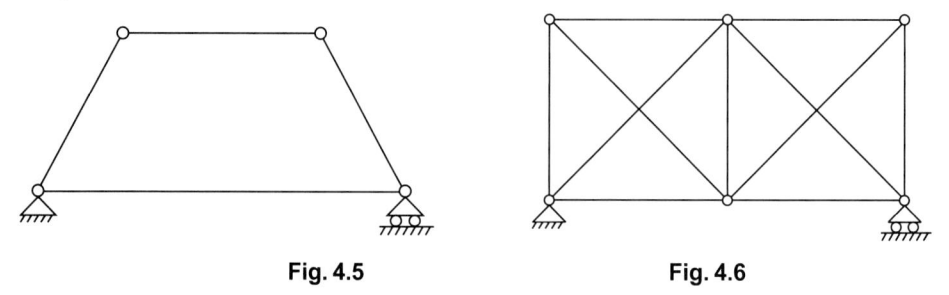

Fig. 4.5   **Fig. 4.6**

A frame is said to be **redundant**, if the number of members in it are more than that required for a perfect frame. Such frames cannot be analysed by using the equations of equilibrium only. Thus a redundant frame is statically indeterminate frame. Each extra member adds one degree of indeterminancy. For the analysis of such frames consistancy of deformations

ANALYSIS OF PIN-JOINTED PLANE FRAMES

is also to be considered. The truss shown in Fig. 4.6 is a typical redundant truss. In this, one diagonal member in each panel is extra. Hence it is two degree redundant frame.

4.2. ASSUMPTIONS

The following assumptions are made in the analysis of pin jointed trusses:
(i) The ends of the members are pin connected (hinged).
(ii) The loads act only at the joints.
(iii) Self weight of the members are negligible.
(iv) Members are having either uniform cross-sections throughout or if they have varying cross-section the centroid is located along the same longitudinal line.

In reality, the members are connected by bolting, rivetting or by welding. No special care is taken to ensure perfect pin-connections. Assuming pin connected end is quite satisfactory for steel trusses, since the members used are slender. For concrete and steel trusses the results are reasonably satisfactory.

In most of the frames the loads act at the joints. Even if a load is not acting at a joint, it can be replaced by its reaction at the joints and local bending effect on the member. The frame may be analysed for the joint loads and the local bending effect taken care in the design.

In most of the trusses, the self weight is really small compared to the loads they carry. Hence self weight of the members may be neglected.

It is the duty of construction engineer to see that the centroid of all cross-sections of a member lie along a single axis so that the member is held in equilibrium by the two forces at its ends.

4.3. NATURE OF FORCES IN MEMBERS

The members of a pin jointed frames are subjected to either tensile or compressive forces. A typical truss *ABCDE* loaded at joint *E* is shown in Fig. 4.7 (*a*). The member *BC* is subjected to compressive force *C* as shown in Fig. 4.7(*b*). Effect of this force on the joint *B* (or *C*) is equal and opposite to force *C* as shown in Fig. 4.7(*b*). The member *AE* is subjected to tensile force *T*. Its effect on the joints *A* and *E* are as shown in Fig. 4.7(*b*).

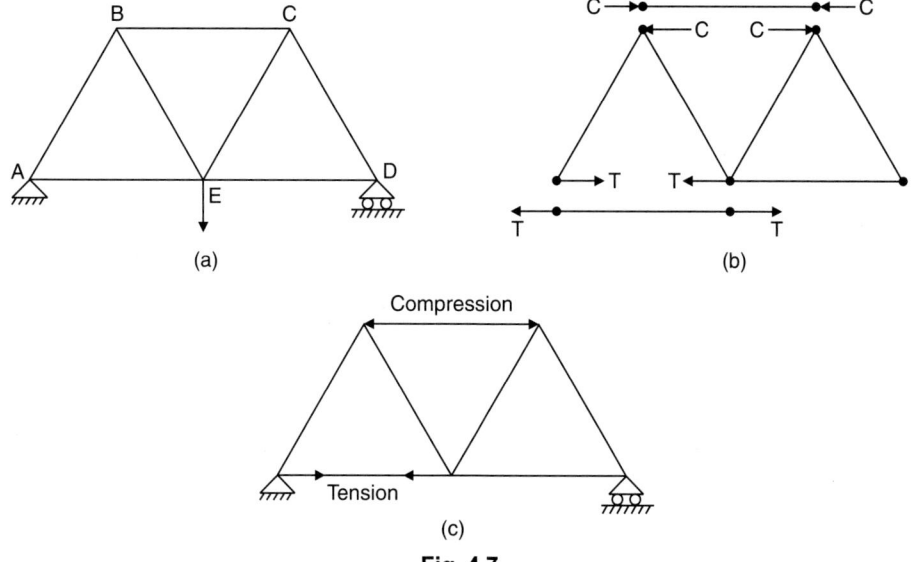

Fig. 4.7

During analysis, we mark the forces on the joints but not forces in the member [Ref. Fig. 4.7 (c)]. Hence, it may be noted that two arrows going away from each other in the figure of frame, show compressive force in the member and two arrows coming towards each other represent the tensile force in the member.

4.4. METHODS OF ANALYSIS

The following four methods are available for the analysis of pin connected plane frames :
(i) Method of joint
(ii) Method of section
(iii) Method of tension coefficient
(iv) Graphical method.

The first three are analytical methods and are dealt in this chapter. For graphical method any book on graphic static may be referred. However nowadays graphical methods are almost given up since computer aided analysis has become popular.

4.5. METHOD OF JOINT

At any joint, the forces in the members meeting and loads acting, constitute a system of concurrent forces. Hence two equations of equilibrium can be formed at each joint. To start with a joint is selected where there are only two unknown forces. Many times such a joint can be identified only after finding the reactions at supports considering the equilibrium of entire frame. Then using two equations of equilibrium for the system of forces at joint the two unknown forces can be found. Then next joint is selected for the analysis where there are now only two unknown forces. Thus the analysis proceeds from joint to joint to get the forces in all the members.

It may be noted that if there are j number of joints, $2j$ number of equations can be formed. In general, there will be three reactions in a determinate plane frames. The force in each member is u nknown. Hence if there are 'm' number of members, the total number of unknowns are $m + 3$. A problem can have unique solution, if there are as many equations as there are unknowns. Hence a pin jointed plane frame analysis is determinate, if

$$2j = m + 3 \qquad \qquad ...(4.2)$$

This equation is same as eqn. (4.1), which was derived on the consideration of a perfect frame. Hence a perfect frame is a determinate frame. If $m > 2j - 3$, then the number of unknowns is more than the number of equations. Hence, a redundant frame is statically indeterminate frame. If $m \leq 2j - 3$, then the number of equations are more than the number of unknowns. Since there can be more than one solution for such case, it shows instability of the frame. Hence a deficient frame is not a stable structure. The method of joint is illustrated with the examples below :

Example 4.1. *Find the forces in all the members of the truss shown in Fig. 4.8 (a). Tabulate the results.*

Solution.

Step 1. Determine the inclination of all inclined members. In this problem

$$\tan \theta = \frac{3}{3} = 1.$$

$$\therefore \qquad \theta = 45°.$$

Step 2. Look for a joint where there are only two unknowns. If such a joint is not available, determine the reactions at supports and then the unknowns at one

ANALYSIS OF PIN-JOINTED PLANE FRAMES

of the support may reduce to only two. In this case at joint C, there are only two unknown forces. *i.e.*, forces in members CB and CD, say F_{CB} and F_{CD}.

[**Note:** Usually in cantilever type trusses, we find such joints without need to find support reactions.]

Step 3. Now at joint C there are two equations of equilibrium and two unknown forces. Hence the unknown forces can be found.

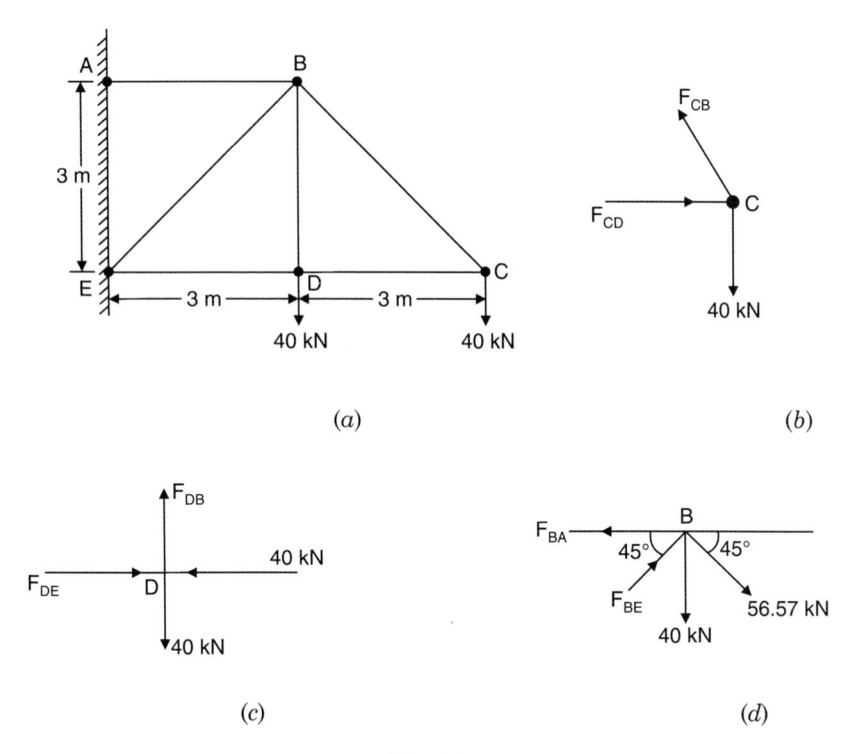

Fig. 4.8

At joint C [Ref. Fig. 4.8 (b)], $\Sigma F_V = 0$
shows that the force F_{CB} should act away from joint C so that its vertical component balances the vertical downward load.

$$F_{CB} \sin 45° - 40 = 0$$
$$\therefore \quad F_{CB} = 56.57 \text{ kN.}$$

Now, $\Sigma F_H = 0$ indicates that F_{CD} should act towards C.

$$F_{CD} - F_{CB} \cos 45° = 0$$
$$\therefore \quad F_{CD} = F_{CB} \cos 45° = 56.57 \cos 45°$$
$$\therefore \quad F_{CD} = 40 \text{ kN.}$$

[**Note :** If assumed direction of member force is wrong, the value of member force will be negative. Then the direction is to be corrected and proceeded further].

Step 4. On the diagram of the truss, mark arrow on the members near the joint analysed to indicate the forces on the joint. At the other end, mark the arrows in the reverse direction. In the present case, arrows are marked at joint C in members CB and CD. Then reversed directions are marked in the members CB and CD near joint B and D.

Step 5. Look for the next joint for analysis where there are only two unknown forces. In this case, joint D is selected for further analysis. Referring to Fig. 4.8 (c)

$$\Sigma F_V = 0 \rightarrow$$
$$F_{DB} = 40 \text{ kN}$$
and
$$\Sigma F_H = 0 \rightarrow$$
$$F_{DE} = 40 \text{ kN}.$$

Step 6. Repeat step 4 and 5 till forces in all members are determined.

In the present case, joint B can be selected for further analysis. Referring to Fig 4.8 (d),

$$\Sigma F_V = 0 \rightarrow$$
$$F_{BE} \sin 45° - F_{BD} - F_{BC} \sin 45° = 0$$

[Note, $F_{BD} = F_{DB}$ and $F_{BC} = F_{CB}$ in magnitude]

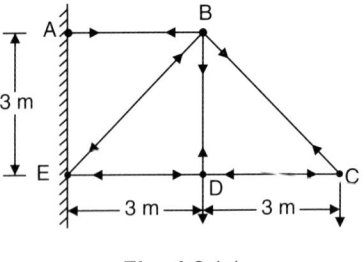

$\therefore \quad F_{BE} \sin 45° - 40 - 56.57 \sin 45° = 0$
i.e. $\quad F_{BE} = 113.14$ kN.
$\Sigma F_H = 0 \rightarrow$
$- F_{BA} + F_{BE} \cos 45° + F_{BC} \cos 45° = 0$
or $\quad F_{BA} = 113.14 \cos 45° + 56.57 \cos 45°$
$= 120$ kN.

Fig. 4.8 (e)

The direction of these forces are marked on the diagram. Now the analysis is complete since the forces in all the members are found.

Step 7. Determine the nature of forces in the members and tabulate the results. Note that the arrows marked on a member are towards joints the member is in compression and if the arrows are going away from the joint [Ref. Fig. 4.8 (e)], the member is in tension as discussed in Art. 4.3. In this case,

Member	Magnitude of Force in kN	Nature
AB	120	Tension
BC	56.57	Tension
CD	40	Compression
DE	40	Compression
BE	113.14	Compression
BD	40	Tension

Example 4.2. *Determine the forces in all the members of the truss shown in Fig. 4.9 (a) and indicate the magnitude and nature of the forces on the diagram of the truss. All inclined members are at 60° to horizontal and length of each member is 2 m.*

Solution. In this problem we don't find a joint with only two unknown forces straight way. Hence let us first find the support reactions.

Considering the equilibrium of entire frame.

$$\Sigma M_A = 0 \rightarrow$$
$$- R_D \times 4 + 40 \times 1 + 60 \times 2 + 50 \times 3 = 0$$
$\therefore \quad R_D = 77.5$ kN
$\Sigma F_H = 0 \rightarrow H_A = 0$

ANALYSIS OF PIN-JOINTED PLANE FRAMES

∴ Reaction at A is vertical.

$$\Sigma F_V = 0 \rightarrow$$
$$R_A + R_D - 40 - 60 - 50 = 0$$
$$R_A = 150 - 77.5, \qquad \text{since } R_D = 77.5 \text{ kN}$$
$$= 72.5 \text{ kN}.$$

Now at joint A there are only two unknown forces. Hence consider the equilibrium of joint A [Fig. 4.9 (b)]

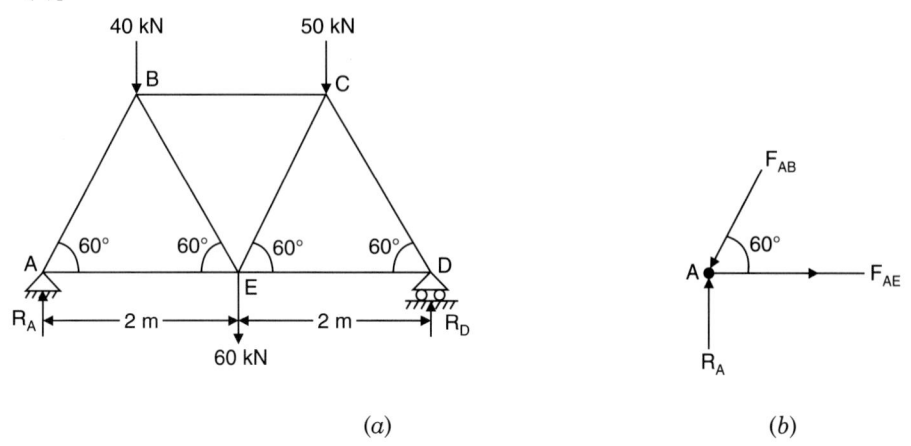

Fig. 4.9. (a), (b)

$$\Sigma F_V = 0 \rightarrow$$
$$F_{AB} \sin 60° - R_A = 0$$

∴ $$F_{AB} = \frac{R_A}{\sin 60°} = \frac{72.5}{\sin 60°} = 83.72 \text{ kN (Compression)}.$$

$$\Sigma F_H = 0 \rightarrow$$
$$F_{AE} - F_{AB} \cos 60° = 0$$

∴ $$F_{AE} = 83.72 \cos 60° = 41.86 \text{ kN (Tension)}$$

Joint D: $\Sigma F_V = 0 \rightarrow$
$$- F_{DC} \sin 60° + R_D = 0$$

∴ $$F_{DC} = \frac{R_D}{\sin 60°} = \frac{77.5}{\sin 60°} = 89.5 \text{ kN (Compression)}.$$

$$\Sigma F_H = 0 \rightarrow$$
$$- F_{DE} + F_{DC} \cos 60° = 0.$$

∴ $$F_{DE} = F_{DC} \cos 60° = 89.5 \cos 60° = 44.75 \text{ kN (Tension)}$$

Joint B: Referring to Fig. 4.9 (d)

$$\Sigma F_V = 0 \rightarrow$$
$$F_{AB} \sin 60° - 40 - F_{BE} \sin 60° = 0$$

∴ $$F_{BE} = \frac{F_{AB} \sin 60° - 40}{\sin 60°} = \frac{72.5 - 40}{\sin 60°} = 37.53 \text{ kN (Tension)}$$

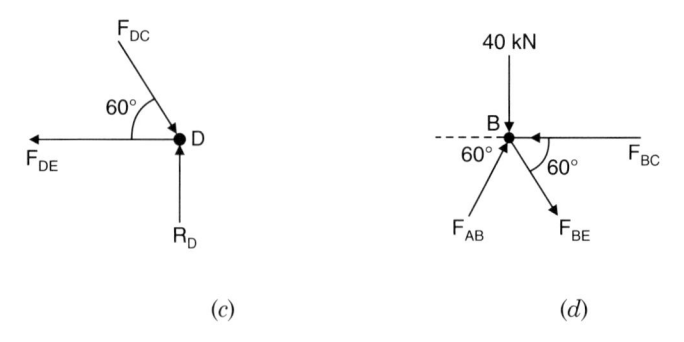

Fig. 4.9. (c), (d)

$$\Sigma F_H = 0 \rightarrow$$
$$F_{AB} \cos 60° + F_{BE} \cos 60° - F_{BC} = 0$$
$$\therefore \quad F_{BC} = 83.72 \cos 60° + 37.53 \cos 60° = 60.62 \text{ kN (Compression)}.$$

Joint C: Referring to Fig. 4.9 (e),
$$\Sigma F_V = 0 \rightarrow$$
$$F_{DC} \sin 60° - 50 - F_{CE} \sin 60° = 0$$
$$\therefore \quad F_{CE} = \frac{F_{DC} \sin 60° - 50}{\sin 60°} = \frac{89.50 \sin 60° - 50}{\sin 60°} = 31.76 \text{ kN (Tension)}$$

Now the forces in all the members are known. If joint E is analysed it will give check for the analysis. The results are shown on diagram of the truss in Fig. 4.9 (f).

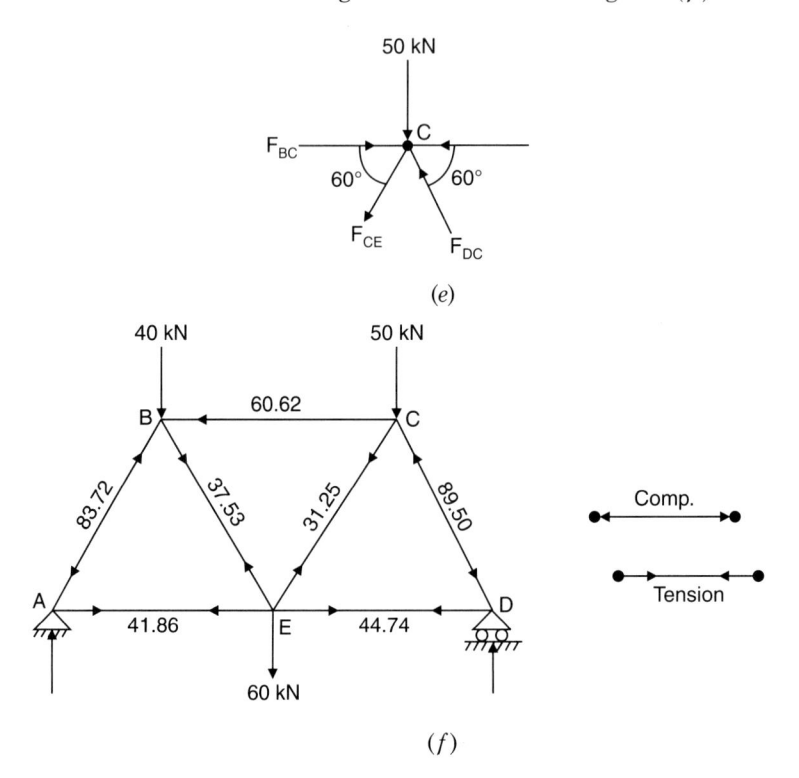

Fig. 4.9. (e), (f)

ANALYSIS OF PIN-JOINTED PLANE FRAMES

Example 4.3. *Analyse the truss shown in Fig. 4.10 (a).*

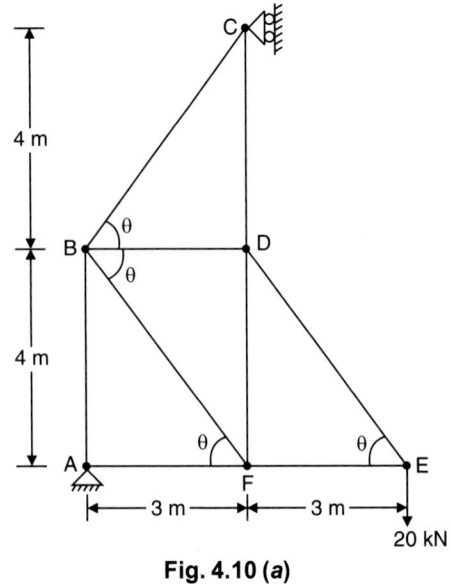

Fig. 4.10 (a)

Solution. Inclined members are making angle θ with horizontal, where

$$\tan \theta = \frac{4}{3}$$

∴ $\theta = 53.13°$.

As soon as a joint is analysed the forces on the joints are marked on members in Fig. 4.10 (b).

Joint E : Referring to Fig. 4.10 (c)

$$\Sigma F_V = 0 \quad \rightarrow$$
$$F_{ED} \sin \theta - 20 = 0.$$

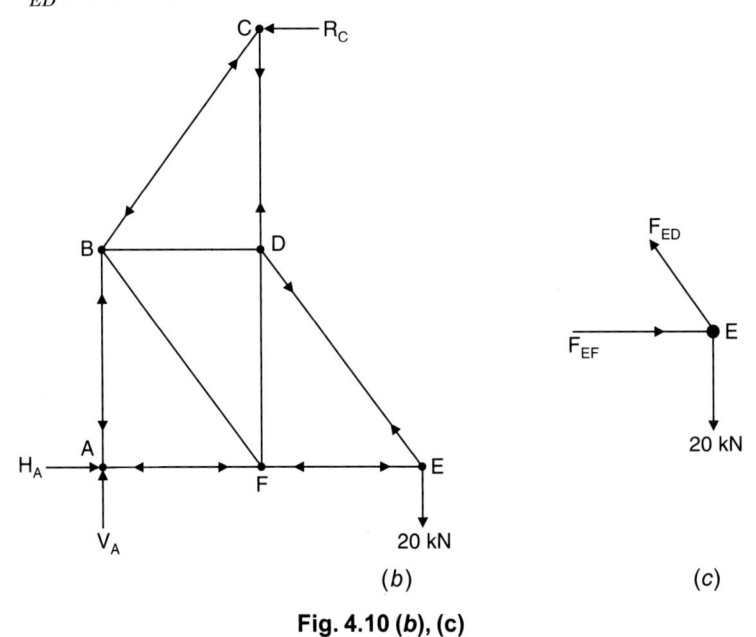

(b)

(c)

Fig. 4.10 (b), (c)

$$F_{ED} = \frac{20}{\sin 53.13°} = 25 \text{ kN (Tension)}$$

$$\Sigma F_H = 0 \rightarrow$$
$$F_{EF} - F_{ED} \cos \theta = 0.$$
$$F_{EF} = 25 \cos 53.13° = 15 \text{ kN (Compression)}$$

At this stage no other joint is having only two unknown forces. Hence no further progress is possible. Let us find the reactions at supports considering the equilibrium of entire truss. Let the reactions be as shown in Fig. 4.10 (b). Then

$$\Sigma M_A = 0 \rightarrow$$
$$-R_C \times 8 + 20 \times 6 = 0$$
$$\therefore \quad R_C = 15 \text{ kN}$$
$$\Sigma F_V = 0 \rightarrow$$
$$V_A - 20 = 0 \quad \text{or} \quad V_A = 20 \text{ kN}.$$
$$\Sigma F_H = 0, \quad H_A - R_C = 0$$
$$\therefore \quad H_A = R_C = 15 \text{ kN}$$

Joint A: Referring to Fig. 4.10 (d).

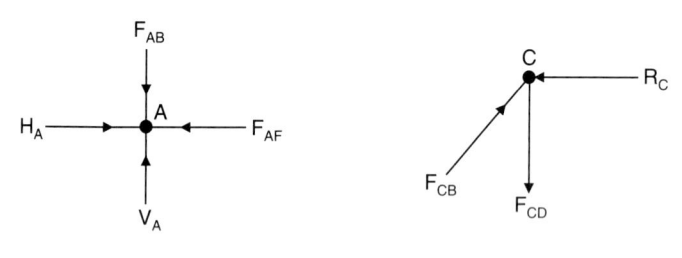

Fig. 4.10 (d)   Fig. 4.10 (e)

$$\Sigma F_V = 0 \rightarrow$$
$$-F_{AB} + V_A = 0.$$
$$\therefore \quad F_{AB} = V_A = 20 \text{ kN (Compression)}$$
$$\Sigma F_H = 0 \rightarrow$$
$$H_A - F_{AF} = 0.$$
$$\therefore \quad F_{AF} = H_A = 15 \text{ kN (Compression)}$$

Joint C: Referring to Fig. 4.10 (e)

$$\Sigma F_H = 0 \rightarrow$$
$$F_{CB} \cos \theta - R_C = 0.$$
$$F_{CB} = \frac{R_C}{\cos \theta} = \frac{15}{\cos 53.13°} = 25 \text{ kN (Compression)}$$

$$\Sigma F_V = 0 \rightarrow$$
$$F_{CB} \sin 53.13° - F_{CD} = 0$$
$$\therefore \quad F_{CD} = 25 \sin 53.13° = 20 \text{ kN (Tension)}$$

ANALYSIS OF PIN-JOINTED PLANE FRAMES

Joint B: Referring to Fig. 4.10 (f)

$$\Sigma F_V = 0 \rightarrow$$

$$F_{BF} \sin 53.13° - F_{BC} \sin 53.13° + F_{AB} = 0.$$

$$F_{BF} = \frac{F_{BC} \sin 53.13° - F_{AB}}{\sin 53.13°} = \frac{25 \sin 53.13° - 20}{\sin 53.13°} = 0$$

$$\Sigma F_H = 0 \rightarrow$$

$$F_{BD} - F_{BF} \cos 53.13° - F_{BC} \cos 53.13° = 0$$

$$F_{BD} = 0 + 25 \cos 53.13° = 15 \text{ kN (Tension)}$$

Joint F: Referring to Fig. 4.10 (g)

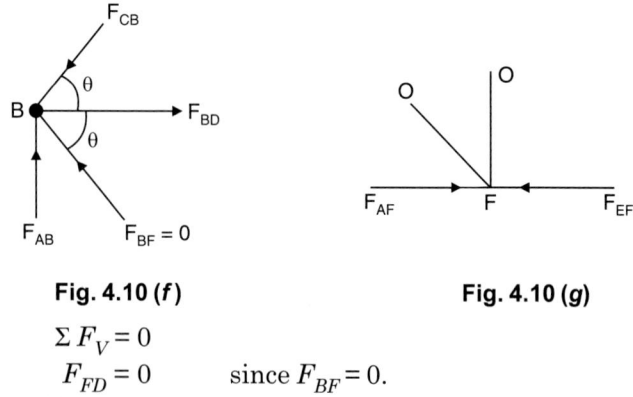

Fig. 4.10 (f)   **Fig. 4.10 (g)**

$$\Sigma F_V = 0$$

$$F_{FD} = 0 \qquad \text{since } F_{BF} = 0.$$

The magnitude of forces are marked on Fig. 4.10 (b).

A member with arrows towards joint is in compression and a member with arrows away from joints is in tension.

Note: When three members are meeting at an unloaded joint and out of them two are collinear, then the force in third member will be zero. Such situations are illustrated in Figs. 4.10 (h) and (i).

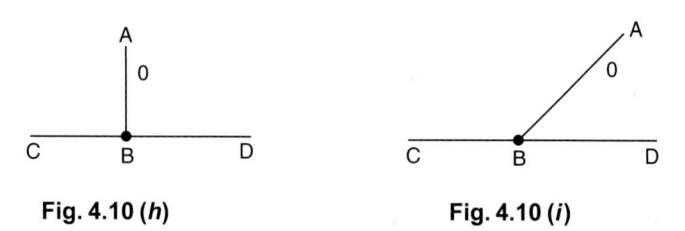

Fig. 4.10 (h)   **Fig. 4.10 (i)**

Example 4.4. *Find the forces in all the members of the truss shown in Fig 4.11 (a).*

Solution.

$$\tan \theta_1 = 4/6 \qquad \therefore \quad \theta_1 = 33.69°$$

$$\theta_2 = \tan^{-1} \frac{8}{3} \times \frac{1}{2} = 53.13°$$

$$\theta_3 = \tan^{-1} 4/3 = 53.13°$$

Thus $\qquad \theta_2 = \theta_3 = 53.13°$

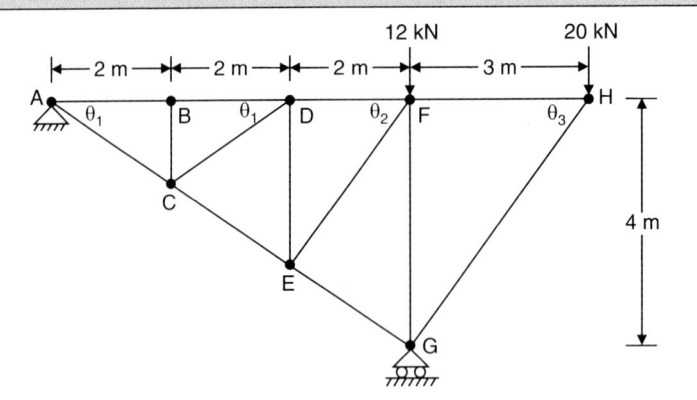

Fig. 4.11 (a)

Joint by joint analysis is carried out as given below and the joint forces are marked in Fig. 4.11 (b). Then the nature of the force in each member is determined.

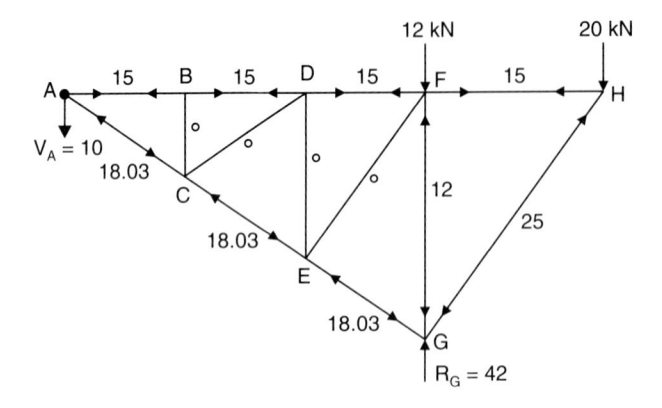

Fig. 4.11 (b)

Joint H: $\Sigma F_V = 0 \rightarrow$
$$F_{HG} \sin \theta_3 = 20$$

∴ $$F_{HG} = \frac{20}{\sin 53.13°} = 25 \text{ kN (Compression)}$$

$$\Sigma F_H = 0 \rightarrow$$
$$F_{HF} - F_{HG} \cos \theta_3 = 0$$

∴ $$F_{HF} = 25 \cos 53.13° = 15 \text{ kN (Tension)}$$

Now, reactions at supports are to be found.
$$\Sigma M_A = 0 \rightarrow$$
$$-R_G \times 6 + 20 \times 9 + 12 \times 6 = 0$$

∴ $$R_G = 42 \text{ kN}$$
$$\Sigma F_V = 0 \rightarrow$$
$$V_A + R_G - 12 - 20 = 0$$
$$V_A = 32 - 42, \quad \text{since } R_G = 42 \text{ kN}$$
$$= -10 \text{ kN}$$
$$= 10 \text{ kN} \downarrow$$

ANALYSIS OF PIN-JOINTED PLANE FRAMES

$$\Sigma F_H = 0 \rightarrow \quad H_A = 0.$$

Hence reactions are as shown in Fig. 4.11 (b)

Joint A: $\quad \Sigma F_V = 0 \rightarrow$
$$F_{AC} \sin \theta_1 - 10 = 0$$

$$\therefore \quad F_{AC} = \frac{10}{\sin 33.69°} = 18.03 \text{ kN (Compression)}$$

$$\Sigma F_H = 0 \rightarrow$$
$$F_{AB} - F_{AC} \cos \theta_1 = 0$$
$$\therefore \quad F_{AB} = 18.03 \cos 33.69° = 15 \text{ kN (Tension)}$$

Joint B: $\quad \Sigma F_V = 0 \rightarrow F_{BC} = 0$
$$\Sigma F_H = 0 \rightarrow F_{CD} = F_{BA} = 15 \text{ kN (Compression)}$$

Joint C:
Σ Forces normal to $AC = 0 \rightarrow$
$$F_{CD} = 0, \quad \text{since } F_{BC} = 0.$$
Σ Forces parallel to $AE \rightarrow$
$$F_{CE} = F_{AC} = 18.03 \text{ kN (Compression)}$$

Joint D:
$$\Sigma F_V = 0 \rightarrow F_{DE} = 0$$
$\Sigma F_H = 0 \rightarrow \quad F_{DF} = F_{BD} = 15 \text{ kN (Tension)}$

Joint E:
Σ Forces normal to $CG = 0 \rightarrow$
$$F_{EF} = 0$$
Σ Forces parallel to $CG \rightarrow$
$$F_{EG} = F_{CE} = 18.03 \text{ (Compression)}$$

Joint F:
$$\Sigma F_V = 0 \rightarrow$$
$$F_{AG} = 12 \text{ kN (Compression)}$$

All member forces are marked on Fig. 4.11 (b).

Example 4.5. *Analyse the truss shown in Fig. 4.12 (a). All members are 3 m long.*

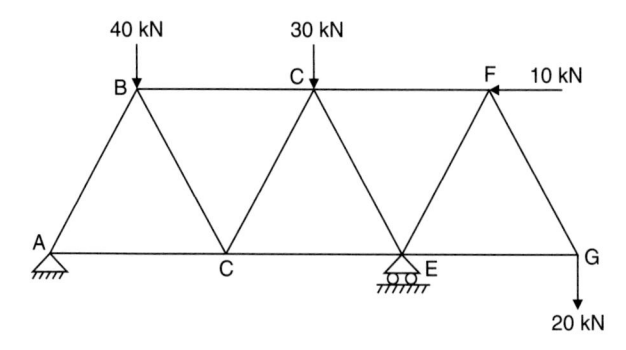

Fig. 4.12 (a)

Solution. Since all members are of same length (3 m), all triangles are equilateral triangles and hence all included angles are 60°. Joint by joint analysis is carried out and the direction of joint forces are marked in Fig. 4.11 (b).

Then nature of the force is determined.

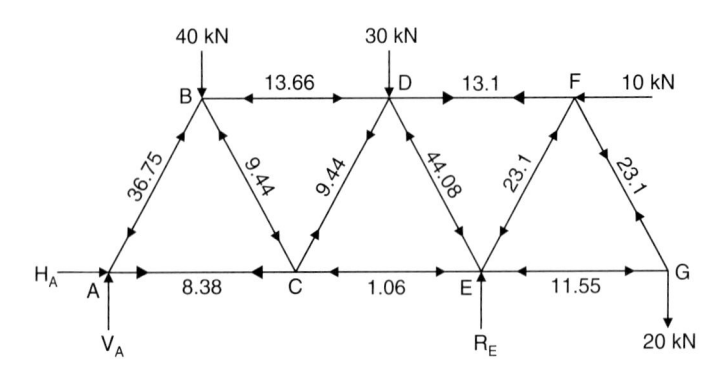

Fig. 4.12 (b)

Joint G: $\Sigma F_V = 0 \rightarrow$
$$F_{GF} \sin 60° = 20$$
∴ $\quad F_{GF} = 23.1$ kN (Tension)
$$\Sigma F_H = 0 \rightarrow$$
$$F_{GE} - F_{GF} \cos 60° = 0$$
∴ $\quad F_{GF} = 23.1 \cos 60° = 11.55$ kN (Compression)

Joint F:
$$\Sigma F_V = 0 \rightarrow$$
$$F_{EF} \sin 60° - F_{GF} \sin 60° = 0.$$
∴ $\quad F_{FE} = F_{GF} = 23.1$ kN (Compression).
$$\Sigma F_H = 0 \rightarrow$$
$$F_{FD} + 10 - F_{GF} \cos 60° - F_{FE} \cos 60° = 0.$$
∴ $\quad F_{FD} = 13.1$ kN (Tension). Since $F_{GF} = F_{FE} = 23.1$ kN.

Now without finding reactions, we can proceed. Hence consider the equilibrium of entire truss.
$$-R_E \times 6 - 10 \times 3 \sin 60° + 40 \times 3 \cos 60° + 30 (3 + 3 \cos 60°) + 20 \times 9 = 0$$
∴ $\quad R_E = 58.17$ kN
$$\Sigma F_V = 0 \rightarrow$$
$$V_A - 40 - 30 - 20 + R_E = 0.$$
∴ $\quad V_A = 38.83$ kN
$$\Sigma F_H = 0$$
$$H_A = 10 \text{ kN}$$

Joint A: $\Sigma F_V = 0 \rightarrow$
$$F_{AB} \sin 60° - 31.83 = 0.$$
∴ $F_{AB} = 36.75$ kN (Compression).
$$\Sigma F_H = 0 \rightarrow$$
$$F_{AC} - F_{AB} \cos 60° + H_A = 0.$$
∴ $\quad F_{AC} = 36.75 \cos 60° - 10$
$$= 8.38 \text{ kN (Tension)}$$

Joint B: $\Sigma F_V = 0 \rightarrow$

$F_{BC} \sin 60° + F_{AB} \sin 60° - 40 = 0$

$F_{BC} \sin 60° = 40 - F_{AB} \sin 60° = 40 - 36.75 \sin 60°$

∴ $F_{BC} = 9.44$ kN (Compression)

$\Sigma F_H = 0 \rightarrow$

$F_{AB} \cos 60° - F_{BC} \cos 60° - F_{BD} = 0$

∴ $F_{BD} = 36.75 \cos 60° - 9.44 \cos 60° = 13.66$ kN (Compression)

Joint C: $\Sigma F_V = 0 \rightarrow$

$F_{CD} \sin 60° - F_{BC} \sin 60° = 0$

$F_{CD} = F_{BC} = 9.44$ kN (Tension)

$\Sigma F_H = 0 \rightarrow$

$F_{CD} \cos 60° + F_{BC} \cos 60° - F_{AC} - F_{CE} = 0.$

∴ $F_{CE} = 9.44 \cos 60° + 9.44 \cos 60° - 8.38$

$= 1.06$ kN (Compression)

Joint D: $\Sigma F_V = 0 \rightarrow$

$F_{DE} \sin 60° - F_{CD} \sin 60° - 30 = 0$

$$F_{DE} = \frac{30 + F_{CD} \sin 60°}{\sin 60°} = \frac{30 + 9.44 \sin 60°}{\sin 60°}$$

$= 44.08$ kN (Compression).

Forces are indicated in Fig. 4.12 (b).

4.6. METHOD OF SECTION

In the method of section, after determining the support reactions, a section line is drawn passing through not more than three members in which forces are not known, such that the frame is cut into two separate parts. Each part should be in equilibrium under the action of loads, reactions and the forces in the members that are cut by the section line. Equilibrium of any one of these two parts is considered and the unknown forces in the members cut by the section line are determined. The system of forces acting on either part of truss constitutes a non-concurrent force system. Since there are only three independent equations of equilibrium, there should be only three unknown forces. Hence, in this method it is an essential requirement that the section line should pass through not more than three members in which forces are not known and it should separate the frame into two parts.

Under the following two situations the method of section is preferred over the method of joint :

(i) In the analysis of large truss in which forces in only few members are required.

(ii) If method of joint fails to start or proceed with analysis for not getting a joint with only two unknown forces.

The method of section is illustrated with the examples 4.6 to 4.9. Examples 4.6 and 4.7 are the cases in which method of section is advantageous since forces in only few members are required. Examples 4.8 and 4.9 are the cases in which method of joints fail to start/proceed to get the solution. In practice the frames may be analysed partly by the method of joint and partly by the method of section as illustrated in example 4.9.

Example 4.6. *Determine the forces in the members FH, HG and GI in the truss shown in Fig. 4.13 (a). Each load is 10 kN and all triangles are equilaterals with sides equal to 4 m.*

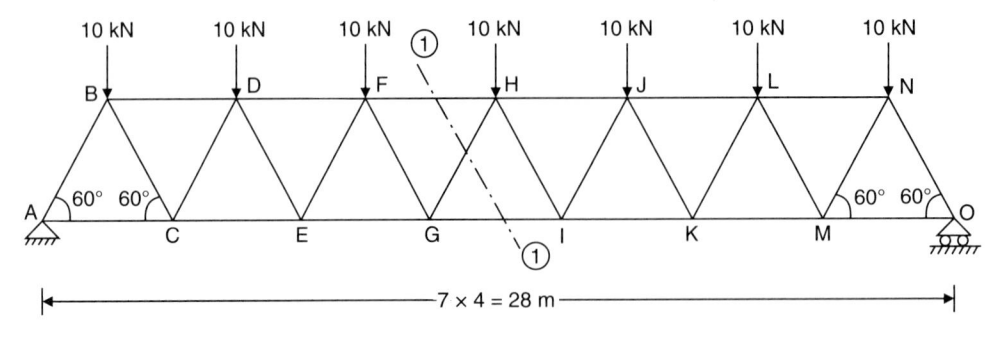

Fig. 4.13 (a)

Solution. Due to symmetry,

$$R_A = R_B = \frac{1}{2} \times \text{total load}$$

$$= \frac{1}{2}(10 \times 7) = 35 \text{ kN}$$

There is no horizontal component of reaction at A.

Take section (1) – (1) which cuts the members *FH, GH* and *GI* and separate the truss into two parts. Consider the equilibrium of left hand side part as shown in Fig. 4.13 (b). (Prefer the part in which number of forces are less).

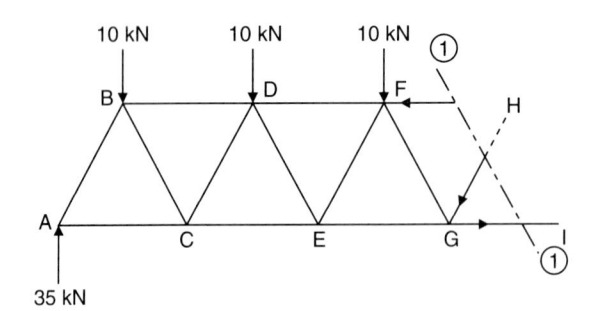

Fig. 4.13 (b)

$$\Sigma M_G = 0 \rightarrow$$
$$-F_{FH} \times 4 \sin 60° + 35 \times 12 - 10 \times 10 - 10 \times 6 - 10 \times 2 = 0$$
∴ $\quad F_{FH} = 69.28$ kN (Compression)
$$\Sigma F_V = 0 \rightarrow$$
$$35 - 10 - 10 - 10 - F_{GH} \sin 60° = 0$$
∴ $\quad F_{GH} = 5.77$ kN (Compression)
$$\Sigma F_H = 0 \rightarrow$$
$$F_{GI} - F_{FH} - F_{GH} \cos 60° = 0$$
∴ $\quad F_{GI} = 69.28 + 5.77 \cos 60° = 72.17$ kN (Tension).

ANALYSIS OF PIN-JOINTED PLANE FRAMES

Example 4.7. *Find the magnitude and nature of forces in the members $U_3 U_4$, $L_3 L_4$ and $U_4 L_3$ of the loaded truss shown in Fig. 4.14 (a).*

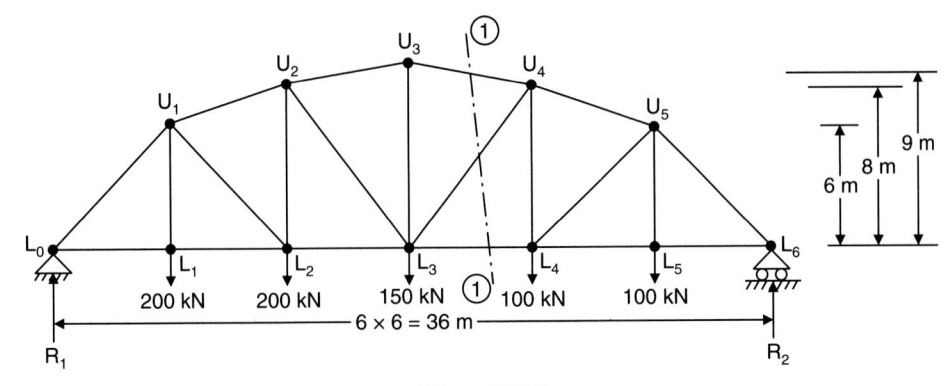

Fig. 4.14 (a)

Solution. Considering the equilibrium of entire truss.

$$\Sigma M_{L_0} = 0 \rightarrow$$

$200 \times 6 + 200 \times 12 + 150 \times 18 + 100 \times 24 + 100 \times 30 - R_2 \times 36 = 0$

∴ $\quad R_2 = 325$ kN

$$\Sigma F_V = 0 \rightarrow$$

$R_1 + R_2 - 200 - 200 - 150 - 100 - 100 = 0$
$R_1 = 200 + 200 + 150 + 100 + 100 - 325$, Since $R_2 = 325$ kN
$\quad = 425$ kN.

Take the section (1) – (1) and consider the equilibrium of right hand side part. Referring to Fig. 4.14 (b).

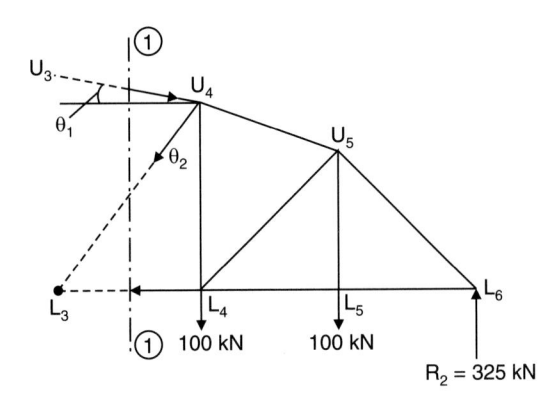

Fig. 4.14 (b)

$\theta_1 = \tan^{-1} 1/6 = 9.46°$
$\theta_2 = \tan^{-1} 6/8 = 36.87°$.

$$\Sigma M_{U_4} = 0 \rightarrow$$

$F_{L_3 L_4} \times 8 + 100 \times 6 - 325 \times 12 = 0$

∴ $\quad F_{L_3 L_4} = 412.5$ kN (Tension)

$$\Sigma M_{L_3} = 0 \rightarrow$$

100 FUNDAMENTALS OF ENGINEERING MECHANICS

$$F_{U_4U_3} \cos\theta_1 \times 8 + F_{U_3U_4} \sin\theta_1 \times 6 + 100 \times 6 + 100 \times 12 - 325 \times 18 = 0$$

$$\therefore \quad F_{U_3U_4} = 456.2 \text{ kN (Compression)}$$

$$\Sigma F_H = 0 \rightarrow$$

$$F_{U_3U_4} \cos\theta_1 - F_{L_3L_4} - F_{U_4L_3} \sin\theta_2 = 0$$

$$456.2 \cos 9.46° - 412.5 - F_{U_4L_3} \sin 36.87° = 0$$

$$\therefore \quad F_{U_4L_3} = 62.49 \text{ kN (Tension)}.$$

Example 4.8. *Find the forces in the members (1), (2), and (3) of French truss shown in Fig. 4.15 (a).*

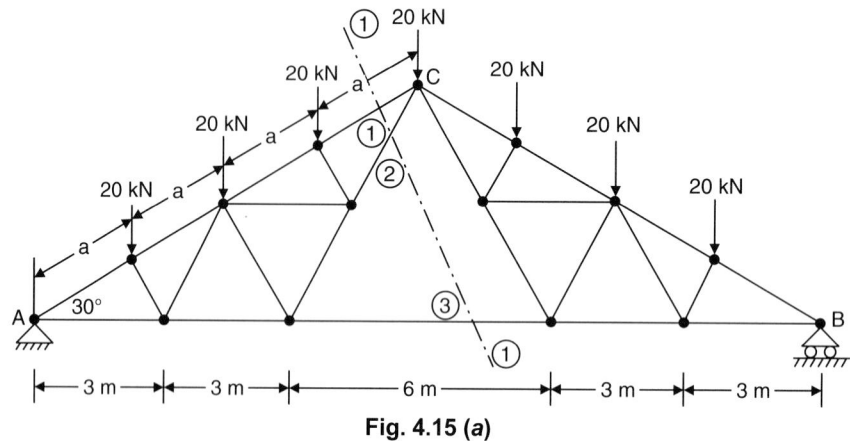

Fig. 4.15 (a)

Solution. Due to symmetry,

$$R_A = R_B = \frac{1}{2} \times 20 \times 7 = 70 \text{ kN}.$$

Drop perpendicular CE on AB.

$$CE = AE \tan 30° = 5.196 \text{ m}.$$
$$DE = 3 \text{ m}.$$

$$\therefore \quad \theta = \tan^{-1} \frac{5.196}{3} = 60°$$

Fig. 4.15 (b)

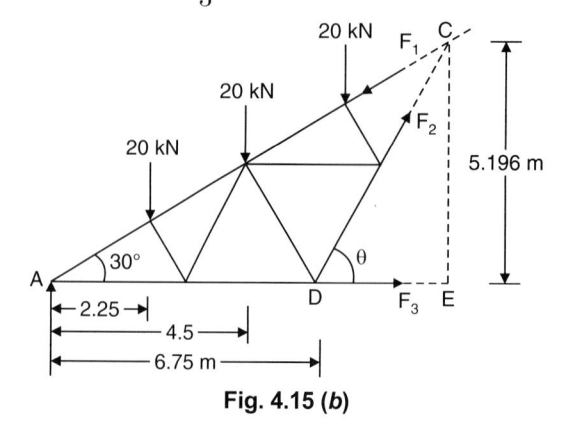

ANALYSIS OF PIN-JOINTED PLANE FRAMES 101

20 kN loads are at equidistances. From $\triangle ACE$, we can make out that horizontal distances of loads from A are

$$\frac{9}{4} = 2.25 \text{ m}, \quad \frac{9 \times 2}{4} = 4.5 \text{ m and } \quad \frac{9 \times 3}{4} = 6.75 \text{ m}.$$

Take section (1) – (1) and consider left hand side portion.
Now, $\Sigma M_A = 0 \rightarrow$
$20 \times 2.25 + 20 \times 4.5 + 20 \times 6.75 - F_2 \sin 60° \times 6 = 0$
$\therefore \quad F_2 = 51.96 \text{ kN}$

[**Note:** F_2 is resolved at D in vertical and horizontal components and then moment about A is calculated].

$$\Sigma F_V = 0 \rightarrow$$
$R_A - 20 - 20 - 20 - F_1 \sin 30° + F_2 \sin 60° = 0$
$70 - 20 - 20 - 20 - F_1 \sin 30° + 51.96 \sin 60° = 0$
$\therefore \quad F_1 = 110 \text{ kN (Compression)}$
$$\Sigma F_H = 0 \rightarrow$$
$F_3 + F_2 \cos 60° - F_1 \cos 30° = 0$
$F_3 = -51.96 \cos 60° + 110 \times \cos 30°$
$= 69.28 \text{ kN (Tension)}$

[**Note:** In this problem, method of joint alone cannot give complete solution.]

Example 4.9. *Find the forces in all the members of the symmetric truss, shown in Fig. 4.16 (a).*

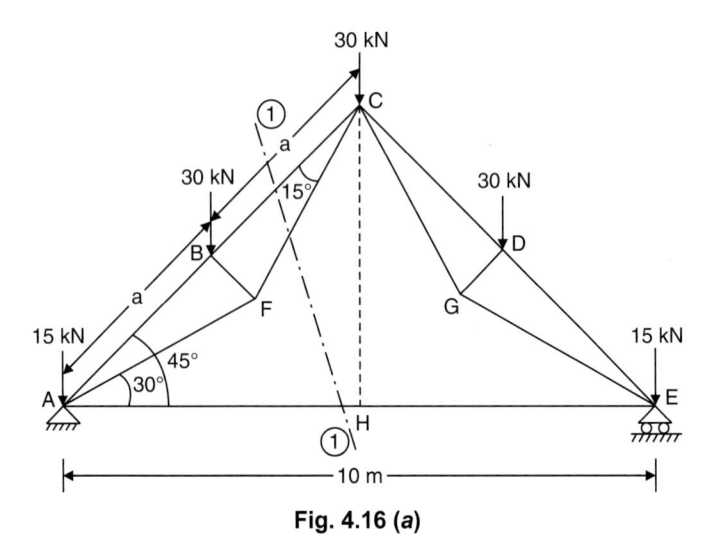

Fig. 4.16 (a)

Solution. Due to symmetry the reactions are equal.

$$\therefore \quad R_A = R_E = \frac{1}{2} \times \text{total load}$$
$$= \frac{1}{2}(15 + 30 + 30 + 30 + 15) = 60 \text{ kN}.$$

Drop perpendicular CH to AE.
$\angle ACH = 45°$
$\therefore \quad \angle FCH = 45° - 15° = 30°$ *i.e.*, FC makes $30°$ with vertical.

It is not possible to find a joint with only two unknown forces. Consider section (1) – (1). For left hand side of the frame, 4.16(b)

$$\Sigma M_C = 0 \rightarrow$$
$$-F_{AE} \times 5 + 60 \times 5 - 15 \times 5 - 30 \times 2.5 = 0.$$
$$\therefore \quad F_{AE} = 30 \text{ kN (Tension)}$$

Assuming the directions for F_{FC} and F_{BC} as shown in Fig. 4.16(b)

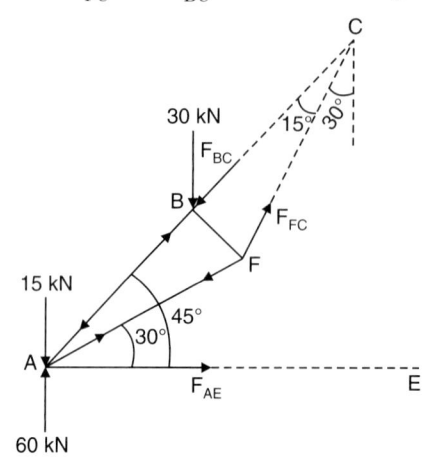

Fig. 4.16 (b)

$$\Sigma F_V = 0 \rightarrow$$
$$F_{FC} \cos 30° - F_{BC} \cos 45° + 60 - 15 - 30 = 0$$
$$0.866 \, F_{FC} - 0.707 \, F_{BC} = -15 \quad \text{...(i)}$$
$$\Sigma F_H = 0 \rightarrow$$
$$F_{FC} \sin 30° - F_{BC} \sin 45° + F_{AE} = 0.$$
$$0.5 \, F_{FC} - 0.707 \, F_{BC} = -30 \quad \text{...(ii)}$$
Since $\quad F_{AE} = 30 \text{ kN}$

Subtracting eqn. (ii) from eqn. (i), we get
$$0.366 \, F_{FC} = 15$$
$$\therefore \quad F_{FC} = 40.98 \text{ kN (Tension)}$$

Substituting this value in eqn. (i),
$$0.866 \times 40.98 - 0.707 \, F_{BC} = -15$$
$$\therefore \quad F_{BC} = 71.41 \text{ kN (Compression)}$$

Assumed directions of F_{FC} and F_{BC} are correct.

Now we can proceed with method of joint to find the forces in other members. Since it is a symmetric truss, analysis of half the truss is sufficient. Other values may be written by making use of symmetry.

Joint B :

ΣForces normal to $AC = 0 \rightarrow$
$$F_{BF} - 30 \cos 45° = 0$$
or $\quad F_{BF} = 21.21 \text{ kN (Compression)}$

ΣForces parallel to $AC = 0 \rightarrow$
$$F_{AB} - F_{BC} - 30 \sin 45° = 0$$

ANALYSIS OF PIN-JOINTED PLANE FRAMES

$$F_{AB} = 71.41 + 21.21 = 92.62 \text{ kN (Compression)}$$

Joint A: $\Sigma F_V = 0 \rightarrow$

$$F_{AF} \sin 30° - F_{AB} \sin 45° - 15 + 60 = 0.$$

$\therefore$
$$F_{AF} = \frac{92.62 \sin 45° - 45}{\sin 30°} = 40.98 \text{ kN (Tension)}$$

The results are tabulated below :

Member	Force in kN
AB and E	– 92.62
BC and DC	– 71.41
BF and DG	– 21.21
AF and EG	40.98
FC and GC	40.98
AE	30

+ means tension and – means compression.

4.7 METHOD OF TENSION COEFFICIENT

In 1924, Prof. Muller Breslau extended method of joint systematically to three dimensional problems. This systematic approach is now known as 'Method of Tension Coefficient'. In this method tensile forces are considered as +ve forces and hence a negative force is a compressive force. In this article the tension coefficient method for plane trusses is presented which can be easily extended to three dimensional trusses.

Figure 4.17 shows a typical member of length L_{ij} oriented at $\theta_{x,ij}$ with x-axis and $\theta_{y,ij}$ with y-axis. Let the tensile force in the member be F_{ij}.

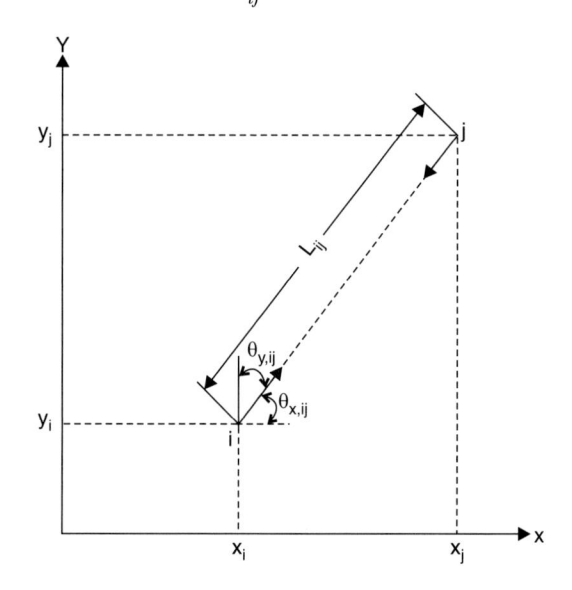

Fig. 4.17

The coordinators of points i and j be (x_i, y_i) and (x_j, y_j) respectively. Then,

$$L_{ij} = \sqrt{(x_j - x_i)^2 + (y_j - y_i)^2}$$

$$\cos \theta_{x,ij} = \frac{x_j - x_i}{L_{ij}}$$

∴ Component of the force F_{ij} in x-direction, acting on joint i, is

$$= F_{ij} \cos \theta_{x,ij}$$
$$= F_{ij} \frac{x_j - x_i}{L_{ij}}$$
$$= t_{ij}(x_j - x_i)$$

where $t_{ij} = \dfrac{F_{ij}}{L_{ij}}$ is the tension coefficient of the member ij. Thus the tension coefficient of a member is defined as tensile force of the member divided by the length of the member. It may be noted that the term has no physical meaning but it is only a notation.

Similarly it may be shown that the component of member force in y-direction at joint i is

$$= F_{ij} \cos \theta_{y,ij}$$
$$= F_{ij} \frac{y_j - y_i}{L_{ij}}$$
$$= t_{ij}(y_j - y_i)$$

Note that the term t_{ij} and t_{ji} are the same.

Now consider the equilibrium of joint A shown in Fig. 4.18 where a number of members are meeting. Let the external loads acting at the joint be X_A and Y_A.

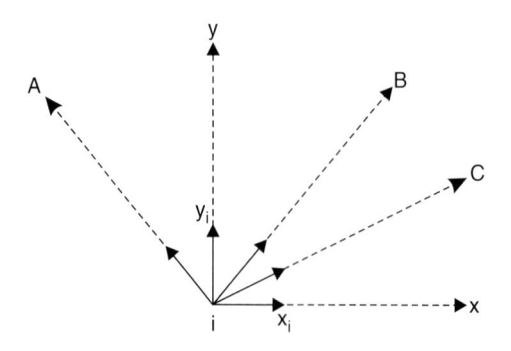

Fig. 4.18

The equilibrium condition in x-direction is,

$$\sum X = 0$$

i.e. $X_i + t_{iA}(x_A - x_i) + t_{iB}(x_B - x_i) + t_{iC}(x_C - x_i) = 0$

i.e. $X_i + \sum t_{ij}(x_j - x_i) = 0$ ...(4.3)

Similarly, it can be shown that,

$$\sum Y = 0, \text{ gives}$$

$$Y_i + \sum t_{ij}(y_j - y_i) = 0 \qquad \text{...(4.4)}$$

Note that in eqns. (4.3) and (4.4) summation is over the number of members meeting at joint i and j refers to the far end joints of the members.

In the method of tension coefficient, the eqns. (4.3) and (4.4) are applied joint by joint to get tension coefficient of members. By multiplying these coefficients with the length of the member, the forces in the members are then calculated. If the quantity works out to be positive, it is tensile force and if it is negative quantity it is compressive force.

In case of space trusses it may be noted that,

$$L_{ij} = \sqrt{(x_j - x_i)^2 + (y_j - y_i)^2 + (z_j - z_i)^2} \qquad \text{...(4.5)}$$

and there will be additional equation of equilibrium

$$Z_i + \sum t_{ij}(z_j - z_i) = 0 \qquad \text{...(4.6)}$$

The following examples, illustrate the method:

Example 4.10 *Analyse the plane truss shown in Fig. 4.8 (example 4.1) by the method of tension coefficient.*

Solution. Taking joint E as the origin, EC as x-axis and EA as y-direction, the coordinates of various joints in the truss are

A (0, 3), B (3, 3), C (6, 0), D (3, 0) and E (0, 0).
$Y_D = -40$ kN, $Y_C = -40$ kN,
Length of various members are,

$$L_{AB} = \sqrt{(3-0)^2 + (3-3)^2} = 3\text{m}$$

$$L_{BC} = \sqrt{(6-3)^2 + (0-3)^2} = 3\sqrt{2}\text{ m}$$

$$L_{CD} = \sqrt{(3-6)^2 + (0-0)^2} = 3\text{m}$$

$$L_{DE} = \sqrt{(0-3)^2 + (0-0)^2} = 3\text{m}$$

$$L_{BD} = \sqrt{(3-3)^2 + (0-3)^2} = 3\text{m}$$

$$L_{BE} = \sqrt{(0-3)^2 + (0-3)^2} = 3\sqrt{2}\text{ m}$$

Now we can proceed with forming and solving equations of equilibrium joint by joint as was done in case of method of joint.

Joint C: $\sum X = 0 \rightarrow$

$$0 + t_{CB}(3-6) + t_{CD}(3-6) = 0$$

$$t_{CB} = -t_{CD} \qquad \text{...(i)}$$

$$\sum Y = 0 \rightarrow$$

$$-40 + t_{CB}(3-0) + t_{CD}(0-0) = 0.$$

$$\therefore \qquad t_{CB} = \frac{40}{3} = 13.333$$

$$\therefore \qquad F_{CB} = t_{CB} \times L_{CB} = \frac{40}{3} \times 3\sqrt{2} = 56.57 \text{ kN}.$$

$\therefore$ From eqn. (i), $t_{CD} = -13.333$

$$\therefore \qquad F_{CD} = -13.333 \times L_{CD} = -13.333 \times 3 = -40 \text{ kN}.$$

Joint D: $\qquad \sum X = 0 \rightarrow$

$0 + t_{DC}(6-3) + t_{DE}(0-3) + t_{DB}(3-3) = 0$

i.e. $\qquad t_{DE} = t_{DC}$

$\qquad \qquad = -13.333 \qquad$ [Note t_{DC} is same as t_{CD}]

$\therefore \qquad F_{DE} = t_{DC} \times L_{DC}$

$\qquad \qquad = -13.333 \times 30 = -40 \text{ kN}.$

$$\sum Y = 0 \rightarrow$$

$-40 + t_{DC}(0-0) + t_{DE}(0-0) + t_{DB}(3-0) = 0$

$$\therefore \qquad t_{DB} = \frac{40}{3} = 13.333$$

$$\therefore \qquad F_{DB} = t_{DB} L_{DB} = \frac{40}{3} \times 3 = 40 \text{ kN}$$

Joint B:

$$\sum X = 0 \rightarrow$$

$0 + t_{BC}(6-3) + t_{BD}(3-3) + t_{BE}(0-3) + t_{BA}(0-3) = 0$

$\qquad 3 t_{BC} - 3 t_{BE} - 3 t_{BA} = 0$

$\qquad t_{BC} - t_{BE} - t_{BA} = 0 \qquad \qquad \qquad ...(ii)$

$$\sum F_y = 0 \rightarrow$$

$0 + t_{BC}(0-3) + t_{BD}(0-3) + t_{BE}(0-3) + t_{BA}(3-3) = 0$

$\qquad -3 t_{BC} - 3 t_{BD} - 3 t_{BE} = 0$

$\therefore \qquad t_{BE} = -(t_{BC} + t_{BD})$

$\qquad \qquad = -(13.333 + 13.333) = -26.667$

$\therefore \qquad F_{BE} = t_{BE} \times L_{BE}$

$\qquad \qquad = -26.667 \times 3\sqrt{2}$

$\qquad \qquad = -113.14 \text{ kN}$

From eqn. (ii),

$13.333 - (-26.667) - t_{BA} = 0$

$\therefore \qquad t_{BA} = 40$

$\therefore \qquad F_{BA} = t_{BA} \times L_{BA}$

$\qquad \qquad = 40 \times 3 = 120 \text{ kN},$

ANALYSIS OF PIN-JOINTED PLANE FRAMES

Hence member forces are as shown in the table below:

Member	Force in kN	Member	Force in kN
AB	120	DE	– 40.0
BC	56.67	EB	– 113.14
CD	– 40.0	BD	40.0

Example 4.11 *Analyse the truss shown in Fig. 4.19 by the method of tension coefficient and determine the forces in all the members.*

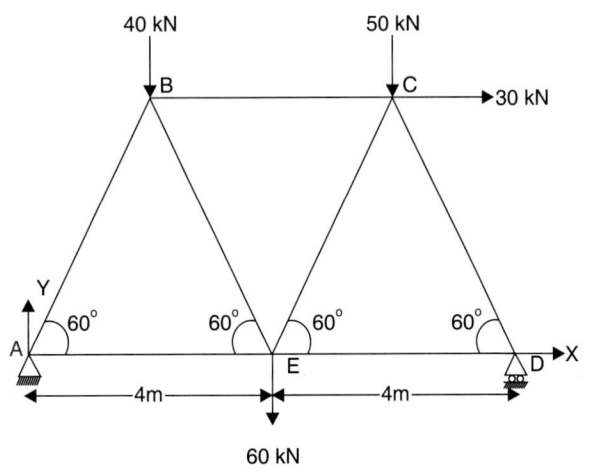

Fig. 4.19

Solution. In this problem, first the support reactions are to be found. Let the support reactions be X_A and Y_A at A and X_D, Y_D at D.

Since at D the support is roller, $X_D = 0$.

Consider entire structure.

Taking moment about A, we get

$$-Y_D \times 8 + 40 \times 2 + 50 \times 6 + 30 \times 4 \sin 60° + 60 \times 4 = 0$$

$$\therefore \quad Y_D = 90.49 \text{ kN}.$$

$\Sigma F_x = 0$, gives $X_A + 30 = 0$ or $X_A = -30$ kN.

$\Sigma F_y = 0$, gives $Y_A - 40 - 50 - 60 + 90.49 = 0$

i.e. $Y_A = 59.51$ kN.

Taking A as origin and x,y axes as shown in Fig. 4.19, the coordinates of various joints are A (0,0), B (2, 3.464), C (6, 3.464), D (8,0) and E (4, 0).

Length of each member is 4 m.

Consider joint by joint equilibrium equations.

Joint A:

$$\Sigma F_y = 0 \rightarrow$$

$$59.51 + t_{AB}(3.464 - 0) + t_{AE}(0 - 0) = 0$$

$$\therefore \quad t_{AB} = -\frac{59.51}{3.464} = -17.18$$

$\therefore \quad F_{AB} = t_{AB} \times L_{AB} = -17.18 \times 4 = -68.72 \text{ kN.}$

$\sum F_x = 0 \rightarrow$

$-30.0 + t_{AB}(2-0) + t_{AE}(4-0) = 0$

$4 t_{AE} = 30 - 2 t_{AB} = 30 - 2(-17.18)$

$= 64.36$

$t_{AE} = 16.09$

$\therefore \quad F_{AE} = t_{AE} \times L_{AE} = 16.09 \times 4 = 64.36 \text{ kN.}$

Joint B :

$\sum F_y = 0 \rightarrow$

$-40 + t_{BA}(0 - 3.464) + t_{BC}(3.464 - 3.464) + t_{BE}(0 - 3.464) = 0$

$-40 - 3.464 t_{BA} - 3.464 t_{BE} = 0$

$t_{BE} = -\dfrac{40}{3.464} - t_{BA}$

$= -11.547 - (-17.18) = 5.637$

$\therefore \quad F_{BE} = t_{BE} \times L_{BE} = 5.631 \times 4 = 22.5 \text{ kN.}$

$\sum F_x = 0 \rightarrow$

$0 + t_{BA}(0-2) + t_{BC}(6-2) + t_{BE}(4-2) = 0$

$\therefore \quad 4 t_{BC} = 2 t_{BA} - 2 t_{BE}$

$\therefore \quad t_{BC} = 0.5(-17.18 - 5.637) = -11.409$

$\therefore \quad F_{BC} = t_{BC} \times L_{BC} = -11.409 \times 4 = 45.63 \text{ kN.}$

Joint C:

$\sum F_y = 0 \rightarrow$

$-50 + t_{CB}(3.464 - 3.464) + t_{CD}(0 - 3.464) + t_{CE}(0 - 3.464) = 0$

$-50 - 3.464 t_{CD} - 3.464 t_{CE} = 0$

i.e. $\quad t_{CD} + t_{CE} = -14.434$ ...(i)

$\sum F_x = 0 \rightarrow$

$30 + t_{CB}(2-6) + t_{CD}(8-6) + t_{CE}(4-6) = 0$

$30 - 4 t_{CB} + 2 t_{CD} - 2 t_{CE} = 0$

$t_{CD} - t_{CE} = -\dfrac{1}{2}[30 - 4 \times (-11.409)]$

$= -37.818$ ...(ii)

Adding eqns. (i) and (ii), we get

$2 t_{CD} = -[14.434 + 37.818]$

$t_{CD} = -26.126$

$\therefore \quad F_{CD} = t_{CD} \times L_{CD} = -26.126 \times 4 = -104.50 \text{ kN.}$

ANALYSIS OF PIN-JOINTED PLANE FRAMES

and
$$t_{CE} = -14.434 - t_{CD} = -14.434 - (-26.126)$$
$$= 11.692$$

∴ $$F_{CE} = t_{CE} \times L_{CE} = 11.692 \times 4 = 46.77 \text{ kN}$$

Joint D:
$$\sum F_x = 0 \rightarrow$$
$$0 + t_{DC}(6-8) + t_{DE}(4-8) = 0$$

i.e.
$$t_{DE} = -0.5 \, t_{DC} = -0.5 \, (-26.126)$$
$$= 13.063$$

∴
$$F_{DE} = t_{DE} \times L_{DE} = 13.063 \times 4$$
$$= 52.25 \text{ kN}$$

The results are shown in the table below:

Member	Force in kN	Member	Force in kN
AB	– 68.72	DE	52.25
BC	– 45.63	EA	64.36
CD	– 104.50	EB	22.5
		EC	46.77

+ ve means tension   – ve means compression

Example 4.12 *Analyse the truss shown in Fig. 4.20 by the method of tension coefficient and determine the forces in all the members.*

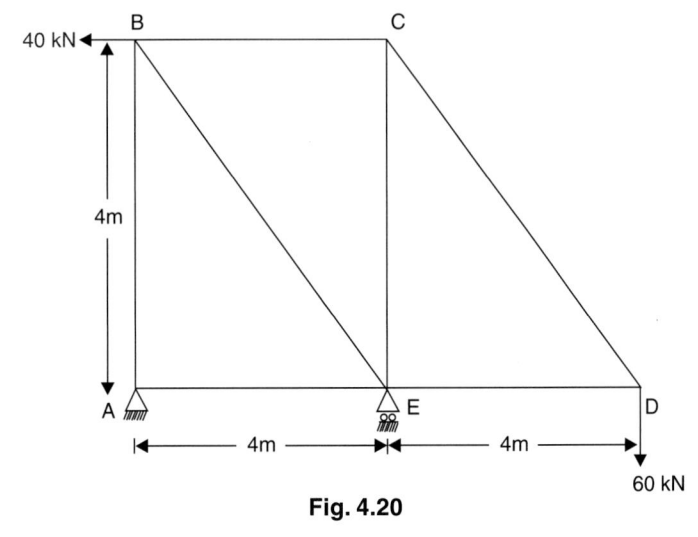

Fig. 4.20

Solution. Considering entire truss, we get
$$\sum M_A = 0 \rightarrow$$
$$Y_E \times 4 - 60 \times 8 + 40 \times 4 = 0$$
∴ $$Y_E = 80 \text{ kN}$$

$$\Sigma F_x = 0 \rightarrow$$
$$X_A - 40 = 0 \quad \text{or} \quad X_A = 40 \text{ kN}.$$
$$\Sigma F_y = 0 \rightarrow$$
$$Y_A + 80 - 60 = 0. \quad \text{or} \quad Y_A = -20 \text{ kN}.$$

Taking A as origin and x, y, axes as shown in Fig. 4.20, the coordinates of various joints are A (0, 0), B (0, 4), C (4, 4), D (8, 0) and E (4, 0).

The length of various members are

$$AB = CE = AE = ED = 4 \text{ m and } BE = CD = 4\sqrt{2} \text{ m}$$

Now consider the equilibrium of joints one by one.

Joint A:

$$\Sigma F_y = 0 \rightarrow$$
$$-20 + t_{AB} (4 - 0) + t_{AE} (0 - 0) = 0$$
$$\therefore \quad t_{AB} = 5.0$$
$$\therefore \quad F_{AB} = t_{AB} \times L_{AB} = 5.0 \times 4 = 20 \text{ kN}$$
$$\Sigma F_x = 0 \rightarrow$$
$$40 + t_{AB} (0 - 0) + t_{AE} (4 - 0) = 0$$
$$t_{AE} = -10$$
$$\therefore \quad F_{AE} = t_{AE} \times L_{AE} = -10 \times 4 = -40 \text{ kN}$$

Joint B :

$$\Sigma F_y = 0 \rightarrow$$
$$0 + t_{BA} (0 - 4) + t_{BC} (4 - 4) + t_{BE} (0 - 4) = 0$$
$$\therefore \quad t_{BE} = -t_{BA} = -(5 - 0) = -5.0$$
$$\therefore \quad F_{BE} = t_{BE} \times L_{BE} = -5.0 \times 4\sqrt{2} = -28.28 \text{ kN}$$
$$\Sigma F_x = 0 \rightarrow$$
$$-40 + t_{BA} (0 - 0) + t_{BC} (4 - 0) + t_{BE} (4 - 0) = 0$$
$$t_{BC} = 10 - t_{BE} = 10 - (-5.0) = 15$$
$$\therefore \quad F_{BC} = t_{BC} \times L_{BC} = 15 \times 4 = 60 \text{ kN}$$

Joint C :

$$\Sigma F_y = 0 \rightarrow$$
$$0 + t_{CB} (4 - 4) + t_{CD} (0 - 4) + t_{CE} (0 - 4) = 0$$
$$\therefore \quad t_{CD} + t_{CE} = 0 \qquad \qquad \ldots(i)$$
$$\Sigma F_x = 0 \rightarrow$$
$$0 + t_{CB} (0 - 4) + t_{CD} (8 - 4) + t_{CE} (4 - 4) = 0$$
$$\therefore \quad t_{CD} = t_{CB} = 15$$
$$\therefore \quad F_{CD} = t_{CD} \times L_{CD} = 15 \times 4\sqrt{2} = 56.57 \text{ kN}$$

ANALYSIS OF PIN-JOINTED PLANE FRAMES

Substituting the value of t_{CD} in eqn. (i), we get
$$15 + t_{CE} = 0 \quad \text{i.e.} \quad t_{CE} = -15$$
$$\therefore \quad F_{CE} = t_{CE} \times L_{CE} = -15 \times 4 = -60 \text{ kN}$$

Joint D :
$$\sum F_x = 0 \rightarrow$$
$$0 + t_{DC}(8-4) + t_{DE}(8-4) = 0$$
$$\therefore \quad t_{DE} = -t_{DC} = -15$$
$$\therefore \quad F_{DE} = t_{DE} \times L_{DE} = -15 \times 4 = -60 \text{ kN}$$

The results are tabulated below:

Member	Force in kN	Member	Force in kN
AB	20	DE	– 60
BC	60	EA	– 40
CD	56.57	EC	– 60
		EB	– 28.28

+ ve means tension   – ve means compression

Important Definitions

1. **A pin jointed frame** is a structure made of slender members pin connected at ends and capable of taking loads at joints.
2. A frame, in which all members lie in a single plane, is called **plane frame.**
3. If all the members of a frame do not lie in a single plane, they are called as space frame.
4. A pin jointed frame, which has got just sufficient number of members to resist the loads without undergoing appreciable deformation in shape is called a **perfect frame.**
5. A frame is said to be **deficient** if the number of members in it are less than that required for a perfect frame. These frames do not retain their shapes when loaded.
6. A frame is said to be **redundant**, if the number of members in it are more than that required for a perfect frame. Such frames cannot be analysed by using the equations of equilibrium only.

Important Formulae

In general the relationship between number of members (m) and number of joints (j) in a perfect frame is given by
$$m = 2j - 3$$

PROBLEMS FOR EXERCISE

4.1 to 4.17: Determine the forces in all the members of the frames shown in Figs. 4.17 to 4.33. Indicate the nature of the forces also. (Tension as + ve and compression as – ve)

4.1 [Ans. $F_{AB} = +67.5$ kN; $F_{BC} = +15$ kN ; $F_{CD} = -25$ kN ; $F_{DE} = -30$ kN ;
$F_{EF} = -105$ kN; $F_{AE} = +62.5$ kN ; $F_{BE} = -62.5$ kN ; $F_{BD} = +25$ kN]

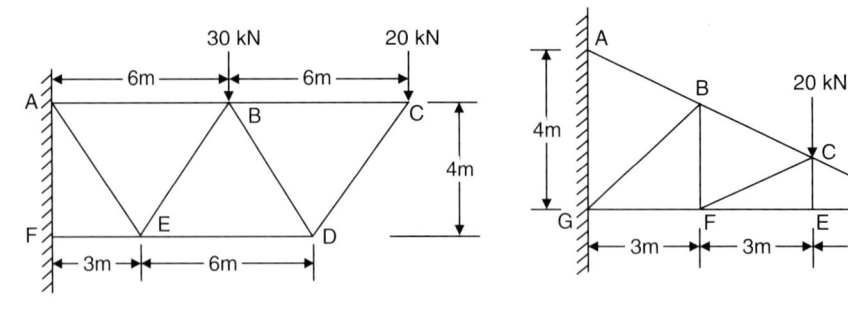

Fig. 4.17 (Prob. 4.1)

Fig. 4.18 (Prob. 4.2)

4.2 [**Ans.** $F_{AB} = +82.0738$ kN; $F_{BC} = +73.866$ kN; $F_{CD} = 49.2443$ kN; $F_{DE} = -45$ kN;
$F_{EF} = -45$ kN; $F_{FG} = -67.5$ kN ; $F_{BG} = -10.0$ kN; $F_{FC} = +24.622$ kN; $F_{CE} = 0$; $F_{BF} = 10$ kN]

4.3 [**Ans.** $F_{AC} = F_{CE} = E_{EG} = +193.1852$ kN; $F_{BD} = F_{DE} = F_{FG} = -193.1852$ kN; All others are zero members]

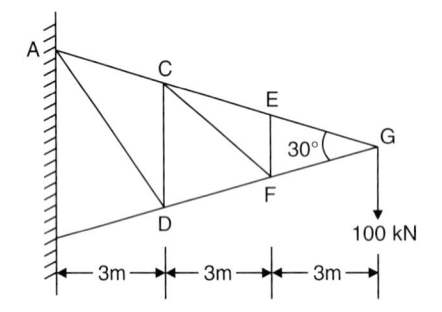

Fig. 4.19 (Prob. 4.3)

Fig. 4.20 (Prob. 4.4)

4.4 [**Ans.** $F_{EC} = +447.2136$ kN; $F_{CA} = +400$ kN; $F_{AB} = -447.2136$ kN; $F_{BD} = -400$ kN;
$F_{CD} = 0$; $F_{CB} = -200$ kN]

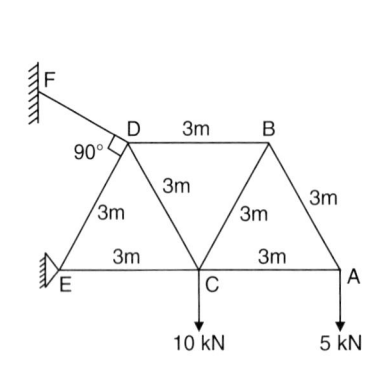

4.5 [**Ans.** $F_{DB} = F_{BA} = +5.7735$ kN; $F_{BC} = F_{DE} = -5.7738$ kN; $F_{AC} = -2.8868$ kN;
$F_{CE} = -14.4338$ kN; $F_{DC} = +17.3205$ kN; $F_{DF} = +20.0$ kN]

4.6 [**Ans.** $F_{AB} = -30$ kN; $F_{AC} = -160$ kN; $F_{BC} = +50$ kN; $F_{BD} = -200$ kN; $F_{CD} = -50$ kN;
$F_{CE} = -120$ kN; $F_{DF} = -266.67$ kN; $F_{DE} = +83.33$ kN]

Fig. 4.21 (Prob. 4.5)

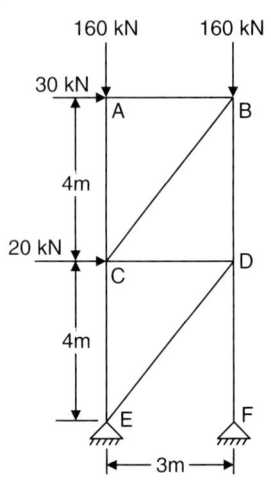

Fig. 4.22 (Prob. 4.6)

ANALYSIS OF PIN-JOINTED PLANE FRAMES

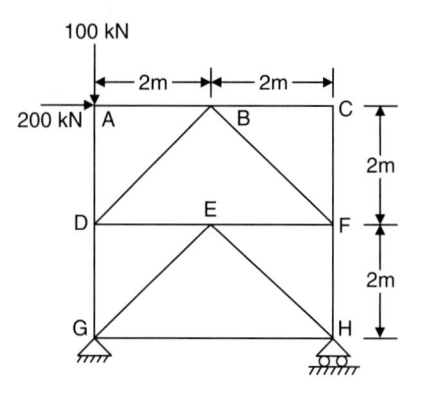

4.7 [Ans. $F_{AB} = -200$ kN; $F_{AD} = -100$ kN; $F_{BC} = F_{CF} = 0$; $F_{BD} = 100\sqrt{2}$ kN; $F_{BF} = -100\sqrt{2}$ kN; $F_{DE} = -100$ kN; $F_{DG} = 0$; $F_{EF} = +100$ kN; $F_{EH} = -100\sqrt{2}$ kN; $F_{EG} = +100\sqrt{2}$ kN; $F_{GH} = +100$ kN]

Fig. 4.23 (Prob. 4.7)

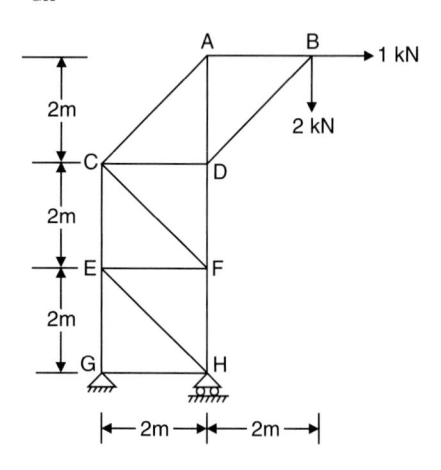

4.8 [Ans. $F_{BD} = -2\sqrt{2}$ kN; $F_{BA} = +3$ kN; $F_{AC} = +3\sqrt{2}$ kN; $F_{AD} = -3$ kN; $F_{DC} = -2$ kN; $F_{DF} = -5$ kN; $F_{CF} = -\sqrt{2}$ kN; $F_{CE} = +6$ kN; $F_{FE} = +1$ kN; $F_{FH} = -4$ kN; $F_{EH} = -\sqrt{2}$ kN; $F_{EG} = +5$ kN; $F_{GH} = +1$ kN]

Fig. 4.24 (Prob. 4.8)   Fig. 4.25 (Prob. 4.9)

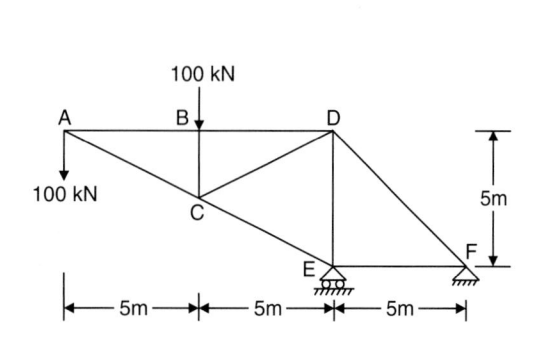

4.9 [Ans. $F_{AC} = -100\sqrt{5}$ kN; $F_{AB} = +200$ kN; $F_{BD} = +200$ kN; $F_{BC} = -100$ kN; $F_{CD} = +50\sqrt{5}$ kN; $F_{CE} = -150\sqrt{5}$ kN; $F_{DE} = +35.0$ kN; $F_{DF} = 300\sqrt{2}$ kN; $F_{EF} = -300$ kN]

4.10 [Ans. $F_{AB} = +5\sqrt{2}$ kN; $F_{AC} = -5$ kN; $F_{BC} = -5$ kN; $F_{BD} = 5$ kN; $F_{CD} = +15\sqrt{2}$ kN; $F_{CE} = -20$ kN; $F_{DE} = -15.0$ kN; $F_{DF} = +20\sqrt{2}$ kN; $F_{EH} = -15$ kN; $F_{EF} = -20$ kN; $F_{FG} = +30\sqrt{2}$ kN; $F_{FH} = +10\sqrt{2}$ kN]

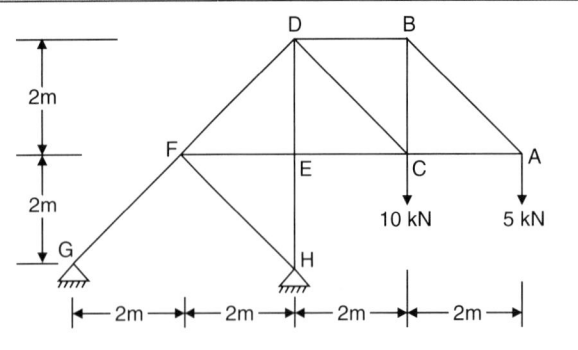

Fig. 4.26 (Prob. 4.10)

Fig. 4.27 (Prob. 4.11)

4.11. [Ans. $F_{AB} = -15$ kN; $F_{AC} = +12\sqrt{2}$ kN; $F_{BD} = -27.5\sqrt{2}$ kN; $F_{BC} = -12.5\sqrt{2}$ kN; $F_{CE} = 0$; $F_{CD} = +25$ kN; $F_{ED} = -27.5\sqrt{2}$ kN]

4.12. [Ans. $F_{AB} = -17.32$ kN; $F_{AC} = +5$ kN; $F_{BC} = -20$ kN; $F_{BD} = -17.32$ kN; kN; $F_{CD} = +20$ kN; $F_{CE} = -15$ kN; $F_{DE} = -30$ kN]

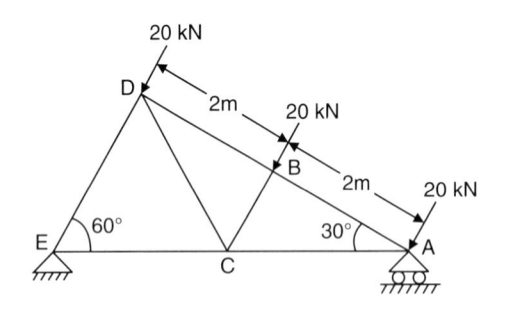

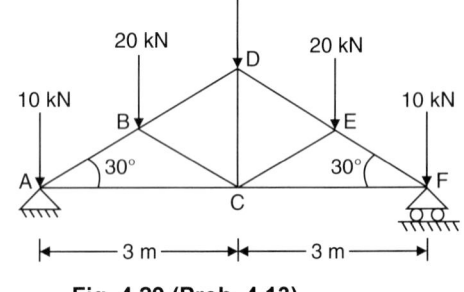

Fig. 4.28 (Prob. 4.12)

Fig. 4.29 (Prob. 4.13)

4.13. [Ans. $F_{AB} = 60$ kN; $F_{AC} = +51.96$ kN; $F_{BC} = -20$ kN; $F_{BD} = -40$ kN; kN; $F_{CD} = +40$ kN; symmetry]

4.14. [Ans. $F_{AC} = -4.5\sqrt{13}$ kN; $F_{AB} = +13.5$ kN; $F_{BC} = +6$ kN; $F_{BD} = +13.5$ kN; $F_{CD} = -0.5\sqrt{13}$ kN; $F_{CE} = -4\sqrt{10}$ kN; $F_{DE} = 8$ kN]

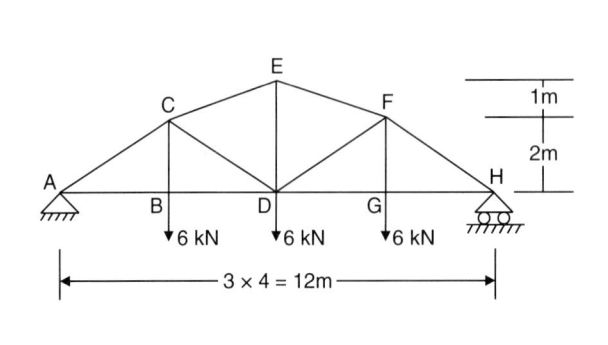

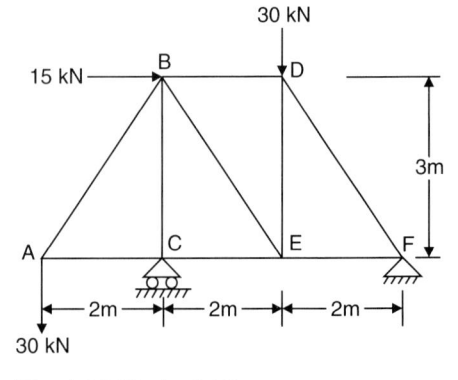

Fig. 4.30 (Prob. 4.14)

Fig. 4.31 (Prob. 4.15)

4.15. [Ans. $F_{AB} = +10\sqrt{13}$ kN; $F_{AC} = -20$ kN; $F_{CB} = -48.75$ kN; $F_{CE} = -20$ kN; $F_{CD} = -7.5$ kN; $F_{BE} = +6.25\sqrt{13}$ kN; $F_{DE} = 18.75$ kN; $F_{DF} = -3.75\sqrt{13}$ kN; $F_{FE} = -7.5$ kN]

4.16. [Ans. $F_{AB} = 16.91$ kN; $F_{AF} = +31.55$ kN; $F_{BF} = +23.91$ kN; $F_{BD} = -23.91$ kN; $F_{BC} = +40$ kN; $F_{CD} = -40$ kN; $F_{DE} = -63.1$ kN; $F_{DF} = +23.91$ kN; $F_{EF} = +31.55$ kN]

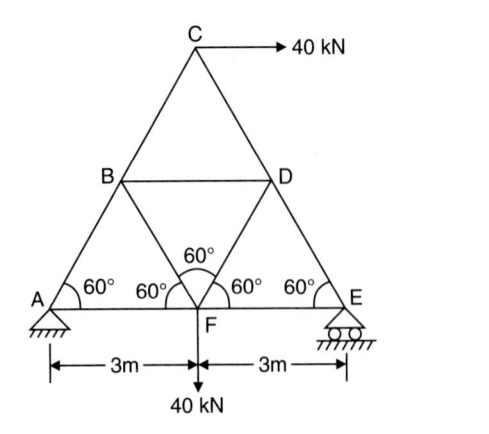

Fig. 4.32 (Prob. 4.16)   Fig. 4.33 (Prob. 4.17)

4.17. [Ans. $F_{AC} = -67.48$ kN; $F_{AB} = +53.99$ kN; $F_{BC} = +10$ kN; $F_{CD} = -8.33$ kN; $F_{CE} = -59.15$ kN; $F_{EF} = -24.5$ kN; $F_{ED} = +52.81$ kN; $F_{FD} = +47.21$ kN; $F_{FG} = -34.64$ kN; $F_{DG} = +47.32$ kN]

4.18. Find the force in the member *FG* of the triangular Howe truss shown in Fig. 4.34.

(**Hint.** Take section (*A'*) – (*A'*) and find force in *FD*. Then analyse joint *F*)

[Ans. + 28 kN]

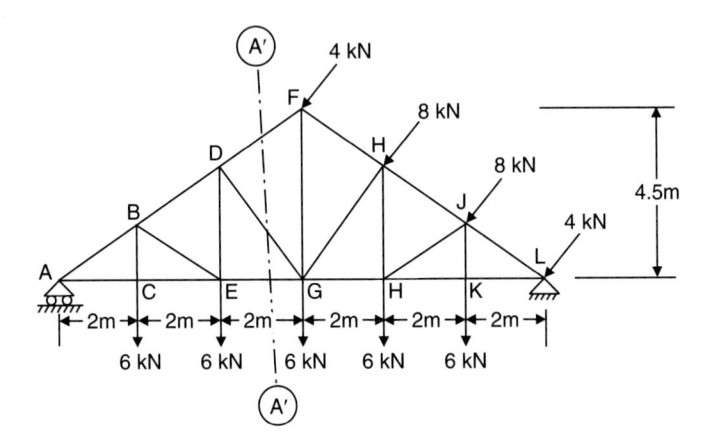

Fig. 4.34 (Prob. 4.18)

4.19. Determine the forces in the members *AB, AC, DF* and *CE* of the scissors truss shown in Fig. 4.35.

(**Hint.** Find reaction R_A and analyse joint *A*. Take section (*A'*) – (*A'*) and find forces in *DF* and *CE*).

[Ans. $F_{AB} = -6.25$ W ; $F_{AC} = 4.51$ W ; $F_{DF} = -3.75$ W ; F_{CE} + 2.75 W]

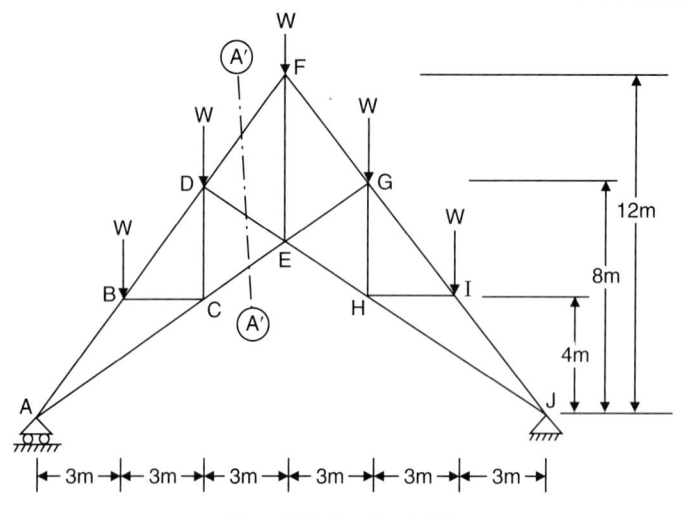

Fig. 4.35 (Prob. 4.19)

4.20. Find the force in member *KL* of the French truss shown in Fig. 4.36.

(**Hint.** Take section $(A') - (A')$ and find F_{LE} and F_{DE}. From joint D find F_{DL}. Then analyse joint L to get F_{KL}.) [**Ans.** + 41.96 kN]

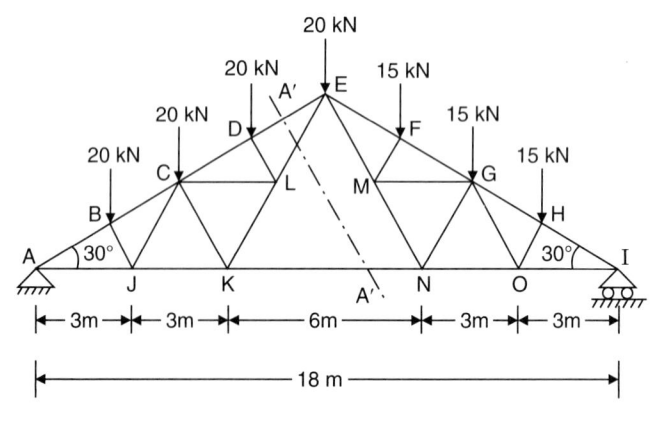

Fig. 4.36 (Prob. 4.20)

5
Friction

When a body moves or tends to move over another body, a force opposing the motion develops at the contact surfaces. This force which opposes the movement or the tendency of movement is called **frictional force** or simply **friction**. Friction is due to the resistance to motion offered by minutely projecting particles at the contact surfaces. In this chapter, the concepts related to friction are explained and the laws of friction presented. Application of these laws to many engineering problems including wedge and rope/belt are illustrated.

5.1. FRICTIONAL FORCE

As defined above frictional force is the resistance offered by minutely projecting particles of a body when it moves over another body. Frictional force has a remarkable property of adjusting itself in magnitude to the force producing or tending to produce the motion so that motion is prevented.

However, there is a limit beyond which the magnitude of this force cannot increase. If the applied force is more than this limit, there will be movement of one body over the other. This limiting value of frictional force when the motion is impending, is known as **Limiting Friction**. It may be noted that when the applied force is less than the limiting friction, the body remains at rest and such frictional force is called **Static Friction**, which will be having any value between zero and the limiting friction. If the value of applied force exceeds the limiting friction, the body starts moving over the other body and the frictional resistance experienced by the body while moving is known as **Dynamic Friction**. Dynamic friction is found to be less than limiting friction. Dynamic friction may be classified into the following two:

(a) **Sliding Friction**

(b) **Rolling Friction.**

Sliding friction is the friction experienced by a body when it slides over the other body and the rolling friction is the friction experienced by a body when it rolls over a surface. It is experimentally found that the magnitude of limiting friction bears a constant ratio to the normal reaction between the two surfaces and this ratio is called **Coefficient of Friction**. Thus, referring to Fig. 5.1 in which the block has a impending motion,

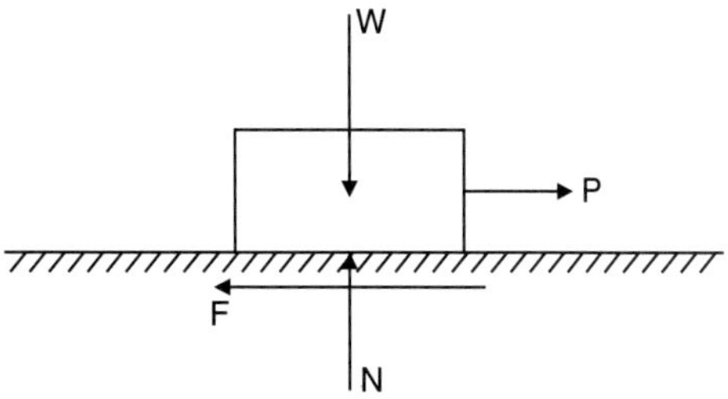

$$\text{Coefficient of friction} = \frac{F}{N}$$

Fig. 5.1

where F is limiting friction and N is normal reaction between the contact surfaces. Coefficient of friction is denoted by μ. Thus

$$\mu = \frac{F}{N}$$

5.2. LAWS OF FRICTION

The principles discussed in Art. 5.1 are mainly due to the experimental studies by Coulomb (1781) and by Morin (1831) on solids in dry conditions. These principles constitute the laws known as Coulomb's laws of friction/laws of dry friction/laws of solid friction. These laws are listed below:

(1) The force of friction always acts in a direction opposite to that in which body tends to move.

(2) Till the limiting value is reached, the magnitude of friction is exactly equal to the force which tends to move the body.

(3) The magnitude of the limiting friction bears a constant ratio to the normal reaction between the two surfaces of contact and this ratio is called coefficient of friction.

(4) The force of friction depends upon the roughness/smoothness of the surfaces.

(5) The force of friction is independent of the area of contact between the two surfaces.

(6) After the body starts moving, the dynamic friction comes into play, the magnitude of which is less than that of limiting friction and it bears a constant ratio with normal force. This ratio is called **coefficient of dynamic friction.**

5.3. ANGLE OF FRICTION, ANGLE OF REPOSE AND CONE OF FRICTION

The above three terms used in this chapter are defined and explained below.

Angle of Friction

Consider the block shown in Fig. 5.2 resting on a horizontal surface and subjected to horizontal pull P. Let F be the frictional force developed and N the normal reaction. Thus at contact surface the reactions are F and N. They can be combined graphically to get the reaction R which acts at angle θ to normal reaction. This angle θ, called the angle of friction is given by

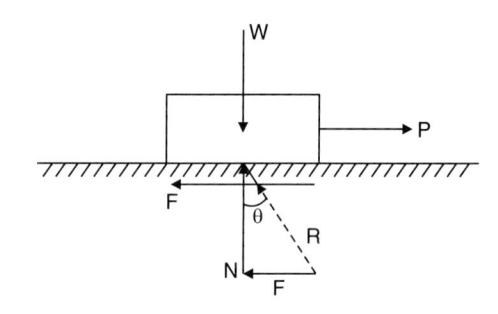

$$\tan \theta = \frac{F}{N}$$

Fig. 5.2

As P increases, F increases and hence θ also increases. θ can reach the maximum value α when F reaches limiting value. At this stage

$$\tan \alpha = \frac{F}{N} = \mu \qquad \ldots(5.1)$$

and this value of α is called **Angle of Limiting Friction**. Hence the angle of limiting friction may be defined as *the angle between the resultant reaction and the normal to the plane on which the motion of the body is impending.*

Angle of Repose

It is well known that when grains (food grain, soil, sand etc.) are heaped, there exists a limit for the inclination of the surface. Beyond this limiting inclinations the grains start rolling down. This limiting angle upto which the grains repose (sleep) is called the angle of repose.

Consider the block of weight W resting on an inclined plane which makes an angle θ with the horizontal as shown in Fig. 5.3. When θ is small the block will rest on the plane. If θ is increased gradually a stage is reached at which the block start sliding down the plane. The angle θ for which motion is impending, is called the angle of repose. Thus *the maximum inclination of the plane on which a body, free from external forces, can repose (sleep)* is called **Angle of Repose**.

Consider the equilibrium of the block shown in Fig. 5.3. Since the surface of contact is not smooth, not only normal reaction, but frictional force also develops. Since the body tends to slide downward, the frictional force will be up the plane.

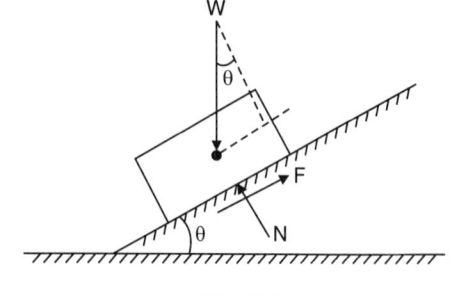

Fig. 5.3

ΣForces normal to the plane = 0, gives
$$N = W \cos \theta \qquad \text{...(5.2)}$$
ΣForces parallel to the plane = 0, gives
$$F = W \sin \theta \qquad \text{...(5.3)}$$
Dividing eqn. (5.3) by eqn. (5.2), we get
$$\tan \theta = \frac{F}{N} \qquad \text{...(5.4)}$$

If ϕ is the value of θ when motion is impending, frictional force will be limiting friction and hence
$$\tan \phi = \frac{F}{N}$$
$$= \mu$$
$$= \tan \alpha$$
or $\qquad \phi = \alpha$

Thus the value of angle of repose is same as the value of limiting angle of repose.

Cone of Friction

When a body is having impending motion in the direction of force P, the frictional force will be limiting friction and the resultant reaction R will make limiting angle α with the normal as shown in Fig. 5.4. If the body is having impending motion in some other direction, the resultant reaction makes limiting frictional angle α with the normal to that direction. Thus when the direction of force P is gradually changed through 360°, the resultant R generates a right circular cone with semi-central angle equal to α.

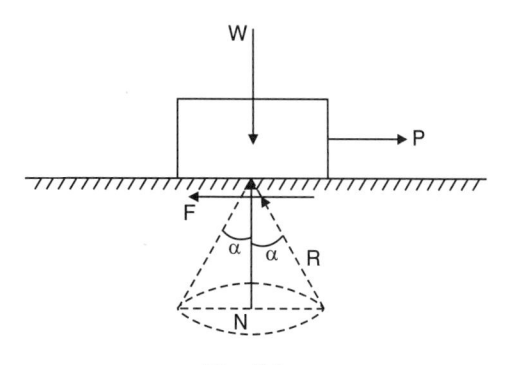

Fig. 5.4

If the resultant R is on the surface of this inverted right circular cone whose semi-central angle is limiting frictional angle (α) the motion of the body is impending. If the resultant is within this cone the body is stationary. This *inverted cone with semi-central angle α equal to limiting frictional angle α, is called* **Cone of Friction.**

Example 5.1. *Block A weighing 1000 N rests over block B which weighs 2000 N as shown in Fig. 5.5(a). Block A is tied to wall with a horizontal string. If the coefficient of friction between blocks A and B is 0.25 and between B and floor is 1/3, what should be the value of P to move the block (B), if*

(a) P is horizontal.

(b) P acts at 30° upwards to horizontal ?

Solution.

(a) When P is horizontal:

The free body diagrams of the two blocks are shown in Fig. 5.5(b). It may be noted that the frictional forces F_1 and F_2 are to be marked in the opposite directions of impending relative motion. Considering block A,

$$\Sigma V = 0 \rightarrow$$
$$N_1 - 1000 = 0 \quad \text{or} \quad N_1 = 1000 \text{ N}$$

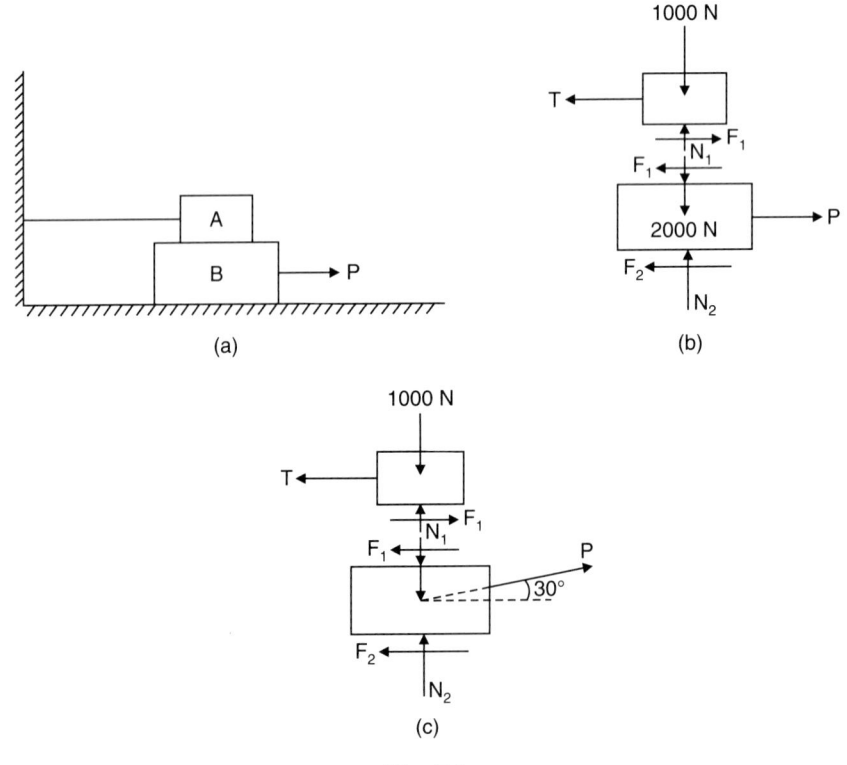

Fig. 5.5

Since F_1 is limiting friction,

$$\frac{F_1}{N_1} = \mu = 0.25$$

∴ $\quad F_1 = 0.25\, N_1 = 0.25 \times 1000 = 250$ N.

FRICTION

$\Sigma H = 0 \rightarrow$
$F_1 - T = 0$

or $\qquad T = F_1 = 250$ N

Consider equilibrium of block B.

$\Sigma V = 0 \rightarrow$
$N_2 - 2000 - N_1 = 0$

or $\qquad N_2 = 2000 + N_1 = 2000 + 1000 = 3000$ N

Since F_2 is limiting friction,

$$F_2 = \mu N_2 = \frac{1}{3} \times 3000 = 1000 \text{ N}$$

$\Sigma H = 0 \rightarrow$
$P - F_1 - F_2 = 0$

∴ $\qquad P = F_1 + F_2 = 250 + 1000$
∴ $\qquad \mathbf{P = 1250 \text{ N} \quad Ans.}$

(b) When P is inclined:

Free body diagram for this case is shown in Fig. 5.5(c).

As in the previous case here also,

$N_1 = 1000$ N

and $F_1 = 250$ N. Consider the equilibrium of block B.

$\Sigma V = 0 \rightarrow$
$N_2 - 2000 - N_1 + P \sin 30° = 0$
$N_2 + P \sin 30° = 2000 + N_1$
$N_2 + 0.5 P = 2000 + 1000$

∴ $\qquad N_2 = 3000 - 0.5P$

From law of friction,

$$F_2 = \frac{1}{3} N_2 = \frac{1}{3} (3000 - 0.5P)$$

$$= 1000 - \frac{0.5}{3} P$$

$\Sigma H = 0 \rightarrow$
$P \cos 30° - F_1 - F_2 = 0$

$$P \cos 30° - 250 - \left(1000 - \frac{0.5}{3} P\right) = 0$$

$$P\left(\cos 30° + \frac{0.5}{3}\right) = 1250$$

or $\qquad \mathbf{P = 1210.43 \text{ N} \quad Ans.}$

Example 5.2. *What should be the value of θ in Fig. 5.6(a) that will make the motion of 900 N block down the plane to impend ? The coefficient of friction for all contact surfaces is 1/3.*

Solution. 9000 N block is on the verge of moving downward. Hence frictional forces F_1 and F_2 [Ref. Fig. 5.6 (b)] act up the plane on 900 N block. Free body diagram of the blocks is as shown in Fig. 5.6(b).

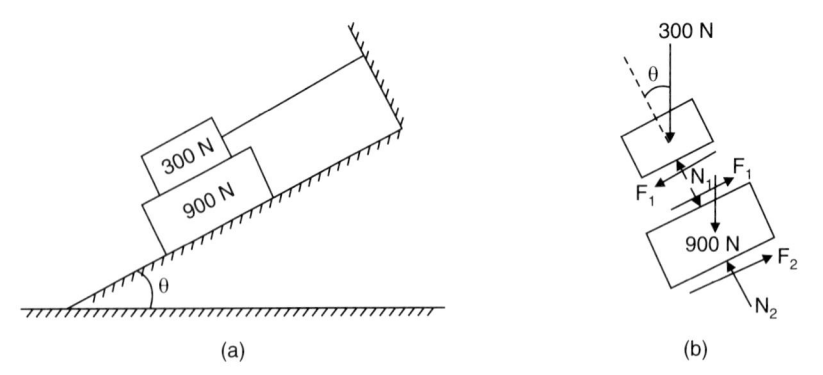

Fig. 5.6

Consider equilibrium of 300 N block.
Σ Forces normal to plane = 0, gives
$$N_1 - 300 \cos \theta = 0$$
or
$$N_1 = 300 \cos \theta \qquad \ldots(1)$$
From law of friction, $F_1 = 1/3 \, N_1 = 1/3 \times 300 \cos \theta$
$$= 100 \cos \theta \qquad \ldots(2)$$
Consider equilibrium of 900 N block:
Σ Forces normal to the plane = 0, gives
$$N_2 - N_1 - 900 \cos \theta = 0$$
$\therefore$
$$N_2 = N_1 + 900 \cos \theta$$
$$= (300 + 900) \cos \theta$$
$$= 1200 \cos \theta \qquad \ldots(3)$$
From law of friction,
$$F_2 = \frac{1}{3} N_2 = \frac{1}{3} \times 1200 \cos \theta = 400 \cos \theta \qquad \ldots(4)$$
ΣForces parallel to plane = 0, gives
$$F_1 + F_2 - 900 \sin \theta = 0$$
$$100 \cos \theta + 400 \cos \theta = 900 \sin \theta$$
$\therefore$
$$\tan \theta = \frac{5}{9}$$
$\therefore$
$\theta = 29.05°$ **Ans.**

Example 5.3. *A block weighing 500N just starts moving down a rough inclined plane when supported by a force of 200N acting parallel to the plane in upward direction. The same block is on the verge of moving up the plane when pulled by a force of 300N acting parallel to the plane. Find the inclination of the plane and the coefficient of friction between the inclined plane and the block.*

Solution. Free body diagram of the block when it just start moving down is shown in Fig. 5.7(a). The direction of frictional force is upward in this case since the direction of impending motion is downward. Since it is limiting case

FRICTION

$$\frac{F}{N} = \mu$$

ΣForce perpendicular to plane = 0, gives

$$N - 500 \cos \theta = 0 \quad \text{or} \quad N = 500 \cos \theta \qquad \ldots(1)$$

From law of friction,

$$F_1 = \mu N = 500\, \mu \cos \theta \qquad \ldots(2)$$

ΣForces parallel to plane = 0, gives

$$200 + F_1 - 500 \sin \theta = 0$$

i.e.
$$200 = 500 \sin \theta - F_1$$
$$200 = 500 \sin \theta - 500\, \mu \cos \theta \qquad \ldots(3)$$

When the block starts moving up the plane when 300 N pull is applied, frictional force F_2 is downward. Free body diagram for this case is shown in Fig. 5.7(b). In this case,

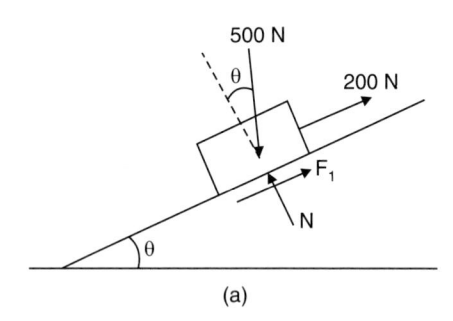

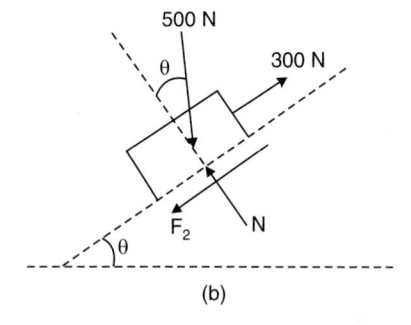

Fig. 5.7

ΣForces normal to plane = 0, gives

$$N - 500 \cos \theta = 0 \quad \text{or} \quad N = 500 \cos \theta \qquad \ldots(4)$$

From the law of friction,

$$F_2 = \mu N = 500\, \mu \cos \theta \qquad \ldots(5)$$

ΣForces parallel to plane = 0, gives

$$300 - F_2 - 500 \sin \theta = 0$$

or
$$300 = F_2 + 500 \sin \theta$$
$$= 500\, \mu \cos \theta + 500 \sin \theta \qquad \ldots(6)$$

Adding eqns. (3) and (6), we get

$$500 = 1000 \sin \theta$$

or
$$\sin \theta = 0.5$$

$\therefore$ **$\theta = 30°$ Ans.**

Substituting it in eqn. (6), we get

$$300 = 500\, \mu \cos 30° + 500 \sin 30°$$

i.e.
$$500\, \mu \cos 30° = 300 - 500 \times 0.5 = 50$$

or
$$\mu = \frac{50}{500 \cos 30°} = \mathbf{0.11547} \quad \textbf{Ans.}$$

Example 5.4. *What is the value of P in the system shown in Fig. 5.8(a) to cause the motion of 500N block to the right side ? Assume the pulley is smooth and the coefficient of friction between other contact surfaces is 0.20.*

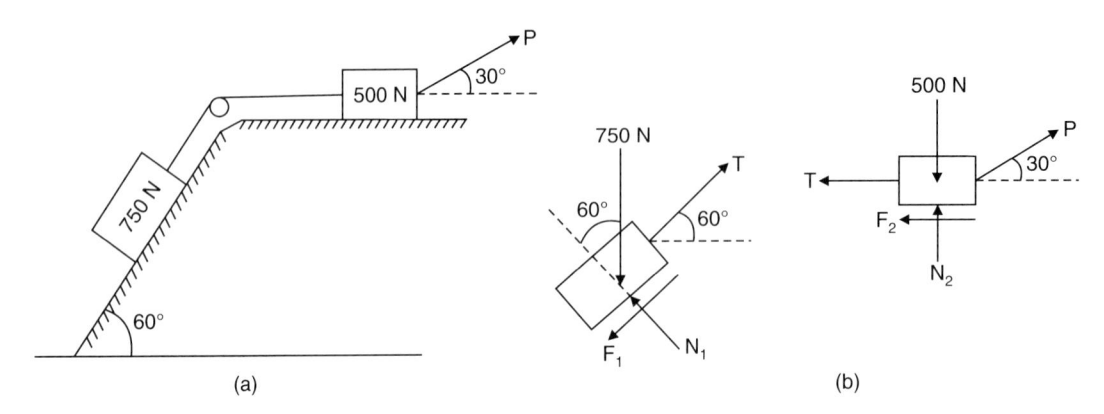

Fig. 5.8

Solution. Free body diagrams of the blocks are as shown in Fig. 5.8(b). Consider the equilibrium of 750N block.

ΣForces normal to the plane = 0, gives

$N_1 - 750 \cos 60° = 0$ or $N_1 = 375$ N.

Since the motion is impending

$F_1 = \mu N_1 = 0.2 \times 375 = 75$ N.

ΣForces parallel to the plane = 0, gives

$T - F_1 - 750 \sin 60° = 0$

∴ $T = F_1 + 750 \sin 60° = 75 + 750 \sin 60° = 724.52$ N.

Consider the equilibrium of 500 N block.

$\Sigma V = 0 \rightarrow$

$N_2 - 500 + P \sin 30° = 0$

or $\qquad N_2 + 0.5P = 500$

i.e., $\qquad N_2 = 500 - 0.5P$

From law of friction,

$F_2 = 0.2 N_2 = 0.2(500 - 0.5P) = 100 - 0.1P$

$\Sigma H = 0 \rightarrow$

$P \cos 30° - T - F_2 = 0$

$P \cos 30° - 724.52 - (100 - 0.1P) = 0$

∴ $P(\cos 30° + 0.1) = 724.52 + 100 = 824.52$

∴ $\qquad\qquad$ **P = 853.52 N   Ans.**

Example 5.5. *Two blocks connected by a horizontal link AB are supported on two rough planes as shown in Fig. 5.9(a). The coefficient of friction between the block A and horizontal surface is 0.4. The limiting angle of friction between block B and inclined plane is 20°. What is the smallest weight W of the block A for which equilibrium of the system can exist, if the weight of block B is 5 kN ?*

FRICTION

Solution. Free body diagrams for block A and B are as shown in Fig. 5.9(b).

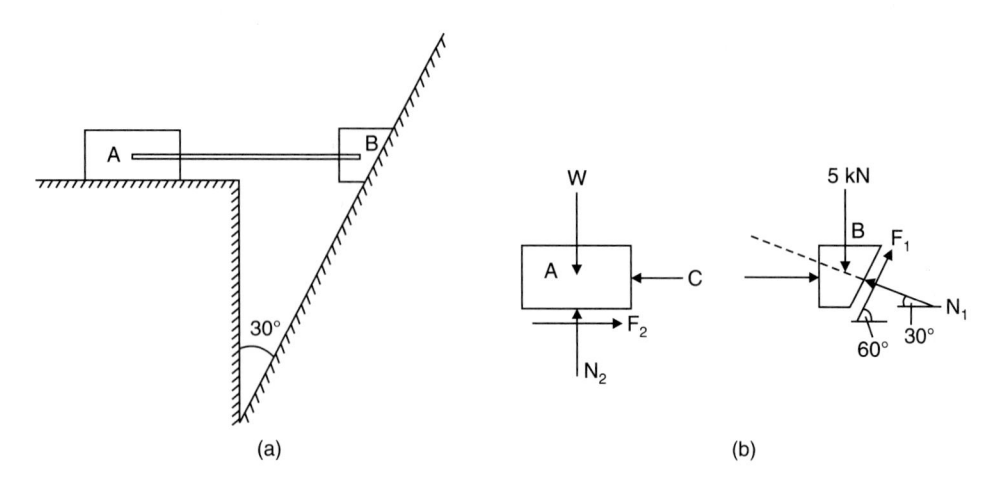

Fig. 5.9

Consider the equilibrium of block B.
From law of friction,
$$F_1 = N_1 \tan 20°$$ [since $\mu = \tan 20°$]
$$\Sigma V = 0 \;\rightarrow$$
$$N_1 \sin 30° + F_1 \sin 60° - 5 = 0$$
$$0.5 N_1 + N_1 \tan 20° \sin 60° = 5$$
∴ $$N_1 = 6.133 \text{ kN}$$
∴ $$F_1 = 6.133 \tan 20° = 2.232 \text{ kN}$$
$$\Sigma H = 0 \;\rightarrow$$
$$C + F_1 \cos 60° - N_1 \cos 30° = 0$$
∴ $$C = 6.133 \cos 30° - 2.232 \cos 60°$$
$$= 4.196 \text{ kN.}$$

Now consider the equilibrium of block A.
$$\Sigma H = 0 \;\rightarrow$$
$$F_2 - C = 0$$
or $$F_2 = C = 4.196 \text{ kN}$$
From law of friction
$$F_2 = \mu N_2 = 0.4 N_2$$
∴ $$N_2 = \frac{F_2}{0.4} = \frac{4.196}{0.4} = 10.49 \text{ kN}$$
$$\Sigma V = 0 \;\rightarrow$$
$$W - N_2 = 0$$
∴ $$W = N_2 = 10.49 \text{ kN} \quad \text{Ans.}$$

Example 5.6. *Two blocks A and B weighing 2000N each are to be held from slipping by the thrust of two weightless link reds each of which is connected by pin joints at one end to the blocks and interconnected by a pin joint at other end O_2 and subjected to a horizontal force P required to keep the blocks from slipping as shown in Fig. 5.10(a). Coefficient of friction is 0.25 for all contact surfaces.*

Solution. Let C_1 be the force in link AO and C_2 be the force in link OB. The free body diagrams of blocks A and B and the hinge O are shown in Fig. 5.10(b). Block A is on the verge of slipping. From the law of friction,

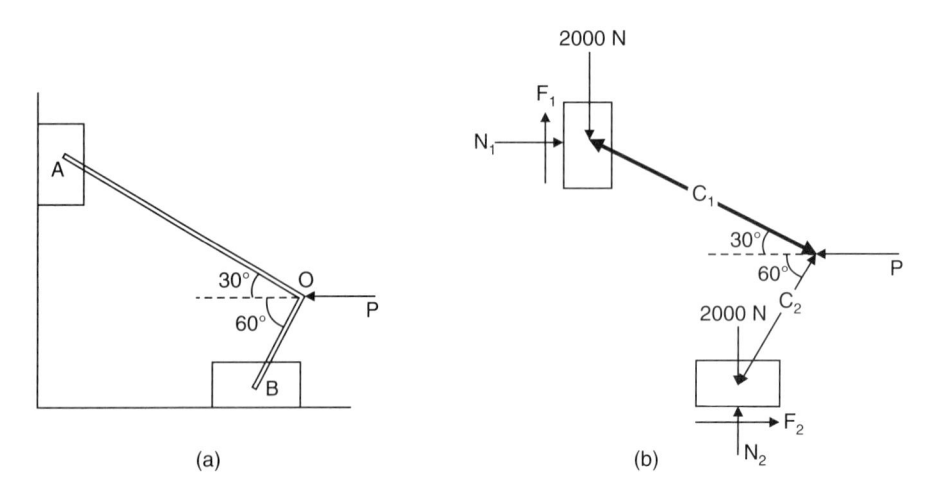

Fig. 5.10

$$F_1 = \mu N_1 = 0.25 N_1$$

Consider the equilibrium of block A.

$$\Sigma H = 0 \rightarrow$$
$$N_1 - C_1 \cos 30° = 0, \text{ or } N_1 = C_1 \cos 30°$$
$$\Sigma V = 0 \rightarrow$$
$$F_1 + C_1 \sin 30° - 2000 = 0$$
$$0.25 N_1 + 0.5 C_1 = 2000$$
$$0.25 C_1 \cos 30° + 0.5 C_1 = 2000$$

∴ $$C_1 = \frac{2000}{0.25 \cos 30° + 0.5} = 2791.32 \text{ N}.$$

Applying Lami's theorem to the equilibrium of the joint O, we get

$$\frac{P}{\sin 90°} = \frac{C_2}{\sin 150°} = \frac{C_1}{\sin 120°}$$

∴ **P = 3223.14 N Ans.**
and **C = 1611.57 N Ans.**

The above solution holds good, provided block B is not slipping. To verify this, consider the equilibrium of block B.

$$\Sigma H = 0 \rightarrow$$
$$F_2 - C_2 \cos 60° = 0$$

∴ $$F_2 = 1611.57 \cos 60° = 805.79 \text{ N}$$
$$\Sigma V = 0 \rightarrow$$
$$N_2 - 2000 - C_2 \sin 60° = 0$$

∴ $$N_2 = 2000 + 1611.57 \sin 60° = 3393.60 \text{ N}$$

Hence limiting friction $= \mu N_2 = 0.25 \times 3393.6 = 842.92$ N.

FRICTION

Since the actual frictional force F_2 developed is less than the limiting frictional force, block B is stationary and hence **P = 3223.14 N is correct answer.**

[**Note:** If F_2 calculated is more than the limiting friction μN_2, there is no possibility of maintaining equilibrium condition in the position shown.]

Example 5.7. *Two planes AC and BC inclined at 60° and 30° to the horizontal meet at C as shown in Fig. 5.11(a). A block of weight 1000N rests on the inclined plane BC and is tied by a rope passing over a pulley to a block weighing W newtons and resting on the plane AC. If the coefficient of friction between the block and plane BC is 0.28 and that between the block and the plane AC is 0.20, find the least and the greatest value of W for the equilibrium of the system.*

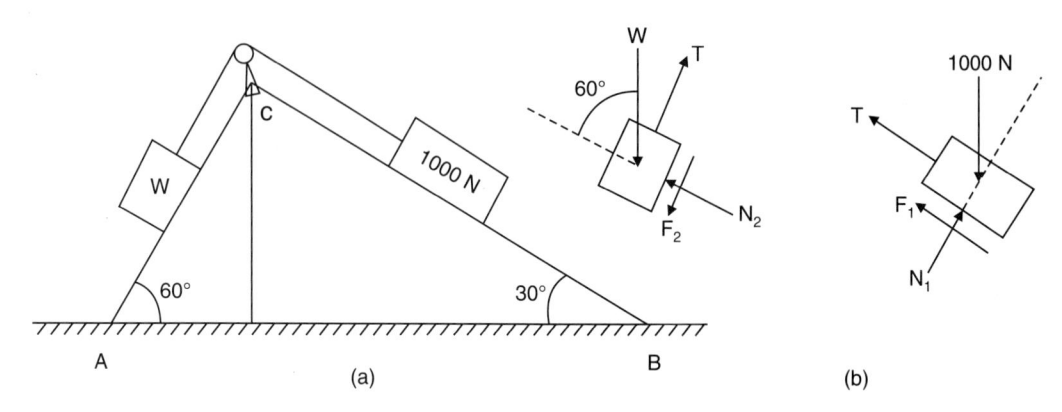

Fig. 5.11

Solution. (*a*) **For the least value of W:**

In this case motion of 1000 N block is impending downward. For this case free body diagram of blocks are as shown in Fig. 5.11(*b*).

Consider the equilibrium of 1000 N block:

ΣForces normal to the plane = 0, gives

$\qquad N_1 - 1000 \cos 30° = 0 \quad \therefore \quad N_1 = 866.03$ newton.

From the law of friction,

$\qquad F_1 = \mu N_1 = 0.28 \times 866.03 = 242.49$ newton.

ΣForces parallel to the plane = 0

$\qquad T + F_1 - 1000 \sin 30° = 0$

$\therefore \qquad T = -242.49 + 500 = 257.51$ newton.

Now consider the equilibrium of block weighing W:

ΣForces normal to the plane = 0, gives

$\qquad N_2 - W \cos 60° = 0 \quad \therefore \quad N_2 = 0.5 W$

$\therefore \qquad F_2 = \mu N_2 = 0.2 \times 0.5 W = 0.1 W$

ΣForces parallel to the plane = 0, gives

$\qquad T - F_2 - W \sin 60° = 0$

$\qquad 257.51 - 0.1 W - W \sin 60° = 0$

$$W = \frac{257.51}{(0.1 + \sin 60°)} = 266.57 \quad \textbf{Ans.}$$

(*b*) **For greatest value of W:**

In this case 1000N block will be on the verge of moving up the plane. The free body diagram for this case is as shown in Fig. 5.11(*c*).

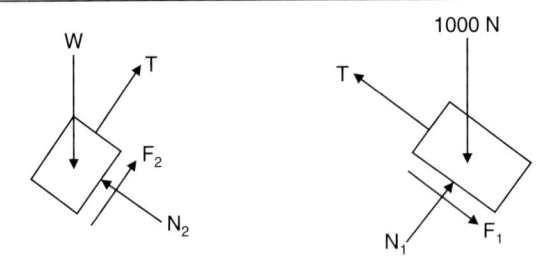

Fig. 5.11(c)

Consider the equilibrium of block weighing 1000N
$$N_1 = 866.03 \text{ newton}$$
and $F_1 = 242.49$ newton, as in the previous case.
ΣParallel to plane = 0, gives
$$T - 1000 \sin 30° - F_1 = 0$$
$\therefore \qquad T = 1000 \sin 30° + 242.49 = 742.49$ newton.

Consider the equilibrium of block weighing W.
ΣForces normal to plane = 0, gives
$$N_2 - W \cos 60° = 0 \quad \therefore \quad N_2 = 0.5\ W$$
$$F_2 = \mu N_2 = 0.2 \times 0.5\ W = 0.1\ W$$
ΣForces parallel to plane = 0, gives
$$T + F_2 - W \sin 60° = 0$$
$$742.49 + 0.1 W - W \sin 60° = 0$$
$\therefore \qquad W = \dfrac{742.49}{(0.1 + \sin 60°)} = 969.28$ newton **Ans.**

Example 5.8. *Two blocks A and B each weighing 1500N are connected by a uniform horizontal bar weighing 1000N. If the angle of limiting friction for all contact surfaces is 15°, find the force P directed parallel to the 60° inclined plane that will cause motion impending to the right [Ref. Fig. 5.12 (a)].*

Solution. Free body diagrams of block A, beam AB and block B are as shown in Figs. 5.12(b), (c) and (d) respectively.

From vertical equilibrium condition for AB, it may be found that 500N force is transferred at A and B, which may be directly added to self weights of the blocks.

Now consider equilibrium of block A.
$$\Sigma V = 0 \quad \rightarrow$$
$$N_A \cos 30° + F_A \sin 30° - (1500 + 500) = 0$$
But from law of friction, $F_A = N_A \tan 15°$
$\therefore \qquad N_A (\cos 30° + \tan 15° \sin 30°) = 2000$
$\therefore \qquad N_A = 2000$ newton
$\therefore \qquad F_A = 2000 \tan 15° = 535.90$ newton.
$$\Sigma H = 0 \quad \rightarrow$$
$$- C + N_A \sin 30° - F_A \cos 30° = 0$$
$\therefore \qquad C = N_A \sin 30° - F_A \cos 30°$
$\qquad\qquad = 2000 \sin 30° - 535.90 \cos 30° = 535.90$ newton.

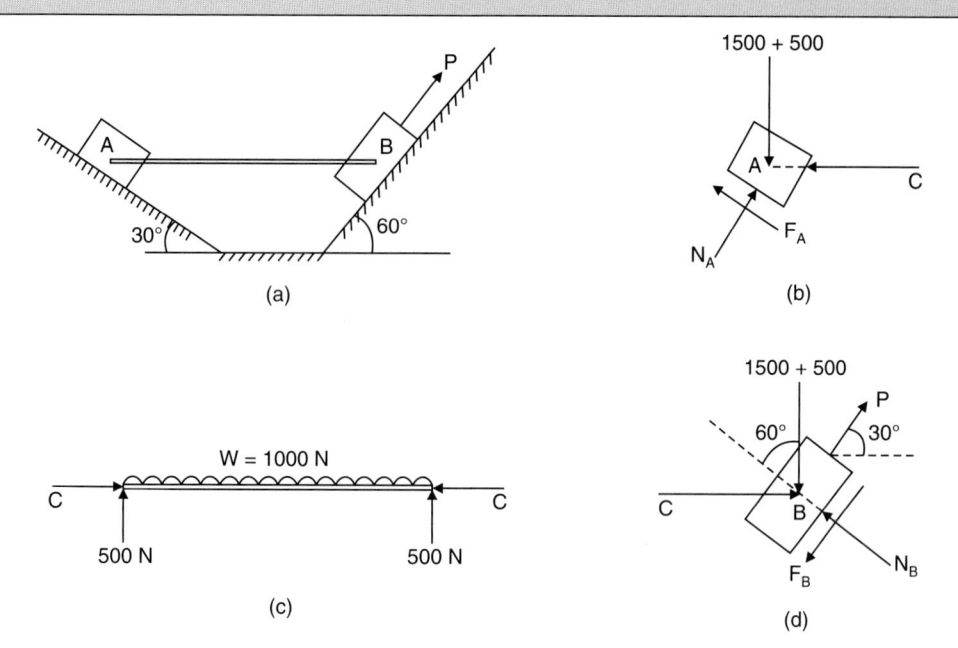

Fig. 5.12

Consider the equilibrium of block B.

ΣForces normal to the inclined plane = 0, gives

$$N_B - 2000 \cos 60° - C \cos 30° = 0$$
$$N_B = 2000 \cos 60° + 535.90 \cos 30° = 1464.10 \text{ newton}$$

∴
$$F_B = N_B \tan 15° = 1464.90 \tan 15°$$
$$= 392.30 \text{ newton.}$$

ΣForces parallel to the inclined plane = 0, gives

$$P - F_B - 2000 \sin 60° + C \sin 30° = 0$$
$$P = 392.30 + 2000 \sin 60° + 535.90 \sin 30°$$
$$\mathbf{P = 1856.40 \ Ans.}$$

5.4. WEDGES

Wedges are small pieces of materials with two of their opposite surfaces not parallel. They are used to slightly lift heavy blocks, machinery, precast beams etc. for final alignments or to make place for inserting lifting devices. The weight of the wedge is very small compared to the weight lifted. Hence, in all problems, the self weight of wedge is neglected.

In the analysis, instead of treating normal reaction and frictional force independently, it is advantageous to consider their resultant. If F is limiting friction, then resultant makes limiting frictional angle α with the normal. Its direction should be marked correctly. For this, it should be noted that the tangential component R is the frictional force and it always acts opposite to impending motion. The analysis procedure is illustrated with two examples below:

Example 5.9. *Determine the minimum force required to move the wedge shown in Fig. 5.13(a). The angle of friction for all contact surfaces is 15°.*

Solution. As wedge is driven, it moves towards left and block moves upwards. Force P required to move the system is minimum when the motion is impending and hence at this stage limiting frictional force acts. Hence resultant makes limiting angle of 15° with normal. The free body diagrams for block and wedge are shown in Fig. 5.13(b). The forces on block and wedge are redrawn in Figs. 5.13(c) and (d) so that Lami's theorem can be applied conveniently. Applying Lami's theorem to the system of forces on block, we get

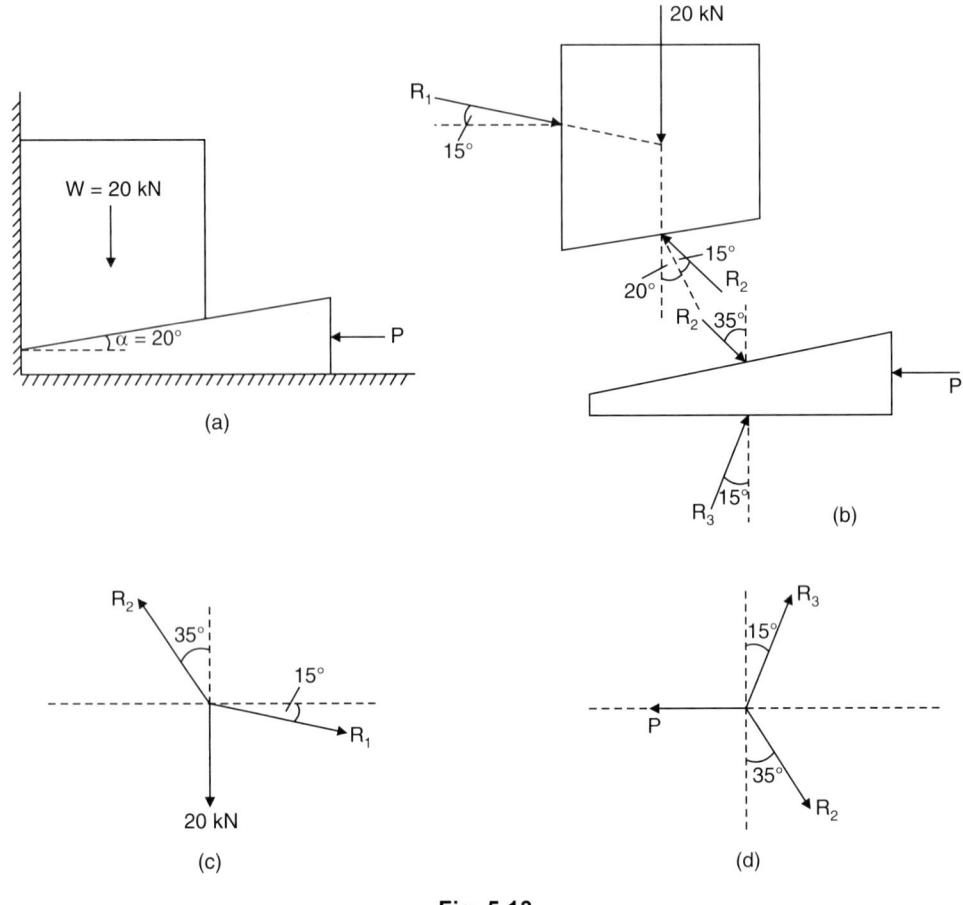

Fig. 5.13

$$\frac{R_1}{\sin 145°} = \frac{R_2}{\sin 75°} = \frac{20}{\sin 140°}$$

∴ $R_1 = 17.847$ kN

and $R_2 = 30.047$ kN.

Applying Lami's theorem to system of forces on the wedge, we get,

$$\frac{P}{\sin 130°} = \frac{R_2}{\sin 105°}$$

∴ **P = 23.835 kN** **Ans.**

Example 5.10. *The block C, weighing 160 kN is to be raised by means of driving wedges A and B as shown in Fig. 5.14(a). Find the value of force P for impending motion of the block upwards, if coefficient of friction is 0.25 for all contact surfaces. Self weight of wedges may be neglected.*

FRICTION

Solution. Let ϕ be the angle of limiting friction. Then
$$\tan \phi = 0.25 \quad \therefore \quad \phi = 14.036°$$

Free body diagrams of wedges A and B, and block C are as shown in Fig. 5.14(b). The problem being symmetric, the forces R_1 and R_2 on wedges A and B are equal. The system of forces on block C and wedge A are shown in the form convenient for applying Lami's theorem in Figs. 5.14(c) and (d).

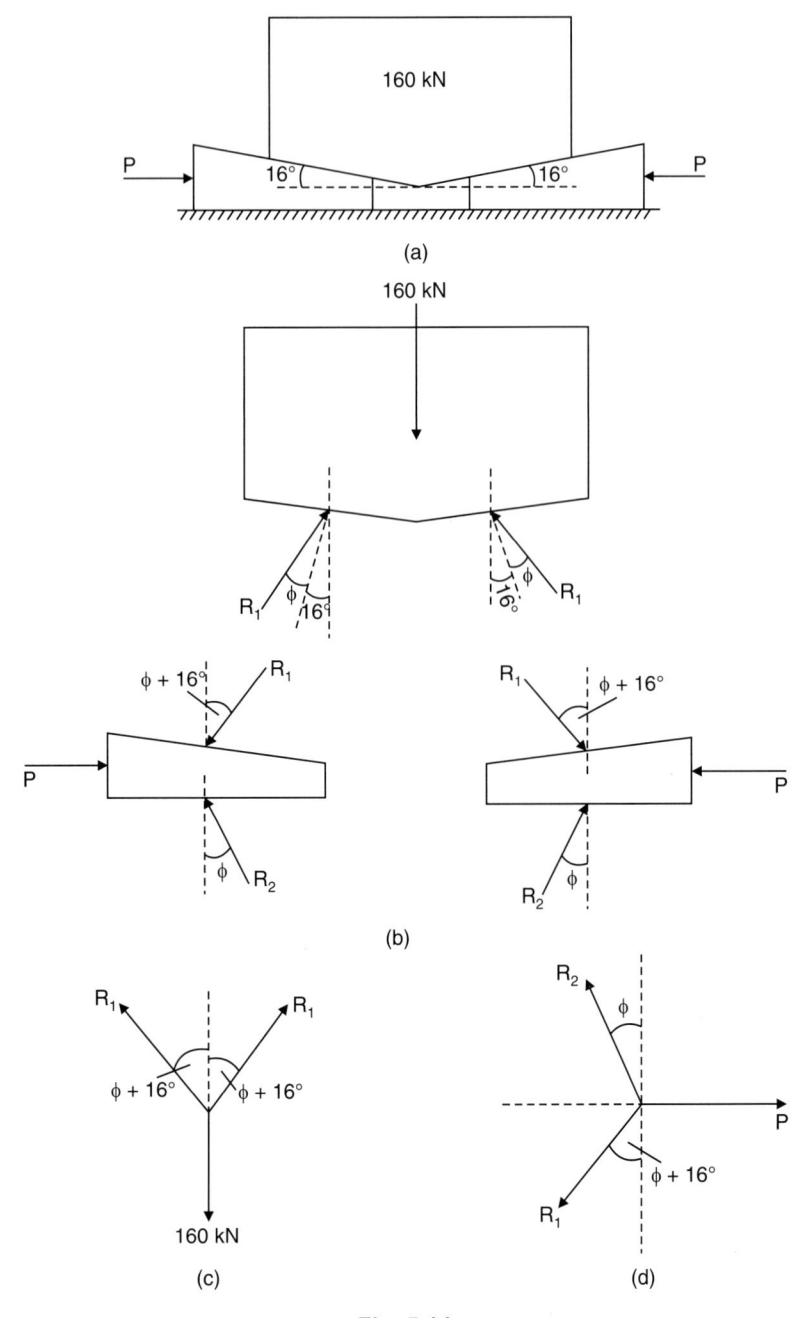

Fig. 5.14

Consider the equilibrium of block C:
$$\frac{R_1}{\sin(180° - 16° - \phi)} = \frac{160}{\sin 2(\phi + 16°)}$$

i.e. $\quad\dfrac{R_1}{\sin 149.96°} = \dfrac{160}{\sin 60.072°}$, since $\phi = 14.036°$

$\therefore\quad R_1 = 92.41$ kN.

Consider the equilibrium of wedge A:
$$\frac{P}{\sin(180° - \phi - \phi - 16°)} = \frac{R_1}{\sin(90° + \phi)}$$

$\therefore\quad$ **P = 66.256 kN Ans.**

5.5. PROBLEMS INVOLVING NON-CONCURRENT FORCE SYSTEMS

There are many practical problems of non-concurrent force systems involving friction. In these cases, apart from law of friction, three equations of equilibrium are to be used. The method of solving such problems is illustrated below with typical problems.

Example 5.11. *A ladder of length 4 m, weighing 200 N is placed against a vertical wall as shown in Fig. 5.15(a). The coefficient friction between the wall and the ladder is 0.2 and that between the floor and the ladder is 0.3. In addition to self weight, the ladder has to support a man weighing 600 N at a distance of 3 m from A. Calculate the minimum horizontal force to be applied at A to prevent slipping.*

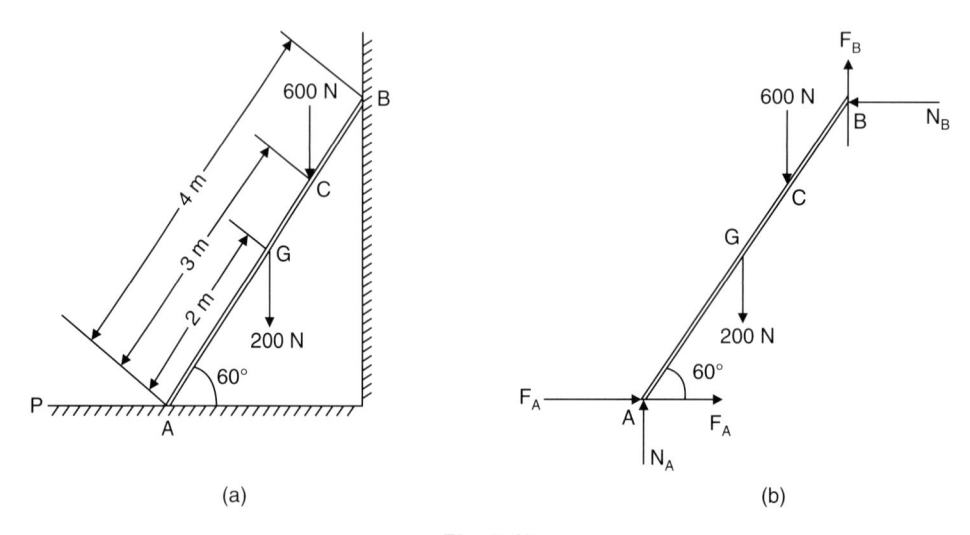

Fig. 5.15

Solution. The free body diagram of the ladder is as shown in Fig. 5.15(b).

$\Sigma M_A = 0$, gives
$$N_B \times 4 \sin 60° + F_B \times 4 \cos 60° - 600 \times 3 \cos 60° - 200 \times 2 \cos 60° = 0$$

Dividing throughout by 4 and rearranging, we get,
$$0.866\, N_B + 0.5\, F_B = 275 \qquad \ldots(1)$$

From the law of friction,
$$F_B = 0.2\, N_B \qquad \ldots(2)$$

FRICTION

Substituting this in eqn. (1), we get

$N_B(0.866 + 0.5 \times 0.2) = 275$

$\therefore \quad N_B = 284.68$ newton ...(3)

$\therefore \quad F_B = 0.2 \times 284.68 = 56.934$ newton ...(4)

$\Sigma V = 0 \rightarrow$

$N_A - 200 - 600 + F_B = 0$

$\therefore \quad N_A = 200 + 600 - 56.934 = 743.066$ newton

$\therefore \quad F_A = 0.3 \, N_A = 0.3 \times 743.066 = 222.92$ newton.

$\Sigma H = 0 \rightarrow$

$P + F_A - N_B = 0$

$\therefore \quad P = N_B - F_A = 284.68 - 222.92$

i.e. $\quad\quad\quad \mathbf{P = 61.76\ newton.\ Ans.}$

Example 5.12. *The ladder shown in Fig. 5.16(a) is 6m long and is supported by a horizontal floor and vertical wall. The coefficient of friction between the floor and the ladder is 0.25 and between wall and the ladder is 0.4. The self weight of the ladder is 200N and may be considered as concentrated at G. The ladder also supports a vertical load of 900N at C which is at a distance of 1m from B. Determine the least value of α at which the ladder may be placed without slipping. Determine the reactions developed at that stage.*

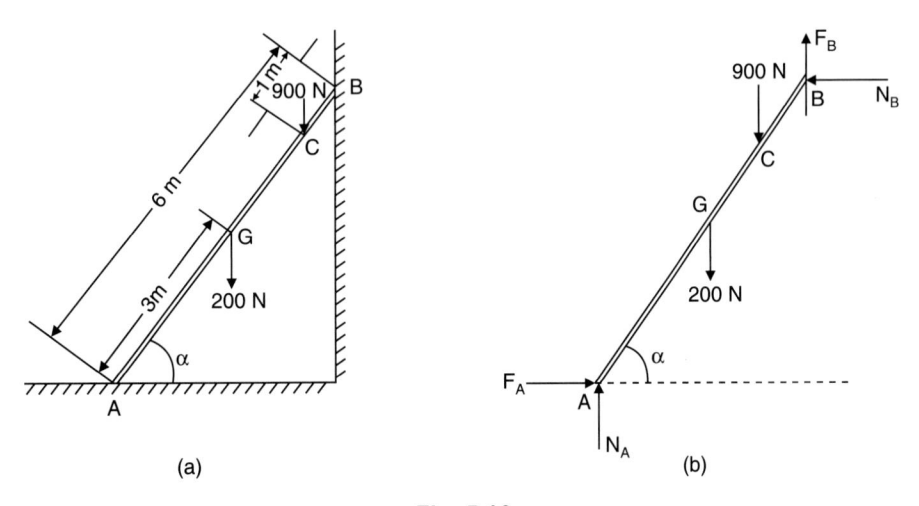

Fig. 5.16

Solution. Free body diagram of ladder for this case is as shown in Fig. 5.16(b). From the law of friction,

$F_A = 0.25 \, N_A$ ...(1)

and $\quad F_B = 0.4 \, N_B$ ...(2)

$\Sigma V = 0 \rightarrow$

$N_A - 200 - 900 + F_B = 0$

i.e. $\quad N_A + 0.4 \, N_B = 1100$ ...(3)

$\Sigma H = 0 \rightarrow$

$F_A - N_B = 0$

i.e. $\quad 0.25 \, N_A = N_B$ ...(4)

From eqns. (3) and (4), we get
$$N_A + 0.4 \times 0.25\, N_A = 1100$$
∴ $\quad\quad\quad\quad\quad N_A = \mathbf{1000\ newton}\quad\textbf{Ans.}$
∴ $\quad\quad\quad\quad\quad F_A = 0.25\, N_A = 0.25 \times 1000 = \mathbf{250\ newton}\quad\textbf{Ans.}$
From eqn. (4), $N_B = 0.25\, N_A = 0.25 \times 1000 = \mathbf{250\ newton}\quad\textbf{Ans.}$
Substituting it in eqn. (2), we get
$$F_B = 0.4 \times 250 = \mathbf{100\ newton}\quad\textbf{Ans.}$$
$$\Sigma M_A = 0 \ \rightarrow$$
$$N_B \times 6 \sin\alpha + F_B \times 6 \cos\alpha - 200 \times 3 \cos\alpha - 900 \times 5 \cos\alpha = 0$$
i.e., $\quad\quad 250 \times 6 \sin\alpha + 100 \times 6 \cos\alpha - 600 \cos\alpha - 4500 \cos\alpha = 0.$
i.e., $\quad\quad 1500 \sin\alpha = (-600 + 600 + 4500) \cos\alpha$
∴ $\quad\quad\quad\quad \tan\alpha = 3$
Hence $\quad\quad\quad \alpha = \mathbf{71.563°}\quad\textbf{Ans.}$

Example 5.13. *A horizontal bar AB of length 3m and weighing 500N is lying in a trough as shown in Fig. 5.17(a). Find how close to end A and B a load of 600N can be placed safely, if coefficient of friction between contact surfaces is 0.2.*

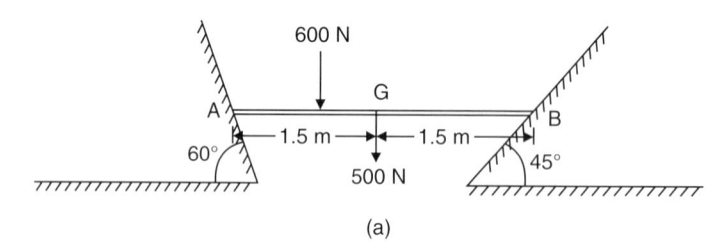

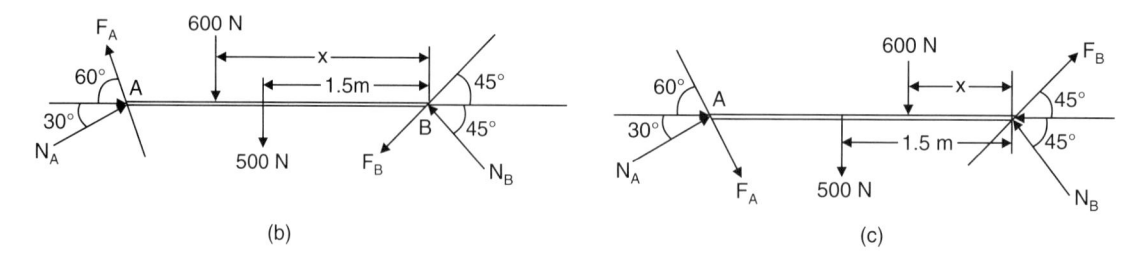

Fig. 5.17

Solution. When the load is close to end A, the end A will slip down and end B will slip up. For this impending motion the free body diagram is as shown in Fig. 5.17(b). From law of friction
$$F_A = 0.2\, N_A \quad\quad\quad\quad\quad\quad\quad\quad\quad\quad\quad\quad\quad\quad\quad\quad ...(1)$$
and $\quad\quad\quad\quad F_B = 0.2\, N_B \quad\quad\quad\quad\quad\quad\quad\quad\quad\quad\quad\quad\quad\quad\quad ...(2)$
$$\Sigma V = 0 \ \rightarrow$$
$$N_A \sin 30° + F_A \sin 60° + N_B \sin 45° - F_B \sin 45° = 1100$$
$$N_A (\sin 30° + 0.2 \sin 60°) + N_B (\sin 45° - 0.2 \sin 45°) = 1100$$
$$0.6732\, N_A + 0.5657\, N_B = 1100 \quad\quad\quad\quad ...(3)$$
$$\Sigma H = 0 \ \rightarrow$$

$N_A \cos 30° - F_A \cos 60° - N_B \cos 45° - F_B \cos 45° = 0$
$N_A (\cos 30° - 0.2 \cos 60°) = N_B (\cos 45° + 0.2 \cos 45°)$
∴ $\quad N_A = 1.1077 \, N_B$ ...(4)

Substituting this value of N_A in eqn. (3), we get
$0.6732 \times 1.1077 \, N_B + 0.5657 \, N_B = 1100$
∴ $\quad N_B = 838.79$ newton
∴ $\quad N_A = 1.1077 \times 839.79 = 929.13$ newton
$\Sigma M_B = 0 \quad \rightarrow$
$600x + 500 \times 1.5 - F_A \sin 60° \times 3 - N_A \sin 30° \times 3 = 0$
$600x = -750 + 0.2 \sin 60° \times 929.13 \times 3 + 929.13 \sin 30° \times 3$
∴ **x = 1.877 m from B. Ans**

When the load is close to end B, the end B may slide down and end A up. Let x be the distance of load from the end B, when motion is impending. For this situation the free body diagram is as shown in Fig. 5.17(a).
$\Sigma V = 0 \quad \rightarrow$
$N_A \sin 30° - F_A \sin 60° + N_B \sin 45° + F_B \sin 45° = 1100$
$N_A (\sin 30° - 0.2 \sin 60°) + N_B (\sin 45° + 0.2 \sin 45°) = 1100$
$0.3268 \, N_A + 0.8485 \, N_B = 1100$ ...(5)
$\Sigma H = 0 \quad \rightarrow$
$N_A \cos 30° + F_A \cos 60° - N_B \cos 45° + F_B \cos 45° = 0$
$N_A (\cos 30° + 0.2 \cos 60°) = N_B (\cos 45° - 0.2 \cos 45°)$
∴ $\quad N_A = 0.5856 \, N_B$ ...(6)

Substituting it in (5), we get
$\quad N_B = 1057.82$ newton.
and hence, $\quad N_A = 0.5856 \times 1057.82 = 619.67$ newton.
$\Sigma M_B = 0 \quad \rightarrow$
$600x + 500 \times 1.5 - N_A \sin 30° \times 3 + F_A \sin 60° \times 3 = 0$
$600x = -750 + 619.67 \sin 30° \times 3 - 0.2 \times 6.19.67 \sin 30° \times 3$
∴ $\quad x = -0.237$ m.

It means, the motion will be impending when the load is at 0.237 m to the right of B, which is not possible case of loading for the bar. Hence load can be placed even on point B safely.

Thus 600 N load can be placed anywhere between B and a point 1.85 m from B.

5.6. ROPE FRICTION

The transmission of power by means of belt or rope drives is possible because of friction, which exists between the wheels and the belt. Similarly band brakes stop the rotating discs because of friction between the belts and the discs. All along the contact surface the frictional resistance develops. Hence the tension in the rope is more on the side it is pulled and is less on the other side. Accordingly, the two sides of the rope may be called as tight side and slack side. The relationship between the forces on slack side and tight side can be derived as explained below:

Figure 5.18(a) shows a load W being pulled up by a force P over a fixed drum. Let the force on slack side be T_1 and on tight side be T_2 [Ref. Fig. 5.18(b)]. T_2 is more than T_1 because frictional force develops between drum and the rope as shown in Fig. 5.18(c). Let θ be the angle of contact between rope and the drum. Now consider an elemental length of rope as shown in Fig. 5.18(d).

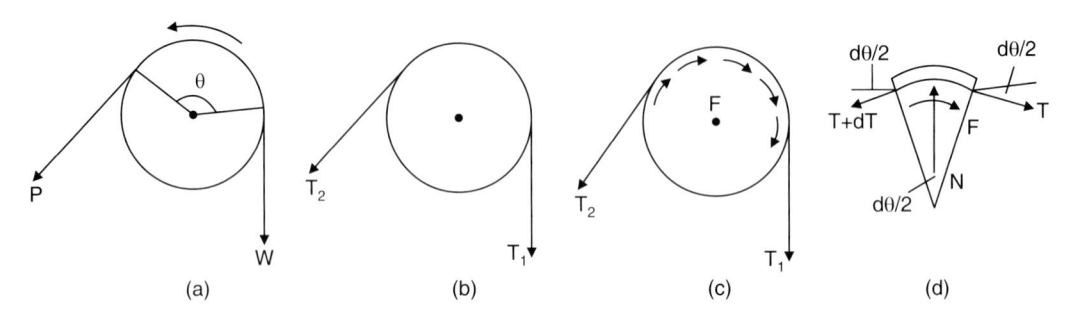

Fig. 5.18

Let T be the force on slack side and $T + dT$ on tight side. There will be normal reaction N on the rope in the radial direction and frictional force $F = \mu N$ in the tangential direction. Then

ΣForces in radial direction = 0, gives

$$N - T \sin \frac{d\theta}{2} - (T + dT) \sin \frac{d\theta}{2} = 0$$

Since $d\theta$ is very small angle, $\sin \frac{d\theta}{2} = \frac{d\theta}{2}$.

$\therefore \quad N - T \dfrac{d\theta}{2} - (T + dT) \dfrac{d\theta}{2} = 0$

i.e., $\quad N = \left(T + \dfrac{dT}{2} \right) d\theta \qquad \qquad …(1)$

From the law of friction,

$$F = \mu N = \mu \left(T + \frac{dT}{2} \right) d\theta \qquad \qquad …(2)$$

where μ is coefficient of friction.

ΣForces in tangential direction = 0, gives

$$(T + dT) \cos \frac{d\theta}{2} = F + T \cos \frac{d\theta}{2}$$

Since $d\theta$ is very small angle, $\cos \dfrac{d\theta}{2} = 1$

$\therefore \quad T + dT = F + T \quad$ or $\quad dT = F \qquad \qquad …(3)$

From eqns. (2) and (3), we get

$$dT = \mu \left(T + \frac{dT}{2} \right) d\theta$$

FRICTION

Neglecting small quantity of higher order, we get
$$dT = \mu T \, d\theta$$
or
$$\frac{dT}{T} = \mu \, d\theta$$

Integrating both sides over 0 to θ, we get
$$\int_{T_1}^{T_2} \frac{dT}{T} = \int_0^\theta \mu \, d\theta$$

$$\left[\log T\right]_{T_1}^{T_2} = \mu \left[\theta\right]_0^\theta$$

∴
$$\log \frac{T_2}{T_1} = \mu\theta$$

i.e.,
$$\frac{T_2}{T_1} = e^{\mu\theta}$$

or
$$T_2 = T_1 \, e^{\mu\theta} \qquad \ldots(4)$$

[**Note:** θ should be in radians.]

Example 5.14. *A rope making $1\frac{1}{4}$ turns around a stationary horizontal drum is used to support a weight W [Fig. 5.19]. If the coefficient of friction is 0.3 what range of weight can be supported by exerting a 600N force at the other end of the rope ?*

Solution. Angle of contact $\theta = 1.25 \times 2\pi = 2.5\pi$

Case 1. Let the impending motion of the weight be downward. Then
$$T_1 = 600 \text{ newton} \quad \text{and } T_2 = W$$
∴ From law of rope friction,
$$W = 600 \, e^{\mu 2.5\pi} = 600 \, e^{0.3 \times 2.5\pi} = 600 \, e^{0.75\pi}$$
$$= 6330.43 \text{ newton.}$$

Case (2). Let the impending motion of weight be upwards. Then
$$T_1 = W \text{ and } T_2 = 600 \text{ newton}$$
∴
$$T_2 = T_1 \, e^{\mu\theta} \text{ gives}$$
$$600 = W \, e^{0.75\pi}$$
∴
$$W = 56.87 \text{ newton.}$$

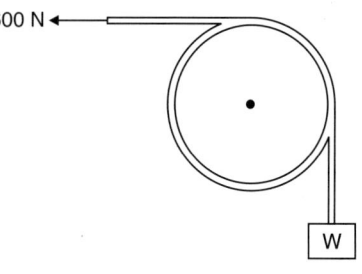

Fig. 5.19

Thus a 600 N force can support a range of loads between 56.87 N to 6330.43 N weight on the other side of the drum.

Example 5.15. *In Fig. 5.20(a), the coefficient of friction between the rope and the fixed drum is 0.2 and between other surfaces of contact is 0.3. Determine the minimum weight W to prevent downward motion of the 1000N block.*

Solution. Since 1000 N block is on the verge of sliding down, the rope connecting it is tight side and the rope connecting W is the slack side. Free body diagrams for W and 1000 N block are as shown in Fig. 5.20(b).

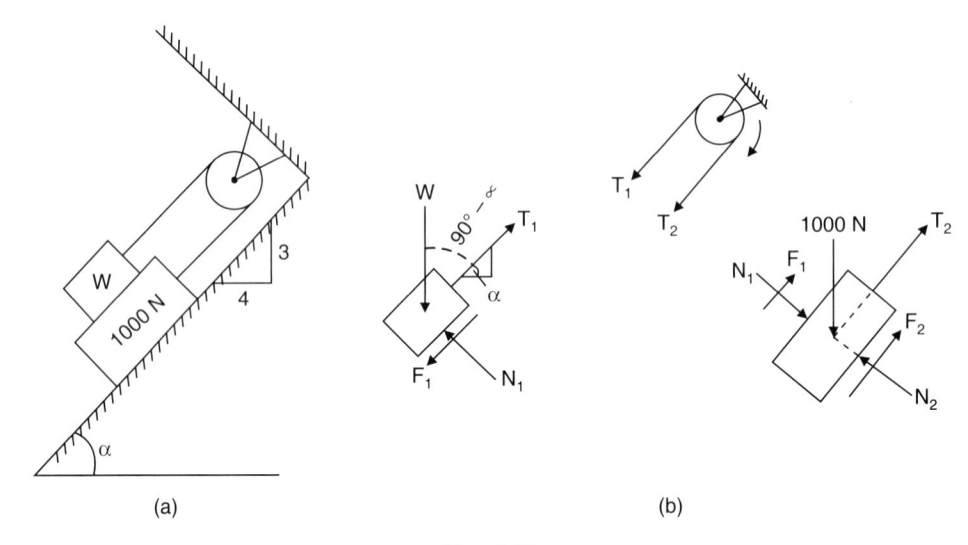

Fig. 5.20

Now, $\tan \alpha = 3/4$ ∴ $\alpha = 36.87°$

Consider the equilibrium of block W

ΣForces perpendicular to plane = 0, gives

$$N_1 - W \cos \alpha = 0 \quad \text{or} \quad N_1 = W \cos 36.87° = 0.8 W \quad \ldots(1)$$

∴ $\quad F_1 = \mu N_1 = 0.3 \times 0.8 W = 0.24 W \quad \ldots(2)$

ΣForces parallel to the plane = 0, gives

$$T_1 - F_1 - W \sin \alpha = 0$$

or $\quad T_1 = F_1 + W \sin 36.87° = 0.24 W + 0.6 W = 0.84 W \quad \ldots(3)$

Angle of contact of rope with pulley = 180° = π radians

From friction equation for rope, we get

$$T_2 = T_1 e^{\mu \theta} = T_1 e^{0.3\pi}$$

i.e., $\quad T_2 = 2.566 T_1$

Substituting the value of T_1 from eqn. (3),

$$T_2 = 2.566 \times 0.84 W = 2.156 W \quad \ldots(4)$$

Now consider the equilibrium of 1000 N block.

ΣForces perpendicular to the plane = 0, gives,

$$N_2 - N_1 - 1000 \cos \alpha = 0$$

$$N_2 - N_1 = 800, \quad \text{since } \cos \alpha = \cos 36.87° = 0.8$$

Substituting the value of N_1 from eqn. (1), we get

$$N_2 = N_1 + 800 = 0.8 W + 800$$

∴ $\quad F_2 = 0.3 N_2 = 0.3 \times 0.8 W + 0.3 \times 800 = 0.24 W + 240 \quad \ldots(5)$

ΣForces parallel to the plane = 0, gives

$$F_1 + F_2 - 1000 \sin \alpha + T_2 = 0$$

FRICTION

i.e. $\quad 0.24\ W + 0.24\ W + 240 - 1000 \sin 36.87° + 2.156\ W = 0$

∴ $\quad$ **W = 136.57 newton   Ans.**

Example 5.16. *A torque of 300 N-m acts on a brake drum shown in Fig. 5.21(a). If the brake band is in contact with the brake drum through 250° and the coefficient of friction is 0.3, determine the force P applied at the end of the brake lever for the position shown in the figure.*

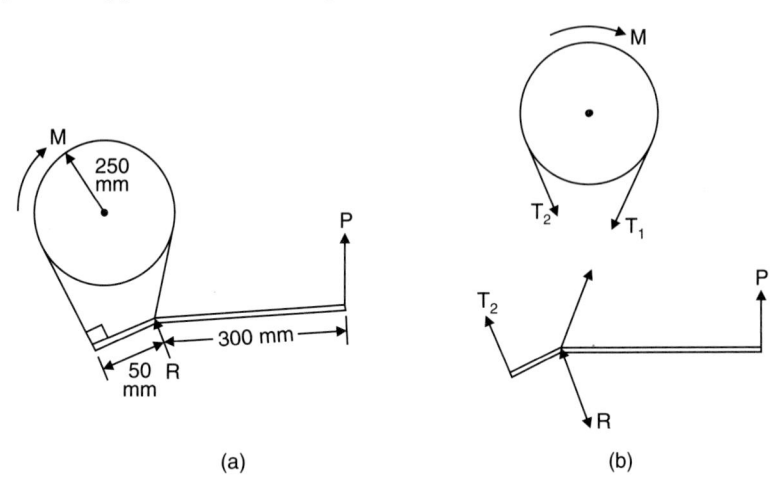

Fig. 5.21

Solution. Figure 5.21(b) shows the free body diagram of brake drum and the lever arm.

Now, $\qquad \theta = 250 \times \dfrac{\pi}{180}$ radians, $r = 250$ mm.

$\mu = 0.3$

∴ $\qquad \mu\theta = 0.3 \times 250 \times \dfrac{\pi}{180} = 1.309$

From rope friction equation,

$$T_2 = T_1 e^{\mu\theta} = T_1 e^{1.309} = 3.7025\ T_1$$

Now, $\quad (T_2 - T_1)r = M$

i.e., $\quad (3.7025 - 1)\ T_1 \times 250 = 300 \times 10^3$

∴ $\qquad T_1 = 444.04$ newton

$T_2 = 3.7025 \times 444.04 = 1644.06$ newton.

Consider the equilibrium of lever arm

$$T_2 \times 50 = P \times 300$$

∴ $\qquad \mathbf{P} = \dfrac{1644.06 \times 50}{300} = \mathbf{274.0\ newton\quad Ans.}$

Important Definitions and Formulae

Definitions

1. The maximum value of the frictional force, which develops between two contacting surfaces when the motion is impending is called **limiting friction**.

2. Between two contacting surfaces, the magnitude of limiting friction bears a constant ratio to normal reaction. This ratio is called **coefficient of friction**. Thus $\mu = \dfrac{F}{N}$, where F is limiting friction and N normal reaction.
3. The angle between the resultant reaction and the normal reaction is called **Angle of Friction** i.e. $\phi = \tan^{-1} \dfrac{F}{N} = \tan^{-1} \mu$.
4. The maximum angle at which grains can repose is called **Angle of Repose**.
5. The inverted cone with semi-central angle equal to the angle of friction is called **Cone of Friction**.

Formulae

1. $\mu = \dfrac{F}{N}$, where F is limiting friction.
2. $\phi = \tan^{-1} \dfrac{F}{N} = \tan^{-1} \mu$.
3. In case of rope friction, $T_2 = T_1 e^{\mu\theta}$.

PROBLEMS FOR EXERCISE

5.1 A pull of 180 N applied upward at 30° to a rough horizontal plane was required to just move a body resting on the plane while a push of 220 N applied along the same line of action was required to just move the same body downwards. Determine the weight of the body and the coefficient of friction. **[Ans.** $W = 990$ N; $\mu = 0.1722$**]**

5.2 The block A shown in Fig. 5.22 weighs 2000 N. The cord attached to A passes over a frictionless pulley and supports a weight equal to 800 N. The value of coefficient of friction between A and the horizontal plane is 0.35. Solve for horizontal force P: (1) if motion is impending towards the left, and (2) if the motion is impending towards the right. **[Ans.** (1) 1252.82 N; (2) 132.82 N**]**

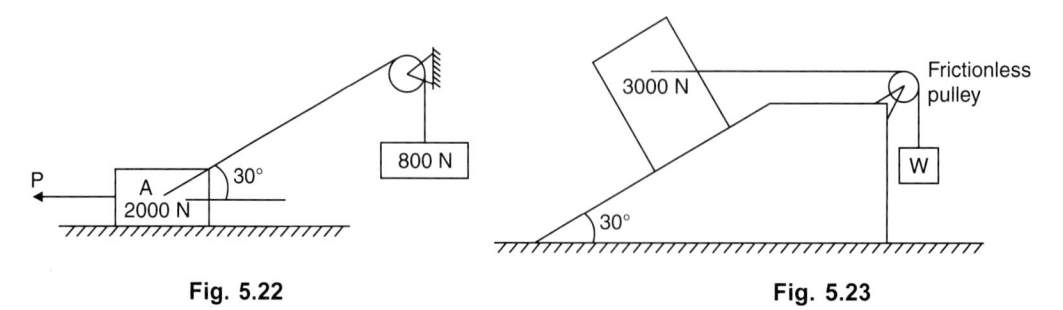

Fig. 5.22

Fig. 5.23

5.3 A 3000 N block is placed on an inclined plane as shown in Fig. 5.23. Find the maximum value of W for equilibrium if tipping does not occur. Assume coefficient of friction as 0.2. **[Ans.** 2636.15**]**

5.4 Find whether block A is moving up or down the plane in Fig. 5.24 for the data given below. Weight of block $A = 300$ N. Weight of block $B = 600$ N. Coefficient of limiting friction between plane AB and block A is 0.2. Coefficient of limiting friction between plane BC and block B is 0.25. Assume pulley as smooth. **[Ans.** The block A is stationary since F developed $< F_{\lim}$**]**

FRICTION

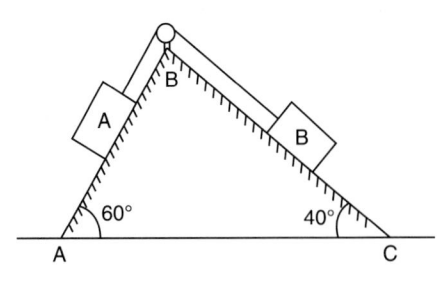

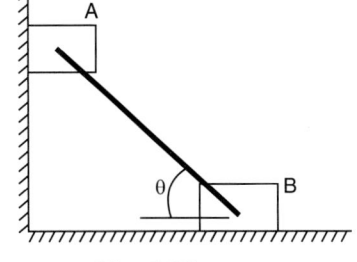

Fig. 5.24

Fig. 5.25

5.5 Two identical blocks A and B are connected by a rod and they rest against vertical and horizontal planes respectively as shown in Fig. 5.25. If sliding impends when $\theta = 45°$, determine the coefficient of friction, assuming it to be the same for both floor and wall. **[Ans. 0.414]**

5.6 Determine the force P required to start the wedge as shown in Fig. 5.26. The angle of friction for all surfaces of contact is 15°. **[Ans. 26.678 kN]**

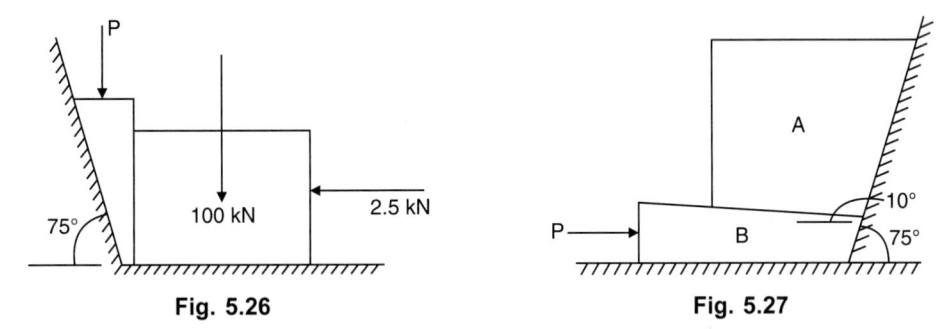

Fig. 5.26

Fig. 5.27

5.7 Two blocks A and B weighing 3 kN, and 15 kN respectively, are held in position against an inclined plane by applying a horizontal force P as shown in Fig. 5.27. Find the least value of P which will induce motion of the block A upwards. Angle of friction for all contact surfaces is 12°.
[Ans. 14.025 kN]

5.8 In Fig. 5.28, C is a stone block weighing 6 kN. It is being raised slightly by means of two wooden wedges A and B with a force P on wedge B. The angle between the contacting surfaces of the wedge is 5°. If coefficient of friction is 0.3 for all contacting surfaces, compute the value of P required to impend upward motion of the block C. Neglect weight of the wedges.
[Ans. 2.344 kN]

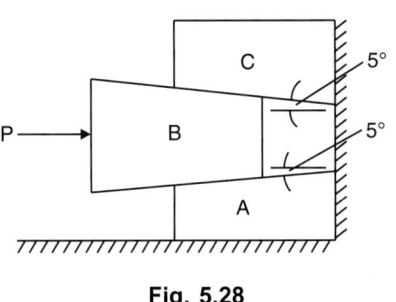

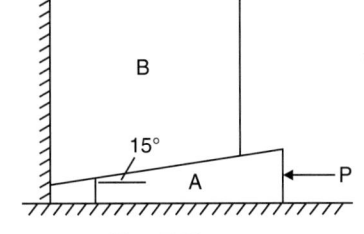

Fig. 5.28

Fig. 5.29

5.9 Find the horizontal force P required to pull the block A of weight 150 N which carries block B of weight 1280 N as shown in Fig. 5.29. Take angle of limiting friction between floor and block A as 14° and that between vertical wall and block B as 13° and coefficient of limiting friction between the blocks as 0.3. **[Ans. $P = 1294.2$ N]**

5.10 The level of a precast beam, weighing 20,000 N is to be adjusted by driving a wedge as shown in Fig. 5.30. If coefficient of friction between the wedge and pier is 0.35 and that between beam and the wedge is 0.25, determine the minimum force P required on the wedge to make adjustment of the beam. Angle of the wedge is 15°.

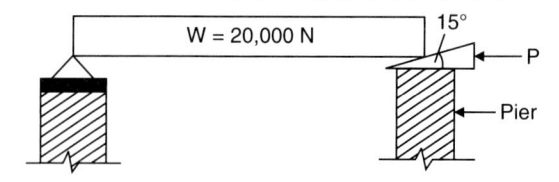

Fig. 5.30

(*Hint:* Vertical component of reaction on wedge at contact with beam = 1/2 vertical load on beam = 10,000 kN) [**Ans.** 9057.4 N]

5.11 A ladder 5 m long rests on a horizontal ground and leans against a smooth vertical wall at an angle of 70° with the horizontal. The weight of the ladder is 300 N. The ladder is on the verge of sliding when a man weighing 750 N stands on a rung 1.5 m high. Calculate the coefficient of friction between the ladder and the floor. [**Ans.** $\mu = 0.1837$]

5.12 A 4 m ladder weighing 200 N is placed against a vertical wall as shown in Fig. 5.31. As a man weighing 800 N, reaches a point 2.7 m from A, the ladder is about to slip. Assuming that the coefficient of friction between the ladder and the wall is 0.2, determine the coefficient of friction between the ladder and the floor. [**Ans.** 0.3548]

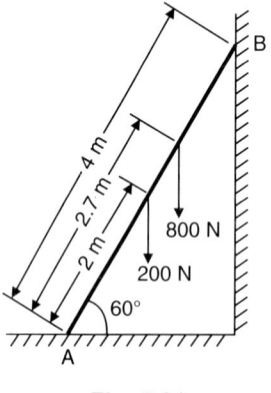

Fig. 5.31

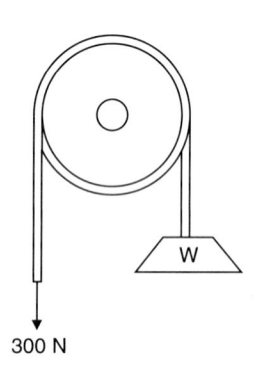

Fig. 5.32

5.13 Determine the maximum weight that can be lowered by a person who can exert a 300 N pull on rope if the rope is wrapped $2\frac{1}{2}$ turns round a horizontal spur as shown in Fig. 5.32. Coefficient of friction between spur and the rope is 0.3. [**Ans.** 33395.33 N]

5.14 Determine the minimum value of W required to cause motion of blocks A and B towards right (ref. Fig. 5.33). Each block weighs 3000 N and coefficient of friction between blocks and inclined planes is 0.2. Coefficient of friction between the drum and rope is 0.1. Angle of wrap over the drum is 90°. [**Ans.** 3065.18 N]

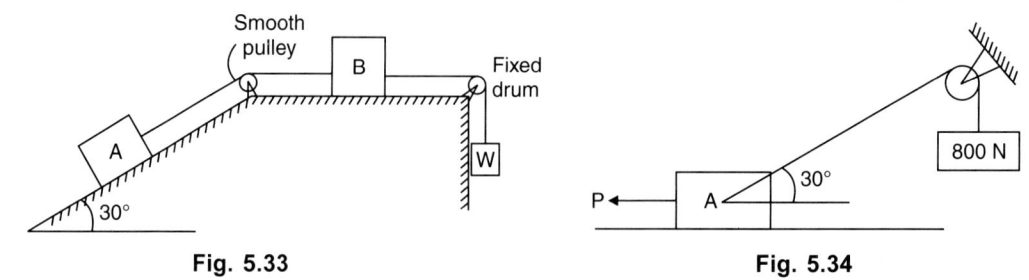

Fig. 5.33

Fig. 5.34

FRICTION

5.15 Block *A* shown in Fig. 5.34 weighs 2000 N. The cord attached to *A* passes over a fixed drum and supports a weight equal to 800 N. The value of coefficient of friction between *A* and the horizontal plane is 0.25 and between the rope and the fixed drum is 0.1. Solve for *P*: (1) if motion is impending towards the left, (2) if the motion is impending towards the right.

[**Ans.** (1) 1230.94 N; (2) 143.0 N]

5.16 The dimension of a brake drum is as shown in Fig. 5.35. Determine the torque *M* exerted on the drum if the load *P* = 50 N. Assume coefficient of kinetic friction between rope and drum to be 0.15.

[**Ans.** 747.685 N-m]

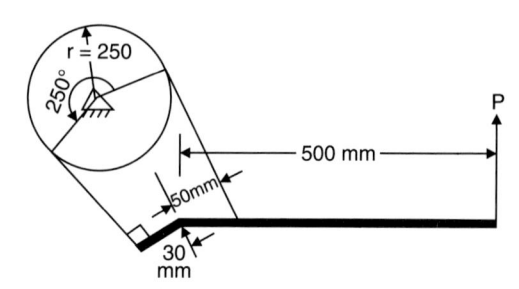

Fig. 5.35

6
Lifting Machines

A **lifting machine** is a device with the help of which heavy loads are lifted by applying small loads in a convenient direction. Pulley used to lift water from a well and screw jacks used to lift motor car are some of the common examples of lifting machines. In this chapter some of the terms connected with lifting machines are explained first followed by the description of the characteristic features of levers, systems of pulleys, wheel and axle, Weston differential pulley block, inclined plane, simple screw jack, differential screw jack and winch crab.

6.1. DEFINITIONS

The terms commonly used while dealing with lifting machines are defined below:

Load: This is the resistance to be overcome by the machine.

Effort: This is the force required to overcome the resistance to get the work done by the machine.

Mechanical Advantage: This is the ratio of load lifted to effort applied. Thus, if W is the load and P is the corresponding effort, then

$$\text{Mechanical Advantage} = \frac{W}{P} \qquad \ldots(6.1)$$

Velocity Ratio: This is the ratio of the distance moved by the effort to the distance moved by the load in the same interval of time. Thus,

$$\text{Velocity Ratio} = \frac{D}{d} \qquad \ldots(6.2)$$

where, D = distance moved by effort

d = distance moved by the load.

Input: The work done by the effort is known as input to the machine. Since work done by a force is defined as the product of the force and the distance moved in the direction of the force,

$$\text{Input} = P \times D \qquad \ldots(6.3)$$

If force P is in newton and distance D is in metre, the unit of input will be N-m. One N-m work is also known as one Joule (J). (for further details refer Ch. XIV).

Output: It is defined as useful work got out of the machine, *i.e.* the work done by the load. Thus,

$$\text{Output} = W \times d \qquad \ldots(6.4)$$

Efficiency: This is defined as the ratio of output to the input. Thus, if we use notation η for efficiency,

LIFTING MACHINES

$$\eta = \frac{\text{Output}}{\text{Input}} = \frac{W \times d}{P \times D} = \frac{W}{P} \times \frac{d}{D}$$

$$= \text{Mechanical Advantage } (MA) \times \frac{1}{\text{Velocity Ratio } (VR)}$$

$$= \frac{MA}{VR} \qquad \ldots(6.5)$$

i.e., $\qquad$ Efficiency $= \dfrac{\text{Mechanical Advantage}}{\text{Velocity Ratio}}$

Ideal Machine: A machine whose efficiency is 1 (*i.e.,* 100%) is called an ideal machine. In other words, in an ideal machine, the output is equal to the input. From eqn. (6.5), in an ideal machine,

$\qquad$ Velocity Ratio = Mechanical Advantage

Ideal Effort: Ideal effort is the effort required to lift the given load by the machine assuming the machine to be ideal.

For ideal machine,
$$VR = MA$$
If P_i is the ideal effort, then
$$VR = \frac{W}{P_i}$$
$\therefore \qquad P_i = \dfrac{W}{VR} \qquad \ldots(6.6)$

Ideal Load: Ideal load is the load that can be lifted using the given effort by the machine, assuming it to be ideal.

For the ideal machine,
$$VR = MA$$
If W_i is the ideal load, then,
$$VR = \frac{W_i}{P}$$
$\therefore \qquad W_i = VR \times P \qquad \ldots(6.7)$

6.2. PRACTICAL MACHINES

In practice, it is difficult to get an ideal machine. Friction exists between all surfaces of contacts of movable parts. Some of the work done by the effort is utilised to overcome frictional resistance. Hence, the useful work done in lifting the load is reduced, resulting in reduction of efficiency.

Let $\qquad P$ = actual effort required
$\qquad P_i$ = ideal effort required
$\qquad W$ = actual load to be lifted
$\qquad W_i$ = ideal load to be lifted

Then, $P - P_i$ is called *effort lost in friction* and $W - W_i$ is called *frictional resistance*.

Now, $$\eta = \frac{MA}{VR} = \frac{W}{P} \times \frac{1}{VR}$$

From eqn. (6.6), $$P_i = \frac{W}{VR}$$

$\therefore$ $$\eta = \frac{P_i}{P}$$

Similarly from eqn. (6.7), $W_i = VR \times P$

$\therefore$ $$\eta = \frac{W}{W_i}$$

Thus, $$\eta = \frac{P_i}{P} = \frac{W}{W_i} \qquad \ldots(6.8)$$

Example 6.1. *In a lifting machine, an effort of 500 N is to be moved by a distance of 20 m to raise a load of 10,000 N by a distance of 0.8 m. Determine the velocity ratio, mechanical advantage and efficiency of the machine. Determine also ideal effort, effort lost in friction, ideal load and frictional resistance.*

Solution. Load, $W = 10{,}000$ N

Effort $P = 500$ N

Distance moved by the effort $D = 20$ m

Distance moved by the load $d = 0.8$ m

Mechanical Advantage,

$$MA = \frac{W}{P} = \frac{10{,}000}{500} = \mathbf{20} \quad \textbf{Ans.}$$

Velocity Ratio, $VR = \dfrac{D}{d} = \dfrac{20}{0.8} = \mathbf{25}$ **Ans.**

Efficiency $= \dfrac{MA}{VR} = \dfrac{20}{25} = 0.8 = \mathbf{80\ percent}$ **Ans.**

Ideal effort, $P_i = \dfrac{W}{VR} = \dfrac{10{,}000}{25} = \mathbf{400\ N}$ **Ans.**

Effort lost in friction $= P - P_i$
$= 500 - 400 = \mathbf{100\ N}$ **Ans.**

Ideal Load, $W_i = P \times VR = 500 \times 25 = \mathbf{12{,}500\ N}$ **Ans.**

Frictional Resistance $= W_i - W$
$= 12{,}500 - 10{,}000$
$= \mathbf{2500\ N}$ **Ans.**

LIFTING MACHINES

6.3. LAW OF MACHINE

The relationship between the load lifted and the effort required in a machine is called the **law of machine.** This is found by conducting experiments in which efforts required for lifting different loads are determined and then load *versus* effort graph as shown in Fig. 6.1 is plotted. This is generally a straight line which does not pass through the origin.

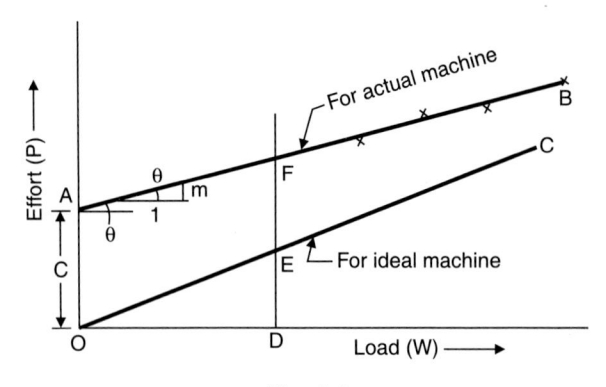

Fig. 6.1

The law of machine can be expressed mathematically in the form:
$$P = mW + C \qquad \ldots(6.9)$$
where, C is the intercept OA and $m = \tan \theta$, the slope of AB. For the ideal machine
$$MA = VR$$
$$\frac{P}{W} = VR$$

This is a straight line relationship passing through the origin and is shown by line OC in Fig. 6.1.

After plotting the law for the actual machine (AB) and the law for ideal machine (OC), it is easy to determine efficiency at any given load. The vertical line DEF corresponding to given load OD is drawn. Then,
$$DE = P_i, \text{ effort required in ideal machine}$$
$$DF = P, \text{ effort required in actual machine}$$
∴ Friction loss = $P - P_i$
$$= DF - DE = EF$$
$$\text{Efficiency, } \eta = \frac{P}{P_i} = \frac{DE}{DF}$$

6.4. VARIATION OF MECHANICAL ADVANTAGE

Mechanical Advantage (MA) is given by:
$$MA = \frac{W}{P}$$
From the law of machine,
$$P = mW + C$$

$$\therefore \quad MA = \frac{W}{mW+C} = \frac{1}{m + \frac{C}{W}} \qquad ...(6.10)$$

As the load increases, $\frac{C}{W}$ which is in denominator, decreases and hence mechanical advantage increases. In limiting case when W tends to infinity, $\frac{C}{W} = 0$ and hence *maximum mechanical advantage equals* $\frac{1}{m}$. The variation of mechanical advantage with respect to load is as shown in Fig. 6.2.

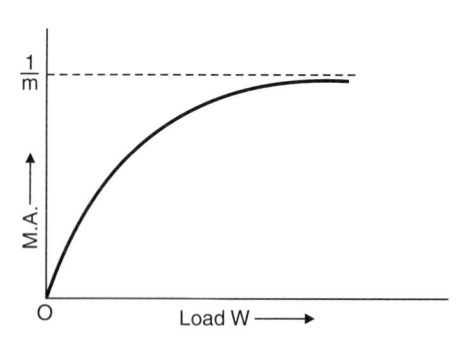

Fig. 6.2

6.5. VARIATION OF EFFICIENCY

From eqn. (6.5), the efficiency of the machine is given by $\eta = \frac{MA}{VR}$. Using the eqn. (6.10),

$$\eta = \frac{1}{VR} \times \frac{1}{m + \frac{C}{W}} \qquad ...(6.11)$$

Since the velocity ratio (*VR*) is constant for a machine, variation of efficiency with load is similar to the variation of mechanical advantage with the load.

The *maximum efficiency* is approached as the load approaches infinity ($W \rightarrow \infty$) and its value is equal to $\frac{1}{VR} \times \frac{1}{m}$. The variation of the efficiency with load is shown in Fig. 6.3.

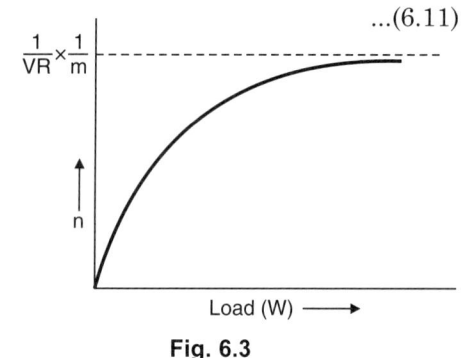

Fig. 6.3

Example 6.2. *In a simple machine, whose velocity ratio is 30, a load of 2400 N is lifted by an effort of 150 N and a load of 3000 N is lifted by an effort of 180 N. Find the law of machine and calculate the load that could be lifted by a force of 200 N. Calculate also:*

(i) The amount of effort wasted in overcoming the friction,

(ii) Mechanical advantage, and

(iii) The efficiency.

Solution. Let the law of machine be
$$P = mW + C$$
In the first case, $P = 150$ N, and $W = 2400$ N
In the second case, $P = 180$ N and $W = 3000$ N
$\therefore \qquad 150 = 2400\, m + C \qquad ...(1)$
$\qquad 180 = 3000\, m + C \qquad ...(2)$

LIFTING MACHINES

Subtracting eqn. (1) from eqn. (2), we get
$$30 = 600\, m$$
$$m = 0.05$$
Substituting this value in eqn. (1), we get
$$150 = 120\, m + C$$
$$\therefore \quad C = 30$$
Hence, **the law of machine is**
$$\mathbf{P = 0.05\, W + 30} \quad \text{Ans.} \qquad \ldots(3)$$
When a force of 200 N is applied:
From the law of machine (3),
$$200 = 0.05\, W + 30$$
$$\therefore \quad \mathbf{W = 3400\ N\ Ans.}$$
Ideal effort is given by:
$$P_i = \frac{W}{VR} = \frac{3400}{30} = 113.33\ N$$
$\therefore$ **Effort wasted in overcoming the friction**
$$= P - P_i = 200 - 113.33$$
$$= \mathbf{86.67\ N} \quad \text{Ans.}$$

Mechanical Advantage $= \dfrac{W}{P} = \dfrac{3400}{200} = \mathbf{17}$ **Ans.**

Efficiency $= \dfrac{MA}{VR} = \dfrac{17}{30} = .5667 = \mathbf{56.67\%}$ **Ans.**

Example 6.3. *In a lifting machine an effort of 150 N raised a load of 7700 N. What is the mechanical advantage? Find the velocity ratio if the efficiency at this load is 60%. If by the same machine, a load of 13,200 N is raised by an effort of 250 N, what is the efficiency? Calculate the maximum mechanical advantage and the maximum efficiency.*

Solution. Effort, $P = 150$ N
Load, $W = 7700$ N

$\therefore$ **Mechanical Advantage:**
$$MA = \frac{W}{P} = \frac{7700}{150} = \mathbf{51.33} \quad \text{Ans.}$$
If the efficiency is 60%,
$$\eta = 0.6$$
$$\eta = \frac{MA}{VR}$$
$$\therefore \quad 0.6 = \frac{51.33}{VR}$$
or
$$VR = \frac{51.33}{0.6} \quad i.e., \quad \mathbf{VR = 85.55} \quad \text{Ans.}$$

When an effort of 250 N raised a load of 13,200 N,

$$MA = \frac{W}{P} = \frac{13,200}{250} = 52.8$$

∴ $$\eta = \frac{MA}{VR} = \frac{52.8}{85.55} = 0.6172$$

i.e., $\eta = \mathbf{61.72\%}$ **Ans.**

Let the law of machine be
$$P = mW + C$$
In the first case, $150 = 7700\,m + C$ ...(1)
In the second case, $250 = 13{,}200\,m + C$ ...(2)
Subtracting eqn. (1) from eqn. (2) we get,
$$100 = 5500\,m$$
∴ $m = 0.01818$

∴ **Maximum mechanical advantage**

$$= \frac{1}{m} = \frac{1}{0.01818} = \mathbf{55}\ \textbf{Ans.}$$

Maximum efficiency $= \dfrac{1}{m} \times \dfrac{1}{VR} = \dfrac{1}{0.01818} \times \dfrac{1}{85.55}$

$= 0.6429 = \mathbf{64.29\%}$ **Ans.**

Example 6.4. *The efforts required for lifting various loads by a lifting machine are tabulated below:*

| Load lifted in N | 100 | 200 | 300 | 400 | 500 | 600 |
| Effort required in N | 16.0 | 22.5 | 28.0 | 34.0 | 40.5 | 46.5 |

*Determine the law of machine. If the velocity ratio is 25, calculate efficiency at each load and plot efficiency **versus** load curve. From this curve, determine the maximum efficiency.*

Solution. Figure 6.4 shows the graph of effort *versus* load. From this figure, $C = 10$ N

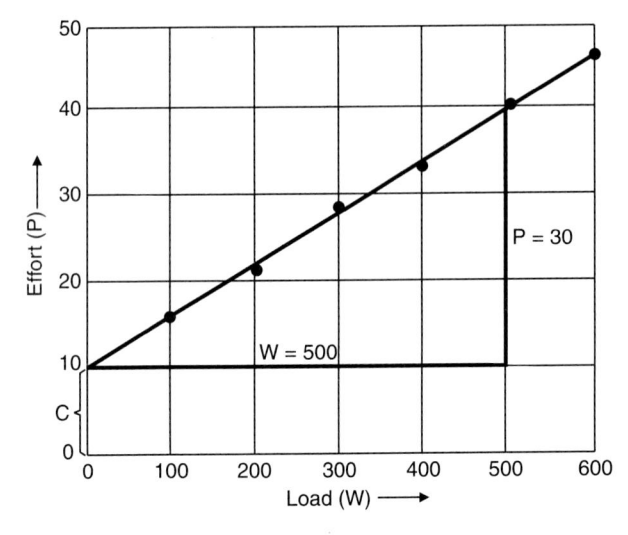

Fig. 6.4

LIFTING MACHINES

and slope $m = \dfrac{30}{500} = 0.06$

∴ **The law of machine is**
$$P = 0.06\,W + 10 \quad \text{Ans.}$$

$$\eta = \dfrac{MA}{VR} = \dfrac{W}{P} \times \dfrac{1}{VR} = \dfrac{W}{P} \times \dfrac{1}{25} = \dfrac{W}{25P}$$

Table below shows the calculation of efficiency for various loads:

Load in N	100	200	300	400	500	600
Effort in N	16.0	22.5	28.0	34.0	40.5	46.5
Efficiency in %	25	36.56	42.86	47.06	49.38	51.61

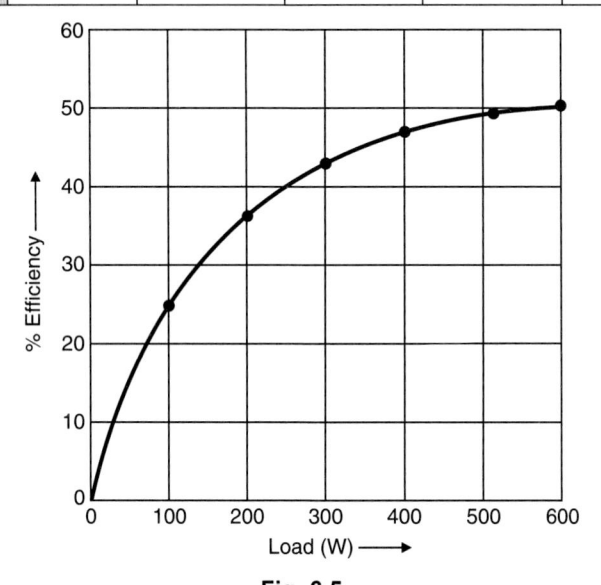

Fig. 6.5

From the graph (Fig. 6.5) maximum efficiency is seen as 50%. Actually if it is plotted for infinitely large load, **maximum efficiency** will be equal to

$$\dfrac{1}{m} \times \dfrac{1}{VR} = \dfrac{1}{0.06} \times \dfrac{1}{25} = 0.6667 = \mathbf{66.67\%} \quad \text{Ans.}$$

6.6. REVERSIBILITY OF A MACHINE

If the removal of effort while lifting results in lowering of the load, the machine is said to be **reversible**. The machine is said to be **self-locking** if the load is not lowered on removal of the effort.

For example, while lifting water from the well, the pot falls back if the effort to pull it up is removed whereas the screw jack used to lift the motor car will hold the car at the same position even if the application of the effort is stopped. Hence, the former is a reversible and later is a self-locking type simple lifting machine.

A simple lifting machine will be reversible or self-locking solely based on its efficiency. It can be shown that a lifting machine is reversible if its efficiency is greater than 50 percent and self-locking if its efficiency is less than 50 percent.

Let W – load being lifted
P – effort required
VR – Velocity Ratio
D – distance moved by the effort
d – distance moved by the load

Then,
$$\text{Input} = P \times D$$
$$\text{Output} = W \times d$$
∴ Work lost in friction $= PD - Wd$

When effort is removed, the load can start moving down if it can overcome the frictional resistance $= PD - Wd$. Hence the condition for the reversibility is:
$$Wd > (PD - Wd)$$
∴
$$2Wd > PD$$

$$\left(\frac{W}{P}\right)\left(\frac{d}{D}\right) > \frac{1}{2}$$

$$MA \times \frac{1}{VR} > \frac{1}{2}$$

i.e.,
$$\eta > \frac{1}{2} \quad \text{or} \quad 50\%.$$

Hence, a machine is reversible if its efficiency is greater than 50%.

Example 6.5. *In a lifting machine in which velocity ratio is 30, a load of 5000 N is lifted with an effort of 360 N. Determine whether it is self-locking or reversible machine. How much is the frictional resistance?*

Solution. $VR = 30$
$W = 5000$ N
$P = 360$ N

$$MA = \frac{W}{P} = \frac{5000}{360} = 13.889$$

Efficiency, $\eta = \dfrac{MA}{VR} = \dfrac{13.889}{30} = 0.4630 = 46.30\%$

Since the efficiency is less than 50%, **it is self-locking machine.   Ans.**
Ideal load, $W_i = P \times VR = 360 \times 30$
$= 10,800$ N
∴ **Frictional resistance** $= W_i - W = 10,800 - 5000 =$ **5,800 N   Ans.**

6.7. PULLEYS

A systematic arrangement of one or more pulleys may provide a simple and convenient lifting machine. In its simplest form, it consists only one pulley over which a rope or chain passes as shown in the Fig. 6.6. In this case, velocity ratio is equal to one since distance moved by effort is equal to the distance moved by the load. It just changes the direction of the applied force.

LIFTING MACHINES

Depending on the arrangement, pulleys are classified as:

1. First order pulley system
2. Second order pulley system
3. Third order pulley system.

At times, it may be difficult or may be detour to find velocity ratio directly. In such cases ideal conditions may be assumed (neglecting friction) and mechanical advantage may be found first. Then applying $VR = MA$ for ideal machine, the velocity ratio is found. This method of determining velocity ratio is used for various pulley systems considered here.

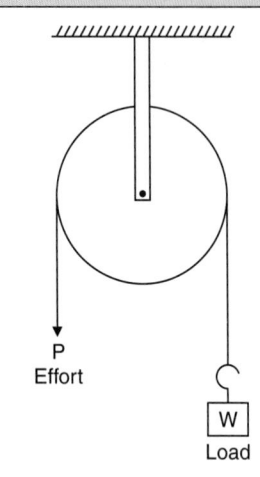

Fig. 6.6

First Order Pulley System

A first order pulley system is shown in the Fig. 6.7. Pulley No. 1 is fixed in position to a support at top. A rope passes over this pulley and one end of this rope is tied to the support at the top, making a loop, in which pulley No. 2 is suspended and effort is applied at the other end. One end of another rope is tied to pulley No. 2 and the other end to the top support which makes a loop in which pulley No. 3 is suspended. Similarly, a number of pulleys can be arranged as shown, when an effort is applied to lift the load except first pulley all other pulleys move vertically. Therefore, first pulley is termed as fixed pulley and the others as movable pulleys.

Let an effort P be applied to lift a load W. In an ideal pulley system (friction = 0), the rope which passes over pulley No. 4 is subjected to a tension $\dfrac{W}{2}$. Then tension in rope which passes over pulley No. 3 is $\dfrac{W}{4}$ and tension in the rope which passes over pulley No. 2 is $\dfrac{W}{8}$.

Hence, an effort equal to $\dfrac{W}{8}$ is required to lift a load W.

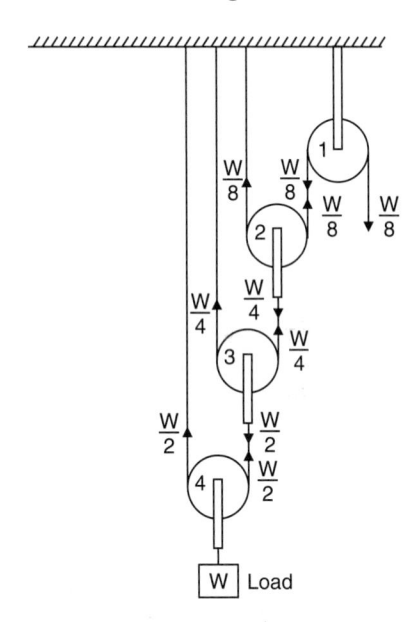

Fig. 6.7

$$\therefore \quad P = \frac{W}{8} \text{ and hence } MA = \frac{W}{P} = 8 = 2^3$$

But in an ideal machine, $VR = MA \Rightarrow VR = 2^3$

It is to be noted that in the system considered, there are three movable pulleys and the velocity ratio is 2^3. If there are only two movable pulleys, then velocity ratio would be 4 (i.e., 2^2). In general, in the first order pulley system, velocity ratio (VR) is given by 2^n, where, n is the number of movable pulleys present in the system. Thus, in first order pulley system.

$$VR = 2^n \qquad \qquad \ldots(6.12)$$

Second Order Pulley System

Figure 6.8 shows a second order pulley system. This system consists of a top pulley block and a bottom pulley block. In a pulley block pulleys may be arranged side by side or may be one below the other as shown in Fig. 6.8. The top pulley block is fixed in position to the top support whereas bottom pulley block can move vertically along with the load which is attached to it. One end of the rope is attached to the hook provided at the bottom of the top pulley block and the effort is applied at the other end. A single rope goes round all the pulleys. Let an effort P be applied to lift a load W.

Neglecting frictional losses, the tension in the rope all along the length is P. Take the section along (1)–(1) and consider the equilibrium of the bottom pulley block.

The load W is lifted using six ropes having equal tension P

$$\therefore \quad W = 6P, \quad \therefore \quad P = \frac{W}{6}, \quad \therefore \quad MA = \frac{W}{P} = 6.$$

But $VR = MA$ in ideal condition

$$\therefore \quad VR = 6.$$

In general, in the second order pulley system velocity ratio is equal to twice the number of movable pulleys in the system.

That is, $\quad VR = 2n \quad$ ...(6.13)

where, n is total number of movable pulley in the system.

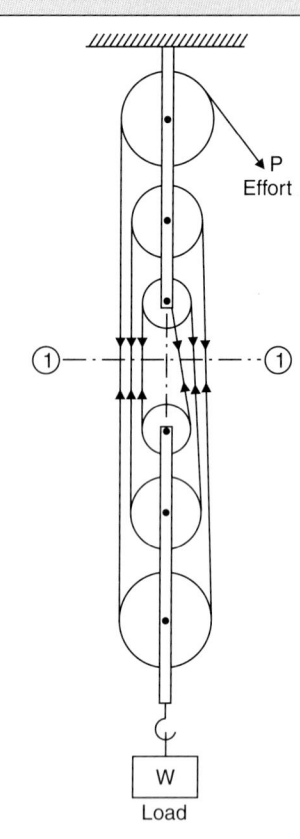

Fig. 6.8

Third Order Pulley System

The arrangement of the pulleys in the third order system is shown in the Fig. 6.9. In this system a pulley (No. 1) is fixed to the top support, over which a rope passes. One end of the rope is attached to a rigid base at the bottom. The other end is attached to a second pulley. Over this pulley another rope passes, whose one end is attached to the same rigid base and the other end to a third pulley as shown. Likewise a series of pulleys can be arranged. The load to be lifted will be attached to the rigid base.

Referring to the Fig. 6.9, let the effort required to lift a load W be P. Then neglecting friction,

Tension in the rope which passes over pulley No. 3 = P

Tension in the rope which passes over pulley No. 2 = $2P$

Tension in the rope which passes over pulley No. 1 = $4P$

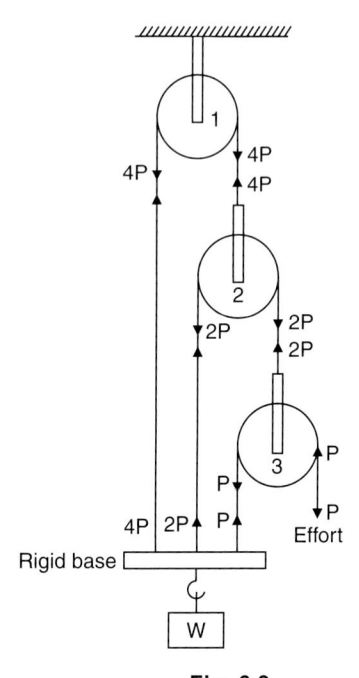

Fig. 6.9

LIFTING MACHINES

∴ A total force of $7P$ is acting on the base.
∴ Lifting force produced on the base = $7P$
Considering the equilibrium of rigid base,

$$7P = W \quad \text{Hence} \quad MA = \frac{W}{P} = 7$$

But in an ideal machine, $VR = MA$, and hence $VR = 7$.

It can be easily seen that, if there are only two pulleys, $VR = 3$ and if there is only one pulley, $VR = 1$. Therefore, in general, for the third order pulley system:

$$VR = 2^n - 1 \qquad \qquad ...(6.14)$$

where, n = number of pulleys.

6.8. WHEEL AND AXLE

This machine consists of an axle A having diameter d and a wheel B having diameter D ($D > d$) co-axially fitted as shown in Fig. 6.10. The whole assembly is mounted on ball bearing so that wheel and axle can be rotated.

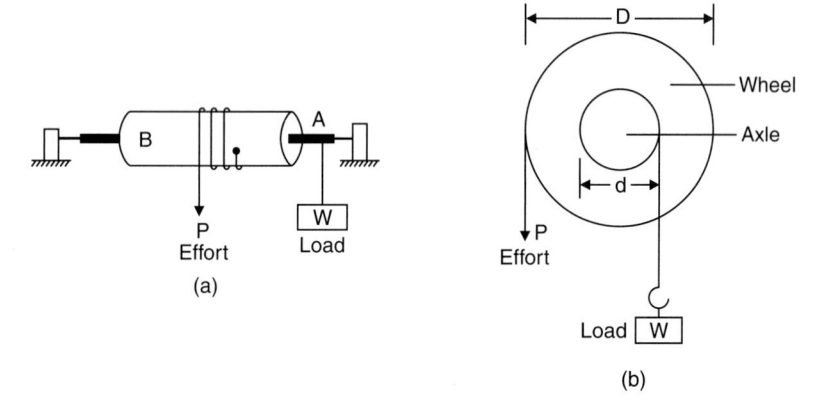

Fig. 6.10

One end of a rope is tied to the pin provided on the wheel and the rope is wound around the wheel. The other end of the rope provides the means for the application of the effort. One end of another rope is tied to the pin provided on the axle and wound around the axle in the opposite direction to that of rope wound to the wheel. To the other end of this rope the load is attached. If the whole assembly is rotated, one rope gets wound up and the other gets unwound.

Suppose the assembly is moved by one complete revolution, then the distance moved by the effort = πD and distance moved by the load = πd

∴ $$VR = \frac{\text{distance moved by effort}}{\text{distance moved by load}}$$

$$= \frac{\pi D}{\pi d}, \quad \text{or} \quad VR = \frac{D}{d} \qquad ...(6.15)$$

6.9. WHEEL AND DIFFERENTIAL AXLE

An improvement over wheel and axle machine is made by using one more wheel of bigger diameter and it is called wheel and differential axle.

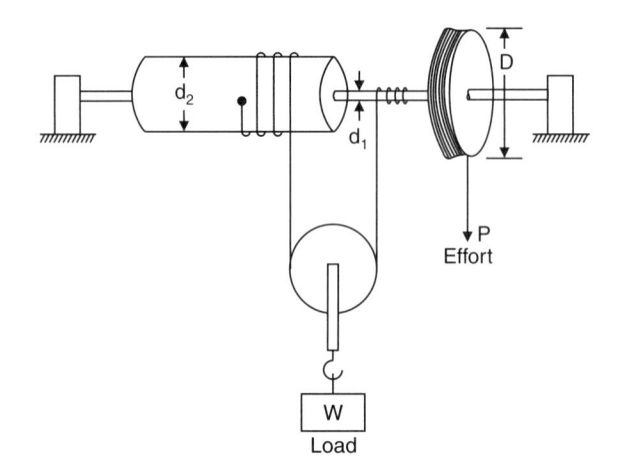

Fig. 6.11

This system consists of a differential axle of diameter d_1 and d_2 ($d_1 < d_2$) and a wheel of diameter D, fixed uniaxially as shown in Fig. 6.11. One end of the rope is tied to the pin provided on the axle portion having diameter d_1 and a part of the rope is wound around it. The other end of the rope is wound around axle on the portion having diameter d_2 in the opposite direction. This pattern of winding forms a loop and a simple pulley is installed in this loop as shown in the figure. The load is attached to this pulley. The second rope is wound to the wheel in such a direction that if it is unwound, the rope around the bigger diameter axle gets wound up and the rope around smaller diameter axle gets unwound. One end of this second rope provides means for application of the effort.

Suppose the whole system makes one complete revolution due to the applied effort, then

Total distance moved by the effort at the differential axle = πD

Length of winding of the rope = πd_2

Length of unwinding of rope = πd_1

∴ Net wound length = $\pi d_2 - \pi d_1$
$= \pi(d_2 - d_1)$

But, the rope is continuous and the load is to be lifted by the pulley block in the loop.

∴ Total height over which pulley is lifted = $\dfrac{\pi(d_2 - d_1)}{2}$

and hence the distance moved by the load = $\dfrac{\pi(d_2 - d_1)}{2}$

∴ $$VR = \dfrac{\pi D}{\dfrac{\pi(d_2 - d_1)}{2}}$$

LIFTING MACHINES

$$VR = \frac{2D}{d_2 - d_1}$$

Hence, velocity ratio in wheel and differential axle is given by:

$$VR = \frac{2D}{d_2 - d_1} \qquad \qquad ...(6.16)$$

6.10. WESTON DIFFERENTIAL PULLEY BLOCK

This is a special type of simple pulley system. It is shown in the Fig. 6.12.

This system consists of two pulley blocks, one at the top attached to the support and the other at the bottom hanging in the chain loop. The top block consists of two wheels of different diameters, but fixed co-axially. The bottom block is a simple pulley to which the load W is attached. An endless chain is wound around the pulley system as shown in the figure. All the wheels are made with teeth so as to accommodate the links of the chain. The chain is essentially used to avoid slipping.

To determine velocity ratio of the system, let us consider pulley block as an ideal machine and determine its mechanical advantage first. In ideal machine $VR = MA$. Let the diameter of the larger wheel of the top block be D and the diameter of the smaller wheel of the top block be d [Fig. 6.13(a)]. Let the effort required to lift the load W be P.

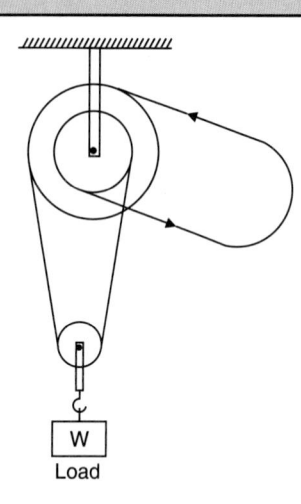

Fig. 6.12

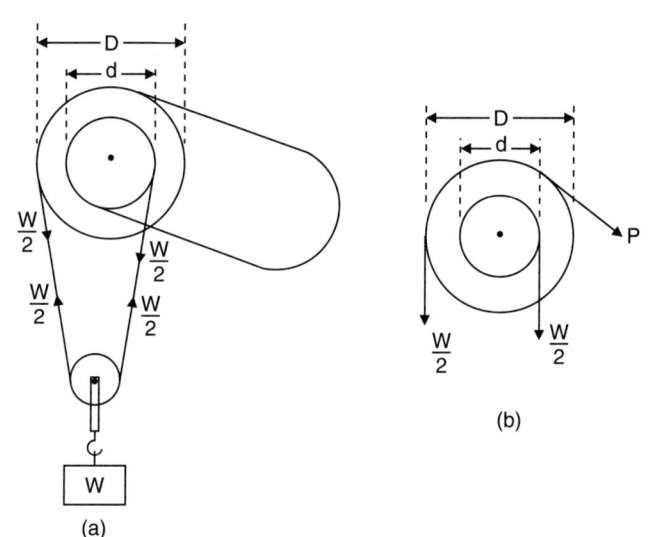

Fig. 6.13

Then the tension in the chain loop in which pulley is hanging is $\dfrac{W}{2}$.

Now, taking moment about the axis of top block [Fig. 6.13(b)],

$$\frac{W}{2} \times \frac{D}{2} = \frac{W}{2} \times \frac{d}{2} + P \times \frac{D}{2}$$

∴ $\dfrac{W}{4}(D-d) = \dfrac{PD}{2}$

$\dfrac{W}{P} = \dfrac{2D}{(D-d)}$

i.e., $MA = \dfrac{2D}{(D-d)}$

In an ideal machine,

$$VR = MA = \dfrac{2D}{(D-d)} \qquad \ldots(6.17)$$

6.11. INCLINED PLANE

Inclined plane is a very simple lifting device. The lift is essentially accomplished with the horizontal displacement.

An inclined plane consists of a plane surface at a definite angle over which the load is to be lifted.

Figure 6.14 shows a typical inclined plane. Here the load is a roller which is to be lifted to a higher elevation. One end of a rope is tied to the roller and the rope is passed over a pulley attached at the top of the inclined plane. At the other end of the rope the effort is applied.

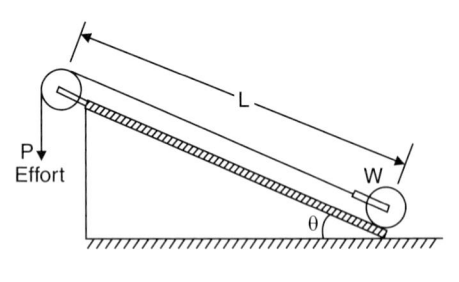

Fig. 6.14

Let the angle of inclination of the plane be θ and the length of the inclined plane be L. Then, if the roller is made to roll from bottom to top, applying an effort P, the load is lifted through a height of $L \sin \theta$. In this process the effort P moves through a distance L vertically downwards.

∴ $VR = \dfrac{\text{distance moved by the effort}}{\text{distance through which load is lifted}}$

$= \dfrac{L}{L \sin \theta} = \dfrac{1}{\sin \theta}$

∴ $VR = \dfrac{1}{\sin \theta} \qquad \ldots(6.18)$

Example 6.6. *In a first order system of pulleys there are three movable pulleys. What is the effort required to raise a load of 6000 N ? Assume efficiency of the system to be 80%.*

If the same load is to be raised using 520 N, find the number of movable pulleys that are necessary.

Assume a reduction of efficiency of 5% for each additional pulley used in the system.

Solution. $VR = 2^n$, where n is the number of movable pulleys.

$VR = 2^3 = 8$

Now, $MA = \eta \times VR = 0.8 \times 8 = 6.4$

LIFTING MACHINES

i.e.,
$$\frac{W}{P} = 6.4$$

∴
$$P = \frac{W}{6.4} = \frac{6000}{6.4}$$

i.e., **P = 937.5 N   Ans.**

In the second case,
$$\text{Effort} = 520 \text{ N}$$
Efficiency $\eta = 0.80 - n_1 \times 0.05$
where n_1 = number of additional pulleys required and equal to $(n - 3)$.
$$MA = \eta \times VR$$

i.e.,
$$\frac{W}{P} = \eta \times VR$$

∴
$$W = P \times \eta \times 2^n$$
$$= P(0.8 - n_1 \times 0.05) \times 2^n$$
$$= P[0.8 - (n - 3) \times 0.05]\, 2^n$$

By going for a trial and error solution, starting with one additional pulley i.e., totally with four pulleys,
$$W = 520\,[0.8 - (4 - 3) \times 0.05]\, 2^4 = 6240 \text{ N} > 6000 \text{ N}$$
i.e., if four pulleys are used, a load of 6240 N can be raised with the help of 520 N effort.

∴ **Number of movable pulleys required = 4   Ans.**

Example 6.7. *What force is required to raise the load W shown in Fig. 6.15 ? Assume efficiency of the system to be 85%.*

Solution. The pulley system shown in the Fig. 6.15 is a variation of the second order pulley system.
$$VR = 2 \times \text{number of movable pulleys} = 6$$
$$MA = \eta \times VR$$
$$= 0.85 \times 6 = 5.1$$

i.e.,
$$\frac{W}{P} = 5.1$$

∴
$$P = \frac{W}{5.1} = \frac{12{,}000}{5.1} = 2352.94 \quad \textbf{Ans.}$$

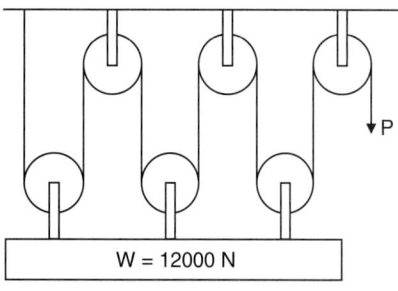

Fig. 6.15

Example 6.8. *Find the pull required to lift the load W shown in Fig. 6.16(a) assuming the efficiency of the system to be 78%.*

Solution. The pulley system shown in Fig. 6.16(a) is a combination of a first order system and a second order system as shown in Figs. 6.16(b) and (c).

Let load W be lifted by a distance x. Consider the first order system portion [Fig. 6.16(b)].

Here there are two movable pulleys. Hence
$$VR = 2^2 = 4$$
In this portion P moves by $4x$.

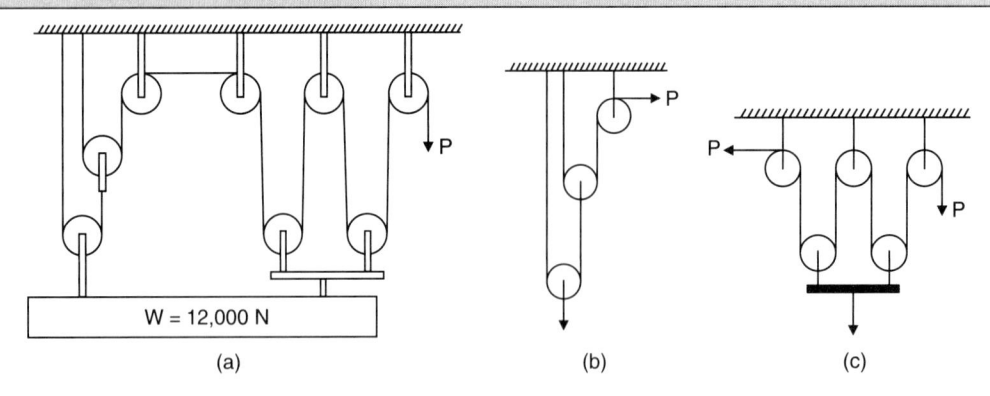

Fig. 6.16

Now, consider the second order pulley system portion [Fig. 6.16(c)]. Here there are two movable pulleys. Hence $VR = 2 \times 2 = 4$.

∴ Distance moved by the effort in this system = $4x$.

Hence, the total distance moved by the effort in the given system = $4x + 4x = 8x$

∴ $$VR = \frac{8x}{x} = 8$$

Now, $MA = \eta \times VR = 0.78 \times 8 = 6.24$

i.e., $$\frac{W}{P} = 6.24$$

∴ $$\mathbf{P} = \frac{12{,}000}{6.24} = \mathbf{1923.08\ N} \quad \textbf{Ans.}$$

Example 6.9. *A lifting machine consists of pulleys arranged in the third order system. There are three pulleys in the system. A load of 1000 N is lifted by an effort of 180 N. Find the efficiency of the machine and the effort lost in friction.*

Solution. For the third order system of pulleys,
$$VR = 2^n - 1$$
where, n is the number of pulleys in the system.

∴ $VR = 2^3 - 1 = 7$

Now, $MA = \eta \times VR$

i.e., $$\frac{W}{P} = \eta \times VR$$

∴ $$\eta = \frac{W}{P} \times \frac{1}{VR} = \frac{1000}{180} \times \frac{1}{7} = 0.7937$$

i.e., $\eta = 79.37\%$

Now,

Ideal effort, $$P_i = \frac{W}{VR} = \frac{1000}{7} = 142.86\ N$$

LIFTING MACHINES

∴ **Effort lost in friction** = $P - P_i$
= 180 − 142.86 = **37.14 N Ans.**

Example 6.10. *What force P is required to raise a load of 2500 N in the system of pulleys shown in Fig. 6.17(a). Assume efficiency of the system to be equal to 70%.*

Solution. Figure 6.17(a) can be split into two simple systems as shown in Figs. 6.17(b) and 6.17(c).

What is shown in Fig. 6.17(b) is a third order pulley system having two pulleys.

∴ $VR = 2^n - 1 = 2^2 - 1 = 3$

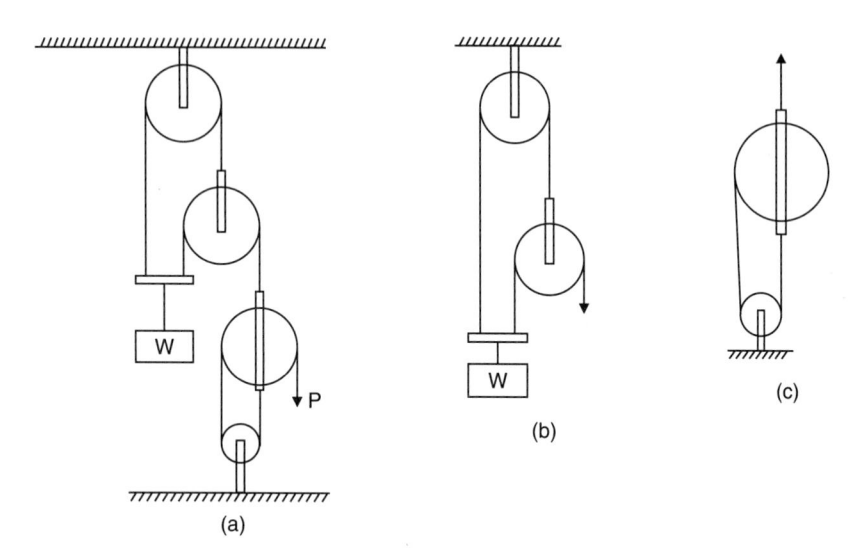

Fig. 6.17

Figure 6.17(c) is also a third order system, having two pulleys.

∴ $VR = 2^2 - 1 = 3$
∴ VR of the whole system = 3 + 3 = 6
Now, $MA = \eta \times VR$

i.e., $\dfrac{W}{P} = \eta \times VR$

∴ $P = \dfrac{W}{\eta \times VR} = \dfrac{2500}{0.7 \times 6}$

i.e., **P = 595.24 N Ans.**

Example 6.11. *In a wheel and axle, diameter of the wheel is 500 mm and that of the axle is 200 mm. The thickness of the cord on the wheel is 6 mm and that of the axle is 20 mm. Find the velocity ratio of the machine. If the efficiency when lifting a load of 1200 N with a velocity of 10 metres per minute is 70%, find the effort necessary.*

Solution. Effective wheel diameter

$$= \dfrac{6}{2} + 500 + \dfrac{6}{2} = 506 \text{ mm}$$

Effective axle diameter = $\dfrac{20}{2} + 200 + \dfrac{20}{2} = 220$ mm.

For a wheel and axle, the velocity ratio is given by $\dfrac{D}{d}$

∴ $\quad VR = \dfrac{512}{220} = 2.33$

Mechanical advantage = Efficiency × velocity ratio
$$= 0.7 \times 2.33 = 1.63$$

$$MA = \dfrac{W}{P}$$

∴ $\quad P = \dfrac{1200}{1.63}$

i.e., $\quad$ **P = 736.2 N Ans.**

Example 6.12. *A load of 20 kN is to be lifted by a differential wheel and axle. It consists of differential axle of 250 mm and 300 mm diameter and the wheel diameter is 800 mm. Find the effort required if the efficiency of the machine is 55%.*

Solution. Differential axle diameters,
$$d_1 = 300 \text{ mm and}$$
$$d_2 = 250 \text{ mm}$$

Wheel diameter, $D = 800$ mm

Load, $\quad W = 20$ kN

Efficiency, $\eta = 55\%$

Velocity ratio, $VR = \dfrac{2D}{d_2 - d_1} = \dfrac{2 \times 800}{300 - 250} = 32$

Mechanical advantage
$$MA = \text{Efficiency} \times \text{velocity ratio}$$
$$= 0.55 \times 32 = 17.6$$

$$MA = \dfrac{W}{P}$$

∴ $\quad P = \dfrac{20{,}000}{17.6} \quad$ i.e., $\quad$ **P = 1136.4 N Ans.**

Example 6.13. *A Weston differential pulley block of diameter 500 mm and 200 mm is used to lift a load of 5000 N. Find the effort required if the efficiency is 60%.*

Solution. Diameter of pulley block $D = 500$ mm, and
$$d = 200 \text{ mm}$$

Load, $W = 5000$ N

Efficiency, $\eta = 60\%$

LIFTING MACHINES 163

$$\text{Velocity ratio} = \frac{2D}{D-d} = \frac{2 \times 500}{500 - 200} = 3.33$$

Mechanical advantage = Efficiency × Velocity ratio
$$= 0.6 \times 3.33 = 2$$

Effort required, $P = \dfrac{W}{MA} = \dfrac{5000}{2}$

i.e., **P = 2500 N   Ans.**

6.12. SCREW JACK

This is a device commonly used to lift heavy loads. Screw jack works on the principle same as that of inclined plane. A typical section of the screw jack is shown in the Fig. 6.18.

The device consists of a nut and a screw. Monolithically cast nut and stand form the body of the jack. The load is carried by the screw head fitted onto the screw as shown in the figure. The body (consisting of nut) is fixed and the screw is rotated by means of a lever.

The axial distance moved by the nut (or by the screw, relative to each other) when it makes one complete revolution is known as **lead of the screw head.** The distance between consecutive threads is called *pitch* (of a screw thread). If the screw is single threaded, then lead of the screw is equal to the pitch. If the screw is double threaded then lead of the screw is twice the pitch.

Let R be the length of the lever and d be the mean diameter of the screw.

Let a load W be lifted using an effort P.

If an effort P is applied at the lever end, it is equivalent to an effort P_1 at the screw [Fig. 6.19(a)] and P_1 is given by the condition:

$$P \times R = P_1 \times \frac{d}{2}$$

∴ $$P_1 = \frac{2PR}{d}$$

Fig. 6.18

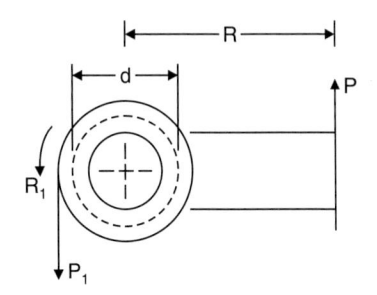

Fig. 6.19(a)

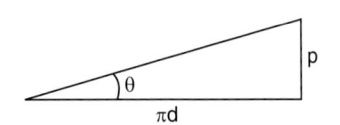

Fig. 6.19(b)

Now, consider one complete revolution of the lever. The load W is lifted up by a distance p equal to the lead of the screw.

This can be compared with that of inclined plane having

inclination $= \tan^{-1} \dfrac{p}{\pi d}$

where, p – lead of the screw
d – mean diameter of screw.

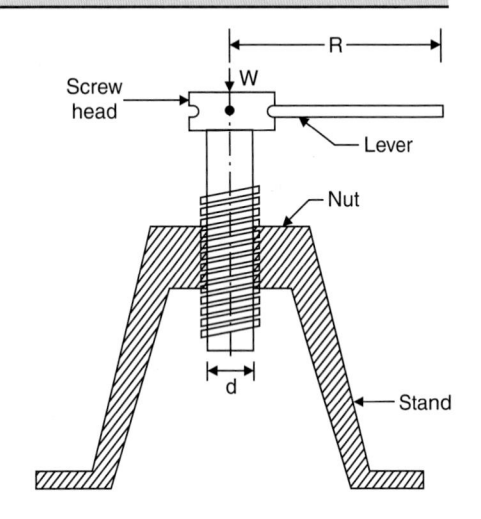

Applying an effort P at the end of the lever is as good as applying an effort P_1 (at the screw) on this inclined plane. [Fig. 6.19(c)].

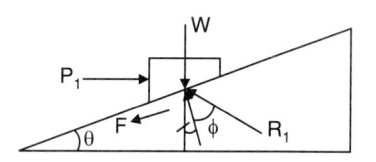

Resolving horizontally *i.e.*, parallel to P_1

$$P_1 = R_1 \sin(\theta + \phi), \qquad \ldots(6.18)$$

where R_1 is resultant reaction and
ϕ is limiting angle of friction. Resolving vertically

$$W = R_1 \cos(\theta + \phi) \qquad \ldots(6.18(a))$$

Fig. 6.19(c)

Dividing eqn. (6.18(a)) by eqn. (6.18(b))

$$\frac{P_1}{W} = W \tan(\theta + \phi)$$

$\therefore \qquad P_1 = W \tan(\theta + \phi)$

But, $\qquad P_1 = \dfrac{2PR}{d}$

$$\frac{2PR}{d} = W \tan(\theta + \phi) \qquad \ldots(6.18(b))$$

$\therefore \qquad P = \dfrac{d}{2R} W \tan(\theta + \phi) \qquad \ldots(6.19)$

We have $\tan \phi = \mu$
where μ is the coefficient of friction.

Then,

$$P = \frac{d}{2R} W \frac{\tan \theta + \tan \phi}{1 - \tan \theta \tan \phi}$$

$$P = \frac{d}{2R} W \frac{\mu + \tan \theta}{1 - \mu \tan \theta} \qquad \ldots(6.19(a))$$

where, $\tan \theta = \dfrac{p}{\pi d}$

If the load is descending, then the friction will be acting in the reverse direction so that the resultant reaction R shifts as shown in Fig. 6.19(d).

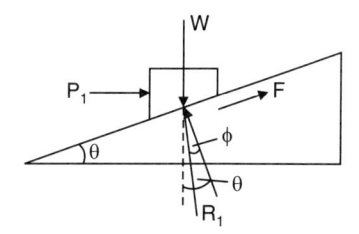

Then eqn. (6.19) changes to

$$P = \frac{d}{2R} W \tan(\theta - \phi) \qquad \ldots(6.19(b))$$

Torque required, $T = PR$

$$= \frac{d}{2} W \tan(\theta + \phi)$$

Fig. 6.19(d)

LIFTING MACHINES

Hence torque required while ascending

$$T = \frac{d}{2} W \tan(\theta + \phi) \qquad \qquad ...(6.20)$$

and torque required while descending

$$T = \frac{d}{2} W \tan(\theta - \phi) \qquad \qquad ...(6.20(a))$$

Now, $\qquad VR = \dfrac{\text{Distance moved by the effort}}{\text{Distance moved by the load}} = \dfrac{2\pi R}{p} \qquad ...(6.21)$

Example 6.14. *A screw jack raises a load of 40 kN. The screw is square threaded having three threads per 20 mm length and 40 mm in diameter. Calculate the force required at the end of a lever 400 mm long measured from the axis of the screw, if the coefficient of friction between screw and nut is 0.12.*

Solution. Screw diameter, $d = 40$ mm

Lead of the screw, $\quad p = \dfrac{20}{3} = 6.667$ mm

Load, $\qquad\qquad\qquad W = 40$ kN
Lever length, $\qquad R = 400$ mm
$\qquad\qquad\qquad\qquad \mu = 0.12$

We have $\qquad P = \dfrac{d}{2R} W \dfrac{\mu + \tan\theta}{1 - \mu \tan\theta}$

and $\qquad\qquad \tan\theta = \dfrac{p}{\pi d} = \dfrac{6.667}{\pi \times 40} = 0.05305$

$\therefore \qquad P = \dfrac{40}{2 \times 400} \times 40{,}000 \left[\dfrac{0.12 + 0.05305}{1 - (0.12 \times 0.05305)} \right]$

i.e., $\qquad\qquad$ **P = 348.32 N Ans.**

Example 6.15. *A screw jack has square threads 50 mm mean diameter and 10 mm pitch. The load on the jack revolves with the screw. The coefficient of friction at the screw thread is 0.05.*

(i) Find the tangential force required at the end of 300 mm lever to lift a load of 6000 N.

(ii) State whether the jack is self-locking. If not, find the torque which must be applied to keep the load from descending.

Solution. $\qquad \tan\theta = \dfrac{p}{\pi d} = \dfrac{10}{\pi \times 50} = 0.0637$

$\therefore \qquad\qquad \theta = 3.6426°$
$\qquad\qquad \tan\phi = 0.05$
$\therefore \qquad\qquad \phi = 2.8624°$

(i) $P = \dfrac{d}{2R} \times W \tan(\theta + \phi)$

$$= \frac{50}{2 \times 300} \times 6000 \tan (3.6426° + 2.8624°)$$

$$P = 57.01 \text{ N} \quad \textbf{Ans.}$$

(ii) We have

$$VR = \frac{2\pi R}{p} = \frac{2\pi \times 300}{10} = 188.496$$

$$MA = \frac{W}{P} = \frac{6000}{57.01} = 105.245$$

$$\text{Efficiency} = \frac{MA}{VA} = \frac{105.245}{188.496} = 0.5583$$

i.e., Efficiency = 55.83% > 50.

Hence the screw jack is not self-locking.

∴ The torque required to keep the load from descending

$$= \frac{d}{2} W \tan(\theta - \phi) = \frac{50}{2} \times 600 \times \tan(3.6426 - 2.8624)$$

$$T = 204.3 \text{ N-mm} \quad \textbf{Ans.}$$

6.13. DIFFERENTIAL SCREW JACK

Differential screw jack is an improvement over simple screw jack. A typical differential screw jack is shown in Fig. 6.20. It consists of two threaded elements A and B. Both A and B have threads in the same direction (right-handed). The element A is a cylinder which has threads on both its outer and inner surfaces. The threads on the outer surface of the element A fits into the nut C which also functions as the base of the whole mechanism. The threads on the element B fit into the threads cut on the inner surface of A. Thus, the element A acts as a screw for the nut C and also as a nut for the element B. With the help of a lever inserted in the holes made on the top of the block D, which is attached to the element B, block D can be rotated. When D is rotated, A rotates with it. Rotation of B is prevented by suitable arrangement.

Let D and d be the mean diameters of the screws A and B, respectively.

Let p_A and p_B be the pitch of the screws A and B, respectively and p_A be greater than p_B.

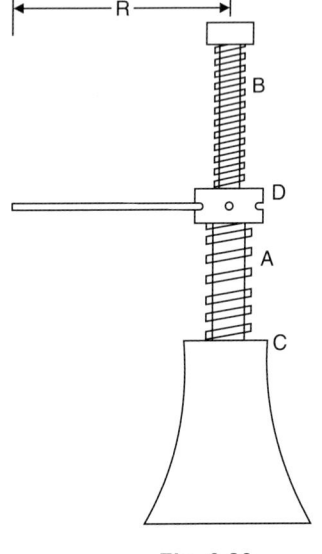

Fig. 6.20

If the lever is rotated through one complete revolution, the height through which the element A moves up = p_A

In the mean time, the element B moves down with respect to C.

The distance through which B comes down = p_B

LIFTING MACHINES

∴ Net height through which load is lifted $= (p_A - p_B)$

Let R be the radial distance (from the centre line of A and B) at which an effort P is applied

Now, $\quad VR = \dfrac{\text{Distance moved by the effort}}{\text{Distance moved by the load}} = \dfrac{2\pi R}{p_A - p_B}$ ...(6.22)

It can be seen from eqn. (6.22) that the velocity ratio in the differential screw jack is increased as compared to that of simple screw jack (eqn. 6.21).

Example 6.16. *The following are the specifications for a differential screw jack:*

(i) *Pitch of smaller screw, 5.0 mm*

(ii) *Pitch of larger screw, 10.0 mm*

(iii) *Lever arm length from centre of screw = 500 mm.*

The screw jack raises a load of 15 kN with an effort of 185 N. Determine the efficiency of the differential screw jack at this load.

If the above jack can raise a load of 40 kN with an effort of 585 N, determine the law of machine.

Solution. Now, $\quad p_A = 10.0$ mm

$\quad p_B = 5.0$ mm

Lever arm length, $\quad R = 500$ mm

$$VR = \dfrac{2\pi R}{p_A - p_B} = \dfrac{2\pi \times 500}{10 - 5.0} = 628.32$$

$$MA = \dfrac{W}{P} = \dfrac{15{,}000}{185} = 81.08 \quad \therefore \quad \eta = \dfrac{MA}{VA} = \dfrac{81.08}{628.32}$$

$$= 0.129 = \textbf{12.9 percent.} \quad \textbf{Ans.}$$

To find law of machine:

Let law of machine be $P = mW + C$

From first case: $\quad 185 = m \times 15{,}000 + C$ ...(1)

From second case: $\quad 585 = m \times 50{,}000 + C$ ...(2)

Subtracting eqn. (1) from eqn. (2), $\quad 400 = 35{,}000\ m.$

or $\quad m = \dfrac{4}{350}$

Substituting this value in eqn. (1) we get,

$\quad 185 = 171.43 + C$

∴ $\quad C = 13.57$ N

∴ **Law of machine is** $P = \dfrac{4}{350}W + 13.57$ **Ans.**

6.14. WINCH CRABS

Winch crabs are lifting machines in which velocity ratio is increased by a gear system. If only one set of gears is used, the winch crab is called a **single purchase winch crab** and if two sets are used it is called **double purchase winch crab.**

Single Purchase Winch Crab

Line diagram of a single purchase winch crab is shown in Fig. 6.21. It consists of a load drum of radius r connected to an axle by gears. The toothed wheel on load drum is called **spur wheel** and the toothed wheel on axle is called **pinion.** Pinion is always smaller in size and it contains less number of teeth as compared to that on the spur wheel. The axle is provided with a handle of arm length R. Let the number of teeth on pinion and spur wheel be T_1 and T_2, respectively. Let the effort be applied at the end of the handle. When one revolution is made, the distance moved by the effort is given by:

$$D = 2\pi R$$

When axle makes one revolution, due to gear arrangement load drum moves by T_1 number of teeth, which means that it makes a revolution of $\dfrac{T_1}{T_2}$.

∴ The distance over which the load moves:

$$d = 2\pi r \times \frac{T_1}{T_2}$$

∴ Velocity ratio,

$$VR = \frac{D}{d} = \frac{2\pi R}{2\pi r \times \dfrac{T_1}{T_2}}$$

i.e.,
$$VR = \frac{R}{r} \times \frac{T_2}{T_1} \qquad \ldots(6.23)$$

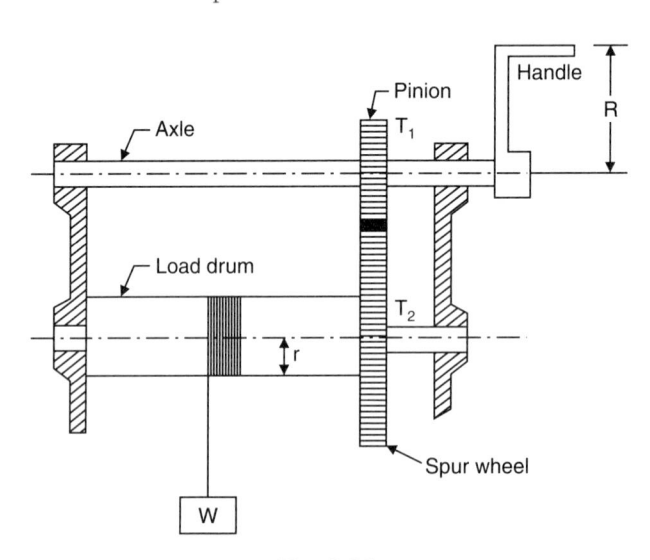

Fig. 6.21

Double Purchase Winch Crab

Velocity ratio of a winch crab can be increased by providing another axle with a pair of pinion and gear as shown in Fig. 6.22. Since two pairs of pinion and gear are used it is called a double purchase winch crab. This is used for lifting heavier loads.

LIFTING MACHINES

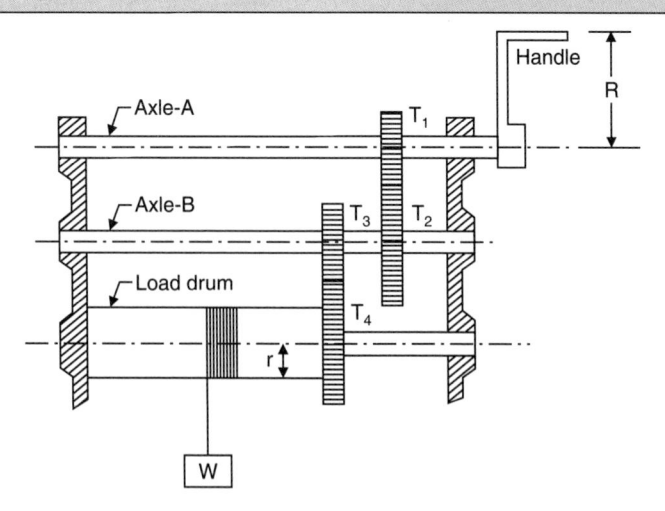

Fig. 6.22

Let the number of teeth on various wheels be T_1, T_2, T_3 and T_4 as shown in Fig. 6.22. Let the handle makes one revolution.

Distance moved by effort P is given by:
$$D = 2\pi R \qquad \ldots(6.24)$$

When axle A makes one revolution, axle B is moved by T_1 teeth, *i.e.*, it makes $\dfrac{T_1}{T_2}$ revolutions. The number of teeth by which spur wheel is moved is $\dfrac{T_1}{T_2} \times T_3$ and hence load drum makes $\left(\dfrac{T_1}{T_2}\right) \times \left(\dfrac{T_3}{T_4}\right)$ revolutions.

∴ The distance moved by the load
$$d = 2\pi r \times \left(\dfrac{T_1}{T_2}\right) \times \left(\dfrac{T_3}{T_4}\right)$$

$$VR = \dfrac{D}{d} = \dfrac{2\pi R}{2\pi r \times \left(\dfrac{T_1}{T_2}\right) \times \left(\dfrac{T_3}{T_4}\right)}$$

i.e.,
$$VR = \dfrac{R}{r} \times \left(\dfrac{T_2}{T_1}\right) \times \left(\dfrac{T_4}{T_3}\right) \qquad \ldots(6.25)$$

Example 6.17. *Following are the specifications of a single purchase crab:*
Diameter of the load drum = 200 mm
Length of lever arm R = 1.2 m
Number of teeth on pinion, T_1 = 10
Number of teeth on spur wheel, T_2 = 100.

Find the velocity ratio of this machine. On this machine efforts of 100 N and 160 N are required to lift the load of 3 kN and 9 kN, respectively. Find the law of the machine and the efficiencies at the above loads.

Solution. Radius of the load drum, $r = \dfrac{200}{2} = 100$ mm

Length of lever arm, $R = 1.2$ m $= 1200$ mm

Velocity ratio of the single purchase crab is given by:

$$VR = \dfrac{R}{r} \times \dfrac{T_2}{T_1} = \dfrac{1200}{100} \times \dfrac{100}{10}$$

VR = 120   Ans.

Let the law of machine be $P = mW + C$

In first case: $P = 100$ N; $W = 3$ kN $= 3000$ N

∴ $100 = m \times 3000 + C$ ...(1)

In the second case: $P = 160$ N; and $W = 9$ kN $= 9000$ N

∴ $160 = m \times 9000 + C$ ...(2)

Subtracting eqn. (1) from eqn. (2), we get

$60 = 6000\, m$

∴ $m = \dfrac{1}{100} = 0.01$

Substituting this value of m in eqn. (1), we get

$$100 = \dfrac{1}{100} \times 3000 + C$$

∴ $C = 70$

Hence, the law of machine is

P = 0.01 W + 70   Ans.

Efficiencies:

In the first case,

$$MA = \dfrac{W}{P} = \dfrac{3000}{100} = 30$$

∴ $\eta = \dfrac{MA}{VR} = \dfrac{30}{120} = 0.25 = \mathbf{25\%}$ **Ans.**

In the second case,

$$MA = \dfrac{W}{P} = \dfrac{9000}{160} = 56.25$$

∴ $\eta = \dfrac{MA}{VR} = \dfrac{56.25}{120}$

i.e., $\eta = 0.4688 = \mathbf{46.88\%}$ **Ans.**

LIFTING MACHINES

Example 6.18. *In a double purchase crab, the pinions have 15 and 20 teeth, while the spur wheels have 45 and 40 teeth. The effort handle is 400 mm long while the effective diameter of the drum is 150 mm. If the efficiency of the winch is 40%, what load will be lifted by an effort of 250 N applied at the end of the handle?*

Solution. $T_1 = 15$; $T_2 = 45$; $T_3 = 20$; $T_4 = 40$
Length of handle, $R = 400$ mm

Radius of the load drum, $r = \dfrac{150}{2} = 75$ mm

$\therefore \qquad VR = \dfrac{R}{r} \times \dfrac{T_2}{T_1} \times \dfrac{T_4}{T_3}$

$\qquad\qquad = \dfrac{400}{75} \times \dfrac{45}{15} \times \dfrac{40}{20} = 32$

Now, $\qquad \eta = \dfrac{MA}{VR}$

$\therefore \qquad 0.40 = \dfrac{MA}{32}$

or $\qquad MA = 12.8$

i.e., $\qquad \dfrac{W}{P} = 12.8$, but $P = 250$ N

$\therefore \qquad W = 12.8 \times 250$

i.e., $\qquad W = 3200$ N

Applied effort lifts a load of 3200 N   Ans.

Important Definitions

1. For load, effort, mechanical advantages, velocity ratio, input, output, efficiency, ideal machine, ideal effort, ideal load—refer page No. 136, 137.
2. **Law of Machine:** The relationship between the load lifted and the effort required in a machine is called law of machine.
3. If the removal of effort while lifting results in lowering of the load, the **machine is** said to be **reversible.** The machine is said to be **self locking** if the load is not lowered on removal of the effort.

Important Formulae

1. Mechanical advantage $= \dfrac{W}{P}$
2. Velocity ratio $= \dfrac{D}{d}$
3. Input $= P \times D$
4. Output $= W \times d$.
5. Efficiency $= \dfrac{MA}{VR} = \dfrac{P_i}{P} = \dfrac{W}{W_i}$

6. In ideal machine, $VR = MA$
7. Ideal effort $= \dfrac{W}{VR}$
8. Ideal load $= W_i = VR \times P$
9. Law of machine: $P = mW + C$
10. Maximum mechanical advantage $= \dfrac{1}{m}$
11. Maximum efficiency $= \dfrac{1}{VR} \times \dfrac{1}{m}$
12. A machine is reversible, if $\eta > 50\%$
13.

Machine	Velocity ratio
(a) First order pulley system	2^n
(b) Second order pulley system	2^n
(c) Third order pulley system	$2^n - 1$
(d) Wheel and axle	$\dfrac{D}{d}$
(e) Wheel and differential axle	$\dfrac{2D}{d_2 - d_1}$
(f) Weston differential pulley block	$\dfrac{2D}{D - d}$
(g) Inclined plane	$\dfrac{1}{\sin \theta}$
(h) Screw jack	$\dfrac{2\pi R}{p}$
(i) Differential screw jack	$\dfrac{2\pi R}{p_A - p_B}$
(j) Winch crabs–single purchase	$\dfrac{R}{r} \times \dfrac{T_2}{T_1}$
–double purchase	$\dfrac{R}{r} \times \dfrac{T_2}{T_1} \times \dfrac{T_4}{T_3}$

PROBLEMS FOR EXERCISE

6.1 In a lifting machine an effort of 1.5 kN is to be moved through a distance of 7.2 m to move a load of 24 kN through a distance of 300 mm. Determine: (1) mechanical advantage (2) velocity ratio, (3) efficiency, (4) ideal effort, (5) effort lost in friction, (6) ideal load, and (7) frictional resistance.
[**Ans.** (1) 16; (2) 24; (3) 66.67%; (4) 1.0 kN; (5) 0.5 kN; (6) 36 kN and (7) 12 kN]

LIFTING MACHINES

6.2 In a lifting machine an effort of 400 N is required to raise a load of 3000 N and an effort of 640 N, to raise a load of 5000 N. How much load can be lifted with an effort of 760 N? If the velocity ratio is 16, determine the efficiency of the machine when an effort of 760 N is applied. Is it a reversible machine?
[**Ans.** $W = 6000$ N; $\eta = 49.34$; It is not reversible]

6.3 The following observations were made in an experiment on a lifting machine:

Load in N	500	1000	1500	2000	2500	3000
Effort in N	26	47	76	95	105	127

Draw the load *versus* effort graph and determine the law of machine. If the velocity ratio is 30 determine the efficiency while lifting a load of 1800 N.

What is the maximum efficiency of the machine?
[**Ans.** $P = 0.04W + 6$; $\eta = 51.28\%$; maximum $\eta = 83.33\%$]

6.4 A lifting machine having velocity ratio 28 starts raising a load of 6420 N on applying an effort of 450 N to it. If suddenly the effort is removed find whether the load comes down or not?
[**Ans.** $\eta = 50.95\%$, hence the load comes down]

6.5 In the first order pulley system having three movable pulleys, how much effort is required to raise a load of 5780 N if the same system raises a load of 1200 N with an effort of 200 N? Assume the efficiency to be constant for the pulley system.
[**Ans.** $P = 963.33$ N]

6.6 For the arrangement of pulleys shown in Fig. 6.23, find the effort required to raise the given load 7280 N. Assume efficiency of the system as 75%.
[**Ans.** $P = 2436.7$ N]

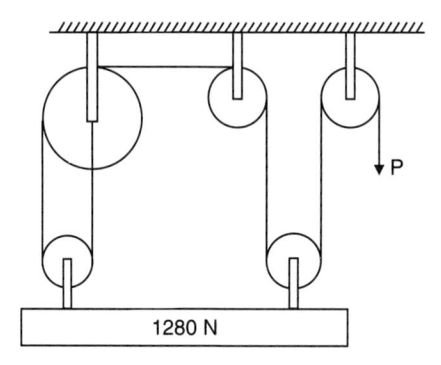

Fig. 6.23

6.7 For the combination of first order and second order pulley system shown in Fig. 6.24 what will be velocity ratio?

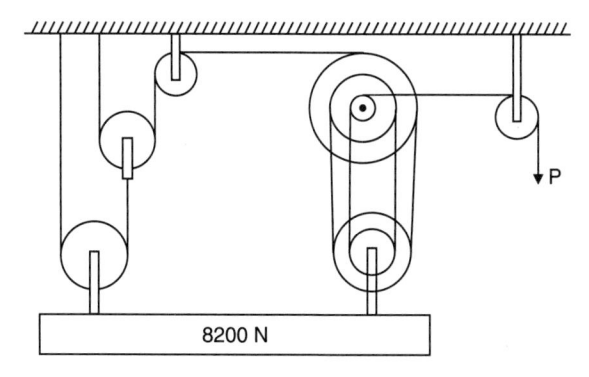

Fig. 6.24

Assuming efficiency to be 80%, calculate what effort is required to raise the load of 8200 N.
[**Ans.** $VR = 8$; $P = 1281.25$ N]

6.8 For a third order pulley system having six movable pulleys, an effort of 720 N is required to raise a load of 30 kN. Calculate the efficiency of the system. [**Ans.** 66.14%]

6.9 For a wheel and axle, the following details are available :
Diameter of the wheel = 540 mm
Diameter of the axle = 270 mm
Thickness of the cord on the wheel = 6 mm
Thickness of the cord on the axle = 20 mm
Calculate the efficiency of the device if an effort of 725 N is required to lift a load of 1000 N. [**Ans.** η = 73.26%]

6.10 A differential axle and wheel consists of a differential axle of 240 mm and 320 mm diameter; and a wheel of diameter 750 mm. Assuming efficiency to be 57%, find the effort required to raise a load of 24 kN. [**Ans.** P = 2.2456 kN]

6.11 A Weston differential pulley block of diameters 400 mm and 800 mm is used to lift a load of 40,000 N. Find the effort required if efficiency of the system is 60%. [**Ans.** 26,667 N]

6.12 The following particulars refer to a screw jack:
(1) Diameter of the screw rod = 62.5 mm
(2) Length of the handle = 250 mm
(3) Pitch of the square threads = 12.5 mm
(4) Coefficient of friction = 0.05
Find the effort required to lift a load of 5000 N. Find also the efficiency of the jack. [**Ans.** P = 71.27 N; η = 55.83%]

6.13 A screw jack carries a load of 10 kN. It has a square threaded screw of pitch 25 mm and mean diameter 60 mm. The coefficient of friction between screw and nut is 0.20. Calculate the torque required to raise the load and the efficiency of the screw. Find also the force required at the end of the handle 500 mm long to lower the load. [**Ans.** T = 102.58 kN-mm; P = 0.2052 kN]

6.14 The following are the specifications of a single purchase crab:
Diameter of the load drum = 220 mm
Length of lever arm, R = 1.0 m
Number of teeth on pinion, T_1 = 10
Number of teeth on spur wheel, T_2 = 120
Determine the velocity ratio of the machine. On this machine, it is found in one trial that an effort of 120 N is required to lift a load 4.2 kN and in another trial an effort of 150 N is required to lift a load of 11.4 kN. Establish the law of machine. Also calculate efficiencies corresponding to 120 N and 150 N efforts.

$$\left(\textbf{Ans. } VR = 109.091; P = \frac{1}{240}W + 102.5; \eta_1 = 32.08\%; \eta_2 = 69.67\%\right)$$

6.15 In a double purchase crab, the pinions have 12 and 18 teeth while the spur wheels have 24 and 36 teeth. The effective diametre of the drum is 142 mm. The effort handle is 450 mm. What load can be lifted by an effort of 220 N applied at the end of handle if efficiency of winch is 42%. [**Ans.** P = 2342.54 N].

7
Transmission of Power

The motion and power from an engine is usually transferred to machine tools by the following means:

    (i) Using belt,

    (ii) Using rope, and

    (iii) Using toothed gear.

In this chapter, each of the above methods of transferring power is discussed.

7.1. BELT DRIVE

Flat leather belts are used for transmission of power. The belt is placed on driver and follower wheels in a state of tension. The belt on the wheel of engine shaft rotates. Due to friction belt also moves with it and it rotates follower wheel. The belt drive may be open type or crossed type as shown in Fig. 7.1 (a) and (b). With open belt drive, if the driver is rotating clockwise, the follower also rotates in clockwise direction, whereas in crossed type follower wheel rotates in counterclockwise direction. Length of the belt required to connect a driver wheel of radius 'r_1' and follower wheel of radius 'r_2' placed at distance 'd' may be calculated as shown below:

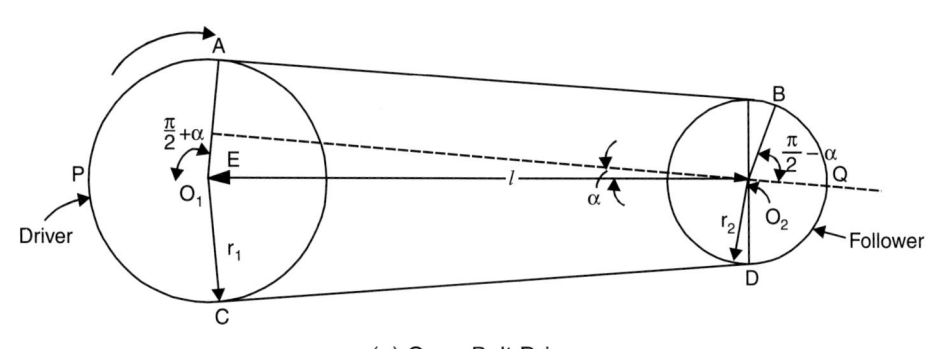

(a) Open Belt Drive

Fig. 7.1 (Contd.)

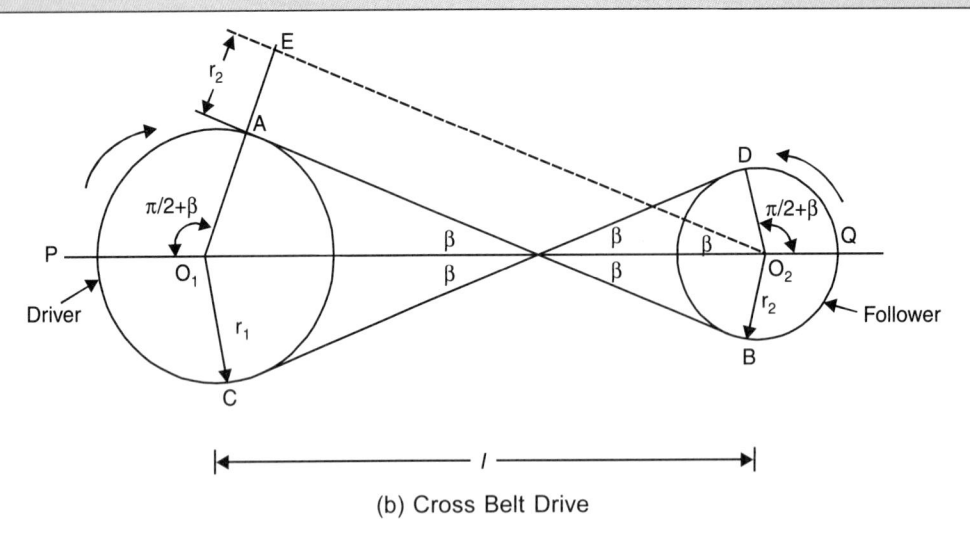

Fig. 7.1. Belt drive

(b) Cross Belt Drive

Length of Belt in Open Drive

Let O_2E be parallel to AB.

Referring to Fig. 7.1 (a), let O_2E be parallel to AB and $\angle O_1 \cdot O_2 E = \alpha$.

$$\therefore \quad \sin \alpha = \frac{O_1 E}{O_1 O_2} = \frac{r_1 - r_2}{l}$$

∴ Length of the belt

$$L = 2 \,(\text{arc } PA + AB + \text{arc } BQ)$$

$$= 2\left[r_1 \left(\frac{\pi}{2} + \alpha\right) + EO_2 + r_2 \left(\frac{\pi}{2} - \alpha\right) \right]$$

$$= \pi (r_1 + r_2) + 2\alpha (r_1 - r_2) + 2EO_2$$

Now, $\quad EO_2 = \sqrt{l^2 - (r_1 - r_2)^2} = l\left[1 - \left(\frac{r_1 - r_2}{l}\right)^2\right]^{1/2}$

Since, $\frac{r_1 - r_2}{l}$ is a small quantity,

$$EO_2 = l\left[1 - \frac{1}{2}\left(\frac{r_1 - r_2}{l}\right)^2\right]$$

$$= l - \frac{1}{2}\frac{(r_1 - r_2)^2}{l}$$

Since, α is small, $\sin \alpha = \alpha = \dfrac{r_1 - r_2}{l}$

$$\therefore \quad L = \pi (r_1 + r_2) + 2\frac{(r_1 - r_2)^2}{l} + 2l - l\frac{(r_1 - r_2)^2}{l^2}$$

TRANSMISSION OF POWER

$$= \pi(r_1 + r_2) + 2l + \frac{(r_1 - r_2)^2}{l} \qquad \ldots (7.1)$$

Length of Belt in Cross Drive

Let EO_2 be parallel to AB, as shown in Fig. 7.1 (b). Angle made by AB with O_1O_2 be β.

Then, $\qquad \angle PO_1A = \angle QO_2B = \dfrac{\pi}{2} + \beta$

and $\qquad \sin \beta = \dfrac{O_1E}{O_1O_2} = \dfrac{r_1 + r_2}{l}$

Since, β is small, $\sin \beta = \beta = \dfrac{r_1 + r_2}{l}$

$$AB = EO_2 = \left[l^2 - (r_1 + r_2)^2\right]^{\frac{1}{2}}$$

$$= l\left[1 - \left(\frac{r_1 + r_2}{l}\right)^2\right]^{1/2}$$

Since, $\dfrac{r_1 + r_2}{l}$ is small,

$$AB \simeq l\left[1 - \frac{1}{2}\left(\frac{r_1 + r_2}{l}\right)^2\right]$$

$\therefore \qquad L = 2\,[\text{arc } PA + AB + \text{arc } QB]$

$$= 2\left[r_1\left(\frac{\pi}{2} + \beta\right) + l\left\{1 - \frac{1}{2}\left(\frac{r_1 + r_2}{l}\right)^2\right\} + r_2\left(\frac{\pi}{2} + \beta\right)\right]$$

$$= \pi(r_1 + r_2) + 2\beta(r_1 + r_2) + 2l - \frac{(r_1 + r_2)^2}{l}$$

Substituting $\dfrac{r_1 + r_2}{l}$ for β, we get,

$$L = \pi(r_1 + r_2) + 2\frac{(r_1 + r_2)^2}{l} + 2l - \frac{(r_1 + r_2)^2}{l}$$

$$= \pi(r_1 + r_2) + 2l + \frac{(r_1 + r_2)^2}{l} \qquad \ldots(7.2)$$

Velocity Ratio

It is defined as the ratio of the speed of the driven wheel to the speed of the driving wheel. Let N_1 be the revolution per minute (rpm) made by driver and N_2 be the rpm of follower. Then angular velocity of driver is $\omega_1 = \dfrac{2\pi N_1}{60}$ and that of follower is $\omega_2 = \dfrac{2\pi N_2}{60}$. If r_1 is the radius of driver and r_2 that of the follower, then the linear velocity of the belt, if there is no slip,

$$v = r_1 \omega_1 = r_2 \omega_2$$

i.e., $\qquad \dfrac{\omega_2}{\omega_1} = \dfrac{r_1}{r_2}$

but
$$\frac{\omega_2}{\omega_1} = \frac{N_2}{N_1}$$

∴ Velocity ratio $= \dfrac{N_2}{N_1} = \dfrac{r_1}{r_2} = \dfrac{d_1}{d_2}$ ...(7.3)

Where d_1 is diameter of driver and d_2 is that of follower. If thickness of belt is not negligible, the linear velocity $v = (r_1 + t)\,\omega_1 = (r_2 + t)\,\omega_2$.

Where t is the thickness of the belt.

Hence,
$$\frac{N_2}{N_1} = \frac{\omega_2}{\omega_1} = \frac{r_1 + \dfrac{t}{2}}{r_2 + \dfrac{t}{2}} = \frac{2r_1 + t}{2r_2 + t} = \frac{d_1 + t}{d_2 + t} \quad ...(7.4)$$

Equations 7.3 and 7.4 are derived for no slippage condition. If there is p_1 percentage slippage in driver wheel and belt and p_2 percentage slippage between follower wheel and the belt, then considering the linear velocity, we find

Velocity of belt at driver end
$$v_b = r_1 \omega_1 \left(1 - \frac{p_1}{100}\right)$$

∴ Velocity of follower, after p_2 percentage slippage
$$r_2 \omega_2 = v_b \left(1 - \frac{p_2}{100}\right)$$
$$= r_1 \omega_1 \left(1 - \frac{p_1}{100}\right)\left(1 - \frac{p_2}{100}\right)$$
$$= r_1 \omega_1 \left(1 - \frac{p_1 + p_2}{100}\right)$$

∴
$$\frac{N_2}{N_1} = \frac{d_1}{d_2}\left(1 - \frac{p_1 + p_2}{100}\right)$$

If 'p' is the total slip during the drive
$$\frac{N_2}{N_1} = \frac{d_1}{d_2}\left(1 - \frac{p}{100}\right) \quad ...(7.5)$$

Stepped Cone Wheel

When the speed of the driven shaft is to be changed very frequently a stepped cone as shown in Fig. 7.2 is used. This situation is very common in machine tools, such as lathe, drilling machine etc. Stepped cone consists of three or more pairs of wheels of different sizes placed adjacent to the one another as shown in the figure. The wheels are having integral casting. One set of wheels are fixed to driving shaft and another set to driven shaft. To facilitate use of the same belt for all the sets, it is obvious that $d_1 + d_2 = d_3 + d_4 = d_5 + d_6$ where d_1, d_3 and d_5 are the diameters of the wheels on driving shaft and d_2, d_4, d_6 those of wheels on driven shaft.

By shifting the belt from one pair to another, the speed of the driven shaft can be altered.

TRANSMISSION OF POWER

Example 7.1. *A shaft running at 250 rpm drives a lathe spindle. The lathe spindle has to run at 100, 125 and 150 rpm. The minimum diameter of the wheel should be 240 mm. Determine the diameters of the other wheels in the system.*

Solution. Let d_1, d_2, d_3 be the diameters of the wheels on speed cone and d_2, d_4, d_6 be the corresponding diameters of speed cone on lathe.

Now, $\quad \dfrac{d_2}{d_1} = \dfrac{N_1}{N_2} = \dfrac{250}{100} = 2.5$

$\therefore \quad d_2 = \dfrac{250}{100} \times 240 = \mathbf{600\ mm\ Ans.}$

$\therefore \quad d_1 + d_2 = 600 + 240 = 840\ mm$

$\dfrac{d_4}{d_3} = \dfrac{250}{125} = 2$

$\therefore \quad d_3 + d_4 = 840\ \text{gives}$

$d_3 = \dfrac{840}{3} = \mathbf{280\ mm\ Ans.}$

$\therefore \quad \mathbf{d_4 = 560\ mm\ Ans.}$

$\dfrac{d_6}{d_5} = \dfrac{250}{150} = \dfrac{5}{3}$

$\therefore \quad d_5 + d_6 = 840 \quad \text{gives}$

$d_5 = \dfrac{840}{\left(1 + \dfrac{5}{3}\right)} = \mathbf{315\ mm\ Ans.}$

$\therefore \quad \mathbf{d_6 = 525\ mm\ Ans.}$

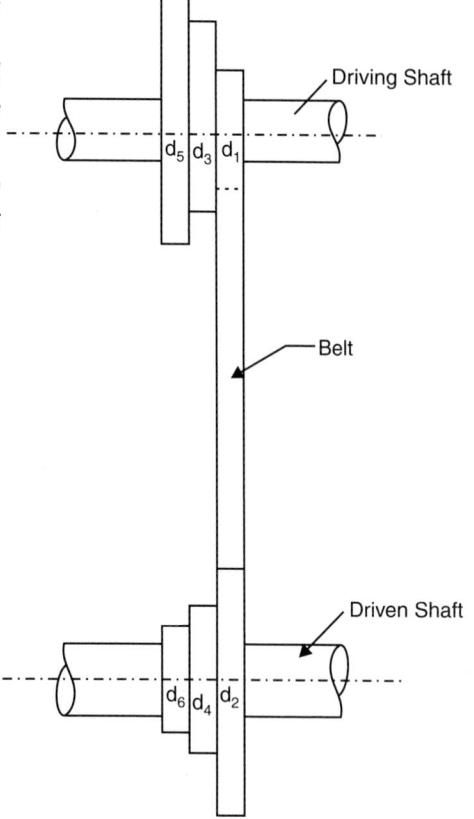

Fig. 7.2. Speed cone wheels

Example 7.2. *An engine, running at 160 rpm drives a line shaft by means of a belt. The diameter of engine wheel is 600 mm and that of the pulley on line shaft is 400 mm. Calculate the speed of the follower shaft if (a) there is no slip (b) if there is a slip of 2.5%.*

Solution.
d_1 = diameter of engine wheel = 600 mm.
d_2 = diameter of wheel on line shaft = 400 mm.
N_1 = speed of engine wheel = 160 rpm.
$N_2 = 2$

$\dfrac{N_2}{N_1} = \dfrac{d_1}{d_2} = \dfrac{600}{400}$, if there is no slip.

$\therefore$ If there is no slip

$N_2 = \dfrac{600}{400} \times 160 = \mathbf{240\ rpm.\ Ans.}$

If there is 2.5% slip $p = \dfrac{2.5}{100}$

$$N_2 = N_1 \times \frac{d_1}{d_2}\left(1 - \frac{p}{100}\right)$$

$$= 160 \times \frac{600}{400}\left(1 - \frac{2.5}{100}\right)$$

$$= \mathbf{234 \text{ rpm. Ans.}}$$

Example 7.3. *A shaft running at 200 rpm drives a lathe spindle which is at a distance 2.5 m. The lathe spindle has to run at 80 rpm. If the diameter of wheel on driver shaft is 240 mm and the drive is by cross belt, determine the length of belt. Assume no slippage.*

Solution.
$$\frac{N_2}{N_1} = \frac{d_1}{d_2}$$

$$\frac{80}{200} = \frac{240}{d_2}$$

$$\therefore \quad d_2 = \frac{240 \times 200}{80} = 600 \text{ mm}$$

$$\therefore \quad r_1 = 120 \text{ mm}, \ r_2 = 300 \text{ mm}, \ l = 2.5 \text{ m} = 2500 \text{ mm}$$

∴ Length of cross belt

$$L = \pi(r_1 + r_2) + 2l + \frac{(r_1 + r_2)^2}{l}$$

$$= \pi(120 + 300) + 2 \times 2500 + \frac{(120 + 300)^2}{2500}$$

$$\mathbf{L = 6390 \text{ mm} = 6.39 \text{ m Ans.}}$$

7.2. TRANSMISSION OF POWER IN BELT DRIVE

The transmission of power by means of belt or rope drives is possible because of friction, that exists between the wheels and the belt. All along the contact surface the frictional resistance develops. Hence, tension in the rope is more on the side it is pulled and is less on the other side. Accordingly, the two sides of the rope may be called as tight side and slack side. The relationship between the two forces (*viz.* on tight side and slack side), can be derived as explained below:

Let the force on slack side be T_1 and on tight side be T_2 (Ref. Fig. 7.3 (*a*)). T_2 is more than T_1 because frictional force develops between the wheel and the belt as shown in Fig. 7.3 (*b*). Let 'θ' be the angle of contact in radians between belt and the wheel. Now consider an elemental length of the belt as shown in Fig. 7.3 (*c*).

Let T be the force on slack side and $T + dT$ be the force on tight side. There will be normal reaction N on the rope in the radial direction and frictional force $F = \mu N$ in the tangential direction. Then,

Σ Forces in radial direction = 0, gives,

$$N - T \sin\frac{d\theta}{2} - (T + dT)\sin\frac{d\theta}{2} = 0$$

TRANSMISSION OF POWER

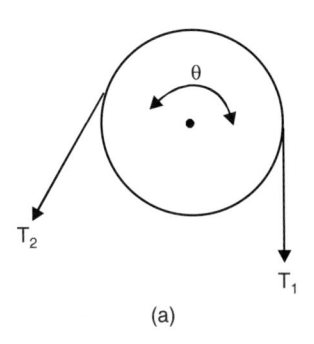

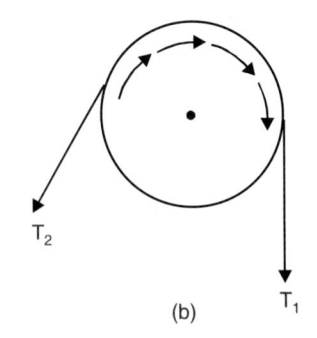

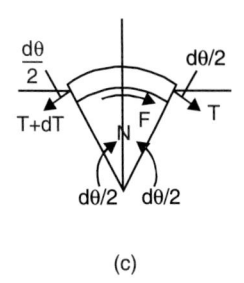

Fig. 7.3

sin $d\theta$ is very small angle,

$$\sin \frac{d\theta}{2} = \frac{d\theta}{2}$$

$$\therefore N - T\frac{d\theta}{2} - (T + dT)\frac{d\theta}{2} = 0$$

or

$$N = \left(T + \frac{dT}{2}\right)d\theta$$

From the law of friction,

$$F = \mu N = \mu\left(T + \frac{dT}{2}\right)d\theta$$

where μ is the coefficient of friction.

Σ Forces in tangential direction = 0, gives,

$$(T + dT)\cos\frac{d\theta}{2} = F + T\cos\frac{d\theta}{2}$$

Since, $d\theta$ is small angle, $\cos\dfrac{d\theta}{2} = 1$

$\therefore \qquad T + dT = F + T$

or

$$dT = F = \mu\left(T + \frac{dT}{2}\right)d\theta$$

Neglecting the small quantity of higher order, we get,

$$dT = \mu T\, d\theta$$

or

$$\frac{dT}{T} = \mu\, d\theta$$

Integrating both sides over 0 to θ, we get,

$$\int_{T_1}^{T_2} \frac{dT}{T} = \int_0^\theta \mu\, d\theta$$

$$[\log T]_{T_1}^{T_2} = \mu[\theta]_0^\theta$$

or $\quad \log T_2 - \log T_1 = \mu\theta$

i.e., $\quad \log \dfrac{T_2}{T_1} = \mu\theta$

$\therefore \quad \dfrac{T_2}{T_1} = e^{\mu\theta} \qquad \qquad \qquad \qquad ...(7.6)$

[**Note:** θ should be in radians.]

The unequal tension exerts a torque on the wheel. Referring to Fig. 7.4 (a) and (b).

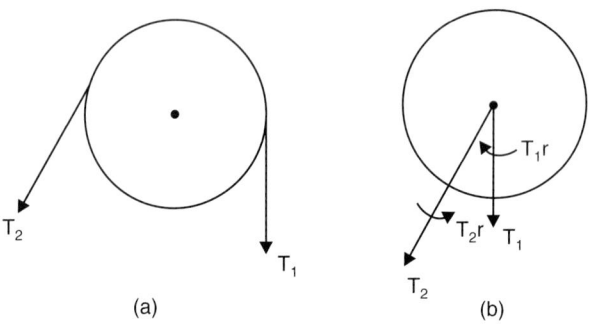

Fig. 7.4

We find T_1 and T_2 can be replaced by the forces T_1 and T_2 in the bearing of the wheel plus a moment of $(T_2 - T_1)\, r$.

If ω is the speed of wheel in rad/sec, then work done on the wheel per second is
$= (T_2 - T_1)\, r\omega = (T_2 - T_1)\, v$.

Where, $v = r\omega$, is linear velocity of the belt.

Since $\quad \dfrac{T_2}{T_1} = e^{\mu\theta}$

We get work done per second

$\quad = \text{Power} = T_1\, (e^{\mu\theta} - 1)\, v \qquad \qquad ...(7.7)$

If the material of the belt has permissible stress of f, then, the maximum tension it can take is $f\, b\, t$ where b is the width of belt and t, its thickness. Then in a design

$\therefore \qquad T_2 = f\, b\, t \qquad \qquad \qquad \qquad ...(7.8)$

Example 7.4. *A rope making* $1\dfrac{1}{4}$ *turns around a stationary horizontal drum is used to support a weight W [Fig. 7.5]. If the coefficient of friction is 0.3 what range of weights can be supported by exerting a 600 N force at the other end of rope?*

Solution. Angle of contact

$$\theta = 1.25 \times 2\pi = 2.5\,\pi$$

Case 1. Let the impending motion of the weight be downward.

Then, $\qquad T_1 = 600\text{ N}, \quad T_2 = W$

TRANSMISSION OF POWER

From the law of rope friction,

$$T_2 = T_1 e^{\mu\theta}$$

We get, $W = 600\, e^{0.3 \times 2.5\pi} = 600\, e^{0.75\pi}$

$= 6330.43$ newton.

Case 2. Let the impending motion of the load W be upward.

Then, $T_1 = W$ and $T_2 = 600$ N

∴ $T_2 = T_1 e^{\mu\theta}$ gives,

$600 = W\, e^{0.75\pi}$

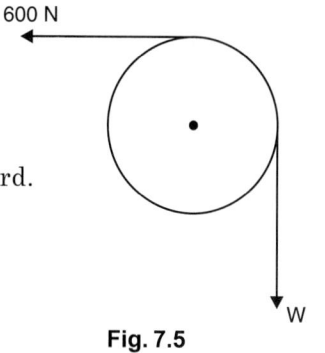

∴ $W = 56.87$ newton.

Fig. 7.5

Thus, **a 600 N force can support a range of load between 56.87 N to 6330.43 N weight. Ans.**

Example 7.5. *An open belt 100 mm wide and 3 mm thick connects two pulleys mounted on a parallel shaft, at 2.5 m apart. The diameter of the larger wheel is 500 mm and that of the smaller is 300 mm. The bigger wheel is rotating at 100 rpm. If the permissible stress in the belt is to be limited to 4.0 N/mm^2, find the maximum power that can be transmitted at this speed. Given, $\mu = 0.3$.*

Solution. Referring to Fig. 4.1 (a), the angle of contact is $\theta = \pi - 2\alpha$,

Where, $\sin\alpha = \dfrac{r_1 - r_2}{t} = \dfrac{1}{2}\dfrac{(500-300)}{2500}$

∴ $\alpha = 2.292$ degrees

$= 0.04$ radians

$\theta = \pi - 2 \times 0.04 = 3.0616$ radians

∴ $T_2 = T_1 e^{\mu\theta} = T_1 \times e^{0.3 \times 3.0616}$

$= 2.505\, T_1$

T_2 may be allowed to go up to

$T_2 = f_1 \times b \times t = 4.0 \times 100 \times 3 = 1200.0$ N

Corresponding, $T_1 = \dfrac{1200.0}{2.505} = 479.04$ N

∴ $P = (T_2 - T_1)\, r\omega$

Now, $r = \dfrac{500}{2}$ and $\omega = \dfrac{2\pi \times 100}{60}$

∴ $P = (12000 - 479.04) \times \dfrac{500}{2} \times \dfrac{2\pi \times 100}{60}$

$= 1.887 \times 10^6$ N-mm/sec

i.e., $P = 1.887$ **kW Ans.**

Example 7.6. *A torque of 300 N-m acts on a brake drum shown in Fig. 7.6 (a). If the brake band is in contact with the brake drum through 250° and the coefficient of friction is 0.3, determine the force P applied at the end of the brake lever for the position shown in the figure.*

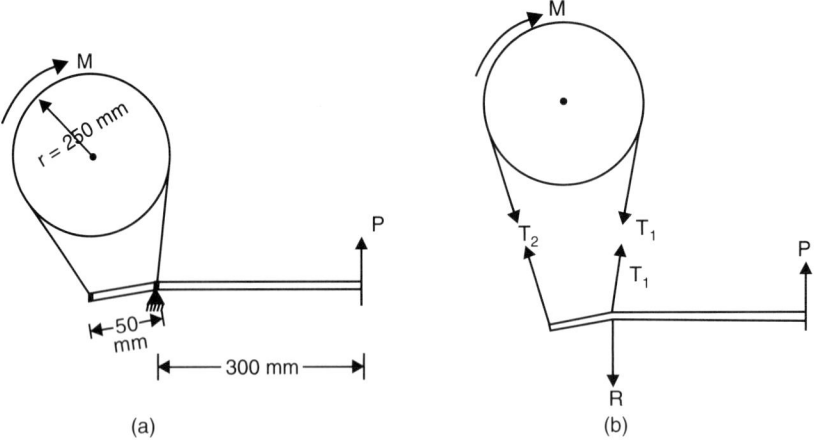

Fig. 7.6

Solution. Figure 7.6 (b) shows the free body diagram of the brake drum and the lever arm.

Now, $\theta = 250 \times \dfrac{\pi}{180}, \quad \mu = 0.3$

$\therefore \quad \mu\theta = 0.3 \times 250 \times \dfrac{\pi}{180} = 1.309$

$r = 250$ mm

From friction equation,

$T_2 = T_1 e^{\mu\theta} = T_1 e^{1.309} = 3.7025 \, T_1$

Now, $(T_2 - T_1) r = $ torque $= M$

$(3.7025 - 1) T_1 \times 250 = 300 \times 10^3$

$\therefore \quad T_1 = 444.04$ newton

$T_2 = 3.7025, \quad T_1 = 3.7025 \times 444.04$

$= 1644.06$ newton

Considering the equilibrium of lever arm,

$T_2 \times 50 = P \times 300$

$\therefore \quad P = \dfrac{1644.06 \times 50}{300} = $ **274.0 newton Ans.**

7.3. CENTRIFUGAL TENSION

In studying forces acting on belt, we have ignored a force acting towards the centre of the circular path. This force may be ignored for the low speed drives, but cannot be ignored for high speed drives.

Consider the belt moving with uniform velocity v along the curved path of radius 'r', as shown in Fig. 7.7 (a). In time dt, let it move from A to B. Since, the body is having

TRANSMISSION OF POWER

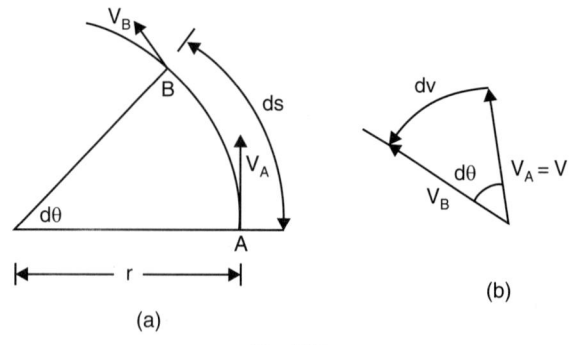

Fig. 7.7

uniform velocity, the tangential velocity v_A at A is equal to the tangential velocity v_B at B in magnitude, say $v_A = v_B = \omega$. However, there is change in velocity dv at right angles to v_A, i.e., in radial inward direction of magnitude dv [Ref. Fig. 7.7 (b)].

Since, $$dv = v\, d\theta = v\, \frac{ds}{r}$$

acceleration in radial inward direction

$$a = \frac{dv}{dt} = \frac{v}{r}\frac{ds}{dt} = \frac{v}{r} \cdot v = \frac{v^2}{r}$$

Thus, when the belt moves with uniform velocity v along a curved path of radius r, it has a radial inward acceleration of magnitude $\frac{v^2}{r}$. Hence, a belt of elemental mass, δm experiences a radial inward force of magnitude $c = \delta m\, \frac{v^2}{r}$ as shown in Fig. 7.8.

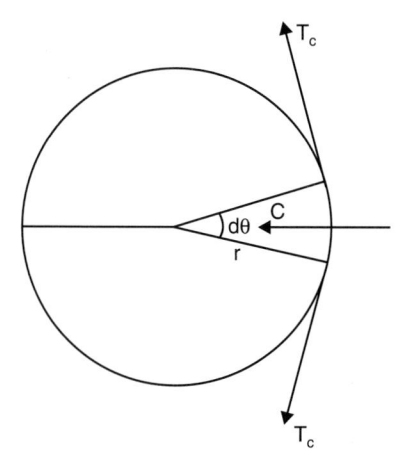

Fig. 7.8

If m is the mass per unit length of belt,
$$\delta m = mr\, d\theta$$

$\therefore$
$$c = mr\, d\theta\, \frac{v^2}{r} = mv^2\, d\theta$$

This centrifugal force C can be supplied by equal pulls T_c at either end of the elemental length of the belt.

$$\therefore \quad C = 2\,T_c \cos\left(\frac{\pi}{2} - \frac{\delta\theta}{2}\right) = 2\,T_c \sin\frac{\delta\theta}{2}$$

$$= 2\,T_c \times \frac{\delta\theta}{2}, \quad \text{Since, } \delta\theta \text{ is small}$$

$$= T_c\,\delta\theta$$

Thus, $\quad T_c\,\delta\theta = C = mv^2\,\delta\theta$

or $\quad T_c = mv^2$ ...(7.9)

Thus, we find the belt on tight side and slack side experience additional force $T_c = mv^2$.

$\therefore$ Tension on slack side $= T_1 + T_c$ ...(7.10 (a))

and Tension on tight side $= T_2 + T_c$ ...(7.10 (b))

$\therefore$ Maximum tension that we can apply on the belt is $T = T_2 + T_c$ but not T_2. If the permissible stress in the belt is f, then,

$$f\,b\,t = T_2 + T_c \qquad\qquad ...(7.11)$$

The term $T_c = mv^2$ is known as centrifugal tension in the belt.

However, it may be noted that the expression for power $P = (T_2 - T_1)\,v$ is not affected because additional term T_c in tight side and slack side tension cancel each other.

Example 7.7. *An open belt 150 mm wide and 8 mm thick connects two wheels mounted on parallel shafts 2.5 m apart. The diameter of the larger pulley is 480 mm and that of the smaller is 320 mm. The bigger wheel is rotating at 800 rpm. If the permissible stress on the belt section is 3 N/mm², calculate the maximum power that can be transmitted at this speed,*

(a) neglecting centrifugal tension,

(b) considering centrifugal tension, if the density of leather is 1100 kg/m³.

Take $\mu = 0.3$

Solution. Referring to Fig. 7.1 (a),

$$\sin\alpha = \frac{r_1 - r_2}{l} = \frac{\left(\dfrac{480}{2} - \dfrac{320}{2}\right)}{2500} = 0.032$$

Since, it is small angle $\alpha = 0.032$ radians.

$\therefore$ Angle of contact $\theta = \pi - 2\alpha = \pi - 2 \times 0.032 = 3.078$

$\therefore \quad T_2 = T_1\,e^{\mu\theta} = T_1\,e^{0.3 \times 3.078}$

$\qquad\qquad = 2.5178\,T_1$

(a) Neglecting centrifugal tension:

Maximum force permitted is given by

$$f\,b\,t = T_2$$
$$3 \times 150 \times 8 = T_2$$

$\therefore \qquad T_2 = 3600$ N

Corresponding, $T_1 = \dfrac{3600}{2.5178} = 1489.82$ N

∴ Maximum power transmitted

$$P = (T_2 - T_1)\, v = (T_2 - T_1)\, r\omega$$
$$= (3600 - 1429.82) \times \dfrac{480}{2} \times \dfrac{2\pi \times 800}{60}$$
$$= 43.634 \times 10^6 \text{ N-mm/sec}$$

i.e., **P = 43.634 kW Ans.**

(b) If centrifugal tension is considered.

Mass of belt per m length = $1100 \times 0.15 \times 0.008 = 1.30$ kg/m.

Velocity, $v = r\omega = \dfrac{480}{2} \times \dfrac{2\pi \times 800}{60}$
$$= 20106 \text{ mm/sec}$$
$$= 20.106 \text{ m/sec}$$

∴ $T_c = mv^2 = 1.32 \times 20.106^2$
$$= 533.62 \text{ N}$$

[**Note:** 1 newton force is defined as the force which makes 1 kg mass to move with 1m/sec velocity. Hence, m should be in kg and v should be in m/sec to get the force in newton].

∴ Maximum force that can be permitted on the belt.

$$T = T_2 + T_c = f\,b\,t$$
$$= 3 \times 150 \times 8 = 3600 \text{ N}$$

i.e., $T_2 + 533.62 = 3600$

∴ $T_2 = 3066.38$ N

Corresponding, $T_1 = \dfrac{3066.38}{2.5178} = 1217.88$ N

∴ The maximum torque that can be transferred

$$= (T_2 - T_1)\, v$$
$$= (3066.38 - 1217.88) \times 20.106$$
$$= 37165.9 \text{ N-m/sec}$$

P = 37.166 kW Ans.

7.4. ROPE DRIVE

If power to be transmitted is high, better friction grip can be achieved by using ropes in a groove as shown in Fig. 7.9.

Let the angle of groove be 2α. Reaction to the rope at contact points be R_1.

∴ The resultant of these two reactions is vertical

$$R = 2R_1 \sin \alpha$$

Frictional force at each contact point
$$= \mu R_1$$

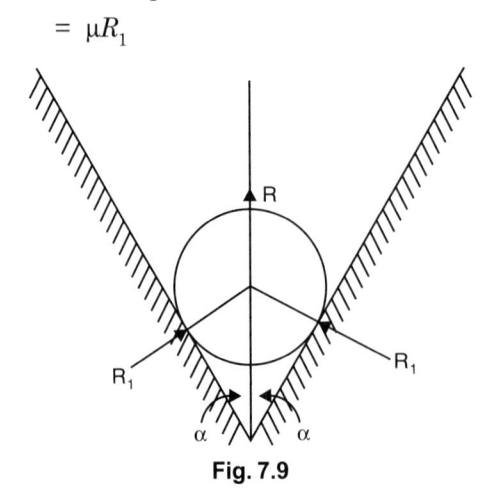

Fig. 7.9

∴ Total frictional force $= 2\mu R_1 = \dfrac{2\mu R}{2\sin\alpha}$

$\qquad\qquad\qquad\quad = \mu R \operatorname{cosec} \alpha$

∴ If we consider the elemental length as in Art 7.2, we get

$$\int_{T_1}^{T_2} \frac{dT}{T} = \int_0^\theta \mu \operatorname{cosec} \alpha\, d\theta$$

∴ We get, $\qquad \dfrac{T_2}{T_1} = e^{\mu\, \operatorname{cosec}\, \alpha\, \times\, \theta}$ ...(7.12)

Where θ is the angle of contact in radians.

If centrifugal tension is to be taken it is equal to $T_c = mv^2$ as earlier.

Example 7.8. *A rope 36 mm in diameter drives a grooved pulley at a speed of 20 m/sec. The angle of lap of the rope on the pulley is 220° and the angle of groove for the pulley is 60°. If the permissible tension in the rope is 1.5 N/mm², calculate the maximum power that can be transmitted in this speed. Take, coefficient of friction μ = 0.3 and the mass of the rope per metre length is 0.9 kg.*

Solution. $T_c = mv^2 = 0.9 \times 20^2 = 360$ N

Maximum permissible force

$$T = f \times \frac{\pi}{4} d^2 = 1.5 \times \frac{\pi}{4} \times 36^2$$

$\qquad\qquad = 1562.8$ N

∴ $\qquad T_2 = T - T_c = 1562.8 - 360$

$\qquad\qquad = 1202.8$ N

∴ $\qquad T_2 = T_1\, e^{\mu\, \operatorname{cosec}\, 30\, \times\, \theta}$

$\qquad\qquad = T_1\, e^{0.3\, \times\, \operatorname{cosec}\, 30°} \times \dfrac{220 \times \pi}{180}$

$$= T_1 \, e^{2.3038}$$
$$= 10.0125 \, T_1$$

∴ $$T_1 = \frac{1202.8}{10.0125} = 120.13 \text{ N}$$

∴ Power $= (T_2 - T_1) \, v = (1202.8 - 120.13) \times 20$
$$= 21653 \text{ N-m/sec}$$
$$\mathbf{P = 21.653 \text{ kW Ans.}}$$

7.5. GEAR DRIVE

In belt and rope drive certain amount of slip is unavoidable. Hence, these drives are not suitable for precision machines like an ordinary clock. In such machines toothed gear drives are used which provide slipless drive. Gear drives can transmit power between the shafts when their ages are parallel, intersecting or even they are neither parallel nor intersecting. For this, different types of gears as listed below are used:

(i) Spur gear
(ii) Bevel gear
(iii) Worm and worm wheel
(iv) Rack and pinion, and
(v) Gear trains.

Spur Gear

If the axes of the driving and driven shafts are parallel and coplanar, spur gears are used. Figure 7.10 shows a typical spur gear drive.

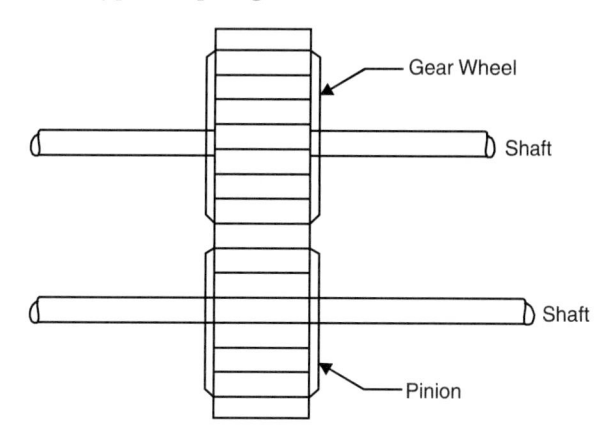

Fig. 7.10 Spur gear

In this, teeth of the gear wheels are parallel to the axes. Teeth of the spur gears are cut on the circumference of the cylindrical discs. The bigger wheel is known as gear wheel whereas the smaller one is known as a pinion. The circle joining the contact points of all teeth is known as pitch circle.

Bevel Gear

If the axes of the two shafts are inclined to one another and intersect when produced bevel gears are used. Figure 7.11 shows a typical bevel gear. The teeth of the bevel gears are cut on the conical surface. If two bevel gears have their axes at right angles to each other and are of equal size they are called mitre gears.

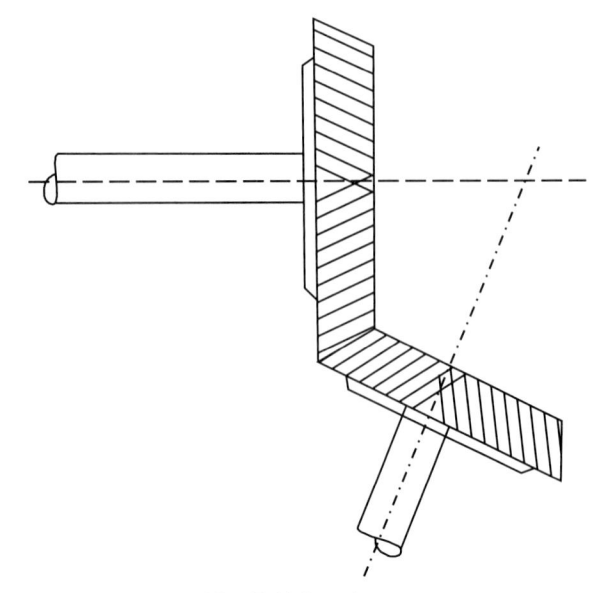

Fig. 7.11 Bevel gear

Worm and Worm Wheel

If the driving and driven shafts are at right angles but not intersecting, worm and worm wheels are used. Figure 7.12 shows a typical worm and worm wheel drive. It consists

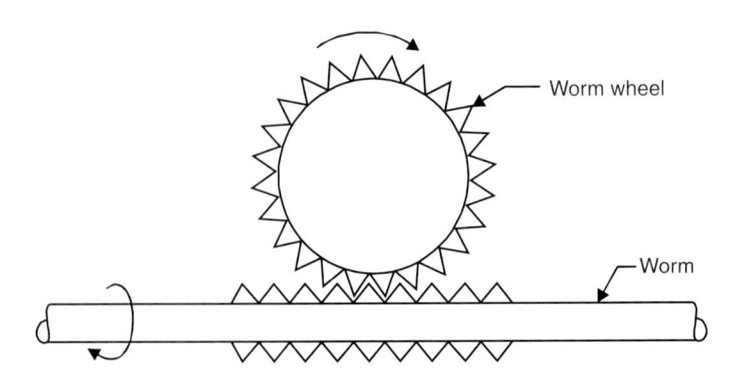

Fig. 7.12 Worm and worm wheel

of a worm which may have one or more helical threads and a worm wheel which engages with the worm. They are used in machine tools when a high velocity ratio is required.

Rack and Pinion Gear

These drives are used when a rotary motion is to be converted to a linear motion. It consists of a rack and pinion as shown in Fig. 7.13, which can engage each other. Rack is

TRANSMISSION OF POWER

a rectangular bar with a series of teeth. One can consider it as a spur gear of infinite drive.

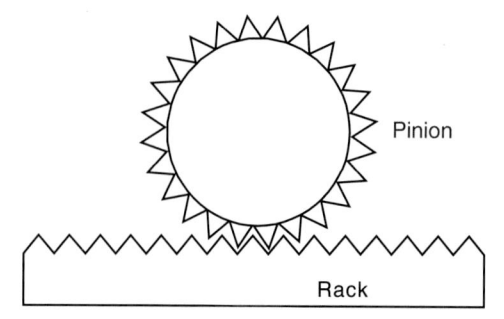

Fig. 7.13 Rack and pinion drive

Gear Trains

A gear train is a combination of a set of gear wheels through which the motion is transmitted. The gear wheels used may be spur or bevel. The gear trains of the following two types are commonly used:

(a) Simple gear train

(b) Compound gear train.

(a) Simple gear train: Figure 7.14 shows a typical simple gear train.

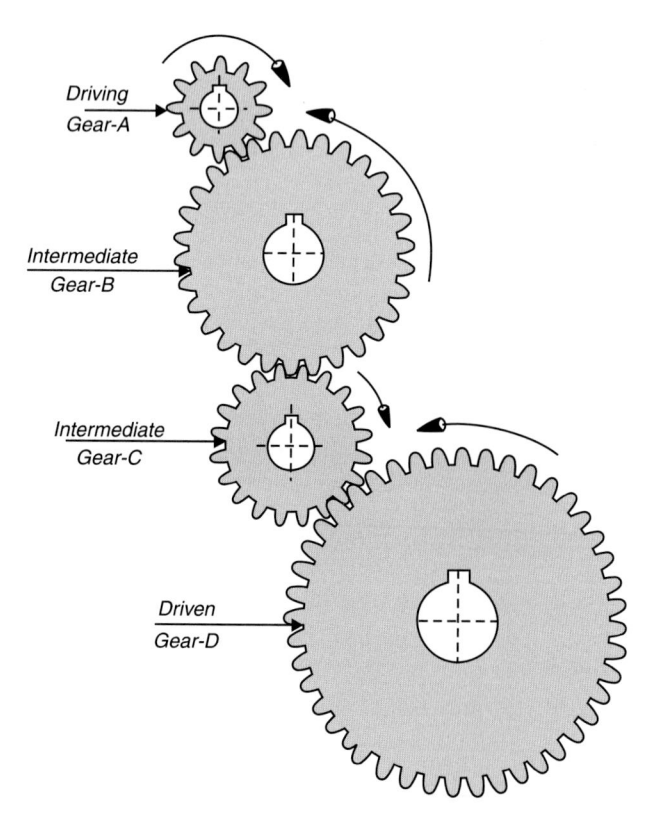

Fig. 7.14 Simple gear train

In this figure A is driving gear and D is the driven gear. Between these two, there are two more gears B and C. All gears are mounted on separate shafts. The intermediate gears B and C are called *idler gears*. The idler gears change the direction of the driven wheel. Even number of idler gears will rotate the driven wheel in the direction opposite to that of driven wheel. Odd number of idler gears rotate the driven wheel in the same direction as that of driving wheel.

(b) Compound gear train. Figure 7.15 shows a typical compound gear train.

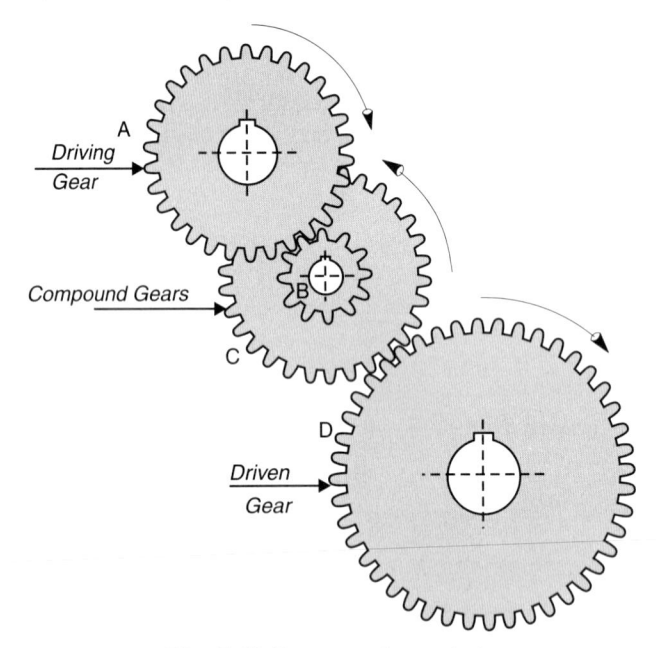

Fig. 7.15 Compound gear train

In this system, each intermediate shaft carries two gears which are keyed to it. Thus in the figure, B and C are mounted on a single shaft. Since, B and C are mounted on the same shaft they rotate at the same speed.

Velocity Ratio of Gear Drives

Let T_1 and T_2 be the number of teeth on a driver gear and on a driven gear respectively. If d_1 and d_2 are the pitch circle diameter of the two gears respectively, then for the slipless fit, the pitch 'p' must be the same in the two gears.

i.e.,
$$p = \frac{\pi d_1}{T_1} = \frac{\pi d_2}{T_2}$$

The term $\frac{d_1}{T_1} = \frac{d_2}{T_2}$ is called the module.

Thus,
$$m = \frac{d_1}{T_1} = \frac{d_2}{T_2} \qquad \ldots(7.13)$$

If N_1 and N_2 are the revolutions per minute of driver and driven gears respectively, then the velocity ratio is

TRANSMISSION OF POWER

$$= \frac{N_2}{N_1} = \frac{d_1}{d_2} = \frac{T_1}{T_2} \qquad \ldots(7.14)$$

Thus, the velocity ratio in the drive is inversely proportional to the number of teeth in the gears.

If the tangential force exerted by the driver at the pitch point is p and the velocity of pitch point is $v = r\omega = \dfrac{r\, 2\pi N}{60}$, then the power transmitted is $p \times v$.

In the design of gears drive, the following two conditions should be satisfied:

$$r_1 + r_2 = l \quad i.e., \quad d_1 + d_2 = 2l \qquad \ldots(1)$$

where, l is the distance between the centres of the two wheels and

$$\frac{N_2}{N_1} = \frac{d_1}{d_2} \qquad \ldots(2)$$

From eqs. (1) and (2) for a given 'l' and velocity ratio d_1 and d_2 can be obtained. But the design is not complete here, since $T_1 = \dfrac{\pi d_1}{p}$ and $T_2 = \dfrac{\pi d_2}{p}$ should be whole number. Hence slight adjustment is to be made in 'l' so that T_1 and T_2 comes out to be whole number.

Example 7.9. *Two parallel shafts are to be connected by spur gearing. The axes of shafts are to be at 540 mm apart as nearly as possible. If one shaft has to run at 100 rpm and the other at 300 rpm, find the number of teeth on each wheel, if the module is to be 8 mm. Determine the exact distance between the two shafts.*

Solution. $\quad d_1 + d_2 = 2l = 2 \times 540 = 1080 \text{ mm} \qquad \ldots(1)$

$$\frac{N_2}{N_1} = \frac{d_1}{d_2} = \frac{300}{100} = 3 \qquad \ldots(2)$$

From eqn. (2), $\quad d_1 = 3d_2$

Substituting it in eqn. (1), we get,

$$4d_2 = 1080 \quad \text{or} \quad d_2 = 270 \text{ mm}$$

$\therefore \qquad d_1 = 3 \times 270 = 810 \text{ mm}$

For a module of 8,

$$\frac{d_1}{T_1} = \frac{d_2}{T_2} = 8$$

$\therefore \qquad T_1 = \dfrac{810}{8} = 101.25$

and $\qquad T_2 = \dfrac{270}{8} = 33.75$

But T_1 and T_2 should be whole numbers with

$$T_1 = 3T_2$$

Let us select $T_1 = 102$ and hence, $T_1 = 34$.

Then, pitch circle diameter are:

$$d_1 = 102 \times 8 = 816 \text{ mm}$$

and $$d_2 = 34 \times 8 = 272 \text{ mm}$$

∴ The exact distance between the shafts

$$l = \frac{1}{2}(d_1 + d_2) = \frac{1}{2}(816 + 272) = 544 \text{ mm}$$

Thus, $T_1 = 102$, $T_2 = 34$ and $l = 544$ **mm Ans.**

Example 7.10. *In a gear drive a power of 1.8 kW is to be transferred by a pinion of 25 teeth of 8 mm module. Calculate the speed of the pinion required, if the pressure between the teeth is to be 480 newtons.*

Solution. Pressure $P = 480$ N. Power transmitted $= 1.8$ kW. If v is the velocity of teeth point,

$$Pv = 1.8 \times 1000$$

$$v = \frac{1800}{P} = \frac{1800}{480} = 3.75 \text{ m/sec}$$

module $$m = 8 = \frac{d}{T} = \frac{d}{25}$$

∴ $d = 200$ mm ∴ $r = 100$ mm $= 0.1$ m

∴ Angular velocity of pinion

$$\omega = \frac{v}{r} = \frac{3.75}{0.100} = 37.5 \text{ rad/sec}$$

If N is the rpm, then,

$$\frac{2\pi N}{60} = \omega = 37.5$$

or $N = 358.1$ **rpm Ans.**

Velocity Ratio in Case of Simple Gear Train

Referring to Fig. 7.14, let N_A, N_B, N_C and N_D be the rpm and T_A, T_B, T_C and T_D be the number of teeth on the wheels A, B, C and D of simple gear train. Then,

(*i*) Since, A drives B

$$\frac{N_B}{N_A} = \frac{T_A}{T_B}$$

(*ii*) B drives C

∴ $$\frac{N_C}{N_B} = \frac{T_B}{T_C}$$

(*iii*) C drives D

∴ $$\frac{N_D}{N_C} = \frac{T_C}{T_D}$$

∴ Velocity ratio between the driving and driven wheels

$$= \frac{N_D}{N_A}$$

TRANSMISSION OF POWER

$$= \frac{N_D}{N_C} \times \frac{N_C}{N_B} \times \frac{N_B}{N_A}$$

$$= \frac{T_C}{T_D} \times \frac{T_B}{T_C} \times \frac{T_A}{T_B}$$

$$= \frac{T_A}{T_D} \qquad \text{...(7.15)}$$

Velocity Ratio in Case of Compound Gear Train

Referring to Fig. 7.15, let N_A, N_B, N_C and N_D be speed in rpm and T_A, T_B, T_C and T_D be teeth on wheels A, B, C and D respectively. Then,

(i) A rotates B

$$\therefore \qquad \frac{N_B}{N_A} = \frac{T_A}{T_B}$$

(ii) $N_B = N_C$, since they are keyed to the same shaft.

(iii) C drives D

$$\therefore \qquad \frac{N_D}{N_C} = \frac{T_C}{T_D}$$

$$\therefore \qquad \text{Velocity ratio} = \frac{N_D}{N_A}$$

$$= \frac{N_D}{N_C} \times \frac{N_B}{N_A}, \quad \text{since,} \quad N_C = N_B$$

$$= \frac{T_C}{T_D} \times \frac{T_A}{T_B} \qquad \text{...(7.16)}$$

Example 7.11. *A gear train consists of 4 gears A, B, C and D and they have 25, 50, 75 and 100 teeth respectively. If A rotates at 160 rpm, determine the speed of driven shaft D*

(a) if intermediate gears B and C are on separate shaft

(b) if B and C are on the same shaft.

Solution. (a) If the intermediate gears are on different shaft, it is a simple gear system.

$$\therefore \qquad \frac{N_D}{N_A} = \frac{T_A}{T_D} = \frac{25}{100}$$

$$\therefore \qquad N_D = \frac{25}{100} \times 160, \quad \text{since,} \quad N_A = 160 \text{ rpm}$$

i.e., $\qquad N_D = $ **40 rpm Ans.**

(b) If the intermediate gears B and C are on the same shaft $N_B = N_C$ and if forms a compound gear train.

$$\therefore \quad \text{Velocity ratio} \quad \frac{N_D}{N_A} = \frac{T_C}{T_D} \times \frac{T_A}{T_B}$$

$$= \frac{75}{100} \times \frac{25}{50}$$

$$\therefore \qquad N_D = \frac{75}{100} \times \frac{25}{50} \times 160$$

i.e., $\qquad N_D = \textbf{60 rpm Ans.}$

Important Formulae

1. Length of belt in open drive

$$L = \pi(r_1 + r_2) + 2l + \frac{(r_1 - r_2)^2}{l}$$

2. Length of belt in cross drive

$$L = \pi(r_1 + r_2) + 2l + \frac{(r_1 + r_2)^2}{l}$$

3. Angular velocity

$$\omega = \frac{2\pi N}{60}$$

4. Velocity ratio:

   (a) $\dfrac{N_2}{N_1} = \dfrac{r_1}{r_2} = \dfrac{d_1}{d_2}$, if belt thickness is negligible.

   (b) $\dfrac{N_2}{N_1} = \dfrac{2r_1 + t}{2r_2 + t} = \dfrac{d_1 + t}{d_2 + t}$, if belt thickness is considered.

   (c) $\dfrac{N_2}{N_1} = \dfrac{d_1}{d_2}\left(1 - \dfrac{p}{100}\right)$, where p is percentage slippage.

5. $\dfrac{\text{Tight side tension}}{\text{Slack side tension}} = \dfrac{T_2}{T_1} = e^{\mu\theta}$

6. Power transmitted $= (T_2 - T_1)\, v$

7. (a) If centrifugal tension is neglected,
$$T = T_2 = f\, b\, t$$
   (b) If centrifugal tension is considered
$$T = T_2 + T_c = f\, b\, t$$
   where $\qquad T_c = mv^2$

8. In case of rope drive in a groove,
$$\frac{T_2}{T_1} = e^{\mu\, \text{cosec}\, \alpha \times \theta}$$

9. In a two gear drive

   module $\qquad m = \dfrac{d_1}{T_1} = \dfrac{d_2}{T_2}$

   and $\qquad \dfrac{N_2}{N_1} = \dfrac{d_1}{d_2} = \dfrac{T_1}{T_2}$

10. In case of simple gear train drive, $\dfrac{N_D}{N_A} = \dfrac{T_A}{T_D}$.

11. In case of compound gear train drive, $\dfrac{N_D}{N_A} = \dfrac{T_C}{T_D} \times \dfrac{T_A}{T_B}$.

Theory Questions

1. Derive the expression for the length of belt in a typical
   (a) Open drive
   (b) Cross drive.
2. Show that in a belt drive the velocity ratio
   (a) $\dfrac{\omega_2}{\omega_1} = \dfrac{N_2}{N_1} = \dfrac{d_1}{d_2}$, if there is no slip
   (b) $\dfrac{\omega_2}{\omega_1} = \dfrac{N_2}{N_1} = \dfrac{d_1}{d_2}\left(1 - \dfrac{p}{100}\right)$, if slip is considered.
3. Derive the expression for the relationship between tight side and slack side tension is a typical belt drive.
4. Explain the term centrifugal tension in a belt drive. Derive the expression for it in a typical belt drive.
5. A rope of diameter d_1 is used in a groove of angle 2α. Find the relationship between tension on tight side and slack side of the rope.
6. State the conditions to be satisfied in a spur gear drive.
7. Derive the expression for velocity ratio in a typical 4 gear (a) Simple gear drive (b) Compound gear drive.

PROBLEMS FOR EXERCISE

7.1 An engine, running at 180 rpm drives a follower shaft by means of a belt. The diameter of the engine pulley is 750 mm and that of the pulley on the follower shaft is 500 mm. Calculate the speed of the follower shaft, if:
   (a) there is no slip
   (b) there is 4 per cent slip. [Ans. (a) 270 rpm, (b) 259.2 rpm]

7.2 A shaft running at 180 rpm drives a spindle which is at a distance 2.4 m. The spindle has to run at 100 rpm. If the diameter of the wheel on the driven shaft is 250 mm and the drive is by open belt, determine the length of the belt. Assume no slippage.
[Ans. $L = 5.904$ m]

7.3 A cross belt 120 mm wide and 8 mm thick connects two pulleys mounted on parallel shafts at 2.5 m apart. The diameter of the larger pulley is 600 mm and that of the smaller is 400 mm. The bigger pulley is rotating at 120 rpm. If the permissible stress in the belt is 2 N/mm², find the maximum power that can be transmitted at this speed. Take $\mu = 0.3$. [Ans. $P = 2.369$ kW]

7.4 The dimensions of a brake drum are as shown in Fig. 7.9. Determine the torque exerted on the drum, if the load $P = 50$ N. Assume the coefficient of kinetic friction between the belt and the drum as 0.15. **[Ans. 747.7 N]**

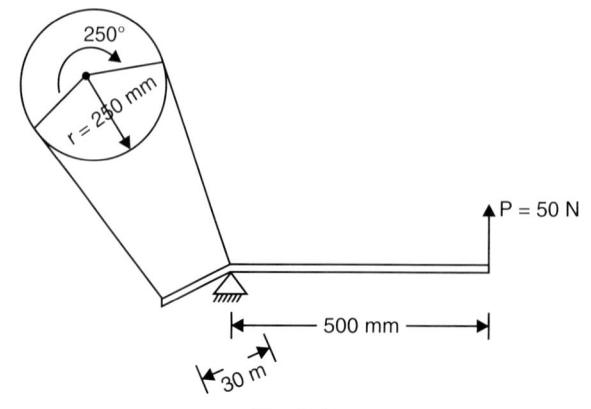

Fig. 7.16

7.5 An open belt 120 mm wide and 8 mm thick connects two pulleys mounted on parallel shafts 2.4 m apart. The diameter of the larger pulley is 540 mm and that of the smaller pulley is 360 mm. The larger pulley is rotating at 850 rpm. If the permissible stress on the belt section is 2.5 N/mm² and the coefficient of friction is 0.3. Calculate the maximum power that can be transmitted at this speed,

(a) neglecting centrifugal tension,

(b) considering the centrifugal tension; if the density of the leather is 1100 kg/m³.

[Ans. (a) 35.704 kW (b) 26.630 kW]

7.6 A rope of 40 mm diameter drives a grooved pulley at a speed of 24 m/sec. The angle of lap of the rope on the pulley is 240° and the angle of the groove for the pulley is 60°. If the permissible tension in the rope is 2 N/mm², calculate the maximum power that can be transmitted in this speed. Take coefficient of friction $\mu = 0.25$. Neglect the centrifugal tension. **[Ans. 52.889 kW]**

7.7 Two parallel shafts are to be connected by spur gearing. The axes of the shafts are to be at 480 mm apart, as nearly as possible. If one shaft has to run at 110 rpm and the other at 330 rpm, find the number of teeth on each wheel, if the module is to be 7.5 mm. Determine the exact distance between the two shafts.

[Ans. $T_1 = 96$, $T_2 = 32$, $l = 480$ mm]

7.8 In a gear drive a power of 2.4 kW is to be transmitted by a pinion of 18 teeth of 8 mm module. Calculate the speed of the pinion required, if the pressure between the teeth is to be 540 newtons. **[Ans. $N = 589.46$ rpm]**

8
Virtual Work Method

So far in this book we saw application of Newtonian mechanics to many engineering mechanics problems. There were many scientists before and after Newton who found solutions to mechanics problems by completely different approach. Jean Bernoulli (350 BC) and Leonardo da Vinci (1435) are the names associated with this school of thought. They imagined a small displacement of the bodies and found out the work done by various forces in the system and said if the body is in equilibrium then the total work done by the system should be zero. This approach is called 'virtual work method'. The word *virtual* is used since in reality there is no actual displacement. The term *virtual* is defined by Webster as 'being in essence of effect but not in fact.'

The work done is virtual if the displacements are virtual or forces acting are virtual. Hence in virtual work approach we may have the following two approaches:

(*i*) Principles of virtual displacements,
(*ii*) Principles of virtual forces.

In this chapter we consider the principle of virtual displacements.

8.1. WORK DONE BY FORCES AND MOMENTS

Work done by a force = force × distance moved in the direction of force.

In Fig. 8.1, various forces acting on a particle are shown. If the particle moves a distance s in x direction, from A to B, then the work done by various forces are as given below:

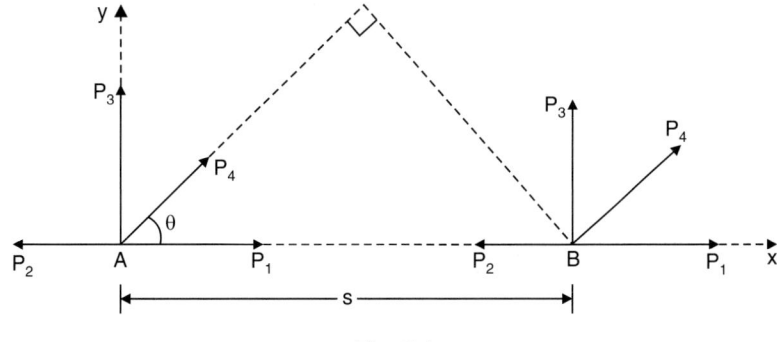

Fig. 8.1

Forces	Work done
P_1	$P_1 s$
P_2	$P_2(-s) = -P_2 s$
P_3	$P_3 \times 0 = 0$
P_4	$P_4 s \cos\theta$

Similarly if the particle moves from A to C in vertical direction (Fig. 8.2), the work done by various forces is as given below:

Forces	Work done
P_1	$P_1 \times 0 = 0$
P_2	$P_2 \times 0 = 0$
P_3	$P_3 \times s' = P_3 s'$
P_4	$P_4 \times s' \sin\theta = P_4 s' \sin\theta$

If the displacement is at a particular angle, say θ to the direction of the force, as is the case of force P_4 in the above cases, it will be convenient to find the perpendicular and parallel components of the forces to displacement and then find the work done. The component of P_4 in x and y directions are:

$$P_{4x} = P_4 \cos\theta$$
$$P_{4y} = P_4 \sin\theta$$

In Fig. 8.1

Displacement in x direction = s

Displacement in y direction = 0

∴ The work done by force P_4, is
$$= P_{4x} s + P_{4y} 0 = P_{4x} s = Ps \cos\theta$$

Similarly in Fig. 8.2

Displacement in x direction = 0

Displacement in y direction = s'

∴ Hence, work done by P_4
$$= P_{4x} \times 0 + P_{4y} s'$$
$$= P \sin\theta \, s' = P_4 s' \sin\theta$$

Fig. 8.2

On the same lines, we can say work done by a moment is equal to the product of the moment and the angular rotation in the direction of the moment. In Fig. 8.3a, the moments M_1, M_2 and M_3 are acting on a body which rotates by angle θ as shown in Fig. 8.3b.

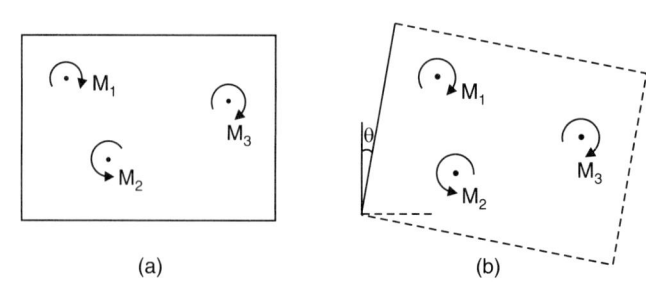

(a)   (b)

Fig. 8.3

VIRTUAL WORK METHOD

Then the work done by various moments are as given below:

Moments	Work done
M_1	$M_1\theta$
M_2	$-M_2\theta$
M_3	$M_3\theta$

1. No work is done by a force if displacement is at right angles to the line of action of force.
2. Work done is – ve if the displacement is in the direction opposite to force.
3. Work done by a moment is – ve if rotation is opposite to the direction of moment.

8.2. THE METHOD

The method consists of the following steps:
1. Give an elemental virtual displacement to the body in equilibrium in suitable direction.
2. Write the work done by each force in the system.
3. Sum up to get total work done by system of forces and equate it to zero to get required equations. Many times such equations are identical to equations of equilibrium. However, the method of getting these equations is completely different in this method.

A number of problems are solved to explain and to illustrate the virtual work method.

8.3. SIGN CONVENTION

In this chapter the positive senses of moments and displacements are as given below:
1. Forces and displacements are +ve if they are in positive directions of coordinates as shown in Fig. 8.4.
2. Moments and rotations are positive if they are in clockwise directions.

Example 8.1. *Determine the resultant of a system of concurrent forces acting on a body as shown in Fig. 8.5a.*

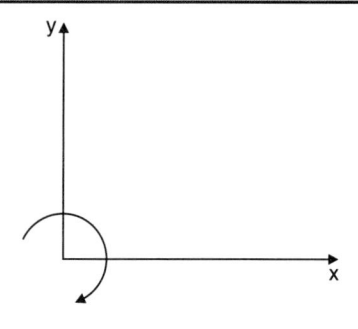

Fig. 8.4

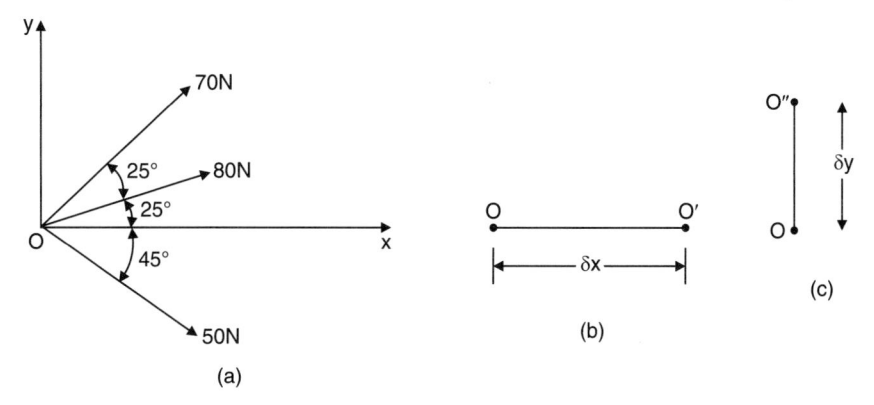

Fig. 8.5

Solution. Let E be the equilibriant of the system which makes anticlockwise angle θ with x-axis. Then the system is in equilibrium with force F. Let us give a virtual displacement of δx in x-direction as shown in Fig. 8.5(b).

Then total virtual work done by the system of given forces is
$$50 \cos 45° \, \delta x + 80 \cos 25° \, \delta x + 70 \cos 50° \, \delta x + E \cos \theta \, \delta x$$

Equating it to zero we get
$$(50 \cos 45° + 80 \cos 25° + 70 \cos 50° + E \cos \theta) \, \delta x = 0$$

Since $\delta x \neq 0$,
$$50 \cos 45° + 80 \cos 25° + 70 \cos 50° + E \cos \theta = 0$$

or
$$- E \cos \theta = 50 \cos 45° + 80 \cos 25° + 70 \cos 50° = 152.86 \text{ N} \qquad \ldots(1)$$

Let us now give virtual displacement δy to point O as shown in Fig. 8.5c. Then virtual work done by the system of forces and the equilibriant are:
$$- 50 \sin 45° \, \delta y + 80 \sin 25° \, \delta y + 70 \sin 50° \, \delta y + E \sin \theta \, \delta y$$

Since the system is in equilibrium with E, we get
$$(- 50 \sin 45° + 80 \sin 25° + 70 \sin 50° + E \sin \theta) \, \delta y = 0$$

Since $\delta y \neq 0$,
$$- 50 \sin 45° + 80 \sin 25° + 70 \sin 50° + E \sin \theta = 0$$

or
$$- E \sin \theta = - 50 \sin 45° + 80 \sin 25° + 70 \sin 50° = 52.07 \sin 25°$$

Resultant is exactly equal but opposite of equillibriant. Hence, eqns. (1) and (2) can be written as
$$R \cos \theta = 152.86 \text{ N}$$

and
$$R \sin \theta = 52.07 \text{ N}$$

∴
$$R = \sqrt{152.86^2 + 52.07^2} = 161.48 \text{ N}$$

and
$$\theta = \tan^{-1} \frac{52.07}{152.86}$$

∴
$$\theta = 18.81°$$

Example 8.2. *Determine the reactions R_A and R_B developed in the simply supported beam shown in Fig 8.6a.*

Solution. Let R_A and R_B be the reactions at supports A and B. Let δy displacement be given to beam at B without giving any displacement at A. (Ref. Fig. 8.6.)

The corresponding displacements for the beam at 20 kN load point and 40 kN load points are $\frac{1}{3} \delta y$ and $\frac{2}{3} \delta y$. By virtual work principle

$$R_A \times 0 - 20 \times \frac{1}{3} \delta y - 40 \times \frac{2}{3} \delta y + R_B \, \delta y = 0$$

or
$$\left(\frac{-20}{3} - \frac{80}{3} + R_B \right) \delta y = 0$$

Since $\delta y \neq 0$
$$\frac{-20}{3} - \frac{80}{3} + R_B = 0$$

or
$$R_B = \frac{20}{3} + \frac{80}{3} = \frac{100}{3} = 33.33 \text{ kN} \quad \textbf{Ans.}$$

VIRTUAL WORK METHOD

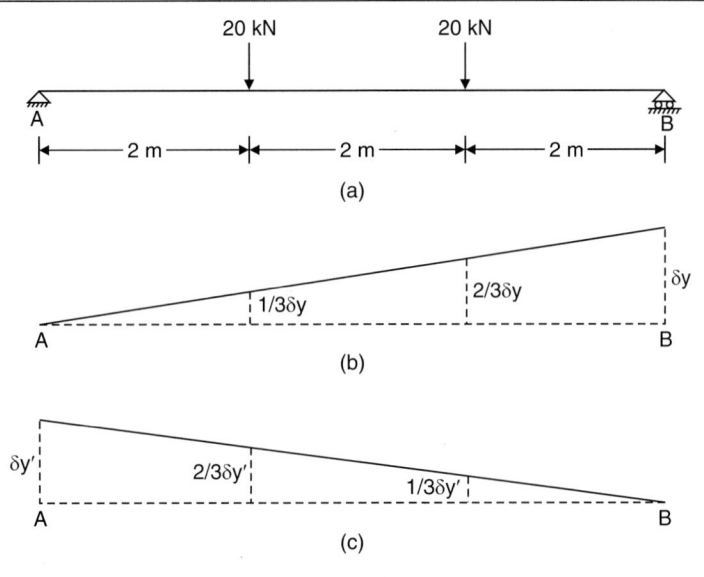

Fig. 8.6

Now let us give virtual displacement to beam as shown in Fig. 8.6c. Since the body is in equilibrium virtual work equation is

$$R_A \, \delta y' - 20 \times \frac{2}{3} \delta y' - 40 \times \frac{1}{3} \delta y' + R_B \times 0 = 0$$

$$\therefore \quad \left(R_A - \frac{40}{3} - \frac{40}{3}\right) \delta y' = 0$$

Since $\delta y \neq 0$,

$$R_A - \frac{40}{3} - \frac{40}{3} = 0$$

or

$$R_A = \frac{80}{3} = 26.667 \text{ kN} \quad \text{Ans.}$$

Example 8.3. *Determine the reactions in the overhanging beam shown in Fig. 8.7a.*

Solution. Let R_A and R_B be the reactions as shown in Fig. 8.7a. Let the virtual displacement δy be given to the beam as shown in Fig. 8.7b. Then the virtual displacement of various loads are as given in the figure. Applying virtual work principle for the beam which is in equilibrium we get

$$20 \left(\frac{\delta y}{8}\right) + R_A \times 0 - 60 \left(\frac{4}{8} \delta y\right) + R_B \frac{6}{8} \delta y - 30 \times \delta y = 0$$

or

$$\left(\frac{20}{8} - \frac{60 \times 4}{8} + R_B \frac{6}{8} - 30\right) \delta y = 0$$

Since $\delta y \neq 0$,

$$\frac{20}{8} - \frac{60 \times 4}{8} + R_B \frac{6}{8} - 30 = 0$$

$$\therefore \quad R_B = 76.667 \text{ kN} \quad \text{Ans.}$$

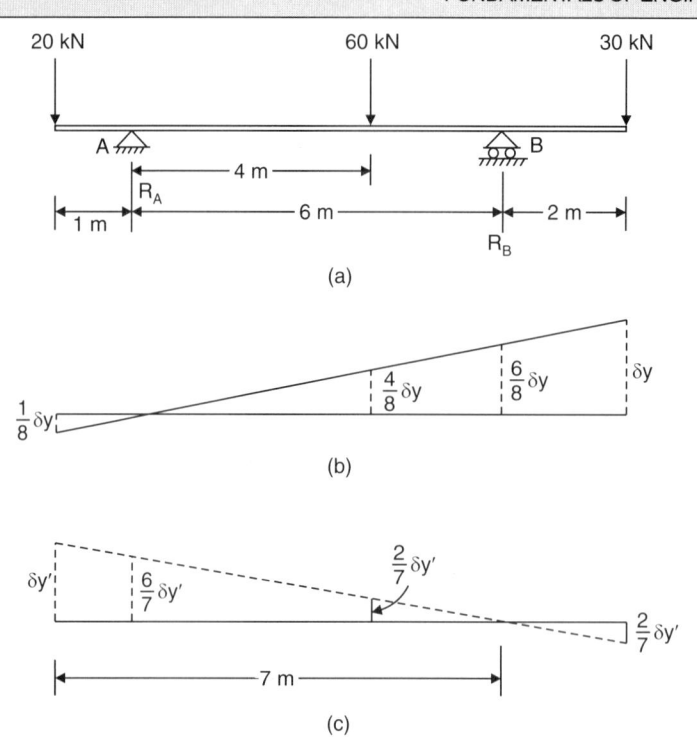

Fig. 8.7

Next, let us give virtual displacement $\delta y'$ to the beam as shown in Fig. 8.7c.

The displacements of various load/reaction points are as shown in the figure. From virtual work principle,

$$-20(\delta y') + R_A \frac{6}{7}\delta y' - 60\left(\frac{2}{7}\delta y'\right) + R_B \times 0 - 30\left(-\frac{2}{7}\delta y'\right) = 0$$

$$\left(-20 + R_A \frac{6}{7} - \frac{60 \times 2}{7} + 30 \times \frac{2}{7}\right)\delta y' = 0$$

Since $\delta y' \neq 0$, $\quad -20 + R_A \frac{6}{7} - \frac{60 \times 2}{7} + 30 \times \frac{2}{7} = 0$

$$R_A = \textbf{33.333 kN} \quad \textbf{Ans.}$$

Example 8.4. *Determine the reaction at A in simply supported beam shown in Fig. 8.8a.*

Solution. Let the virtual displacement δy be given at point A as shown in Fig. 8.8b.

Total uniformly distributed load = $40 \times 2 = 80$ kN

Its centre of gravity is at mid point of AC

Displacement of the load at $A = \delta y$

Displacement of the load at $C = \dfrac{\delta y}{2}$

∴ Displacement of 80 kN load at its centroid

$$= \text{Average of } \delta y \text{ and } \frac{\delta y}{2} = \frac{\delta y + \delta y/2}{2} = 0.75\,\delta y$$

VIRTUAL WORK METHOD

Hence, work done by udl
$$= -40 \times 2(0.75\, \delta y) = -60\, \delta y$$

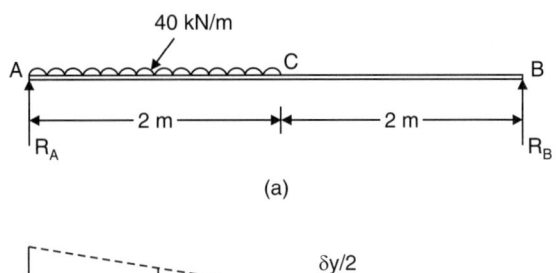

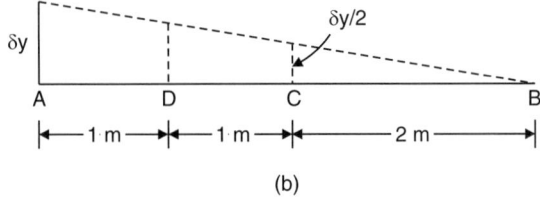

Fig. 8.8

Writing virtual work equation for the system of forces we get
$$R_A\, \delta y - 60\, \delta y + R_B \times 0 = 0$$
$$(R_A - 60)\, \delta y = 0$$

Since, $\delta y \neq 0$
$$R_A = \mathbf{60\ kN} \quad \mathbf{Ans.}$$

Example 8.5. *Determine the reaction at A in the overhanging beam AB as shown in Fig. 8.9 (a).*

Solution. Let the virtual displacement δy be given at point C keeping displacement at B zero as shown in Fig. 8.9b. Let R_A and R_B be the reactions at A and B respectively.

The centroid of udl moves down by $\dfrac{0 + 2\delta y}{2} = \delta y$. Total udl $= 30 \times 6$ kN; downward. Hence virtual work equation is

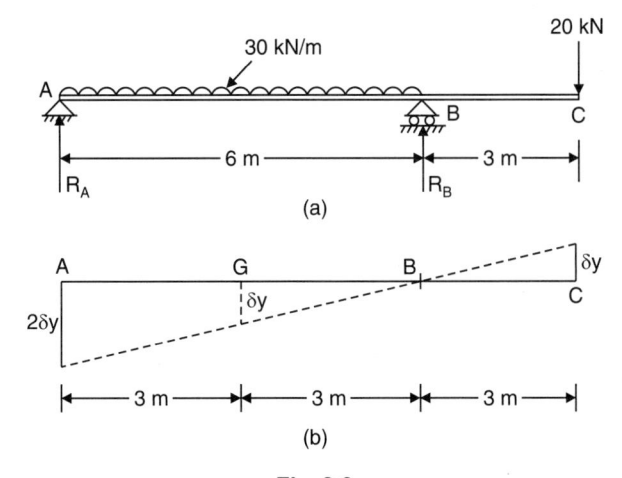

Fig. 8.9

$$R_A(-2\delta y) - 30 \times 6(-\delta y) + R_B \times 0 - 20(\delta y) = 0$$
$$(-2R_A + 180 - 20)\delta y = 0$$
Since $\delta y \neq 0$,
$$-2R_A + 180 - 20 = 0$$
$$\therefore \quad R_A = 80 \text{ kN} \quad \text{Ans.}$$

Example 8.6. *The simply supported beam AB of span 5 m is subjected to a concentrated load and an external moment as shown in Fig. 8.10a. Determine the reaction at B.*

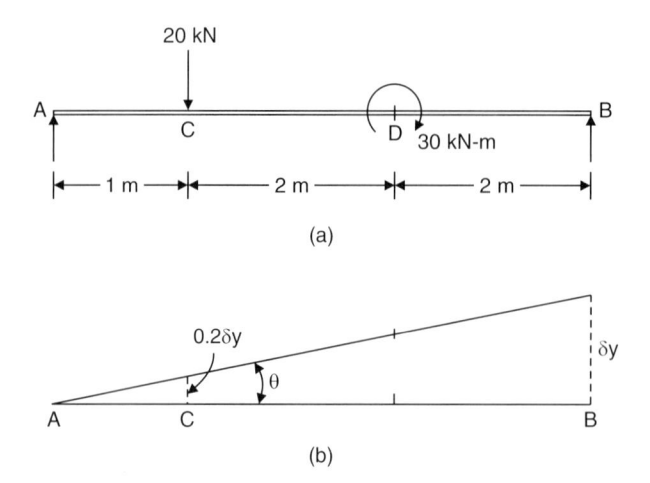

Fig. 8.10

Solution. Let a virtual displacement δy be given at B as shown in Fig. 8.10b.
Displacement at $A = 0$
Displacement at $C = 0.2\,\delta y$
Rotation of moment 30 kNm is,
$$\theta = \frac{-\delta y}{5} = -0.2\,\delta y$$
−ve sign since it is anticlockwise.
Writing virtual work equations, we get
$$R_A \times 0 - 20(0.2\,\delta y) + 30(-0.2\,\delta y) + R_B\,\delta y = 0$$
$$(-4 - 6 + R_B)\,\delta y = 0$$
Since $\delta y \neq 0$, $\quad -4 - 6 + R_B = 0$
or
$$R_B = 10 \text{ kN} \quad \text{Ans.}$$

Example 8.7. *Determine the reaction R_A in the overhanging beam shown in the Fig. 8.11(a) by virtual work method.*

Solution. Let us give virtual displacement δy at A as shown in Fig. 8.11b.
Displacement at A is δy

Displacement at D is $\dfrac{3}{5}\,\delta y = 0.6\,\delta y$.

$\therefore\quad$ Displacement at centroid of udl $= \dfrac{\delta y + 0.6\delta y}{2} = 0.8\,\delta y$

Displacement at $B = 0$

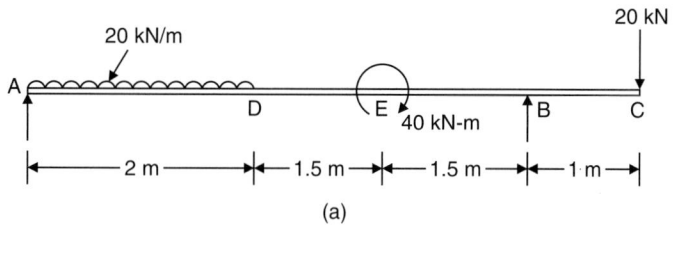

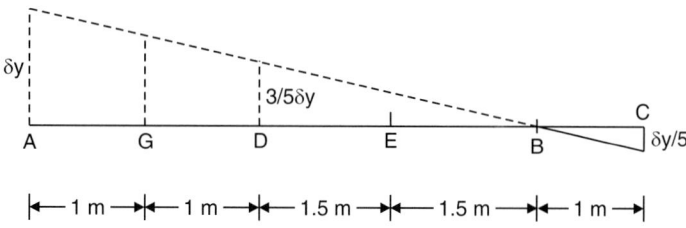

VIRTUAL WORK METHOD

Fig. 8.11

Displacement at C is $-\dfrac{1}{5}\delta y = -0.2\,\delta y$

Rotation of moment $= \dfrac{\delta y}{5} = 0.2\delta y$

Writing the virtual work equation for the system, we get

$$R_A\,\delta y - 20 \times 2\,(0.8\,\delta y) + 40\,(0.2\,\delta y) + R_B \times 0 - 20\,(-0.2\,\delta y) = 0$$
$$(R_A - 32 + 8 + 0 + 4)\,\delta y = 0$$

Since $\delta y \neq 0$

$$R_A - 32 + 8 + 4 = 0 \quad \text{or} \quad R_A = \mathbf{20\ kN}\quad \textbf{Ans.}$$

Example 8.8. *Determine the reaction R_A in the beam as shown in Fig 8.12a*

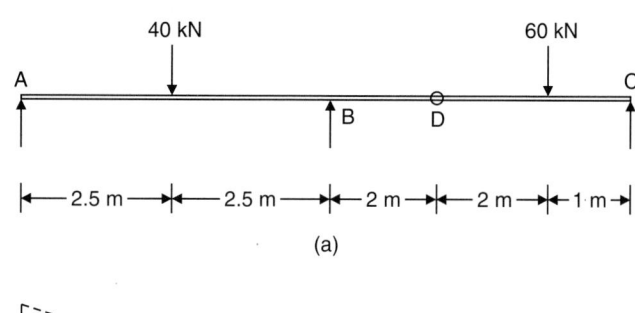

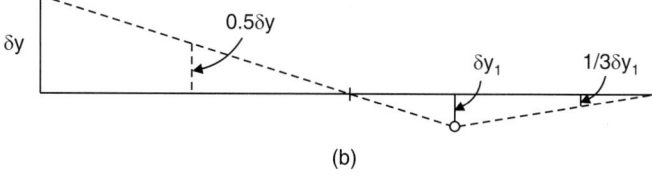

Fig. 8.12

Solution. Let us give virtual displacement δy at A and no displacements at B and C. Since D is hinged, the shape of displaced structure is as shown in Fig. 8.12b.

Now, displacement under 40 kN load = $\dfrac{2.5}{5}\delta y = 0.5\,\delta y$

Displacement at D, $\delta y_1 = -\dfrac{2}{5}\delta y = -0.4\,\delta y$

Displacement of 60 kN load $= -\dfrac{1}{3}\delta y_1 = -\dfrac{1}{3}\times 0.4\,\delta y$

∴ Virtual work equation is

$$R_A\,\delta y - 40\,(0.5\,\delta y) + R_B \times 0 - 60\left(-\dfrac{1}{3}\times 0.4\,\delta y\right) + R_C \times 0 = 0$$

∴ $(R_A - 20 + 8)\,\delta y = 0$

Since $\delta y \neq 0$,

$R_A - 20 + 8 = 0$ or $\quad R_A = \mathbf{12\ kN}$ **Ans.**

Example 8.9. *The surfaces AB and BC shown in Fig 8.13 (a) are smooth. Determine the ratio of loads P and Q for which the system is in equilibrium.*

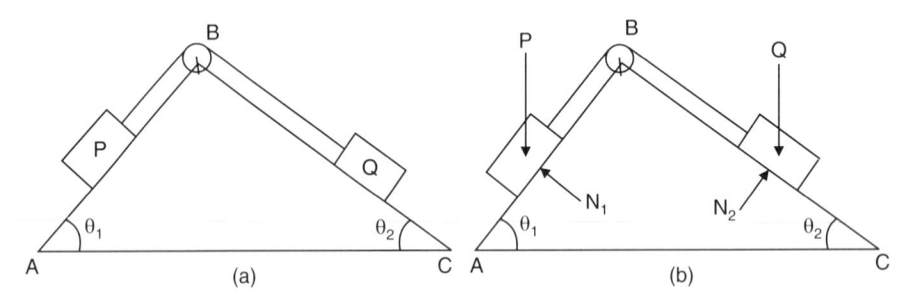

Fig. 8.13

Solution. Since the surfaces AB and BC are smooth frictional forces do not exist. There are normal reactions N_1 and N_2 acting on the blocks as shown in Fig. 8.13 b. If displacement δs is given to block of weight P down the plane block Q moves up the plane BC by δs. There is no movement in the direction of normal to planes.

The work done by various forces in the system are:

$$-P\sin\theta_1\,(-\delta s) + 0 - Q\sin\theta_2\,(\delta s) + 0 = 0$$

i.e., $\quad P\sin\theta_1 = Q\sin\theta_2$

or $\quad \dfrac{P}{Q} = \dfrac{\sin\theta_2}{\sin\theta_1}\quad$ **Ans.**

Example 8.10. *Determine the magnitude of horizontal force P to be applied to a block of weight 1500 N to hold it in the position on a smooth inclined plane AB which makes an angle of 30° with the horizontal (Ref. Fig. 8.14. a).*

Solution. The system of forces acting on the block are as shown in the Fig. 8.14(b).

Since the plane is smooth no frictional force acts. Let the block be given a virtual displacement δs up the plane.

Then displacement in vertical direction $= \delta s\sin\theta$

Then displacement in horizontal direction $= \delta s\cos\theta$

VIRTUAL WORK METHOD

Then displacement in normal to plane = 0

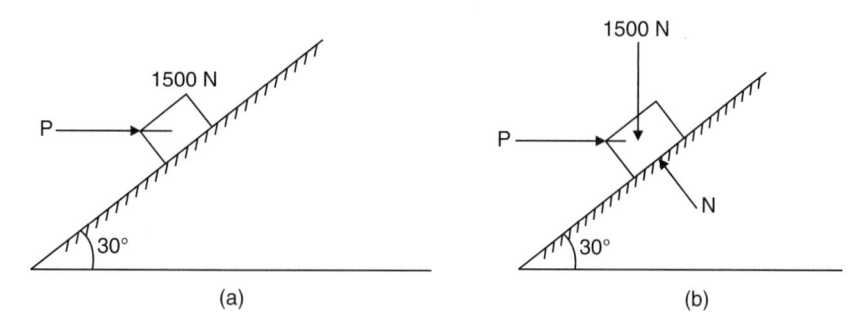

Fig. 8.14

∴ From principle of virtual work,
$$- 1500\ (\delta s \sin 30°) + P\ \delta s \cos 30° + 0 = 0.$$
or $\quad P \cos 30° = 1500 \sin 30°$
$$P = 1500 \tan 30° = \mathbf{866.03\ N} \quad \mathbf{Ans.}$$

Example 8.11. *A block weighing 1000N is to be held on an inclined plane AB which makes angle of 70° with horizontal. For what range of force P it can be kept in equilibrium. Coefficient of friction between the plane and the block is 0.3 (Ref. Fig. 8.15a).*

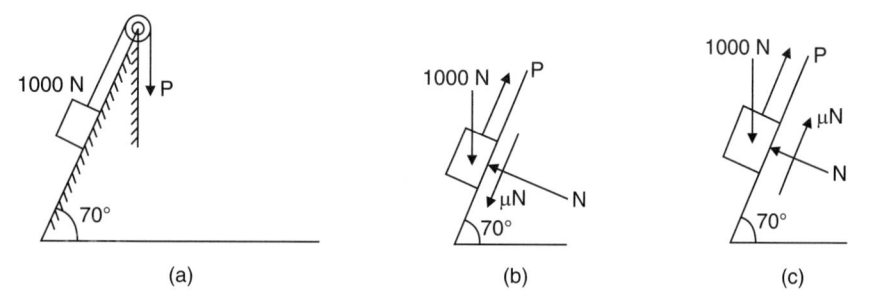

Fig. 8.15

Solution. (*a*) When motion is impending up the plane: The system of forces acting in this situation are shown in Fig. 8.15 (*b*). Frictional force acts down the plane
$$F = \mu N = \mu\ 1000 \cos 70° = 0.3 \times 1000 \cos 70° = 300 \cos 70°$$

Let us give virtual displacement δs up the plane. Let the direction up the plane be taken +ve.

Then from principle of virtual work, we get
$$- 1000 \sin 70°\ (\delta s) - 300 \cos 70°\ (\delta s) + P\ \delta s = 0$$
∴ $\quad - 1000 \sin 70° - 300 \cos 70° + P = 0$
or $\quad P = 1000 \sin 70° + 300 \cos 70° = 1042.3\ N$

(*b*) When motion is impending down the plane. The frictional force acts up the plane as shown in Fig. 8.15 (*c*). Let δs be virtual displacement up the plane. Then virtual work equation is,
$$- 1000 \sin 70°\ \delta s + F\ \delta s + P\ \delta s = 0$$
∴ $\quad - 1000 \sin 70° + 300 \cos 70° + P = 0$
or $\quad P = 1000 \sin 70° - 300 \cos 70° = 837.1\ N$

Hence, the block is in equilibrium for $P = 837.1\ N$ to $1042.3\ N$.

Example 8.12. *Ball 1 shown in Fig. 8.16 weighs 200 N and ball 2 weighs 150 N. They are connected by an inextensible string of length l. Determine the position of the balls if the system is left on the inclined plane shown in the figure. Assume the planes are smooth.*

Solution. Let the system be in equilibrium when string DE makes an angle θ with plane AB as shown in Fig. 8.16 (b).

Now, $DB = l \cos\theta$ and $EB = l \sin\theta$

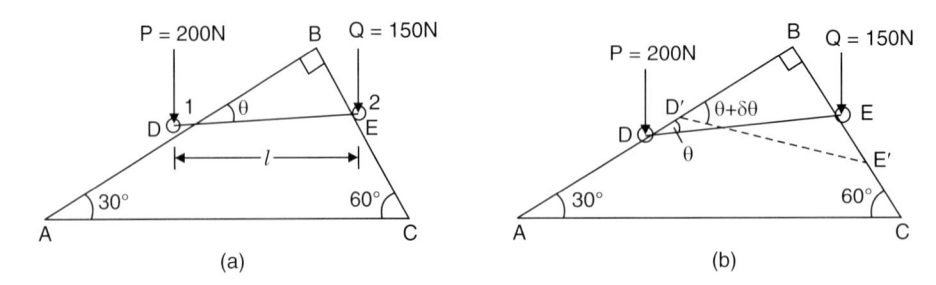

Fig. 8.16

Let a virtual displacement be given to the system of balls so that they move to $D'E'$ i.e., by an angle $\delta\theta$.

Component of P down the plane $= P \sin 30° = 200 \sin 30°$
Component of Q down the plane $= Q \sin 60° = 150 \sin 60°$

Since
$$D_B = l \cos\theta$$
$$\delta(D_B) = -l \sin\theta\, \delta\theta$$

and
$$E_B = l \sin\theta$$
$$\delta(E_B) = l \cos\theta \cdot \delta\theta$$

No displacement normal to the plane.

∴ From principle of virtual work, we get

$$200 \sin 30°\, \delta(DB) + 150 \sin 60°\, \delta(EB) = 0$$
$$200 \sin 30° (-l \sin\theta)\, \delta\theta + 150 \sin 60°\, l \cos\theta\, \delta\theta = 0$$

or
$$\tan\theta = \frac{150 \sin 60°}{200 \sin 30°} = 1.299$$

$$\theta = 52.41° \quad \textbf{Ans.}$$

Example 8.13. *For the system of connected bodies as shown in Fig. 8.17a determine the force P required to make the motion impending to the left. Coefficient of friction for all contacting surfaces is 0.25. Pulleys are smooth.*

Solution. Since motion is impending to the left, frictional forces are acting in the directions shown in Fig. 8.17(b).

$$F_1 = 0.25\, N_1 = 0.25 \times 250 = 62.5 \text{ N}$$
$$F_2 = 0.25\, N_2 = 0.25 \times 1000 \cos 45° = 176.78 \text{ N}$$
$$F_3 = 0.25\, N_3 = 0.25 \times 500 = 12.5 \text{ N}$$

Let us give a virtual displacement towards left. Taking the displacement of the system to left as +ve, virtual work equation is

$$P\delta s - F_1\, \delta s - 1000 \sin 45°\, \delta s - F_2\, \delta s - F_3\, \delta s = 0$$
∴
$$(P - F_1 - 1000 \sin 45° - F_2 - F_3)\, \delta s = 0$$

VIRTUAL WORK METHOD

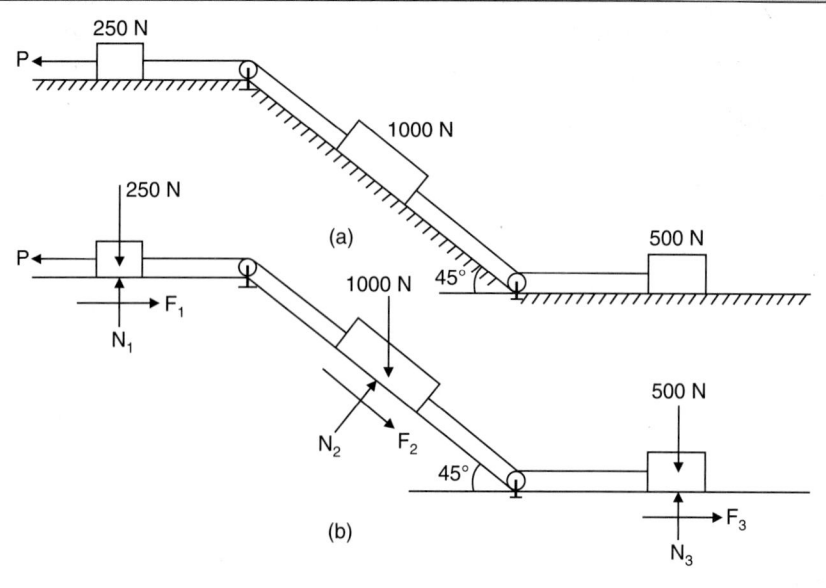

Fig. 8.17

Since $\delta s \neq 0$,

$$P = F_1 + 1000 \sin 45° + F_2 + F_3$$
$$= 62.5 + 1000 \sin 45° + 176.78 + 125$$
$$= \mathbf{1071.38 \, N} \quad \mathbf{Ans.}$$

Example 8.14. *Using the principle of virtual work determine the weight W_B for which the motion of block A is impending up the plane. Given: coefficient of friction between block A and the plane = 0.3 and pulleys are smooth (Ref. Fig. 8.18a)*

Solution.
$$\tan \theta = \frac{3}{4}$$
$$\theta = 36.87°$$
$$F = \mu N = \mu W_A \cos \theta$$
$$= 0.3 \times 200 \cos 36.87° = 48 \, N.$$

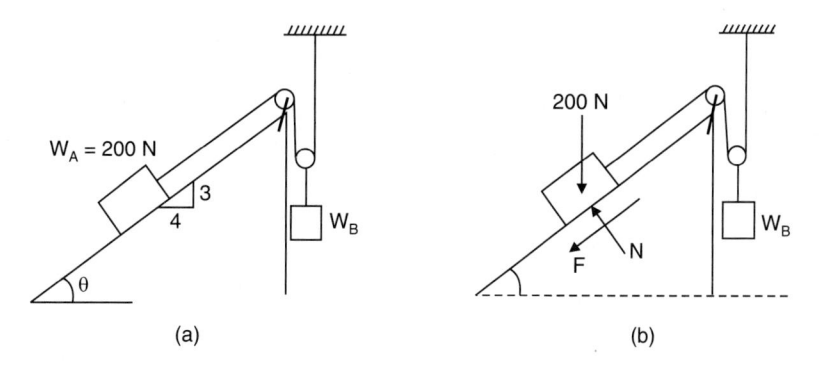

Fig. 8.18

Let us give virtual displacement of δs up the plane to block A.

Component of block A down the plane $= W_A \sin \theta$

Component normal to the plane do not do any work since there is no displacement normal to the plane.

When block A moves by δs, weight B moves down by $\dfrac{\delta s}{2}$.

Hence, virtual work equation is

$$-200 \sin \theta \, \delta s - F \, \delta s + W_B \left(\dfrac{\delta s}{2}\right) = 0$$

or

$$\left(-200 \sin \theta - F + \dfrac{W_B}{2}\right) \delta s = 0$$

Since $\delta s \neq 0$

$$-200 \sin \theta - F + \dfrac{W_B}{2} = 0$$

or
$$W_B = 2(200 \sin \theta + F)$$
$$= 2(200 \sin 36.87° + 48) = \mathbf{336 \ N} \quad \mathbf{Ans.}$$

Example 8.15. *Ladder AB of length 4 m and weight 200 N is held in position by applying force P as shown in Fig. 8.19. Assuming smooth wall and floor, determine the force P. If instead of force P for preventing from slipping, a horizontal rope is tied at a distance of 1 m from A, what is the tension in the rope ?*

Solution. (a) *P* at floor level
$$AB = l = 4 \text{m}$$

Let ladder rotate by angle $\delta\theta$ as shown in Fig. 8.19(b).

Taking C as origin.

Coordinate of $A = x = -4 \sin \theta$
$$\therefore \quad \delta x = -4 \cos \theta \, \delta\theta$$

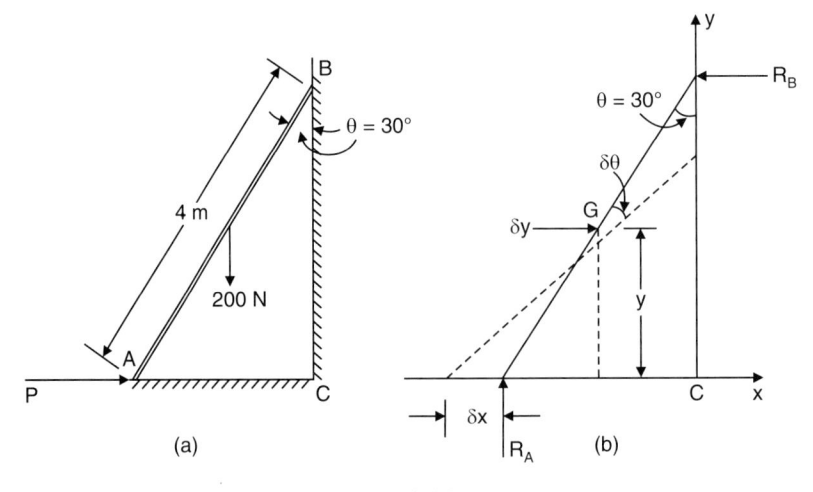

Fig. 8.19

P moves by $-4 \cos \theta \, \delta\theta$

Now, $\qquad y = \dfrac{l}{2} \cos \theta$

$$\delta y = -\frac{l}{2}\sin\theta\,\delta\theta = -2\sin\theta\,\delta\theta \text{ since } l = 4 \text{ m.}$$

The reactions R_A and R_B are not doing any work since there is no movement of their point of applications in the direction they act. Applying principle of virtual work, we get

$$P\,\delta x - 200\,\delta y = 0$$
$$P(-4\cos\theta)\,\delta\theta - 200(-2\sin\theta)\,\delta\theta = 0$$
$$P(-4\cos\theta) + 400\sin\theta = 0$$

or
$$P\frac{400\sin\theta}{4\cos\theta} = 100\tan 30° = \mathbf{57.735 \text{ N}} \quad \textbf{Ans.}$$

(b) If a rope is used instead of force P, the system of forces is as shown in Fig. 8.20a.

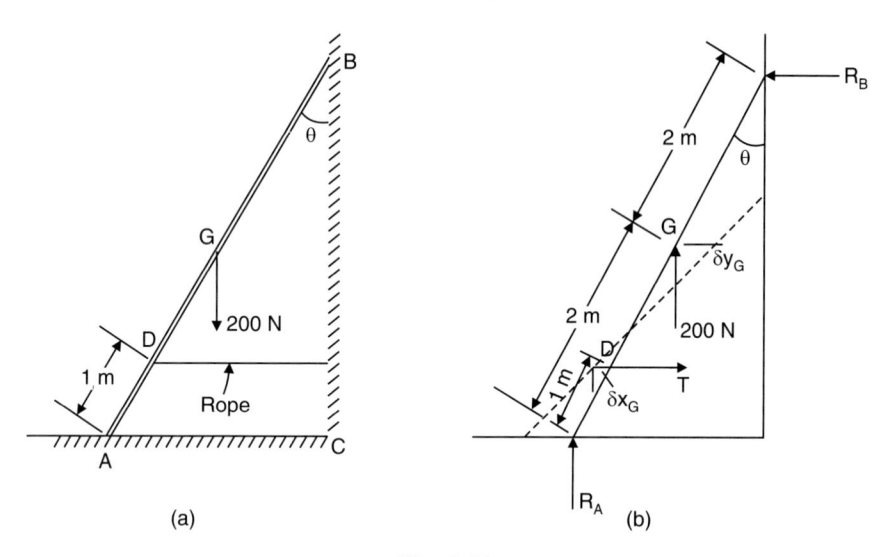

Fig. 8.20

The system of forces in this case are as shown in Fig. 8.20b. Taking C as origin,
$$x_D = -3\sin\theta$$
$$\delta x_D = -3\cos\theta\,\delta\theta$$
$$y_G = 2\cos\theta$$
$$\delta y_G = -2\sin\theta\,\delta\theta$$

There is no displacement in the direction of R_B and R_A. Hence from the principle of virtual work,

$$T(\delta x_D) + 200(\delta y_G) = 0$$
$$T(-3\cos\theta\,\delta\theta) - 200(-2\sin\theta\,\delta\theta) = 0$$
$$(-3T\cos\theta + 400\sin\theta)\,\delta\theta = 0$$

Since $\delta\theta \neq 0$
$$3T\cos\theta = 400\sin\theta$$
$$T = \frac{400}{3}\tan\theta = \frac{400}{3}\tan 30° = \mathbf{76.98 \text{ N}} \quad \textbf{Ans.}$$

Example 8.16. *The ladder shown in Fig. 8.21a is 6 m long and is supported by a horizontal floor and vertical wall. The coefficient of friction between the floor and the ladder is 0.4 and between the wall and the ladder is 0.25. The weight of the ladder is 200 N. The ladder also supports a vertical load of 900 N at C which is at a distance of 1 m from B. Determine the least value of θ for which the ladder may be placed without slipping. Determine the reaction at this stage.*

Solution. The various forces acting on the ladder when it is on the verge of slipping are as shown in Fig. 8.21b. Note frictional forces always oppose impending motions and they are equal to $\mu \times$ normal reaction where μ is coefficient of friction. To calculate virtual works we need the displacement in the direction of forces. Normal reactions N_A and N_B are not going to do any work since displacements in direction of those forces is always zero. Taking O as origin,

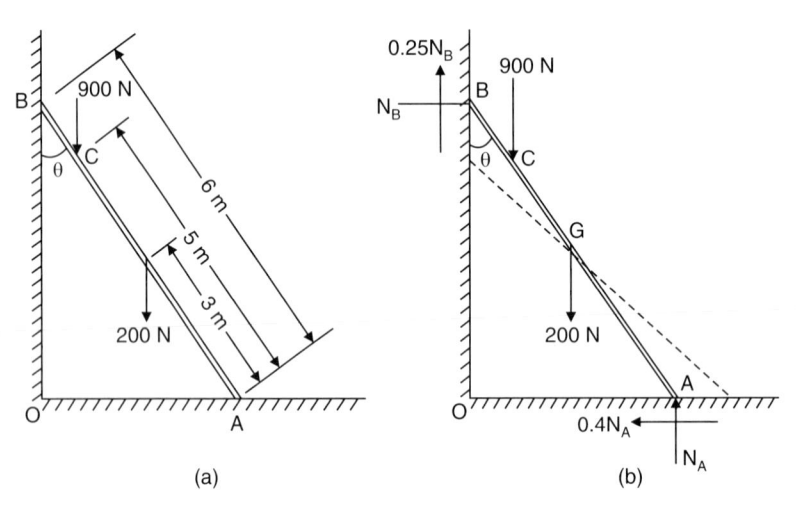

now,
$\quad x_A = 6 \sin \theta \qquad \delta x_A = 6 \cos \theta \, \delta\theta$
$\quad y_G = 3 \cos \theta \qquad \delta y_G = -3 \sin \theta \, \delta\theta$
$\quad y_C = 5 \cos \theta \qquad \delta y_C = -5 \sin \theta \, \delta\theta$
and $\quad y_B = 6 \cos \theta \qquad \delta y_B = -6 \sin \theta \, \delta\theta$

Fig. 8.21

Applying principles of virtual work we get,
$-0.4 \, N_A \, 6 \cos \theta \, \delta\theta - 200 \, (-3 \sin \theta \, \delta\theta) - 900 \, (-5 \sin \theta \, \delta\theta) + 0.25 \, N_B \, (-6 \sin \theta \, \delta\theta) = 0$
Since $\delta\theta \neq 0$,
$\quad -2.4 \, N_A \cos \theta + 600 \sin \theta + 4500 \sin \theta - 1.5 \, N_B \sin \theta = 0 \qquad \ldots(1)$
From the equilibrium conditions,
$\quad \Sigma F_x = 0 \rightarrow 0.4 \, N_A = N_B$
and $\quad \Sigma F_y = 0 \rightarrow N_A + 0.25 \, N_B = 900 + 200 = 1100$
$\quad N_A + 0.25 \times 0.4 \, N_A = 1100$
$\therefore \qquad N_A = 1000 \text{ N}$
Hence $\quad N_B = 0.4 \times 1000 = 400 \text{ N}$
Substituting these values in eqn. (1), we get
$\quad -2.4 \times 1000 \cos \theta + 600 \sin \theta + 4500 \sin \theta - 1.5 \times 400 \sin \theta = 0.$
i.e., $\qquad 4500 \sin \theta = 2400 \cos \theta$

VIRTUAL WORK METHOD 215

or
$$\tan \theta = \frac{2400}{4500}$$
$$\theta = 28.073 \quad \text{Ans.}$$

Example 8.17. *Determine the force in the member FH of the truss shown in Fig 8.22. Each load is 10 kN and all triangles are equilaterals with sides 4 m.*

Solution. Due to symmetry

$$R_A = R_B = \frac{1}{2} \text{ total load} = \frac{1}{2} \times 10 \times 7 = 35 \text{ kN}$$

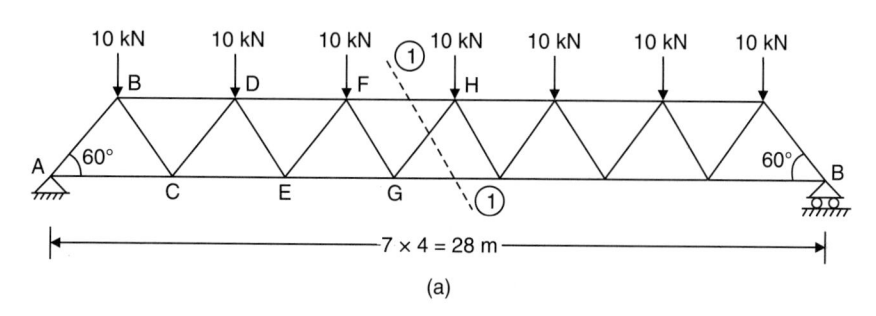

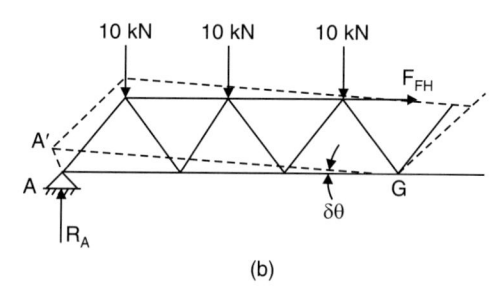

Fig. 8.22

Take section 1-1 as shown in Fig. 8.22 (a) and consider left side portion as shown in Fig. 8.22 (b). Imagine a rotation of $\delta\theta$ about point G.

Vertical displacement of $A = 12\,\delta\theta$
Vertical displacement of $B = 10\,\delta\theta$
Vertical displacement of $D = 6\,\delta\theta$
and Vertical displacement of $F = 2\,\delta\theta$
Horizontal displacement of $F_{FH} = 2 \tan 60°$ towards right

∴ From the principle of virtual displacement, we have

$$R_A\, 12\,\delta\theta - 10 \times 10\,\delta\theta - 10 \times 6\,\delta\theta - 10 \times 2\,\delta\theta + F_{FH}\, 2 \tan 60° = 0$$
$$2 F_{FH} \tan 60° = - R_A\, 12 + 100 + 60 + 20$$
$$= -35 \times 12 + 180 = -240$$

$$F_{FH} = -\frac{240}{2 \tan 60°} = -69.28 \text{ kN}$$

i.e., F_{FH} direction is to the reverted
i.e., F_{FH} is compressive force of **69.28 kN** **Ans.**

Note: The difference between the displacement in this case and the ladder problem. Here the rotation is free rotation about a point. Hence

vertical displacement = horizontal coordinate × δθ

and horizontal displacement = vertical distance × δθ

In ladder problem, the two edges are constrained to move along the specified two edges. In such cases:

Vertical displacement (– y direction) of point D

$$= \frac{d(y_D)}{dy}$$

and horizontal displacement $= \frac{d(x_D)}{dx}$

Example 8.18. *Determine the force developed in the member DF of the truss shown in Fig. 8.23.*

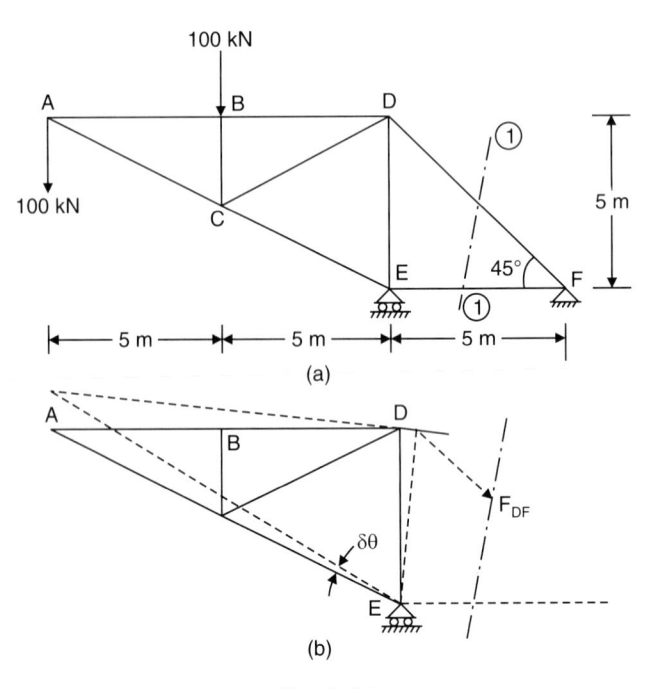

Fig. 8.23

Solution. Take the section 1-1 as shown in Fig. 8.23a and consider left side portion. Imagine a virtual rotation δθ at support E. Then

Vertical displacement of point $A = 10\,\delta\theta$

Vertical displacement of point $B = 5\,\delta\theta$

Vertical displacement of point $D = 0$

Horizontal displacement of point $D = 5\,\delta\theta$

∴ From virtual work principle,

$$-100 \times 10\,\delta\theta - 100 \times 5\,\delta\theta + F_{DF} \cos 45° \times 5\,\delta\theta = 0$$

∴ $F_{DF}\, 5 \cos 45° = 100 \times 10 + 100 \times 5 = 1500$

VIRTUAL WORK METHOD

$$\therefore \qquad F_{DF} = \frac{1500}{5 \cos 45°} = 300\sqrt{2} = 424.26 \text{ kN (tensile)}.$$

Example 8.19. *Neglecting the friction, determine the effort required to lift a load by the screw jack shown in Fig. 8.24. Take pitch of the jack = p.*

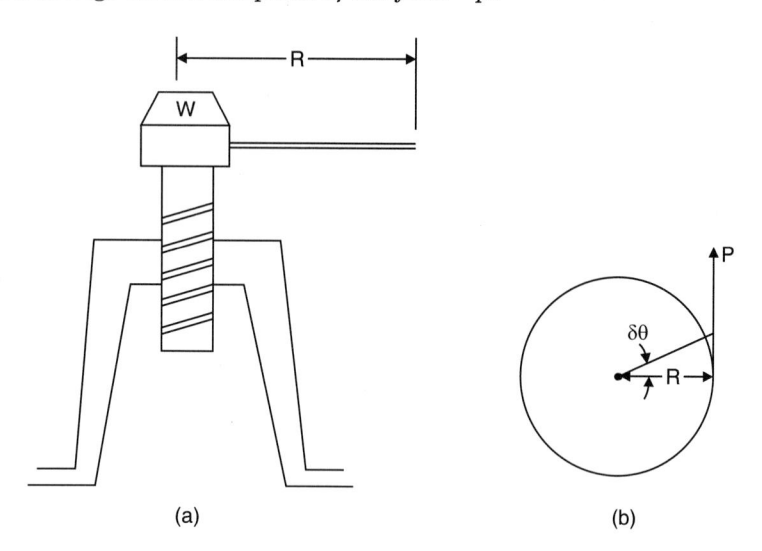

Fig. 8.24

Solution. Let P be the effort required. Let a virtual rotation $\delta\theta$ be given to effort.
$\therefore$ Virtual work done by $P = PR\delta\theta$
For one full rotation of effort, the load moves by distance p (definition of pitch).

$\therefore$ For $\delta\theta$ rotation distance through which weight W moves $= \dfrac{p}{2\pi} \delta\theta$

From virtual work principle, total work done is zero.

$$\therefore \qquad PR\delta\theta = W \frac{p}{2\pi} \delta\theta \quad \text{or} \quad P = \frac{Wp}{2\pi R} \quad \textbf{Ans.}$$

Example 8.20. *Neglecting friction, determine the pressure applied by screw press when effort P is applied (Ref. Fig. 8.25a)*

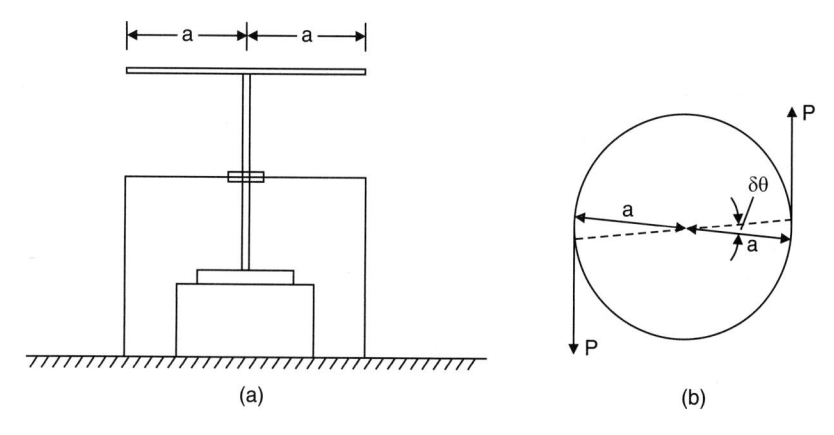

Fig. 8.25

Solution. Let a rotation $\delta\theta$ be given to the screw press by applying efforts P as shown in Fig. 8.25b. Work done by effort.
$$= 2Pa\,\delta\theta = 2Pa\,\delta\theta$$
If p is the pitch of the screw, work done by the load W is
$$= W \times \frac{p}{2\pi}\,\delta\theta$$
Hence from principle of virtual work
$$2\,Pa\,\delta\theta = W\,\frac{p}{2\pi}\,\delta\theta$$
or
$$P = \frac{Wp}{4\pi\,a} \quad \textbf{Ans.}$$

Example 8.21. *A simple linkage consists of two equal bars as shown in Fig. 8.26. When a force P was applied, it was brought into equilibrium by force Q, when bar made angle θ as shown in the figure. Neglecting weights of the bars, determine the relation between θ and F.*

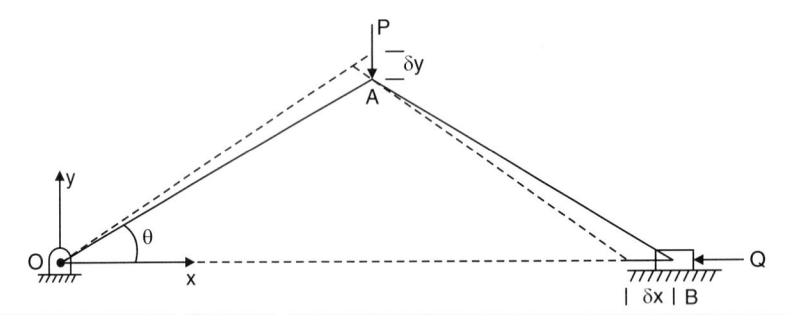

Fig. 8.26

Solution. Let a virtual displacement $\delta\theta$ be given to linkage as shown in the figure. For +ve value of $\delta\theta$ the point P moves up and the force Q moves to the left. Let 'l' be the length of members.

For point A, $\qquad y = l\sin\theta$
$\therefore \qquad\qquad\qquad \delta y = l\cos\theta\,\delta\theta$
For point B, $\qquad x = l\cos\theta + l\cos\theta = 2l\cos\theta$
$\therefore \qquad\qquad\qquad \delta x = -2l\sin\theta\,\delta\theta$

Writing virtual work equation, we get
$$-P\delta y - Q\delta x = 0$$
$$-Pl\cos\theta\,\delta\theta - Q(-2l\sin\theta)\,\delta\theta = 0$$
$\therefore \qquad\qquad Q = \dfrac{P}{2}\cot\theta \quad \textbf{Ans.}$

Example 8.22. *Determine the horizontal reaction developed at joint B, when the rigid frame shown in Fig. 8.27 (a) is loaded by a force F. Neglect weights of bars and use virtual work method.*

Solution. Let a virtual + ve rotation $\delta\theta$ be given to the first arm as shown in the Figure 8.27 (b). 'δy' be vertical displacement of point C and δx be the horizontal displacement of support B, if restraint is removed for horizontal displacement and a force H_B is applied. Selecting cartesian coordinates as shown in the figure,

VIRTUAL WORK METHOD

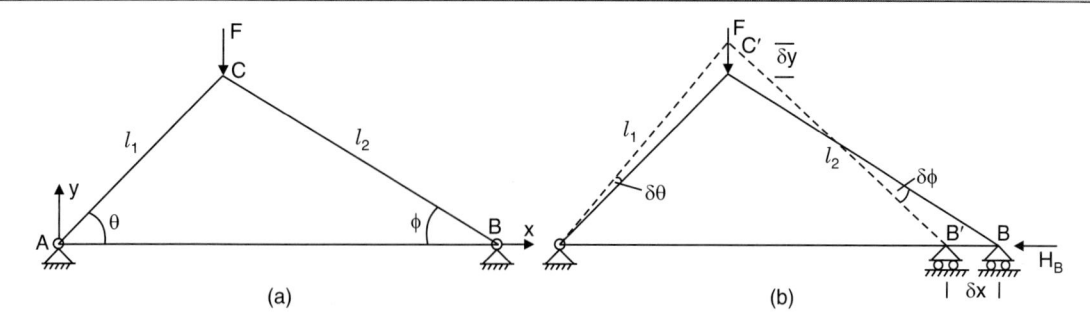

Fig. 8.27

For point C, $y = l_1 \sin \theta = l_2 \sin \phi$ ...(1)

If $\delta\theta$ is the rotation of first bar,

$$\delta y = l_1 \cos \theta \, \delta\theta = l_2 \cos \phi \, \delta\phi \qquad ...(2)$$

$$\therefore \quad \delta\phi = \frac{l_1 \cos \theta}{l_2 \cos \phi} \delta\theta \qquad ...(3)$$

Now for point B,

$$x = l_1 \cos \theta + l_2 \cos \phi$$

$$\therefore \quad \delta x = -l_1 \sin \theta \, \delta\theta - l_2 \sin \phi \, \delta\phi$$

$$= -l_1 \sin \theta \, \delta\theta - l_2 \sin \phi \, \frac{l_1 \cos \theta}{l_2 \cos \phi} \delta\theta$$

$$= \left(-l_1 \sin \theta - l_1 \frac{\sin \phi}{\cos \phi} \cos \theta \right) \delta\theta \qquad ...(4)$$

Writing virtual work equation for the system, we get

$$(-F) \, \delta y - H_B \, \delta x = 0$$

$$(-F) \, l_1 \cos \theta \, \delta\theta - H_B \left(-l_1 \sin \theta - l_1 \frac{\sin \phi}{\cos \phi} \cos \theta \right) \delta\theta = 0$$

$$\therefore \quad H_B = \frac{F \cos \theta}{\sin \theta + \frac{\sin \phi}{\cos \phi} \cos \theta} = \frac{F \cos \theta}{\sin \theta + \tan \phi \cos \theta}$$

Thus $\quad H_B = \dfrac{F \cos \theta}{\sin \theta + \tan \phi \cos \theta}$ **Ans.**

[Check, if $\theta = \phi$, the result will be same as in the previous example].

Example 8.23. *The frame ABCD shown in Fig. 8.28 (a) is a pin jointed plane frame with all the five members of equal length and equal cross-section. If joint C is subjected to a vertical downward load W, determine the force developed in the member BD. Use virtual work method.*

Solution. Let member BD be replaced by force P in it, as shown in Fig. 8.28 (b). Selecting cartesian coordinate system as shown in the figure and taking length of each member as 'l'

$$x_B = -l \cos \theta$$

∴ $\delta x_B = l \sin\theta\, \delta\theta$

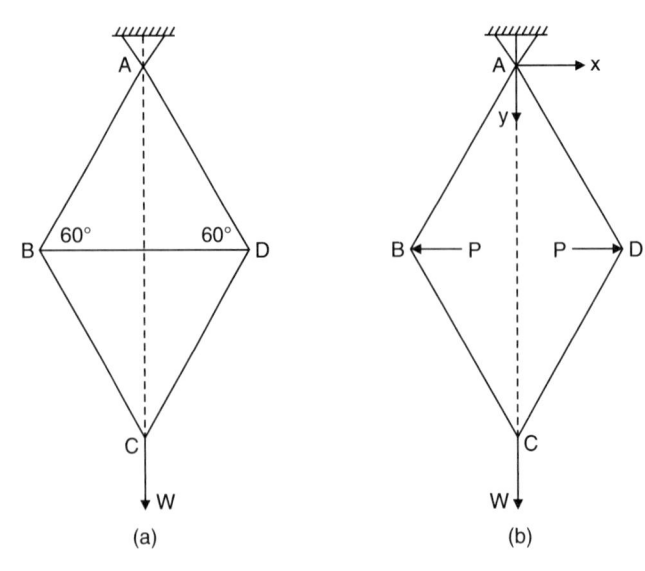

Fig. 8.28

Similarly $\quad x_D = l \cos\theta$
∴ $\quad \delta x_D = -l \sin\theta\, \delta\theta$
$\quad y_C = 2 \times l \sin\theta$
∴ $\quad \delta y_C = 2l \cos\theta\, \delta\theta$

Writing virtual work equation for the system, we get
$$-P\, \delta x_B + P\, \delta x_D + W\, \delta y_C = 0.$$
$$-P(l \sin\theta\, \delta\theta) + P(-l \sin\theta\, \delta\theta) + W\, 2l \cos\theta\, \delta\theta = 0$$

i.e., $\quad P \sin\theta = W \cos\theta$
or $\quad P = W \cot\theta$

Since $\triangle ABD$ is equilateral triangle, $\theta = 60°$

Hence $\quad\quad P = \dfrac{W}{\sqrt{3}}\quad$ **Ans.**

Example 8.24. *Determine the horizontal and vertical component of reactions developed at end B of the pin jointed plane frame loaded as shown in Fig 8.29 (a).*

Solution. (*i*) Horizontal reaction:

Removing the restraint to support *B* and allowing it to roll in *x* direction with a horizontal force H_B, let the angle of rotation of the members of frame be $\delta\theta$.

Now, $\quad y_C = a \sin\theta$
∴ $\quad \delta y_C = a \cos\theta\, \delta\theta$
$\quad y_D = a \sin\theta$
∴ $\quad \delta y_D = a \cos\theta\, \delta\theta$
$\quad x = 6a \cos\theta$
∴ $\quad \delta x = -6a \sin\theta\, \delta\theta$

VIRTUAL WORK METHOD

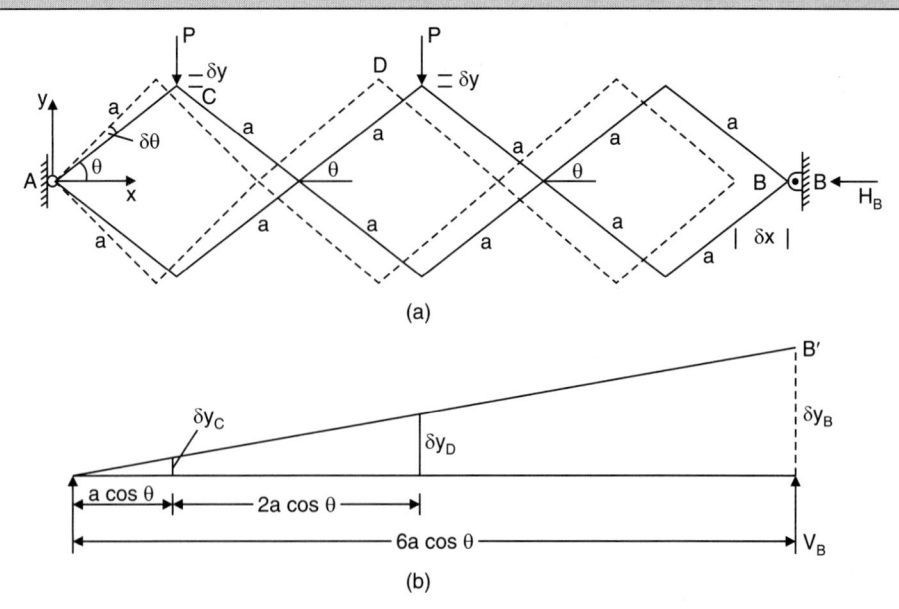

Fig. 8.29

Considering work done by all the active forces, virtual work equation is

$$(-P)\,\delta y_C + (-P)\,\delta y_D + (-H_B)\,\delta x = 0$$

$$-Pa\cos\theta\,\delta\theta - Pa\cos\theta\,\delta\theta - H_B(-6a\sin\theta)\,\delta\theta = 0$$

i.e.,
$$H_B = \frac{2P\cos\theta}{6\sin\theta} = \frac{P}{3}\cot\theta \quad \text{Ans.}$$

(ii) Vertical reaction at B

Let δy be vertical displacement given to joint-B after removing restraint in vertical direction. The corresponding vertical force at B be V_B. (Ref Fig. 8.29 b)

Vertical displacement of C

$$\delta y_C = \frac{a\sin\theta}{6a\sin\theta}\,\delta y_B = \frac{1}{6}\delta y_B$$

Vertical displacement of D:

$$\delta y_D = \frac{3a\sin\theta}{6a\sin\theta}\,\delta y_B = \frac{1}{2}\delta y_B$$

∴ Virtual work equation is,

$$(-P)\,\delta y_C + (-P)\,\delta y_D + V_B\delta y_B = 0$$

$$-P\frac{1}{6}\delta y_B - P\frac{1}{2}\delta y_B + V_B\,\delta y_B = 0$$

∴
$$V_B = \frac{4}{6}P = 2/3\,P. \quad \text{Ans.}$$

Example 8.25. *An inextensible string is subjected to equal loads W at distance L and 2L from end A. If a horizontal force F is applied at other end, the equilibrium position is as shown in Fig 8.30. Express the angles θ_1 and θ_2 with the vertical in terms of F and W.*

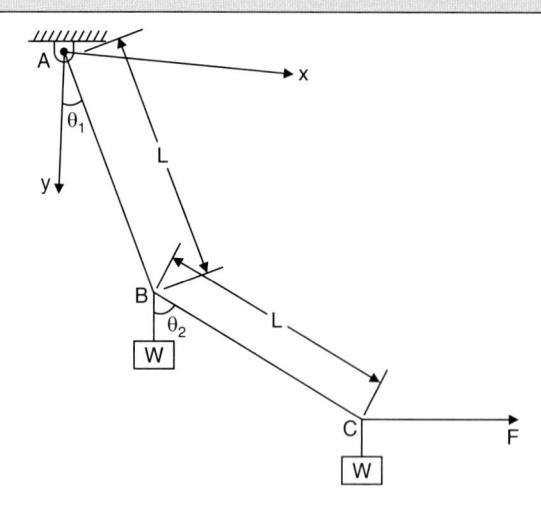

Fig. 8.30

Solution. With respect to the coordinate systems shown in the figure the three active forces are W, W and F. It is a system of two degrees of freedom in which θ_1 and θ_2 can vary independently.

∴ Principle of virtual work is

$$\delta U = \frac{\partial U}{\partial \theta_1} \delta\theta_1 + \frac{\partial U}{\partial \theta_2} \delta\theta_2 = 0$$

Keeping θ_1 constant and varying θ_2, the principle of virtual equation is

$$W\,\delta y_C + F\,\delta x = 0$$

$$W \frac{\partial}{\partial \theta_2}(L\cos\theta_2)\,\delta\theta_2 + F \frac{\partial}{\partial \theta_2}(L\sin\theta_2)\,\delta\theta_2 = 0$$

$$-WL\sin\theta_2 + FL\cos\theta_2 = 0.$$

∴ **$\tan\theta_2 = F/W$ Ans.**

Now, keeping θ_2 constant and varying θ_1,

$$y_B = L\cos\theta_1$$

∴ $$\delta y_B = -L\sin\theta_1\,\delta\theta_1$$

$$y_C = L\cos\theta_1 + L\cos\theta_2$$

Since θ_2 is constant,

$$\delta y_C = -L\sin\theta_1\,\delta\theta_1$$

$$x_C = L\sin\theta_1 + L\sin\theta_2$$

$$\delta x_C = L\cos\theta_1\,\delta\theta_1 \quad \text{(since } \theta_2 \text{ is constant.)}$$

Writing virtual work equation, we get

$$W\,\delta y_B + W\,\delta y_C + F\,\delta x_C = 0$$

i.e., $$W(-L\sin\theta_1)\,\delta\theta_1 + W(-L\sin\theta_1)\,\delta\theta_1 + F(L\cos\theta_1)\,\delta\theta_1 = 0$$

∴ $$\delta\theta_1 = \frac{F}{2W} \quad \text{Ans.}$$

VIRTUAL WORK METHOD

Important Definition

The work done by system of forces acting on a body due to imaginary consistent displacements is called virtual work. The method which uses the condition that, if the body is in equilibrium the total virtual work done by all forces in the system is zero, is called virtual work method.

Equation

Work done = Force × displacement in the direction of the force.
= Fd or $M\theta$

PROBLEMS FOR EXERCISE

8.1 Determine the vertical reaction developed at support B in the beam shown in Fig. 8.31.

[**Ans.** $V_B = 35$ kN use virtual work method].

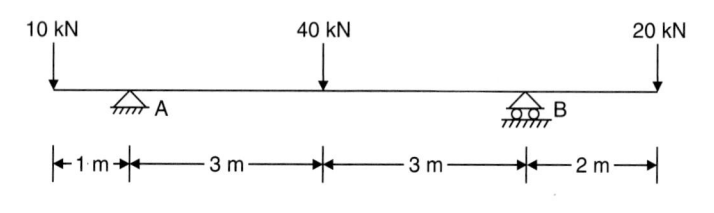

Fig. 8.31

8.2 Determine the reaction V_B in the overhanging beam shown in Fig 8.32 by virtual work method.
[**Ans.** $V_B = 38.33$ kN]

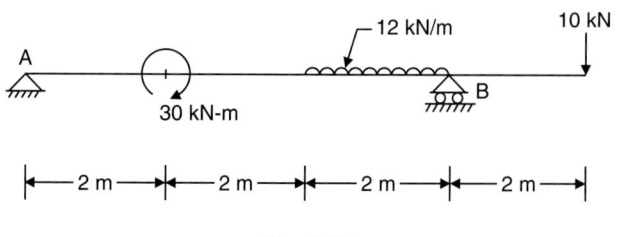

Fig. 8.32

8.3 Using virtual work method determine the force F required to hold the frame in equilibrium when subjected to a force P as shown in Fig 8.33.

$$\left[\textbf{Ans. } F = \frac{P}{2} \tan \theta \right]$$

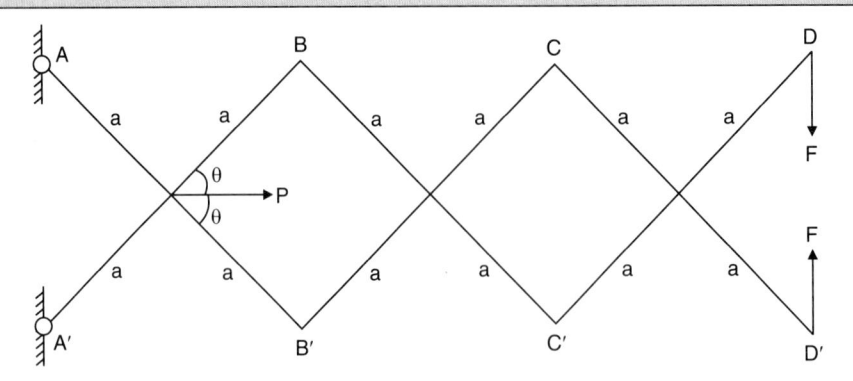

Fig. 8.33

8.4 The frame *ABCDEF* shown in Fig. 8.34 is made up of six identical rods pin connected at ends. The rod *AB* is fixed in horizontal position and the middle points of *AB* and *DE* are connected by an inextensile string so as to keep *DE* also in horizontal position. Determine the tension in string *GH*, if the weight of each rod is *W*. Use virtual work method. [**Ans.** $T = 3W$]

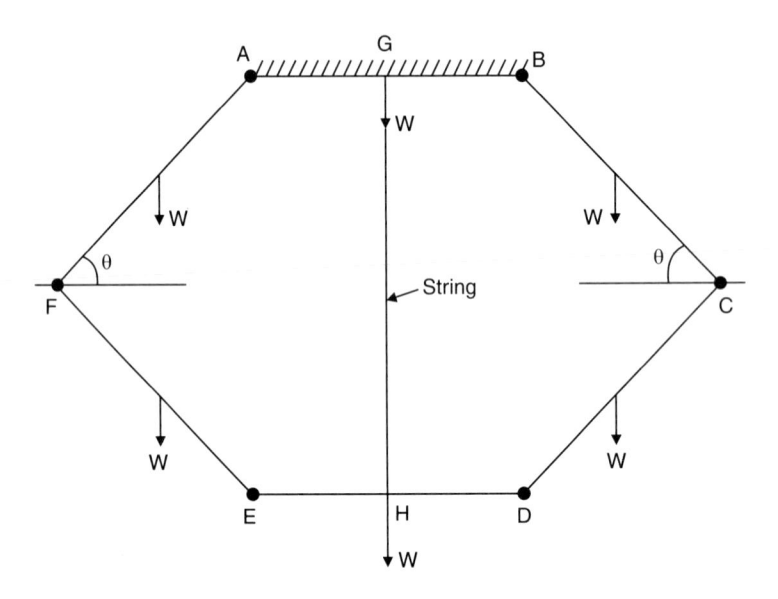

Fig. 8.34

8.5 The double pendulum shown in Fig. 8.35 consists of two links pinned smoothly together and held in position by a horizontal force *F*. If the length of link *AB* is $2L$ and weighs $2W$ while link *BC* is of length *L* and weighs *W*, determine the equilibrium position as defined by θ_1 and θ_2.

$$\left[\textbf{Ans. } \tan \theta_1 = \frac{2W}{P}, \tan \theta_2 = \frac{W}{2P}\right]$$

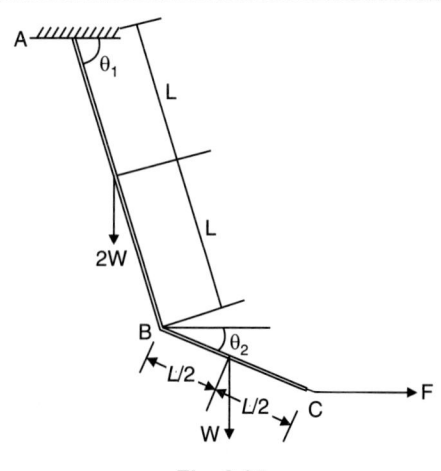

Fig. 8.35

9
Centroid and Moment of Inertia

Centroid and moment of intertia are two important properties of a section, which are required frequently in the analysis of many engineering problems. In this chapter the meaning of centre of gravity and centroid is explained. Procedure for locating centroid of plane figures is illustrated. The term moment of inertia is defined and the method of finding moment of inertia of a given sectional area is explained. Determination of the centroid and the moment of inertia of sections will be of great importance in the study of subjects like strength of materials, structural design and machine design.

9.1. CENTRE OF GRAVITY

Consider a suspended body as shown in Fig. 9.1 The weights of various parts of this body are acting vertically downward. The only upward force is the force in the string. To satisfy the equilibrium condition the resultant weight of the body W must act along the line of the string (1) – (1). Now, if the position is changed and the body is suspended again, it will reach equilibrium in a particular position. Let the line of action of resultant weight be (2) – (2) intersecting line (1) – (1) at G. It is obvious that if the body is suspended in any other position, the line of action of resultant weight W passes through G. This point is called the centre of gravity. Thus, *centre of gravity can be defined as the point through which resultant of force of gravity (Weight) of the body acts.*

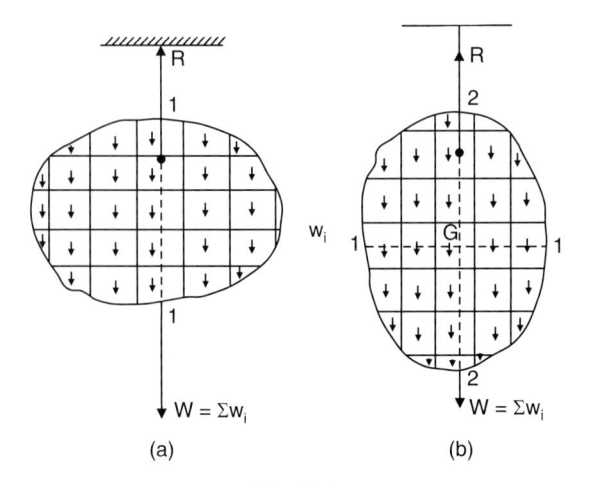

Fig. 9.1

CENTROID AND MOMENT OF INERTIA

9.2. CENTRE OF GRAVITY OF A FLAT PLATE

Consider a flat plate of thickness t as shown in Fig. 9.2. Let W_i be the weight of any elemental portion acting at a point (x_i, y_i). Let W be the total weight of the plate acting at the point $(\bar{x}, \bar{y})$. According to definition of centre of gravity, the point $(\bar{x}, \bar{y})$ is the centre of gravity. Now,

Total weight $\qquad W = \Sigma W_i \qquad$ ...(9.1)

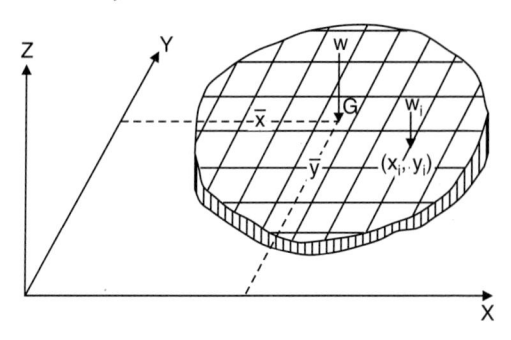

Fig. 9.2

Taking moment about x axis and equating moment of resultant to moment of component forces, we get

$$W\bar{y} = W_1 y_1 + W_2 y_2 + W_3 y_3 + \ldots$$
$$= \Sigma W_i y_i$$

$$\therefore \quad \bar{y} = \frac{\Sigma W_i y_i}{W} \qquad ...(9.2)$$

Similarly, taking moment about y axis we get,

$$W\bar{x} = W_1 x_1 + W_2 x_2 + W_3 x_3 + \ldots$$
$$= \Sigma W_i x_i$$

$$\bar{x} = \frac{\Sigma W_i x_i}{W} \qquad ...(9.3)$$

9.3. CENTROID

Let A_i be the area of the ith element of plate of uniform thickness in the plate. If γ is the unit weight of the material of plate and t its uniform thickness, then

$$W_i = \gamma A_i t$$

$\therefore$ Total weight, $\quad W = \Sigma \gamma A_i t = \gamma t \Sigma A_i$
$$= \gamma t A \qquad ...(9.4)$$

where, $A = \Sigma A_i$ is total area.

From equations 9.2 and 9.3, we get

$$\bar{y} = \frac{\Sigma A_i \gamma t y_i}{\gamma t A} = \frac{\Sigma A_i y_i}{A} \qquad ...(9.5)$$

and $\qquad \bar{x} = \dfrac{\Sigma A_i \gamma t x_i}{\gamma t A} = \dfrac{\Sigma A_i x_i}{A} \qquad$ ...(9.6)

since γ and t are constants.

The terms $\Sigma A_i y_i$ and $\Sigma A_i x_i$ may be considered as moment of area about x axis and y axis. Thus, the distance of the centre of gravity of plate of uniform thickness from an axis can be located by dividing moment of area about that axis by the total area. It will not depend upon the magnitude of thickness of plate and unit weight of material. If the thickness reduces to infinitesimal, then the plate reduces to an area. Still the expressions for finding centre of the area are same as Eqn. 9.5 and 9.6. Centre of gravity is a misnomer for the area. It is to be called as **centroid**. It may be noted that since the moment of area about an axis divided by total area gives the distance of centroid from that axis, the moment of area is zero about any centroidal axis.

9.4. DIFFERENCE BETWEEN CENTRE OF GRAVITY AND CENTROID

From the above discussion we can draw the following differences between centre of gravity and centroid:

(1) The term centre of gravity applies to bodies with mass and weight, and centroid applies to plane areas.

(2) Centre of gravity of a body is a point through which the resultant gravitational force (weight) acts for any orientation of the body whereas centroid is a point in a plane area such that the moment of area about any axis through that point is zero.

9.5. USE OF AXIS OF SYMMETRY

Centroid of an area lies on the axis of symmetry if it exists. This is useful theorem to locate the centroid of an area. This theorem can be proved as follows:

Consider the area shown in Fig. 9.3. In this figure y-axis is the axis of symmetry. From Eqn. 9.6, the distance of centroid from this axis is given by:

$$\frac{\Sigma A_i x_i}{A}$$

Consider the two elemental areas shown in Fig. 9.3, which are equal in size and are equidistant from the axis, but on either side. Now the sum of moments of these areas cancel each other since the areas and distances are the same, but signs of distances are opposite. Similarly, we can go on considering one area on one side of symmetric axis and corresponding image area on the other side, and prove that total moments of area $(\Sigma A_i x_i)$ about the symmetric axis is zero. Hence the distance of centroid from the symmetric axis is zero, *i.e.*, centroid always lies on symmetric axis.

Making use of the symmetry we can conclude that:

(1) Centroid of a circle is its centre (Fig. 9.4);

(2) Centroid of a rectangle of sides b and d is at a distance $\frac{b}{2}$ and $\frac{d}{2}$ from any corner (Fig. 9.5).

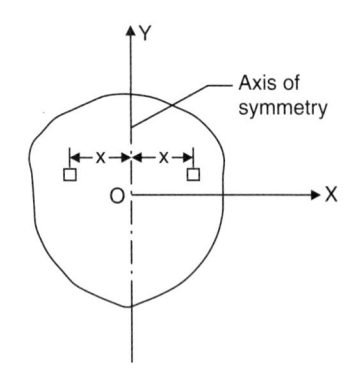

Fig. 9.3

CENTROID AND MOMENT OF INERTIA 229

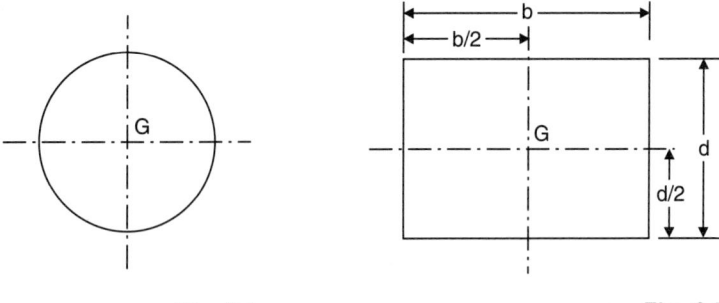

Fig. 9.4

Fig. 9.5

9.6. DETERMINATION OF CENTROID OF SIMPLE FIGURES FROM FIRST PRINCIPLE

For simple figures like triangle and semicircle, we can write general expression for the elemental area and its distance from an axis. Then equations 9.5 and 9.6 reduce to:

$$\bar{y} = \frac{\int y \, dA}{A} \qquad \qquad ...(9.7)$$

$$\bar{x} = \frac{\int x \, dA}{A} \qquad \qquad ...(9.8)$$

The location of the centroid using the above equations may be considered as finding centroid from first principles. Now, let us find centroid of some standard figures from first principles.

Centroid of a Triangle – Consider the triangle ABC of base width b and height h as shown in Fig. 9.6. Let us locate the distance of centroid from the base. Let b_1 be the width of elemental strip of thickness dy at a distance y from the base. Since $\triangle AEF$ and $\triangle ABC$ are similar triangles, we can write:

$$\frac{b_1}{b} = \frac{h-y}{h}$$

$$b_1 = \left(\frac{h-y}{h}\right) b = \left(1 - \frac{y}{h}\right) b$$

∴ Area of the element $= dA = b_1 dy$

$$= \left(1 - \frac{y}{h}\right) b \, dy$$

Area of the triangle $\quad A = \dfrac{1}{2} bh$

∴ From Eqn. (9.7)

$$\bar{y} = \frac{\text{Moment of area}}{\text{Total area}} = \frac{\int y \, dA}{A}$$

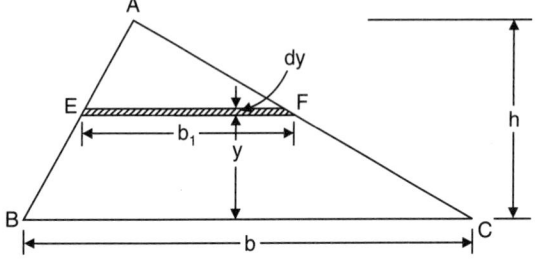

Fig. 9.6

Now, $\int y\,dA = \int_0^h y\left(1 - \frac{y}{h}\right) b\,dy = \int_0^h \left(y - \frac{y^2}{h}\right) b\,dy$

$$= b\left[\frac{y^2}{2} - \frac{y^3}{3h}\right]_0^h = \frac{bh^2}{6}$$

∴ $\bar{y} = \int \frac{y\,dA}{A} = \frac{bh^2}{6} \times \frac{1}{\frac{1}{2}bh}$

∴ $\bar{y} = \frac{h}{3}$

Thus the centroid of a triangle is at a distance $\frac{h}{3}$ from the base (or $\frac{2h}{3}$ from the apex) of the triangle where h is the height of the triangle.

Centroid of a Semicircle – Consider the semicircle of radius R as shown in Fig. 9.7. Due to symmetry centroid must lie on y axis. Let its distance from diametral axis be $\bar{y}$. To find $\bar{y}$, consider an element at a distance r from the centre O of the semicircle, radial width being dr and bound by radii at θ and $\theta + d\theta$.

Area of element $= r\,d\theta\,dr$.

Its moment about diametral axis x is given by:
$r\,d\theta \times dr \times r \sin\theta = r^2 \sin\theta\,dr\,d\theta$

∴ Total moment of area about diametral axis

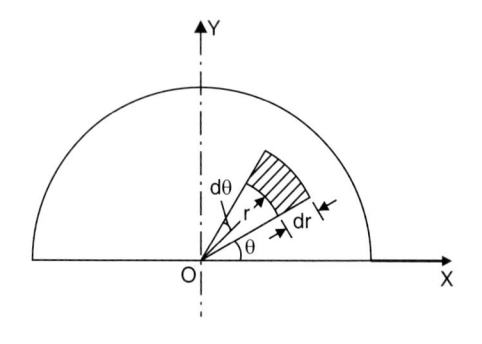

Fig. 9.7

$$= \int_0^\pi \int_0^R r^2 \sin\theta\,dr\,d\theta = \int_0^\pi \left[\frac{r^3}{3}\right]_0^R \sin\theta\,d\theta$$

$$= \frac{R^3}{3}[-\cos]_0^\pi = \frac{R^3}{3}[1+1] = \frac{2R^3}{3}$$

Area of semicircle $A = \frac{1}{2}\pi R^2$

∴ $\bar{y} = \frac{\text{Moment of area}}{\text{Total area}} = \frac{\frac{2R^3}{3}}{\frac{1}{2}\pi R^2} = \frac{4R}{3\pi}$

Thus, the centroid of the semi-circle is at a distance $\frac{4R}{3\pi}$ from the diametral axis.

Centroid of Sector of a Circle – Consider the sector of a circle of angle 2α as shown in Fig. 9.8. Due to symmetry, centroid lies on x axis. To find its distance from the centre O, consider the elemental area shown.

Area of the element $= r\,d\theta\,dr$

Its moment about y axis
$$= rd\theta \times dr \times r \cos\theta = r^2 \cos\theta \, dr \, d\theta$$
∴ Total moment of area about y axis
$$= \int_{-\alpha}^{\alpha} \int_0^R r^2 \cos\theta \, dr \, d\theta$$
$$= \left[\frac{r^3}{3}\right]_0^R [\sin\theta]_{-\alpha}^{\alpha}$$
$$= \frac{R^3}{3} \, 2\sin\alpha$$

Total area of the sector
$$= \int_{-\alpha}^{\alpha} \int_0^R r \, dr \, d\theta$$
$$= \int_{-\alpha}^{\alpha} \left[\frac{r^2}{2}\right]_0^R d\theta$$
$$= \frac{R^2}{2} [\theta]_{-\alpha}^{\alpha} = R^2 \alpha$$

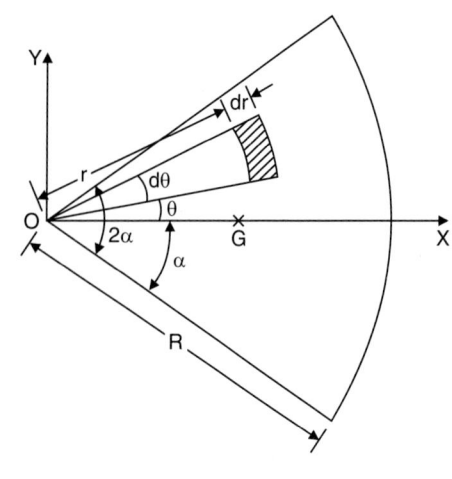

Fig. 9.8

∴ The distance of centroid from centre O
$$= \frac{\text{Moment of area about } y \text{ axis}}{\text{Area of the figure}}$$
$$= \frac{\frac{2R^3}{R}\sin\alpha}{R^2\alpha} = \frac{2R}{3\alpha}\sin\alpha$$

Centroid of Parabolic Spandrel – Consider the parabolic spandrel shown in Fig. 9.9. Height of the element at a distance x from O is $y = kx^2$

Width of element $= dx$

∴ Area of the element $= kx^2 \, dx$

∴ Total area of spandrel $= \int_0^a kx^2 \, dx$

$$\left[\frac{kx^3}{3}\right]_0^a = \frac{ka^3}{3}$$

Moment of area about y axis
$$= \int_0^a kx^2 \, dx \times x = \int_0^a kx^3 \, dx$$
$$= \frac{ka^4}{4}$$

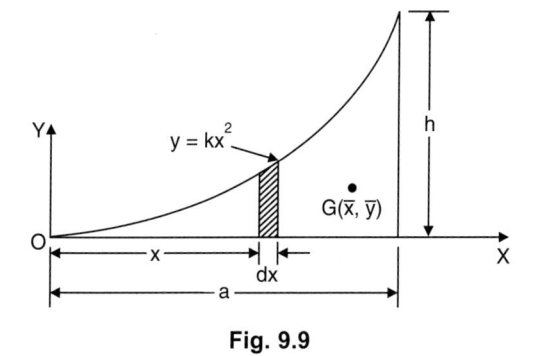

Fig. 9.9

Moment of area about x axis

$$= \int_0^a kx^2 \, dx \frac{kx^2}{2} = \int_0^a \frac{k^2 x^4}{2} dx = \frac{k^2 a^5}{10}$$

$$\therefore \quad \bar{x} = \frac{ka^4}{4} \div \frac{ka^3}{3} = \frac{3a}{4}$$

$$\bar{y} = \frac{k^2 a^5}{10} \div \frac{ka^3}{3} = \frac{3}{10} ka^2$$

From the Fig 9.9, at $x = a$, $y = h$

$$\therefore \quad h = ka^2 \quad \text{or} \quad k = \frac{h}{a^2}$$

$$\therefore \quad \bar{y} = \frac{3}{10} \times \frac{h}{a^2} a^2 = \frac{3h}{10}$$

Thus, centroid of spandrel is located at $\left(\dfrac{3a}{4}, \dfrac{3h}{10}\right)$

Centroids of some common figures are shown in Table 9.1.

Table 9.1 Centroid of Some Common Figures

Shape	Figure	$\bar{x}$	$\bar{y}$	Area
Triangle		—	$\dfrac{h}{3}$	$\dfrac{bh}{2}$
Semicircle		0	$\dfrac{4R}{3\pi}$	$\dfrac{\pi R^2}{2}$
Quarter circle		$\dfrac{4R}{3\pi}$	$\dfrac{4R}{3\pi}$	$\dfrac{\pi R^2}{4}$
Sector of a circle		$\dfrac{2R}{3\alpha} \sin \alpha$	0	αR^2
Parabolic Spandrel		$\dfrac{3a}{4}$	$\dfrac{3h}{10}$	$\dfrac{ah}{3}$

CENTROID AND MOMENT OF INERTIA

9.7. CENTROID OF COMPOSITE SECTIONS

So far, the discussion was confined to locating the centroid of simple figures like rectangle, triangle, circle, semicircle, etc. In engineering practice, use of sections whcih are built up with many simple sections is very common. Such sections may be called as built-up sections or composite sections. To locate the centroid of composite sections, one need not go for the first principle (method of integration). The given composite section can be split into suitable simple figures and then the centroid of each simple figure can be found by inspection or using the standard formulae listed in Table 9.1. Assuming the area of the simple figure as concentrated at its centroid, its moment about an axis can be found by multiplying the area with distance of its certroid from the reference axis. After determining moment of each area about reference axis, the distance of centroid from the axis is obtained by dividing total moment of area by total are of the composite section.

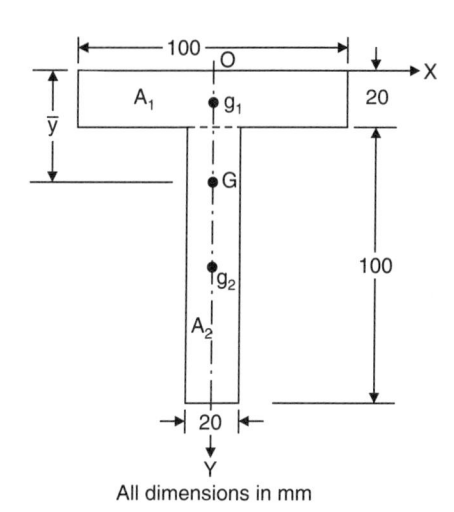

Fig. 9.10

Example 9.1. *Locate the centroid of the T-Section shown in the Fig. 9.10.*

Solution. Selecting the axis as shown in Fig. 9.10, we can say due to symmetry centroid lies on y axis, $i.e.\ \bar{x} = 0$.

Now the given T-section may be divided into two rectangles A_1 and A_2 each of size 100×20 and 20×100. The centroid of A_1 and A_2 are g_1 (0, 10) and g_2 (0, 70), respectively.

∴ The distance of centroid from top is given by:

$$\bar{y} = \frac{\Sigma a_i\, y_i}{\Sigma a_i} = \frac{100 \times 20 \times 10 + 20 \times 100 \times 70}{100 \times 20 + 20 \times 100} = 40 \text{ mm}$$

Hence, **centroid of T-section is on the symmetric axis at a distance 40 mm from the top.** Ans.

Example 9.2. *Find the centroid of the unequal angle 200 × 150 × 12 mm, shown in Fig. 9.11.*

Solution. The given composite figure can be divided into two rectangles:

$$A_1 = 150 \times 12 = 1800 \text{ mm}^2$$
$$A_2 = (200 - 12) \times 12 = 2256 \text{ mm}^2$$

Total area $\quad A = A_1 + A_2 = 4056 \text{ mm}^2$

Selecting the reference axis x and y as shown in Fig. 9.11. The centroid of A_1 is g_1 (75, 6) and that of A_2 is:

$$g_2\left[6, 12 + \frac{1}{2}(200 - 12)\right]$$

i.e., $\qquad g_2$ (6, 106).

∴ $\qquad \bar{x} = \dfrac{\text{Moment of area about } y \text{ axis}}{\text{Total area}}$

$$= \frac{A_1 x_1 + A_2 x_2}{A}$$

$$= \frac{1800 \times 75 + 2256 \times 6}{4056} = 36.62 \text{ mm}$$

$\bar{y} = \dfrac{\text{Moment of area about } x \text{ axis}}{\text{Total area}}$

$$= \frac{A_1 y_1 + A_2 y_2}{A}$$

$$= \frac{1800 \times 6 + 2256 \times 106}{4056} = 61.62 \text{ mm}$$

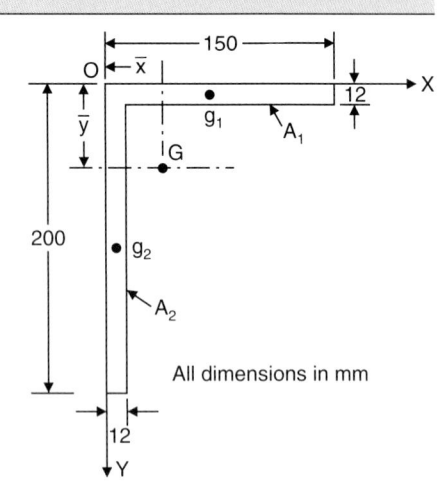

Fig. 9.11

Thus, the centroid is at $\bar{x} = 36.62$ **mm** and $\bar{y} = 61.62$ **mm** as shown in the figure. **Ans.**

Example 9.3. *Locate the centroid of the I-section shown in Fig. 9.12.*

Solution. Selecting the co-ordinate system as shown in Fig. 9.12, due to symmetry centroid must lie on y axis,

i.e., $\qquad \bar{x} = 0$.

Now, the composite section may be split into three rectangles.

$$A_1 = 100 \times 20 = 2000 \text{ mm}^2.$$

Centroid of A_1 from the origin is:

$$y_1 = 30 + 100 + \frac{20}{2} = 140 \text{ mm}$$

Similarly $A_2 = 100 \times 20 = 2000 \text{ mm}^2$

$$y_2 = 30 + \frac{100}{2} = 80 \text{ mm}$$

$A_3 = 150 \times 30 = 4500 \text{ mm}^2$, and

$$y_3 = \frac{30}{2} = 15 \text{ mm}$$

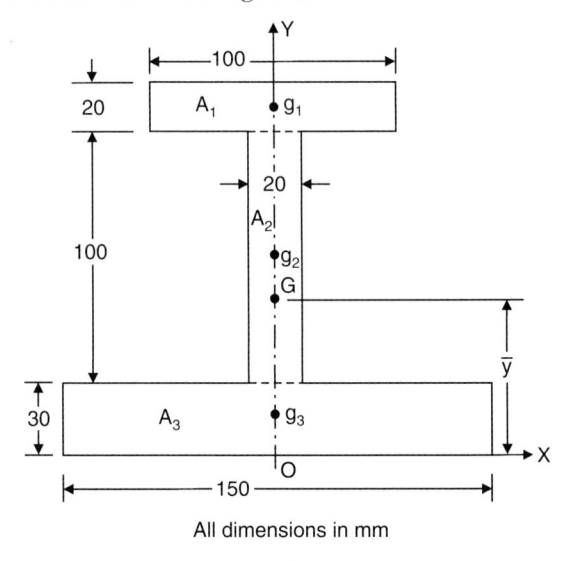

Fig. 9.12

CENTROID AND MOMENT OF INERTIA

$$\bar{y} = \frac{A_1 y_1 + A_2 y_2 + A_3 y_3}{A}$$

$$= \frac{2000 \times 140 + 2000 \times 80 + 4500 \times 15}{2000 + 2000 + 4500} = 59.71 \text{ mm}$$

Thus, **the centroid is on the symmetric axis at a distance 59.71 mm from the bottom** as shown in Fig. 9.12. **Ans.**

Example 9.4. *Determine the centroid of the section of the concrete dam shown in Fig. 9.13.*

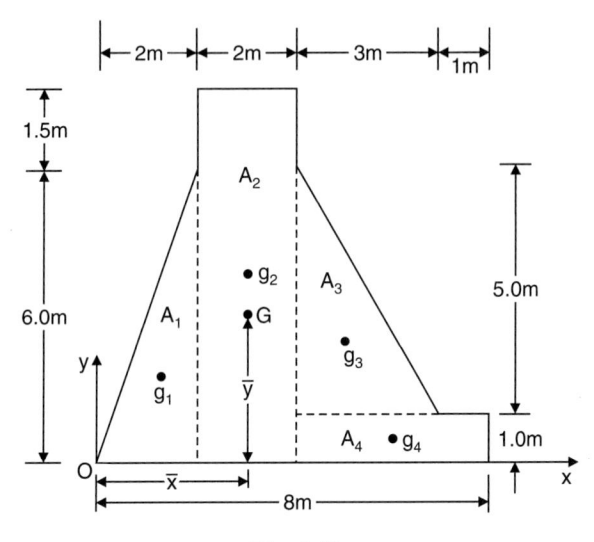

Fig. 9.13

Solution. Let the axis be selected as shown in Fig. 9.13. [Note that it is convenient to take axes in such a way that the centroids of all simple figures are having positive coordinates. If coordinates of any simple figure comes out to be negative, one should be careful in assigning the sign to moment of area of that figure.]

The composite figure can be conveniently divided into two triangles and two rectangles, as shown in Fig. 9.13.

Now,
$$A_1 = \frac{1}{2} \times 2 \times 6 = 6 \text{ m}^2$$
$$A_2 = 2 \times 7.5 = 15 \text{ m}^2$$
$$A_3 = \frac{1}{2} \times 3 \times 5 = 7.5 \text{ m}^2$$
$$A_4 = 1 \times 4 = 4 \text{ m}^2$$
$$A = \text{total area} = 32.5 \text{ m}^2$$

Centroides of simple figures are:
$$x_1 = \frac{2}{3} \times 2 = \frac{4}{3} \text{ m}$$
$$y_1 = \frac{1}{3} \times 6 = 2 \text{ m}$$
$$x_2 = 2 + 1 = 3 \text{ m}$$

$$y_2 = \frac{7.5}{2} = 3.75 \text{ m}$$

$$x_3 = 2 + 2 + \frac{1}{3} \times 3 = 5 \text{ m}$$

$$y_3 = 1 + \frac{1}{3} \times 5 = \frac{8}{3} \text{ m}$$

$$x_4 = 4 + \frac{4}{2} = 6 \text{ m}$$

$$y_4 = 0.5 \text{ m}$$

$$\bar{x} = \frac{A_1 x_1 + A_2 x_2 + A_3 x_3 + A_4 x_4}{A}$$

$$= \frac{6 \times \frac{4}{3} + 15 \times 3 + 7.5 \times 5 + 4 \times 6}{32.5} = 3.523 \text{ m}$$

$$\bar{y} = \frac{A_1 y_1 + A_2 y_2 + A_3 y_3 + A_4 y_4}{A}$$

$$= \frac{6 \times 2 + 15 \times 3.75 + 7.5 \times \frac{8}{3} + 4 \times 0.5}{32.5} = 2.777 \text{ m}$$

The centroid is at $\bar{x} = \mathbf{3.523\ m}$ **Ans.**

and $\bar{y} = \mathbf{2.777\ m}$ **Ans.**

Example 9.5. *Determine the centroid of the area shown in Fig. 9.14 with respect to the axes shown.*

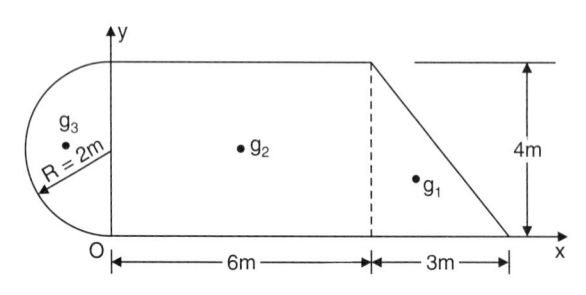

Fig. 9.14

Solution. The composite section is divided into three simple figures, a triangle, a rectangle and a semicircle.

Now, area of triangle $A_1 = \frac{1}{2} \times 3 \times 4 = 6 \text{ m}^2$

Area of rectangle $A_2 = 6 \times 4 = 24 \text{ m}^2$

Area of semicircle $A_3 = \frac{1}{2} \times \pi \times 2^2 = 6.2832 \text{ m}^2$

∴ Total area $A = 36.2832 \text{ m}^2$

The coordinates of centroids of these three simple figures are:

$$x_1 = 6 + \frac{1}{3} \times 3 = 7 \text{ m}$$

CENTROID AND MOMENT OF INERTIA

$$y_1 = \frac{4}{3} \text{ m}$$
$$x_2 = 3 \text{ m}$$
$$y_2 = 2 \text{ m}$$
$$x_3 = \frac{-4R}{3\pi} = -\frac{4 \times 2}{3\pi} = -0.8488 \text{ m}$$
$$y_3 = 2 \text{ m}$$

(Note carefully the sign of x_3).

$$\bar{x} = \frac{A_1 x_1 + A_2 x_2 + A_3 x_3}{A} = \frac{6 \times 7 + 24 \times 3 + 6.2832 \times (-0.8488)}{36.2832}$$

i.e., $\bar{x} = \mathbf{2.995 \text{ m}}$ **Ans.**

$$\bar{y} = \frac{A_1 y_1 + A_2 y_2 + A_3 y_3}{A} = \frac{6 \times \frac{4}{3} + 24 \times 2 + 6.2832 \times 2}{36.2832}$$

i.e. $\bar{y} = \mathbf{1.890}$ **Ans.**

Example 9.6. *In a gusset plate, there are six rivet holes of 21.5 mm diameter as shown in Fig. 9.15. Find the position of the centroid of the gusset plate.*

Solution. The composite area is equal to a rectangle of size 160 × 280 mm plus a triangle of size 280 mm base width and 40 mm height and minus areas of six holes. In this case also the Eqns. 9.5 and 9.6 can be used for locating centroid by treating area of holes as negative. The area of simple figures and their centroids are as shown in Table given below:

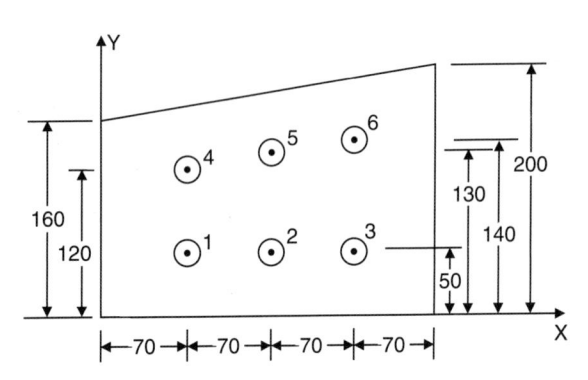

Fig. 9.15

Figure	Area in mm²	x_i in mm	y_i in mm
Rectangle	160 × 280 = 44,800	140	80
Triangle	$\frac{1}{2} \times 280 \times 40 = 5600$	$\frac{2}{3} \times 280$	$160 + \frac{40}{3} = 173.33$
1st hole	$\frac{-\pi \times 21.5^2}{4} = -363.05$	70	50
2nd hole	– 363.05	140	50
3rd hole	– 363.05	210	50
4th hole	– 363.05	70	120
5th hole	– 363.05	140	130
6th hole	– 363.05	210	140

∴ $A = \Sigma A_i = 48,221.70$

∴ $\Sigma A_i x_i = 44{,}800 \times 140 + 5600 \times \frac{2}{3} \times 280 - 363.05\,(70 + 140 + 210 + 70 + 140 + 210)$

$= 70, 12, 371.3 \text{ mm}^3$

$$\bar{x} = \frac{\Sigma A_i x_i}{A} = 145.42 \text{ mm}$$

$\Sigma A_i y_i = 44,800 \times 80 + 5600 \times 173.33 - 363.05(50 \times 3 + 120 + 130 + 140)$
$= 43,58,601 \text{ mm}^3$

$$\bar{y} = \frac{\Sigma A_i y_i}{A} = \frac{43,58,601}{48221.70} = 90.39 \text{ mm}$$

Thus, the coordinates of centroid of composite figure is given by:

$\bar{x} = \mathbf{145.42}$ **mm** Ans.

$\bar{y} = \mathbf{90.39}$ Ans.

Example 9.7. *Determine the coordinates x_c and y_c of the centre of a 100 mm diameter circular hole cut in a thin plate so that this point will be the centroid of the remaining shaded area shown in Fig. 9.16 (All dimensions are in mm).*

Solution. If x_c and y_c are the coordinates of the centre of the circle, centroid also must have the coordinates x_c and y_c as per the condition laid down in the problem. The shaded area may be considered as a rectangle of size 200 mm × 150 mm *minus* a triangle of sides 100 mm × 75 mm and a circle of diameter 100 mm.

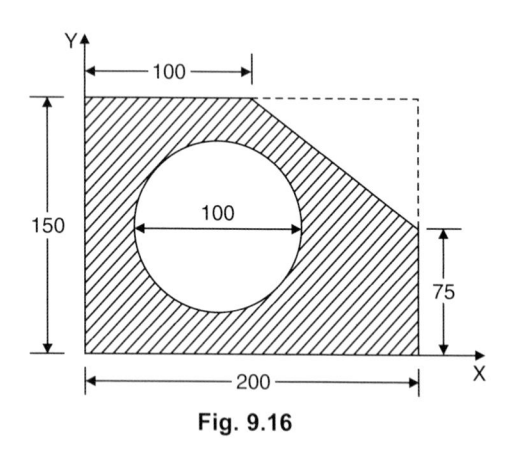

Fig. 9.16

∴ Total area $= 200 \times 150 - \frac{1}{2} \times 100 \times 75 - \left(\frac{\pi}{4}\right) 100^2$

$= 18,396 \text{ mm}^2$

$$\bar{x} = x_c = \frac{200 \times 150 \times 100 - \frac{1}{2} \times 100 \times 75 \times \left[200 - \left(\frac{100}{3}\right)\right] - \frac{\pi}{4} \times 100^2 \times x_c}{18,396}$$

∴ $x_c(18,396) = 200 \times 150 \times 100 - \frac{1}{2} \times 100 \times 75 \times 166.67 - \frac{\pi}{4} \times 100^2 x_c$

$$x_c = \frac{23,75000}{26,250} = \mathbf{90.48 \text{ mm}} \quad \text{Ans.}$$

Similarly,

$18,396 \, y_c = 200 \times 150 \times 75 - \frac{1}{2} \times 100 \times 75 \times (150 - 25) - \frac{\pi}{4} \times 100^2 \, y_c$

∴ $$y_c = \frac{17,81250.0}{26,250} = \mathbf{67.86 \text{ mm}} \quad \text{Ans.}$$

Centre of the circle should be located at (90.48, 67.86) so that this point will be the centroid of the remaining shaded area shown in Fig. 9.16.

Note: The centroid of the given figure will coincide with the centroid of the figure without circular hole. Hence, the centroid of the given figure can be obtained by determining the centroid of the figure without the circular hole.

CENTROID AND MOMENT OF INERTIA

Example 9.8. *Determine the coordinates of the centroid of the plane area shown in Fig. 9.17 with reference to the axes shown. Take x = 40 mm.*

Solution. The composite figure is divided into the following simple figures:

(1) A rectangle $\quad A_1 = (14x) \times (12x) = 168\, x^2$
$x_1 = 7x;\ y_1 = 6x$

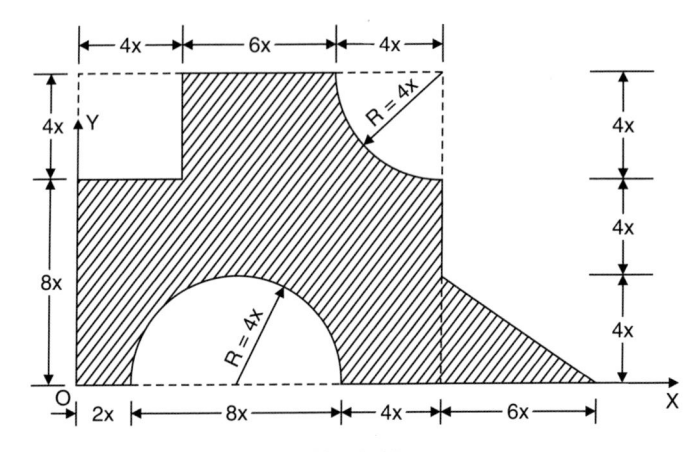

Fig. 9.17

(2) A triangle $\quad A_2 = \dfrac{1}{2}(6x) \times (4x) = 12x^2$
$x_2 = 14x + 2x = 16x$
$y_2 = \dfrac{4x}{3}$

(3) A rectangle to be subtracted
$A_3 = (-4x) \times (4x) = -16x^2$
$x_3 = 2x;\ y_3 = 8x + 2x = 10x$

(4) A semicircle to be subtracted
$A_4 = -\dfrac{1}{2}\pi(4x)^2 = -8\pi x^2$
$x_4 = 6x$
$y_4 = \dfrac{4R}{3\pi} = 4 \times \dfrac{(4x)}{3\pi} = \dfrac{16x}{3\pi}$

(5) A quarter of a circle to be subtracted
$A_5 = -\dfrac{1}{4} \times \pi(4x)^2 = -4\pi x^2$
$x_5 = 14x - \dfrac{4R}{3\pi} = 14x - (4)\left(\dfrac{4x}{3\pi}\right) = 12.3023\, x$
$y_5 = 12x - 4 \times \left(\dfrac{4x}{3\pi}\right) = 10.3023\, x$

Total area $\quad A = 168x^2 + 12x^2 - 16x^2 - 8\pi x^2 - 4\pi x^2$
$= 126.3009\, x^2$

$$\bar{x} = \frac{\Sigma A_r x_i}{A}$$

$\Sigma A_r x_i = 168x^2 \times 7x + 12x^2 \times 16x - 16x^2 \times 2x - 8\pi x^2 \times 6x - 4\pi x^2 \times 12.3023x$
$= 1030.6083 x^3$

∴ $\bar{x} = \dfrac{1030.6083\, x^3}{126.3009\, x^2} = 8.1599\, x = 8.1599 \times 40$ (since $x = 40$ mm)

$= 326.40$ mm

$$\bar{y} = \frac{\Sigma A_i y_i}{A}$$

$\Sigma A_i y_i = 168\, x^2 \times 6x + 12x^2 \times \dfrac{4x}{3} - 16x^2 \times 10x - 8\pi x^2 \times \dfrac{16x}{3\pi} - 4\pi x^2 \times 10.3023x$
$= 691.8708\, x^3$

∴ $\bar{y} = \dfrac{691.8708}{126.3009\, x^2} = 5.4780\, x$

$= 219.12$ mm (since $x = 40$ mm)

Centroid is at **(326.40, 219.12)** **Ans.**

9.8. MOMENT OF INERTIA

Consider the area shown in Fig. 9.18(a). dA is an elemental area with coordinates as x and y. The term $\Sigma y^2\, dA$ is called **moment of inertia** of the area about x axis and is denoted as I_{xx}. Similarly, the moment of inertia about y axis is $I_{yy} = \Sigma x^2\, dA$

In general, if r is the distance of elemental area dA from the axis AB [Fig. 9.18(b)], the sum of the terms $\Sigma r^2\, dA$ to cover the entire area is called moment of inertia of the area about the axis AB. If r and dA can be expressed in general terms, for any element then the sum becomes an integral. Thus,

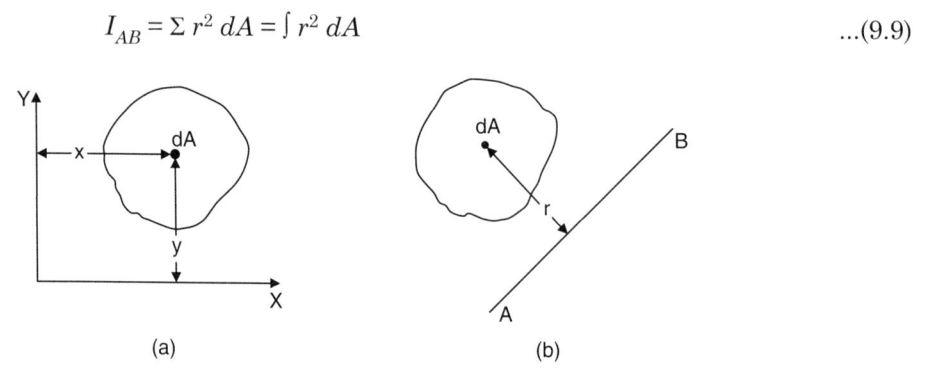

$$I_{AB} = \Sigma r^2\, dA = \int r^2\, dA \qquad \text{...(9.9)}$$

Fig. 9.18

Ther term $r dA$ may be called as moment of area, similar to moment of a force, and hence $r^2 dA$ may be called as moment of moment of area or the second moment of area. Thus, the moment of inertia of a plane figure is nothing but second moment of area. In fact, the term '**second moment of area**' appears to correctly signify the meaning of the expression $\Sigma r^2\, dA$. The term 'moment of inertia' is rather a misnomer. However, term moment of inertia has come to stay for long time and hence it will be used in this book also.

CENTROID AND MOMENT OF INERTIA

Though moment of inertia of plane area is a purely mathematical term, it is one of the important properties of areas. The strength of members subject to bending depends on the moment of inertia of its cross-sectional area. Students will find this property of area very useful when they study subjects like strength of materials, structural design and machine design.

The moment of inertia is a fourth dimensional term since it is a term obtained by multiplying area by the square of the distance. Hence, in SI units, if metre (m) is the unit for linear measurements used then m^4 is the unit of moment of inertia. If millimetre (mm) is the unit used for linear measurements, then mm^4 is the unit of moment of inertia. In MKS system m^4 or cm^4 and FPS system ft^4 or in^4 are commonly used as units for moment of inertia.

9.9. POLAR MOMENT OF INERTIA

Moment of inertia about an axis perpendicular to the plane of an area is known as **polar moment of inertia**. It may be denoted as J or I_{zz}. Thus, the moment of inertia about an axis perpendicular to the plane of the area at O in Fig. 9.19 is called polar moment of inertia at point O, and is given by

$$I_{zz} = \Sigma\, r^2\, dA \qquad \ldots(9.10)$$

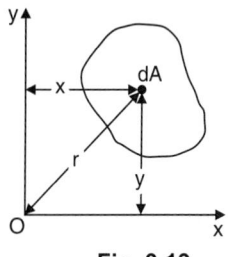

Fig. 9.19

9.10. RADIUS OF GYRATION

Radius of gyration is a mathematical term defined by the relation

$$k = \sqrt{\frac{I}{A}} \qquad \ldots(9.11)$$

where k = radius of gyration
         I = moment of inertia
and A = the cross-sectional area.

Suffixes with moment of inertia I also accompany the term radius of gyration k. Thus, we can have,

$$k_{xx} = \sqrt{\frac{I_{xx}}{A}}$$

$$k_{yy} = \sqrt{\frac{I_{yy}}{A}}$$

$$k_{AB} = \sqrt{\frac{I_{AB}}{A}}$$

and so on.

The relation between radius of gyration and moment of inertia can be put in the form:

$$I = Ak^2 \qquad \ldots(9.12)$$

From the above relation a geometric meaning can be assigned to the term 'radius of gyration.' We can consider k as the distance at which the complete area is squeezed and kept as a strip of negligible width (Fig. 9.20) such that there is no change in the moment of inertia.

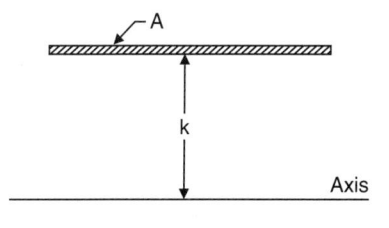

Fig. 9.20

9.11. THEOREMS OF MOMENT OF INERTIA

There are two theorems of moment of inertia:
(1) Perpendicular axis theorem, and
(2) Parallel axis theorem.
These are explained and proved below.

Perpendicular Axis Theorem

The moment of inertia of an area about an axis perpendicular to its plane (polar moment of inertia) at any point O is equal to the sum of moments of inertia about any two mutually perpendicular axis through the same point O and lying in the plane of the area.

Referring to Fig. 9.21, if $z - z$ is the axis normal to the plane of paper passing through point O, as per this theorem,

$$I_{zz} = I_{xx} + I_{yy} \qquad ...(9.13)$$

The above theorem can be easily proved. Let us consider an elemental area dA at a distance r from O. Let the coordinates of dA be x and y. Then from definition:

$$I_{zz} = \Sigma r^2 dA = \Sigma(x^2 + y^2)\, dA$$
$$= \Sigma x^2 dA + \Sigma y^2 dA$$
$$I_{zz} = I_{xx} + I_{yy}$$

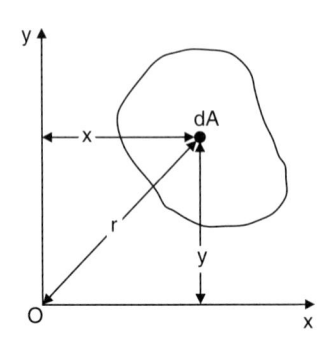

Fig. 9.21

Parallel Axis Theorem

Moment of inertia about any axis in the plane of an area is equal to the sum of moment of inertia about a parallel centroidal axis and the product of area and square of the distance between the two parallel axes. Referring to Fig. 9.22, the above theorem means:

$$I_{AB} = I_{GG} + A\, y_c^2 \qquad ...(9.14)$$

where I_{AB} – moment of inertia about axis AB

I_{GG} – moment of inertia about centroidal axis GG parallel to AB.

A – The area of the plane figure given and

y_c – the distance between the axis AB and the parallel centroidal axis GG.

Proof: Consider an elemental parallel strip dA at a distance y from the centroidal axis (Fig. 9.22).

Then, $I_{AB} = \Sigma(y + y_c)^2\, dA = \Sigma(y^2 + 2yy_c + y_c^2)dA$
$$= \Sigma y^2 dA + \Sigma 2yy_c\, dA + \Sigma y_c^2\, dA$$

Now,
$\Sigma y^2\, dA$ = Moment of inertia about the axis GG
$$= I_{GG}$$
$\Sigma 2yy_c\, dA = 2y_c \Sigma y\, dA$
$$= 2y_c A \frac{\Sigma y dA}{A}$$

Fig. 9.22

In the above term $2y_c A$ is constant and $\dfrac{\Sigma y dA}{A}$ is the distance of centroid from the reference axis GG. Since GG is passing through the centroid itself $\dfrac{y dA}{A}$ is zero and hence the term $\Sigma 2yy_c\, dA$ is zero.

CENTROID AND MOMENT OF INERTIA

Now, the third term,
$$\Sigma y_c^2 dA = y_c^2 \Sigma dA = A y_c^2$$
$$\therefore \quad I_{AB} = I_{GG} + A y_c^2$$

Note: The above equation cannot be applied to any two parallel axes. One of the axes (*GG*) must be centroidal axis only.

9.12. MOMENT OF INERTIA FROM FIRST PRINCIPLES

For simple figures, moment of inertia can be obtained by writing the general expression for an element and then carrying out integration so as to cover the entire area. This procedure is illustrated with the following three cases:

(1) Moment of inertia of rectangle about the centroidal axis
(2) Moment of inertia of a triangle about the base
(3) Moment of inertia of a circle about a diametral axis.

Moment of Inertia of a Rectangle about the Centroidal Axis—Consider a rectangle of width b and depth d (Fig. 9.23). Moment of inertia about the centroidal axis $x-x$ parallel to the short side is required.

Consider an elemental strip of width dy at a distance y from the axis. Moment of inertia of the elemental strip about the centroidal axis xx is:

$$= y^2 \, dA$$
$$= y^2 \, b \, dy$$

$$\therefore \quad I_{xx} = \int_{-\frac{d}{2}}^{\frac{d}{2}} y^2 b \, dy$$

$$= b \left[\frac{y^3}{3} \right]_{-\frac{d}{2}}^{\frac{d}{2}}$$

$$= b \left[\frac{d^3}{24} + \frac{d^3}{24} \right]$$

$$I_{xx} = \frac{bd^3}{12}$$

Fig. 9.23

Moment of Inertia of a Triangle About its Base—Moment of inertia of a triangle with base width b and height h is to be determined about the base AB (Fig. 9.24)

Consider an elemental strip at a distance y from the base AB. Let dy be the thickness of the strip and dA its area. Width of this strip is given by:

$$b_1 = \frac{(h-y)}{h} \times b$$

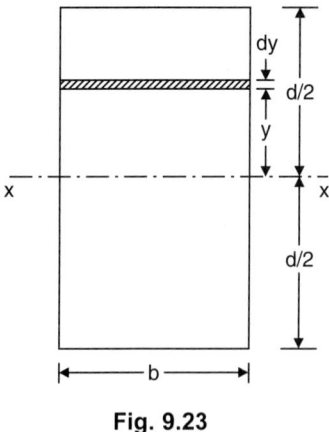

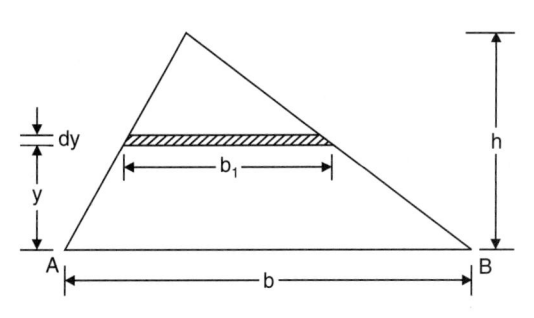

Fig. 9.24

Moment of inertia of this strip about AB
$$= y^2 dA = y^2 b_1 dy$$
$$= y^2 \frac{(h-y)}{h} \times b \times dy$$

∴ Moment of inertia of the triangle about AB,
$$I_{AB} = \int_0^h \frac{y^2 (h-y) b dy}{h} = \int_0^h \left(y^2 - \frac{y^3}{h} \right) b dy$$
$$= \left[\frac{y^3}{3} - \frac{y^4}{4h} \right]_0^h b = b \left[\frac{h^3}{3} - \frac{h^4}{4h} \right]$$
$$I_{AB} = \frac{bh^3}{12}$$

Moment of Inertia of Circle about its Diametral Axis—Moment of inertia of a circle of radius R is required about it's diametral axis as shown in Fig. 9.25.

Consider an element of sides $rd\theta$ and dr as shown in the figure. Its moment of inertia about the diametral axis $x - x$:
$$= y^2 dA$$
$$= (r \sin \theta)^2 \, r \, d\theta \, dr$$
$$= r^3 \sin^2 \theta \, d\theta \, dr$$

∴ Moment of inertia of the circle about $x - x$ is given by
$$I_{xx} = \int_0^R \int_0^{2\pi} r^3 \sin^2 \theta \, d\theta \, dr$$
$$= \int_0^R \int_0^{2\pi} r^3 \frac{(1 - \cos 2\theta)}{2} \, d\theta \, dr$$
$$= \int_0^R \frac{r^3}{2} \left[\theta - \frac{\sin 2\theta}{2} \right]_0^{2\pi} dr$$
$$= \left[\frac{r^4}{8} \right]_0^R [2\pi - 0 + 0 - 0] = \frac{2\pi}{8} R^4$$
$$I_{xx} = \frac{\pi R^4}{4}$$

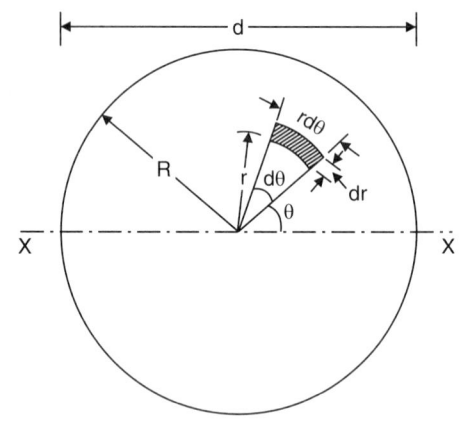

Fig. 9.25

If d is the diameter of the circle, then
$$R = \frac{d}{2}$$
∴
$$I_{xx} = \frac{\pi}{4} \left(\frac{d}{2} \right)^4$$

CENTROID AND MOMENT OF INERTIA

$$I_{xx} = \frac{\pi d^4}{64}$$

9.13. MOMENT OF INERTIA OF STANDARD SECTIONS

Rectangle – Refferring to Fig. 9.26,

(a) $I_{xx} = \dfrac{bd^3}{12}$ as derived in Art. 9.12.

(b) $I_{yy} = \dfrac{db^3}{12}$ can be derived on the same lines.

(c) *About the base AB*, from parallel axis theorem,

$$I_{AB} = I_{xx} + Ay_c^2$$

$$= \frac{bd^3}{12} + bd\left(\frac{d}{2}\right)^2, \quad \text{since } y_c = \frac{d}{2}$$

$$= \frac{bd^3}{12} + \frac{bd^3}{4}$$

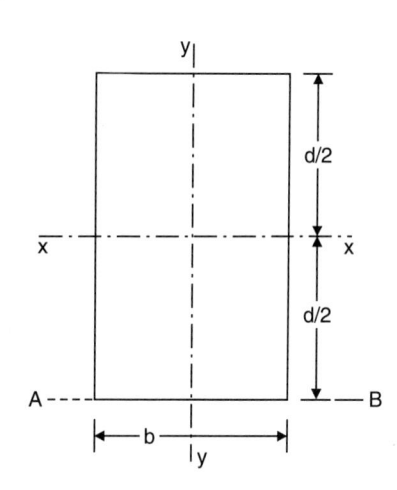

$$= \frac{bd^3}{3}$$

Fig. 9.26

Hollow Rectangular Section – Referring to Fig. 9.27, Moment of inertia I_{xx} = Moment of inertia of larger rectangle – Moment of inertia of hollow portion. That is,

$$= \frac{BD^3}{12} - \frac{bd^3}{12}$$

$$= \frac{1}{12}(BD^3 - bd^3)$$

Triangle – Referring to Fig. 9.28,

(a) About the base:

As found in Art. 9.12,

$$I_{AB} = \frac{bh^3}{12}$$

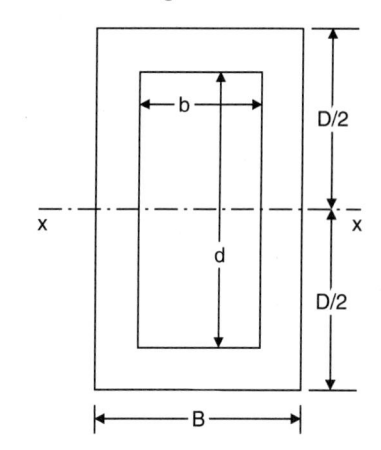

Fig. 9.27

(b) About centroidal axis, $x - x$ parallel to base:

From parallel axis theorem,

$$I_{AB} = I_{xx} + A y_c^2$$

Now, y_c, the distance between the non-centroidal axis AB and centroidal axix $x - x$, is equal to $\dfrac{h}{3}$

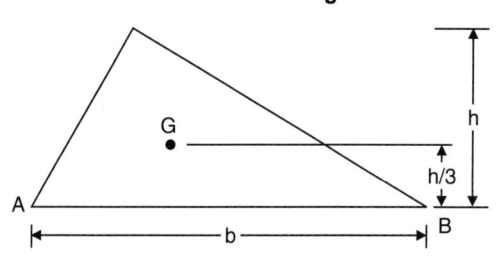

Fig. 9.28

∴ $\dfrac{bh^3}{12} = I_{xx} + \dfrac{1}{2} bh \left(\dfrac{h}{3}\right)^2 = I_{xx} + \dfrac{bh^3}{18}$

∴ $I_{xx} = \dfrac{bh^3}{12} - \dfrac{bh^3}{18} = \dfrac{bh^3}{36}$

Moment of Inertia of a Circle about any diametral axis

$$= \dfrac{\pi d^4}{64} \text{ (as found in Art. 9.12)}$$

Moment of Inertia of a Hollow Circle – Referring to Fig. 9.29,

I_{AB} = Moment of inertia of solid circle of diameter D about AB *minus* Moment of inertia of circle of diameter d about AB. That is,

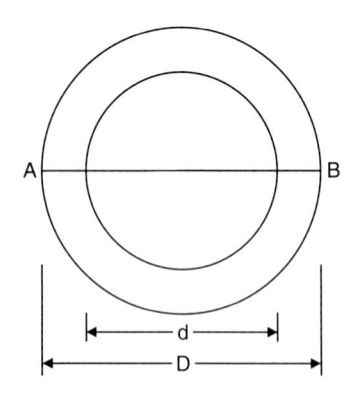

$$= \dfrac{\pi D^4}{64} - \dfrac{\pi d^4}{64}$$

$$= \dfrac{\pi}{64} (D^4 - d^4)$$

Fig. 9.29

Moment of Inertia of a Semicircle – (a) *About Diametral Axis*:

If the limit of integration is put as 0 to π instead of 0 to 2π in the derivation for the moment of inertia of a circle about diametral axis (Ref. art. 9.12), the moment of inertia of a semicircle is obtained. It can be observed that the moment of inertia of a semicircle (Fig. 9.30) about the diametral axis AB:

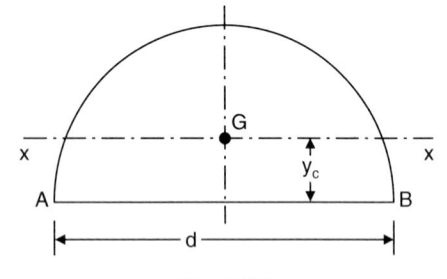

$$= \dfrac{1}{2} \times \dfrac{\pi d^4}{64} = \dfrac{\pi d^4}{128}$$

Fig. 9.30

(b) *About centroidal axis* $x - x$: Now, the distance of centroidal axix y_c from the diametral axis is given by:

$$y_c = \dfrac{4R}{3\pi} = \dfrac{2d}{3\pi}$$

and, $\qquad$ Area $A = \dfrac{1}{2} \times \dfrac{\pi d^2}{4} = \dfrac{\pi d^2}{8}$

From parallel axes theorem,

$$I_{AB} = I_{xx} + A y_c^2$$

$$\dfrac{\pi d^4}{128} = I_{xx} + \dfrac{\pi d^2}{8} \times \left(\dfrac{2d}{3\pi}\right)^2$$

CENTROID AND MOMENT OF INERTIA

$$I_{xx} = \frac{\pi d^4}{128} - \frac{d^4}{18\pi}$$

$$= 0.0068598\, d^4 = 0.11\, R^4$$

Moment of Inertia of a Quarter of a Circle – (a) About the base:

If the limit of integration is put as 0 to $\frac{\pi}{2}$ instead of 0 to 2π in the derivation for moment of inertia of a circle (Ref. art. 9.12), the moment of inertia of a quarter of a circle is obtained. It can be observed that moment of inertia of the quarter of a circle about the base AB:

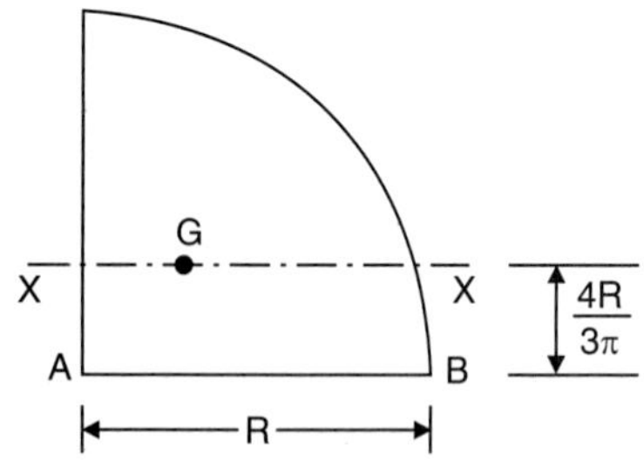

$$= \frac{1}{4} \times \frac{\pi d^4}{64} = \frac{\pi d^4}{256} = \frac{\pi R^4}{16}$$

Fig. 9.31

(b) About centroidal axis $x - x$

Now, the distance of centroidal axix y_c from the base is given by:

$$y_c = \frac{4R}{3\pi} = \frac{2d}{3\pi}$$

and the area

$$A = \frac{1}{4} \times \frac{\pi d^2}{4} = \frac{\pi d^2}{16}$$

From parallel axes theorem,

$$I_{AB} = I_{xx} + A\, y_c^{\,2}$$

$$\frac{\pi d^4}{256} = I_{xx} + \frac{\pi d^2}{16}\left(\frac{2d}{3\pi}\right)^2$$

$$I_{xx} = \frac{\pi d^4}{256} - \frac{d^4}{36\pi} = 0.00343 d^4 = 0.055\, R^4$$

The moment of inertia of common standard sections are presented in Table 9.2.

Table 9.2 Moment of Inertia of Standard Sections

Shape	Axis	Moment of Inertia
Rectangle	(a) Centroidal axis $x - x$	$I_{xx} = \dfrac{bd^3}{12}$
	(b) Centroidal axis $y - y$	$I_{yy} = \dfrac{db^3}{12}$
	(c) $A - B$	$I_{AB} = \dfrac{bd^3}{3}$

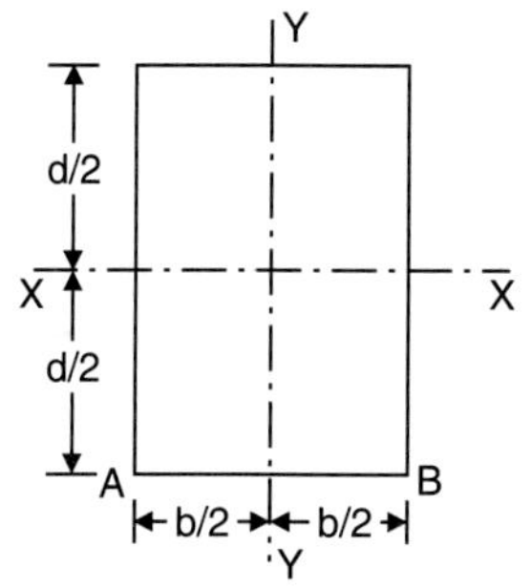

(Contd...)

Hollow Rectangle 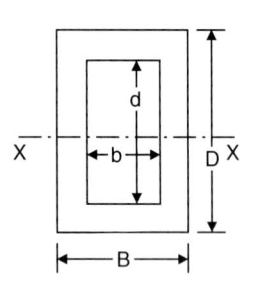	Centroidal axis $x-x$	$I_{xx} = \dfrac{BD^3 - bd^3}{12}$
Triangle	(a) Centroidal axis $x-x$	$I_{xx} = \dfrac{bh^3}{36}$
	(b) Base AB	$I_{AB} = \dfrac{bh^3}{12}$
Circle 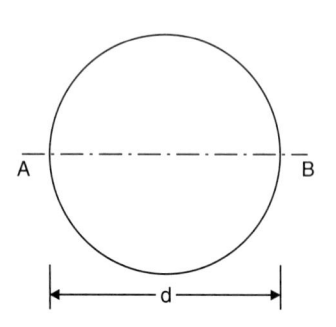	Diametral axis	$I = \dfrac{\pi d^4}{64} = \dfrac{\pi R^4}{4}$
Hollow Circle 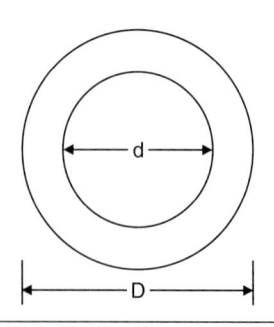	Diametral axis	$I = \dfrac{\pi}{64}(D^4 - d^4)$ $= \dfrac{\pi}{4}(R^4 - r^4)$ where $R = D/2$ and $r = d/2$

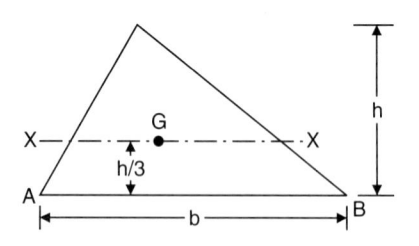

(Contd...)

CENTROID AND MOMENT OF INERTIA 249

Semi Circle	(a) $A - B$	$I_{AB} = \dfrac{\pi d^4}{128}$
	(b) Centroidal axis	$I_{xx} = 0.0068598\, d^4$
		$= 0.11\, R^4$
		where $R = d/2$
Quarter of a Circle	(a) $A - B$	$I_{AB} = \dfrac{\pi d^4}{256}$
	(b) Centroidal axis $x - x$	$I_{xx} = 0.00343\, d^4$
		$= 0.0055\, R^4$
		where $R = d/2$

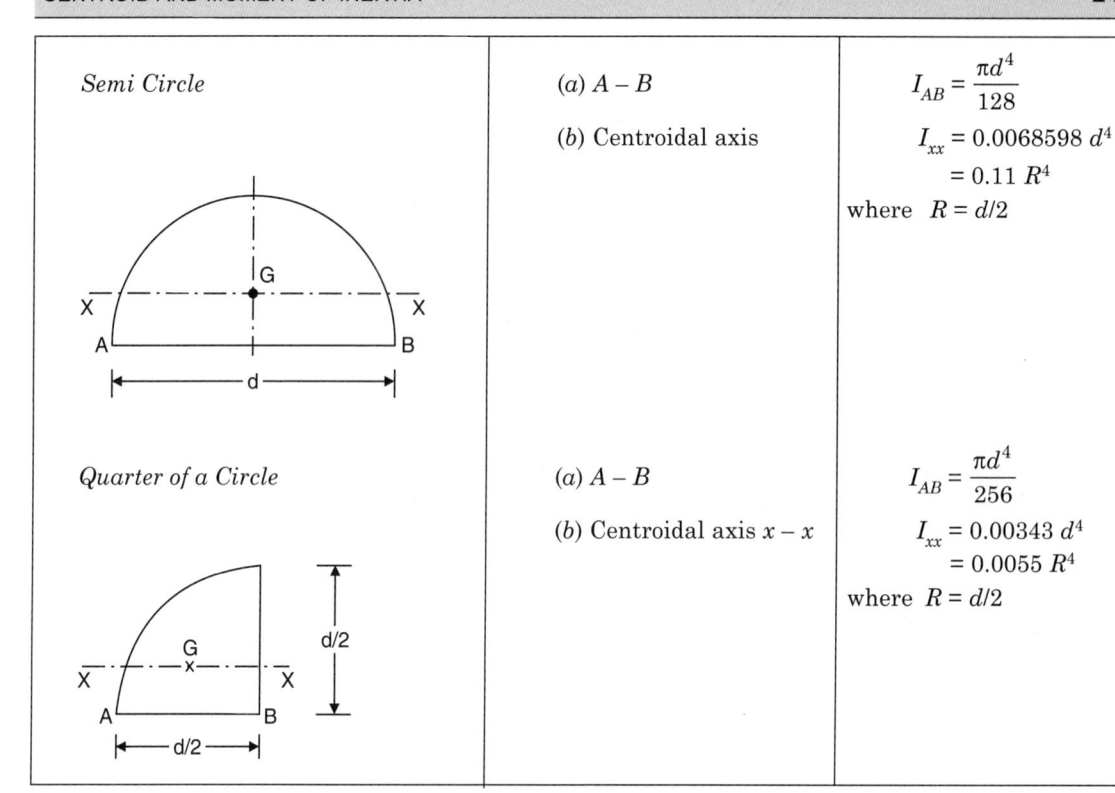

9.14. MOMENT OF INERTIA OF COMPOSITE SECTIONS

Beams and columns having composite sections are commonly used in structures. Moment of inertia of these sections about an axis can be found by the following steps:

(1) Divide the given figure into a number of simple figures.

(2) Locate the centroid of each simple figure by inspection or using standard expressions.

(3) Find the moment of inertia of each simple figure about its centroidal axis. Add the term Ay^2 where A is the area of the simple figure and y is the distance of the centroid of the simple figure from the reference axis. This gives moment of inertia of the simple figure about the reference axis.

(4) Sum up moments of inertia of all simple figures to get the moment of inertia of the composite section.

The procedure given above is illustrated below. Referring to the Fig. 9.32, it is required to find out the moment of inertia of the section about axis $A - B$.

(1) The section in the figure is divided into a rectangle, a triangle and a semi circle. The areas of the simple figures A_1, A_2 and A_3 are calculated.

(2) The centroids of the rectangle (g_1), triangle (g_2) and semi circle (g_3) are located. The distances y_1, y_2 and y_3 are found from the axis AB.

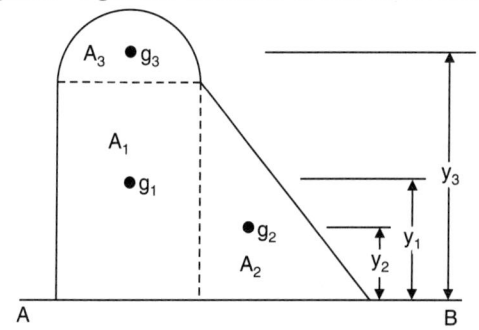

Fig. 9.32

(3) The moment of inertia of the rectangle about its centroid (I_{g1}) is calculated using standard expression. To this, the term $A_1 y_1^2$ is added to get the moment of inertia about the axis AB as:

$$I_1 = I_{g1} + A_1 y_1^2$$

Similarly, the moment of inertia of the triangle ($I_2 = I_{g2} + A_2 y_2^2$) and of semicircle ($I_3 = I_{g3} + A_3 y_3^2$) about axis AB are calculated.

(4) Moment of inertia of the composite section about AB is given by:

$$I_{AB} = I_1 + I_2 + I_3$$
$$= I_{g1} + A_1 y_1^2 + I_{g2} + A_2 y_2^2 + I_{g3} + A_3 y_3^2 \qquad ...(9.15)$$

In most engineering problems, moment of inertia about the centroidal axis is required. In such cases, first locate and centroidal axis as discussed in art. 9.7 and then find the moment of inertia about this axis.

Referring to Fig. 9.33, first the moment of area about any reference axis, say AB, is taken and is divided by the total area of the section to locate centroidal axis $x - x$. Then the distances of centroid of individual figures y_{c1}, y_{c2} and y_{c3} from the axis $x - x$ are determined. The moment of inertia of the composite section about the centroidal axis $x - x$ is calculated using the expression:

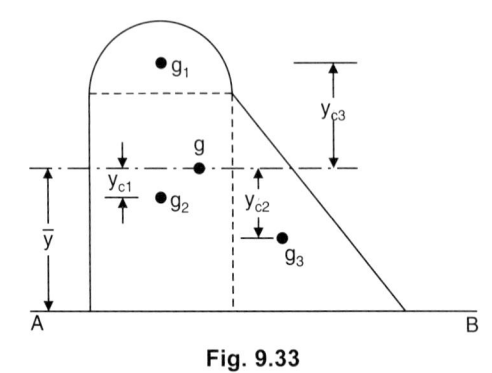

Fig. 9.33

$$I_{xx} = I_{g1} + A_1 y_{c1}^2 + I_{g2} + A_2 y_{c2}^2 + I_{g3} + A_3 y_{c3}^2 \qquad ...(9.16)$$

Sometimes the moment of inertia is found about a convenient axis and then using parallel axis theorem, the moment of inertia about centroidal axis is found.

In the above example, the moment of inertia I_{AB} is found and $\bar{y}$, the distance of CG from axis AB is calculated. Then from parallel axis theorem,

$$I_{AB} = I_{xx} + A \bar{y}^2$$
$$I_{xx} = I_{AB} - A \bar{y}^2$$

where A is the area of composite section.

Example 9.9. *Determine the moment of inertia of the section shown in Fig. 9.34 about an axis passing through the centroid and parallel to the top most fibre of the section.*

Also determine moment of inertia about the axis of symmetry. Hence find radii of gyration.

Solution. The given composite section can be divided into two rectangles as follows:

Area $\qquad A_1 = 150 \times 10 = 1500$ mm^2
Area $\qquad A_2 = 140 \times 10 = 1400$ mm^2
Total Area $\qquad A = A_1 + A_2 = 2900$ mm^2

Due to symmetry, centroid lies on the symmetric axis $y - y$.

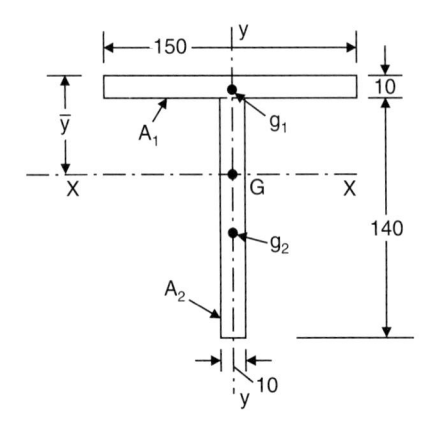

Fig. 9.34

CENTROID AND MOMENT OF INERTIA

The distance of the centroid from the top most fibre is given by:

$$\bar{y} = \frac{\text{Sum of moment of the areas about the top most fibre}}{\text{Total area}}$$

$$= \frac{1500 \times 5 + 1400\,(10 + 70)}{2900} = 41.21 \text{ mm}$$

Referring to the centroidal axes $x - x$ and $y - y$, the centroid of A_1 is g_1 (0.0, 36.21) and that of A_2 is g_2 (0.0, 38.79).

Moment of inertia of the section about $x - x$ axis I_{xx} = Moment of inertia of A_1 about $x - x$ axis + moment of inertia of A_2 about $x - x$ axis.

$$\therefore \qquad I_{xx} = \frac{150 \times 10^3}{12} + 1500\,(36.21)^2 + \frac{10 \times 140^3}{12} + 1400\,(38.79)^2$$

i.e., $\qquad I_{xx} = \mathbf{63{,}72442.5 \text{ mm}^4} \quad \text{Ans.}$

Similarly, $\qquad I_{yy} = \frac{10 \times 150^3}{12} + \frac{140 \times 10^3}{12} = \mathbf{28{,}24166.7 \text{ mm}^4} \quad \text{Ans.}$

Hence, the moment of inertia of the section about an axis passing through the centroid and parallel to the top most fibre is 63,72442.5 mm⁴ and moment of inertia of the section about the axis of symmetry is 28,24166.66 mm⁴.

The radius of gyration is given by:

$$k = \sqrt{\frac{I}{A}}$$

$$\therefore \qquad k_{xx} = \sqrt{\frac{I_{xx}}{A}} = \sqrt{\frac{63{,}72442.5}{2900}}$$

$$k_{xx} = \mathbf{46.88 \text{ mm}} \quad \text{Ans.}$$

Similarly, $\qquad k_{yy} = \sqrt{\frac{28{,}24166.66}{2900}}$

$$k_{yy} = \mathbf{31.21 \text{ mm}} \quad \text{Ans.}$$

Example 9.10. *Determine the moment of inertia of the L-section shown in the Fig. 9.33 about its centroidal axes parallel to the legs. Also find the polar moment of inertia.*

Solution. The given section is divided into two rectangles A_1 and A_2.

$$\text{Area } A_1 = 125 \times 10 = 1250 \text{ mm}^2$$
$$\text{Area } A_2 = 75 \times 10 = 750 \text{ mm}^2$$
$$\text{Total Area} = 2000 \text{ mm}^2$$

First, the centroid of the given section is to be located.

Two reference axes (1) – (1) and (2) – (2) are chosen as shown in Fig. 9.35.

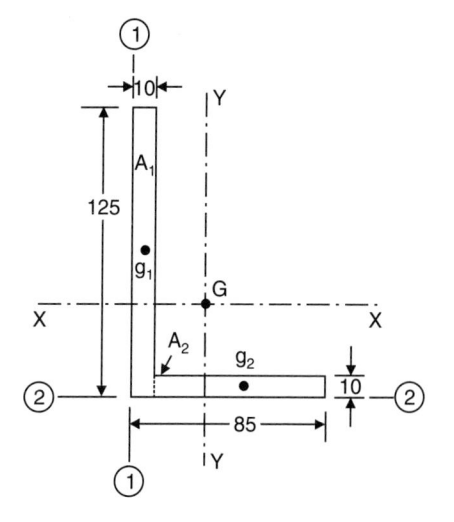

Fig. 9.35

The distance of centroid from the axis (1) – (1)

$$= \frac{\text{sum of moment of areas } A_1 \text{ and } A_2 \text{ about (1)} - (1)}{\text{Total area}}$$

i.e., $\bar{x} = \dfrac{1250 \times 5 + 750\left(10 + \dfrac{75}{2}\right)}{2000}$

$= 20.94$ mm

Similarly, the distance of the centroid from the axis (2) – (2)

$= \bar{y} = \dfrac{1250 \times \dfrac{125}{2} + 750 \times 5}{2000} = 40.94$ mm

With respect to the centroidal axes $x - x$ and $y - y$, the centroid of A_1 is g_1 (15.94, 21.56) and that of A_2 is g_2 (26.56, 35.94).

∴ $\quad I_{xx}$ = Moment of inertia of A_1 about $x - x$ axis
$\quad\quad\quad\quad$ + Moment of inertia of A_2 about $x - x$ axis

∴ $\quad I_{xx} = \dfrac{10 \times 125^3}{12} + 1250 \times 21.56^2 + \dfrac{75 \times 10^3}{12} + 750 \times 39.94^2$

i.e., $\quad I_{xx} = \mathbf{34{,}11298.9 \text{ mm}^4}$ **Ans.**

Similarly,

$I_{yy} = \dfrac{125 \times 10^3}{12} + 1250 \times 15.94^2 + \dfrac{10 \times 75^3}{12} + 750 \times 26.56^2$

i.e., $\quad I_{yy} = \mathbf{12{,}08658.9 \text{ mm}^4}$ **Ans.**

Polar moment of inertia $= I_{xx} + I_{yy}$

$= 34{,}11298.9 + 12{,}08658.9$

$I_{zz} = \mathbf{46{,}19957.8 \text{ mm}^4}$ **Ans.**

Example 9.11. *Determine the moment of inertia of the symmetric I-section shown in Fig. 9.36 about its centroidal axes $x - x$ and $y - y$.*

Also, determine moment of inertia of the section about a centroidal axis perpendicular to $x - x$ axis and $y - y$ axis.

Solution. The section is divided into three rectangles A_1, A_2, A_3.

Area $\quad A_1 = 200 \times 9 = 1800$ mm²
Area $\quad A_2 = (250 - 9 \times 2) \times 6.7$
$\quad\quad\quad\quad = 1554.4$ mm²
Area $\quad A_3 = 200 \times 9 = 1800$ mm²
Total Area $\quad A = 5154.4$ mm²

The section is symmetrical about both $x - x$ and $y - y$ axes. Therefore, its centroid will coincide with the centroid of rectangle A_2.

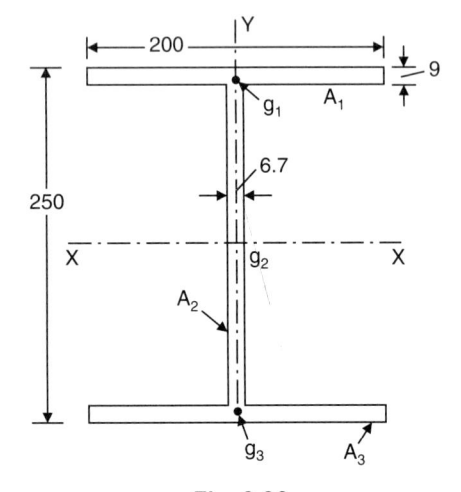

Fig. 9.36

With respect to the centroidal axes $x - x$ and $y - y$, the centroid of rectangle A_1 is g_1 (0.0, 120.5), that of A_2 is g_2 (0.0, 0.0) and that of A_3 is g_3 (0.0, 120.5).

CENTROID AND MOMENT OF INERTIA

I_{xx} = Moment of inertia of A_1 + Moment of inertia of A_2 + Moment of inertia of A_3 about $x-x$ axis

$$I_{xx} = \frac{200 \times 9^3}{12} + 1800 \times 120.5^2 + \frac{6.7 \times 232^3}{12} + 0 + \frac{200 \times 9^3}{12} + 1800 \, (120.5)^2$$

I_{xx} = 5,92,69202 mm^4 Ans.

Similarly,

$$I_{yy} = \frac{9 \times 200^3}{12} + \frac{232 \times 6.7^3}{12} + \frac{9 \times 200^3}{12}$$

I_{yy} = 1,20,05815 mm^4 Ans.

Moment of inertia of the section about a centroidal axis perpendicular to $x-x$ and $y-y$ axis is nothing but polar moment of inertia, and is given by:

$$I_{zz} = I_{xx} + I_{yy} = 5,92,69202 + 1,20,05815$$

I_{zz} = 7,12,75017 mm^4 Ans.

Example 9.12. *Compute the second moment of area of the channel section shown in Fig. 9.38 about centroidal axes $x-x$ and $y-y$.*

Solution. The section is divided into three rectangles A_1, A_2 and A_3.

Area A_1 = 100 × 13.5 = 1350 mm^2
Area A_2 = (400 – 27) × 8.1 = 3021.3 mm^2
Area A_3 = 100 × 13.5 = 1350 mm^2
Total   Area A = 5721.3 mm^2

The given section is symmetric about horizontal axis passing through the centroid g_2 of the rectangle A_2. A reference axis (1) – (1) is chosen as shown in Fig. 9.37.

The distance of the centroid of the section from (1) – (1)

$$= \frac{1350 \times 50 + 3021.3 \times \frac{8.1}{2} + 1350 \times 50}{5721.3} = 25.73 \text{ mm}$$

With reference to the centroidal axes $x-x$ and $y-y$, the centroid of the rectangle A_1 is g_1 (24.27, 193.25), that of A_2 is g_2 (21.68, 0.0) and that of A_3 is g_3 (24.27, 193.25).

∴ I_{xx} = Moment of inertia of A_1, A_2 and A_3 about $x-x$

$$= \frac{100 \times 13.5^3}{12} + 1350 \times 193.25^2 + \frac{8.1 \times 373^3}{12} + \frac{100 \times 13.5^3}{12} + 1350 \times 193.25^2$$

I_{xx} = **1.359 × 10^8 mm^4**   Ans.

Similarly, $I_{yy} = \dfrac{13.5 \times 100^3}{12} + 1350 \times 24.27^2 + \dfrac{373 \times 8.1^3}{12} + 3021.3 \times 21.68^2$

$$+ \frac{13.5 \times 100^3}{12} + 1350 \times 24.27^2$$

I_{yy} = 5276986   Ans.

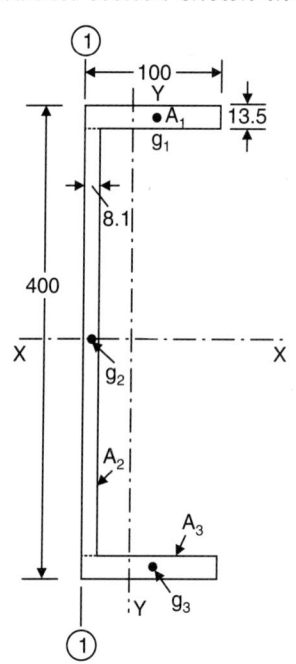

Fig. 9.37

Example 9.13. *Determine the polar moment of inertia about centroidal axes of the I-section shown in the Fig. 9.38. Also determine the radii of gyration with respect to $x-x$ and $y-y$ axes.*

The section is divided into three rectangles as shown in Fig. 9.38.

Area $A_1 = 80 \times 12 = 960$ mm^2
Area $A_2 = (150 - 22) \times 12 = 1536$ mm^2
Area $A_3 = 120 \times 10 = 1200$ mm^2

Total area $A = 3696$ mm^2

Due to symmetry, centroid lies on axis $y - y$. The bottom fibre (1) – (1) is chosen as reference axis to locate the centroid.

The distance of the centroid from (1) – (1)

$$= \frac{\text{Sum of moments of the areas of the rectangles about (1) – (1)}}{\text{Total area of section}}$$

$$= \frac{960 \times (150 - 6) + 1536 \times \left(\frac{128}{2} + 10\right) + 1200 \times 5}{3696} = 69.78 \text{ mm}$$

With reference to the centroidal axes $x - x$ and $y - y$, the centroid of the rectangles A_1 is g_1 (0.0, 74.22), that of A_2 is g_2 (0.0, 4.22) and that of A_3 is g_3 (0.0, 64.78).

$$I_{xx} = \frac{80 \times 12^3}{12} + 960 \times 74.22^2 + \frac{12 \times 128^3}{12} + 1536 \times 4.22^2 + \frac{120 \times 10^3}{12} + 1200 \times 64.78^2$$

$I_{xx} = 1,24,70028$ mm^4

$$I_{yy} = \frac{12 \times 80^3}{12} + \frac{128 \times 12^3}{12} + \frac{10 \times 120^3}{12}$$

$= 19,70432$ mm^4

Polar Moment of Inertia $= I_{xx} + I_{yy}$

$= 1,24,70027 + 19,70432$

$= 1,44,40459$ mm^4 **Ans.**

$\therefore \quad k_{xx} = \sqrt{\frac{I_{xx}}{A}} = \sqrt{\frac{1,24,70027}{3696}}$

$= 58.09$ mm **Ans.**

$k_{yy} = \sqrt{\frac{I_{yy}}{A}} = \sqrt{\frac{19,70432}{3696}}$

$= 23.09$ mm

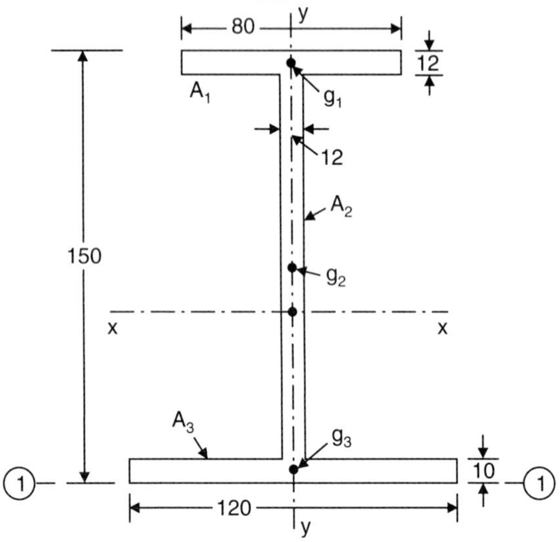

Fig. 9.38

Example 9.14. *Determine the moment of inertia of the built-up section shown in Fig. 9.39 about its centroidal axes $x - x$ and $y - y$.*

Solution. The given composite section may be divided into simple rectangles and triangles as shown in the Fig. 9.39.

Area $A_1 = 100 \times 30 = 3000$ mm^2
Area $A_2 = 100 \times 25 = 2500$ mm^2
Area $A_3 = 200 \times 20 = 4000$ mm^2
Area $A_4 = \dfrac{1}{2} \times 87.5 \times 20 = 875$ mm^2

CENTROID AND MOMENT OF INERTIA 255

$$\text{Area } A_5 = \frac{1}{2} \times 87.5 \times 20 = 875 \text{ mm}^2$$

Total area $A = 11250 \text{ mm}^2$

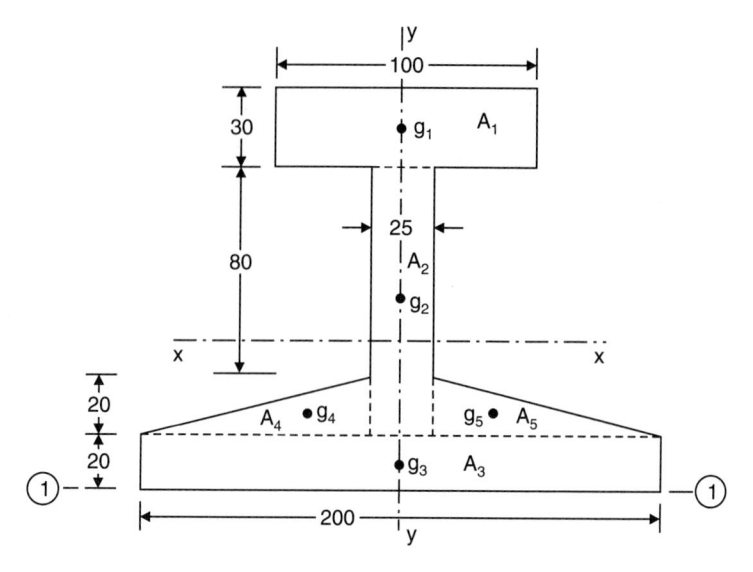

Fig. 9.39

Due to symmetry, centroid lies on the axis $y - y$.

A reference axis (1) – (1) is choosen as shown in the figure.

The distance of the centroidal axis from (1) – (1)

$$= \frac{\text{Sum of moments of areas about (1) – (1)}}{\text{Total area}}$$

$$\bar{y} = \frac{3000 \times 135 + 2500 \times 70 + 4000 \times 10 + 875 \left(\frac{1}{3} \times 20 + 20\right) \times 2}{11250}$$

$= 59.26$ mm

With reference to the centroidal axes $x - x$ and $y - y$, the centroid of the rectangle A_1 is g_1 (0.0, 75.74), that of A_2 is g_2 (0.0, 10.74), that of A_3 is g_3 (0.0, 49.26), the centroid of triangle A_4 is g_4 (41.66, 32.59) and that of A_5 is g_5 (41.66, 32.59).

$$I_{xx} = \frac{100 \times 30^3}{12} + 3000 \times 75.74^2 + \frac{25 \times 100^3}{12} + 2500 \times 10.74^2 + \frac{200 \times 20^3}{12}$$

$$+ 4000 \times 49.26^2 + \frac{87.5 \times 20^3}{36} + 875 \times 32.59^2 + \frac{87.5 \times 20^3}{36} + 875 \times 32.59^2$$

$I_{xx} = 3{,}15{,}43447 \text{ mm}^4$ **Ans.**

$$I_{yy} = \frac{30 \times 100^3}{12} + \frac{100 \times 25^3}{12} + \frac{20 \times 200^3}{12} + \frac{20 \times 87.5^3}{36} + 875 \times 41.66^2$$

$$+ \frac{20 \times 87.5^3}{36} + 875 \times 41.66^2$$

$I_{yy} = 1{,}97{,}45122 \text{ mm}^4$ **Ans.**

Example 9.15. *Determine the moment of inertia of the built-up section shown in the Fig. 9.40 about an axis AB passing through the top most fibre of the section as shown.*

Solution. In this problem, it is required to find out the moment of inertia of the section about an axis AB. So there is no need to find out the position of the centroid.

The given section is split into simple rectangles as shown in Fig. 9.40.

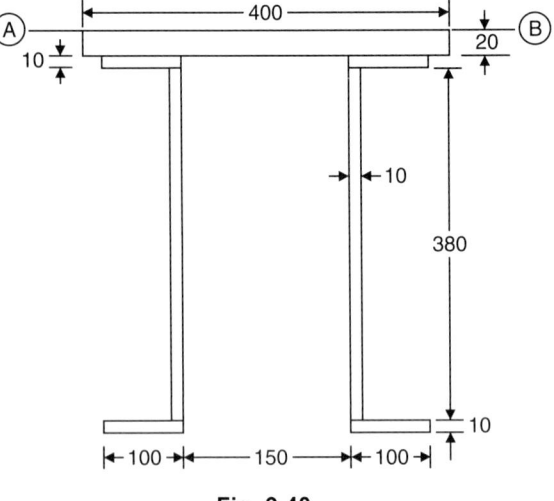

Fig. 9.40

Now,
Moment of Inertia about AB = Sum of moments of intertia of the rectangles about AB

$$= \frac{400 \times 20^3}{12} + 400 \times 20 \times 10^2 + \left[\frac{100 \times 10^3}{12} + 100 \times 10 \times (20+5)^2\right] \times 2$$

$$+ \left[\frac{10 \times 380^3}{12} + 10 \times 380 \times (30+190)^2\right] \times 2$$

$$+ \left[\frac{100 \times 10^3}{12} + 100 \times 10 \times (20+10+380+5)^2\right] \times 2$$

$I_{AB} = 8.06093 \times 10^8$ mm^4 **Ans.**

Example 9.16. *Calculate the moment of inertia of the built-up section shown in Fig. 9.41 about a centroidal axis parallel to AB. All members are 10 mm thick.*

Solutin. The built-up section is divided into six simple rectangles as shown in the figure.

The distance of centroidal axis from AB

$$= \frac{\text{Sum of the moment of areas about } AB}{\text{Total Area}}$$

$$= \frac{\Sigma A_i y_i}{A}$$

Now,
$\Sigma A_i y_i = 250 \times 10 \times 5 + 2 \times 40 \times 10$
$\qquad \times (10 + 20) + 40 \times 10 \times (10+5) +$
$\qquad 40 \times 10 \times 255 + 250 \times 10 \times (10+125)$
$\qquad = 4,82000$ mm^3

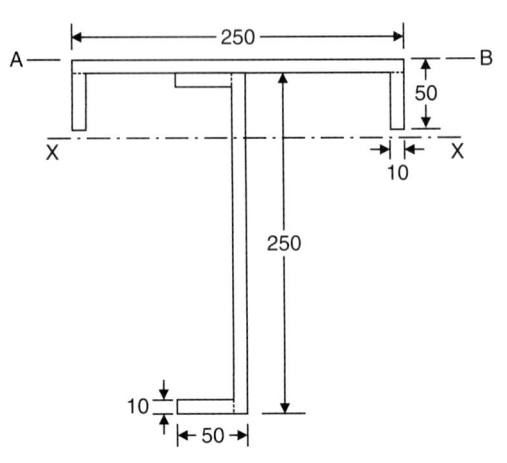

Fig. 9.41

$$A = 2 \times 250 \times 10 + 40 \times 10 \times 4 = 6600 \text{ mm}^2$$

$$\therefore \quad \bar{y} = \frac{\Sigma A_i \, y_i}{A} = \frac{4,82000}{6600} = 73.03 \text{ mm}$$

Now,

$$\begin{Bmatrix} \text{Moment of inertia about the} \\ \text{centroidal axis} \end{Bmatrix} = \begin{Bmatrix} \text{Sum of the moment of inertia of the individual} \\ \text{rectangles} \end{Bmatrix}$$

$$= \frac{250 \times 10^3}{12} + 250 \times 10 \times (73.03 - 5)^2$$

$$+ \left[\frac{10 \times 40^3}{12} + 40 \times 10 \, (73.03 - 30)^2 \right] \times 2$$

$$+ \frac{40 \times 10^3}{12} + 40 \times 10 \, (73.03 - 15)^2$$

$$+ \frac{10 \times 250^3}{12} + 250 \times 10 \, (73.03 - 135)^2$$

$$+ \frac{40 \times 10^3}{12} + 40 \times 10 \, (73.03 - 255)^2$$

$$I_{xx} = 5,03,99395 \text{ mm}^4 \quad \text{Ans.}$$

Example 9.17. *A built-up section of structural steel consists of a flange plate 400 mm × 20 mm, a web plate 600 mm × 15 mm and two angles 150 mm × 150 mm × 10 mm assembled to form a section as shown in Fig. 9.42. Determine the moment of inertia of the section about the horizontal centroidal axis.*

Solution. Each angle is divided into two rectangles as shown in Fig. 9.42.

The distance of the centroidal axis from the bottom fibres of section

$$= \frac{\text{Sum of the moment of the areas about bottom fibres}}{\text{Total area of the section}}$$

$$= \frac{\Sigma A_i \, y_i}{A}$$

Now,

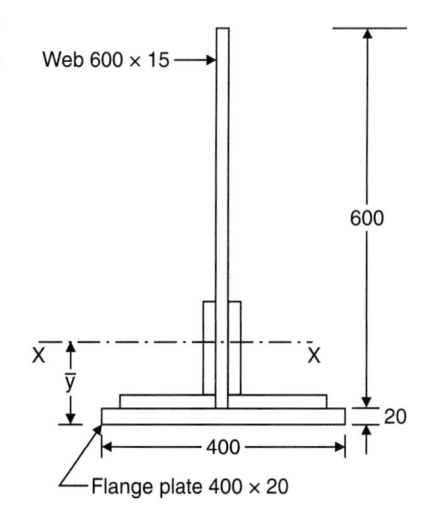

Fig. 9.42

$$\Sigma A_i \, y_i = 600 \times 15 \times \left(\frac{600}{2} + 20 \right) + 140 \times 10 \times (70 + 30) \times 2$$

$$+ 150 \times 10 \times (5 + 20) \times 2 + 400 \times 20 \times 10$$

$$= 3315000 \text{ mm}^3$$

$$A = 600 \times 15 + 140 \times 10 \times 2 + 150 \times 10 \times 2 + 400 \times 20$$

$$= 22,800 \text{ mm}^2$$

$$\therefore \quad \bar{y} = \frac{\Sigma A_i \, y_i}{A} = \frac{3315000}{22,800}$$

Moment of inertia of section about centroidal axis $\Big\} = \Big\{$ Sum of the moment of inertia of the all simple figures about centroidal axis

$$= \frac{15 \times 600^3}{12} + 600 \times 15 \,(145.39 - 320)^2 + \left[\frac{10 \times 140^3}{12} + 1400\,(145.39 - 100)^2\right] \times 2$$

$$+ \left[\frac{150 \times 10^3}{12} + 1500 \times (145.39 - 15)^2\right] 2 + \frac{400 \times 20^3}{12} + 400 \times 20 \times (145.39 - 10)^2$$

$I_{xx} = 7.45156 \times 10^8$ mm^4 **Ans.**

Example 9.18. *Compute the moment of inertia of the 100 mm × 150 mm reactangle shown in Fig. 9.43 about x – x axis to which it is inclined at an angle* $\theta = \sin^{-1}\left(\frac{4}{5}\right)$.

Solution. The rectangle is divided into four triangles as shown in the figure. [The dividing line between A_1 and A_2 is parallel to $x-x$ axis].

Now $= \sin^{-1}\left(\frac{4}{5}\right) = 53.13°$

From the geometry of the Fig. 9.43,
$BK = AB \sin(90° - \theta)$
$\quad = 100 \sin(90° - 53.13°)$
$\quad = 60$ mm
$ND = BK = 60$ mm
$\therefore FD = \dfrac{60}{\sin \theta} = \dfrac{60}{\sin 53.13} = 75$ mm
$\therefore AF = 150 - FD = 75$ mm
Hence $FL = ME = 75 \sin \theta = 60$ mm

$$AE = FC = \frac{AB}{\cos(90 - \theta)} = \frac{100}{0.8}$$
$\quad = 125$ mm

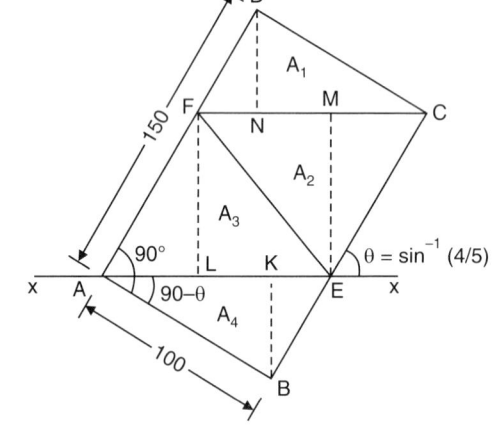

Fig. 9.43

Moment of inertia of the section about $x - x$ axis $\Big\} = \Big\{$ Sum of the moments of intertia of individual triangular areas about $x - x$ axis.

$= I_{DFC} + I_{FCE} + I_{FEA} + I_{AEB}$

$$= \frac{125 \times 60^3}{36} + \frac{1}{2} \times 125 \times 60 \times \left(60 + \frac{1}{3} \times 60\right)^2 + \frac{125 \times 60^3}{36} + \frac{1}{2} \times 125 \times 60 \times \left(\frac{2}{3} \times 60\right)^2$$

$$+ \frac{125 \times 60^3}{36} + \frac{1}{2} \times 125 \times 60 \times \left(\frac{1}{3} \times 60\right)^2 + \frac{125 \times 60^3}{36} + \frac{1}{2} \times 125 \times 60 \times \left(\frac{1}{3} \times 60\right)^2$$

$I_{xx} = 3{,}60{,}00000$ mm^4 **Ans.**

CENTROID AND MOMENT OF INERTIA

Example 9.19. *Find moment of inertia of the shaded area shown in the Fig. 9.44 about the axis AB.*

Solution. The section is divided into a triangle *PQR*, a semicircle *PSQ* having base on axis *AB* and a circle having its centre on axis *AB*.

Now,

Moment of inertia of the section about axis *AB*

$$= \begin{cases} \text{Moment of inertia of triangle } PQR \text{ about } AB \\ + \text{ Moment of inertia of semicircle } PSQ \\ \text{about } AB - \text{moment of inertia of circle about} \\ AB \end{cases}$$

$$= \frac{80 \times 80^3}{12} + \frac{\pi}{128} \times 80^4 - \frac{\pi}{64} \times 40^4$$

$I_{AB} = 42,92979 \text{ mm}^4$ **Ans.**

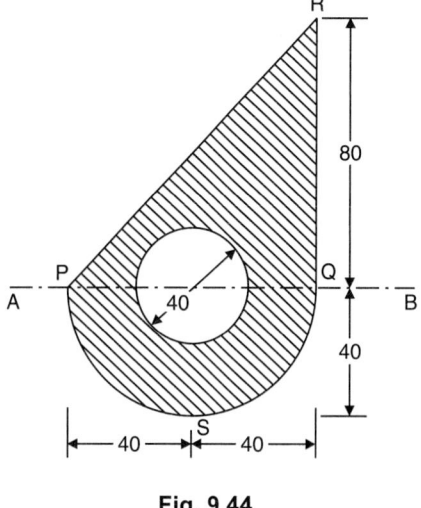

Fig. 9.44

Example 9.20. *Find the second moment of the shaded portion shown in the Fig. 9.45 about its centroidal axis.*

Solution. The section is divided into three simple figures *viz.*, a triangle *ABC*, a rectangle *ACDE* and a semicircle.

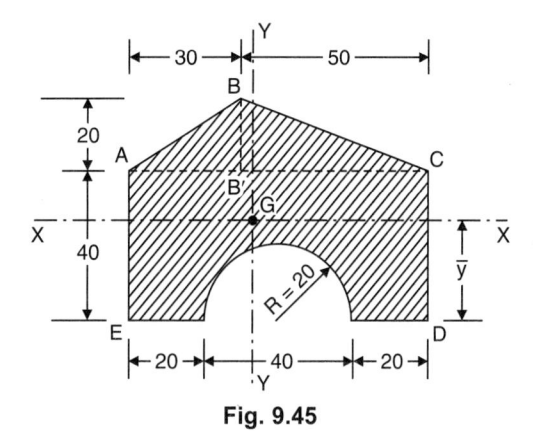

Fig. 9.45

The distance of centroid of the section from base *ED*

$$= \frac{\text{Moment of area of tirangle } ABC + \text{Moment of area of rectangle } ACDE - \text{Moment of semicircle}}{\text{Area of triangle } ABC + \text{Area of rectangle } ACDE - \text{Area of semicircle}}$$

$$\bar{y} = \frac{\frac{1}{2} \times 80 \times 20 \left(\frac{1}{3} \times 20 + 40\right) + 40 \times 80 \times 20 - \frac{1}{2} \times \pi \times 20^2 \times \frac{4 \times 20}{3\pi}}{\frac{1}{2} \times 80 \times 20 + 40 \times 80 - \frac{1}{2} \times \pi \times 20^2} = 28.47 \text{ mm}$$

Similarly, the distance of centroid from side *AE* is given by:

$$\bar{x} = \frac{\frac{1}{2} \times 30 \times 20 \times \frac{2}{3} \times 30 + \frac{1}{2} \times 50 \times 20 \times \left(\frac{1}{3} \times 50 + 30\right) + 40 \times 80 \times 40 - \frac{1}{2} \times \pi \times 20^2 \times 40}{\frac{1}{2} \times 80 \times 20 + 40 \times 80 - \frac{1}{2} \times \pi \times 20^2}$$

$= 39.21 \text{ mm}$

[**Note:** In calcualting moment of area of $\triangle ABC$, it is split into two trinagles.]

Moment of inertia about centroidal $x - x$ axis

$$= \begin{cases} \text{Moment of inertia of triangle } ABC \text{ about } x \\ - x \text{ axis} + \text{Moment of inertia of rectangle} \\ \text{about } x - x \text{ axis} - \text{ moment of inertia semicircle} \\ \text{about } x - x \text{ axis} \end{cases}$$

$$\therefore I_{xx} = \frac{80 \times 20^3}{36} + \frac{1}{2} \times 80 \times 20 \left(60 - \frac{2}{3} \times 20 - 28.47\right)^2 + \frac{80 \times 40^3}{12} + 80 \times 40 \times (28.47 - 20)^2$$

$$- \left[0.0068598 \times 20^4 + \frac{1}{2} \pi \times 20^2 \left(28.47 - \frac{4 \times 20}{3\pi}\right)^2\right]$$

$I_{xx} = 6{,}86944 \text{ mm}^4$ **Ans.**

Similarly, $I_{yy} = \dfrac{20 \times 30^3}{36} + \dfrac{1}{2} \times 20 \times 30 \left(39.21 - \dfrac{2}{3} \times 30\right)^2 + \dfrac{20 \times 50^3}{36} + \dfrac{1}{2} \times 20 \times 50$

$$\times \left[39.21 - \left(30 + \frac{1}{3} \times 50\right)\right]^2 + \frac{40 \times 80^3}{12} + 40 \times 80(39.21 - 40)^2$$

$$- \frac{1}{2} \times \frac{\pi}{64} 40^4 - \frac{1}{2} \times \frac{\pi}{4} \times 40^2 (40 - 39.21)^2$$

$I_{yy} = 1868392 \text{ mm}^4$ **Ans.**

Important Definitions

1. **Centroid** is the point in a plane section such that for any axis through that point moment of area is zero.
2. Moment of inertia of an area about an axis perpendicular to its plane is called **polar moment of inertia**.
3. **Perpendicular axis theorem** states that the moment of inertia of a given area about an axis perpendicular to the plane of area through a point is equal to the sum of moment of inertia of that area about any two mutually perpendicular axes through that point, the axes being in the plane of the area.
4. **Parallel axes theorem** states that the moment of inertia of an area about any axis in the plane of the area is equal to the sum of moment of inertia about the centroidal axis of the area and the product of area and square of the distance of the centroid of the area from the axis.

Important Formulae

1. $\bar{x} = \dfrac{\Sigma A_i x_i}{A} = \dfrac{\int x \, dA}{A}; \; \bar{y} = \dfrac{\Sigma A_i y_i}{A} = \dfrac{\int y \, dA}{A}$
2. Centroid lies on the line of symmetry
3. $I_{xx} = \Sigma A_i y_i^2, \; I_{yy} = \Sigma A_i x_i^2, \; I_{zz} = \Sigma A_i r_i^2$
4. $k = \sqrt{\dfrac{I}{A}}$ or $I = Ak^2$

CENTROID AND MOMENT OF INERTIA 261

5. $I_{zz} = I_{xx} + I_{yy}$
6. $I_{AB} = I_{yy} + Ay_c^2$
7. For finding centroid of standard figures refer table 9.1
8. For finding moment of inertia of standard figures refer table 9.3.

PROBLEMS FOR EXERCISE

9.1. Determine the centroid of the built-up section in Fig. 9.46. Express the coordinates of centroid with respect to x and y axes shown. [**Ans.** $\bar{x}$ = 48.91 mm ; $\bar{y}$ = 61.30 mm]

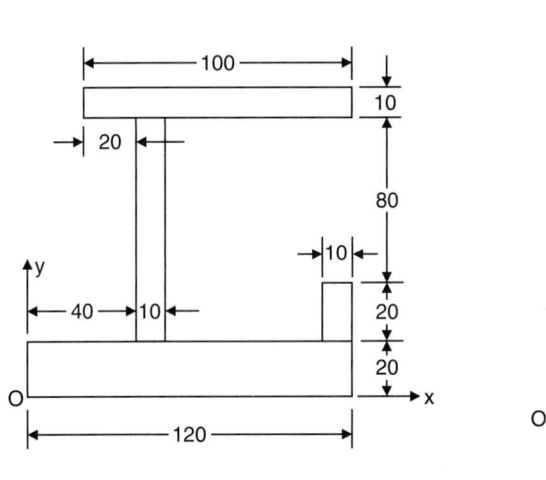

Fig. 9.46

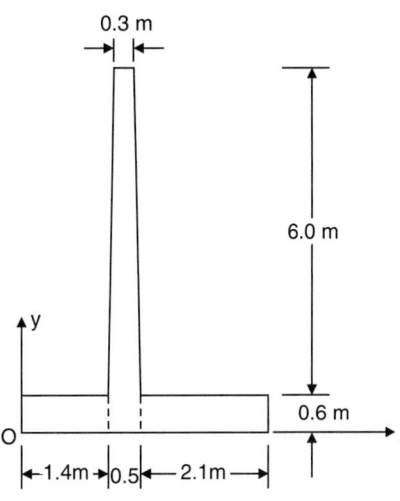

Fig. 9.47

9.2. Determine the centroid of the reinforced concrete retaining wall section shown in Fig. 9.47.
[**Ans.** $\bar{x}$ = 1.848 m; $\bar{y}$ = 1.825 m]

9.3. Find the coordinates of the centroid of the shaded area with respect to the axes shown in Fig. 9.48.
[**Ans.** $\bar{x}$ = 43.98 mm; $\bar{y}$ = 70.15 mm]

Fig. 9.48

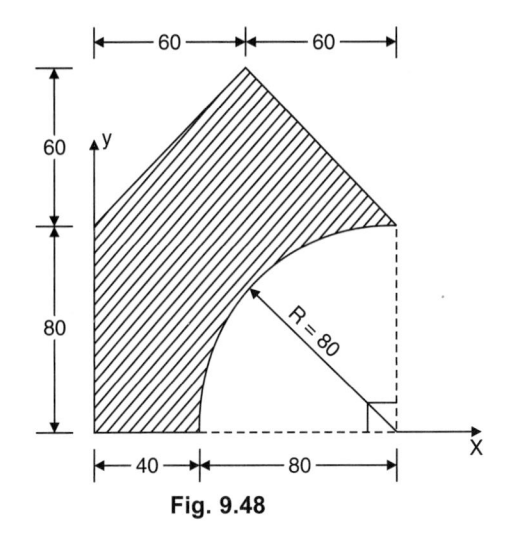

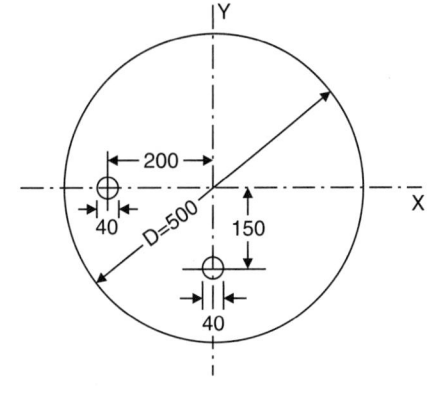

Fig. 9.49

9.4. A circular plate of uniform thickness and of diameter 500 mm as shown in Fig. 9.49 has two circular holes of 40 mm diameter each. Where should a 80 mm diameter hole be drilled so that the centre of gravity of the plate will be at the geometric centre.

[**Ans.** $\bar{x}$ = 50 mm; $\bar{y}$ = 37.5 mm]

9.5. With respect to the coordinate axes x and y locate the centroid of the shaded area shown in Fig. 9.50 [**Ans.** $\bar{x}$ = 97.47 mm; $\bar{y}$ = 70.69 mm]

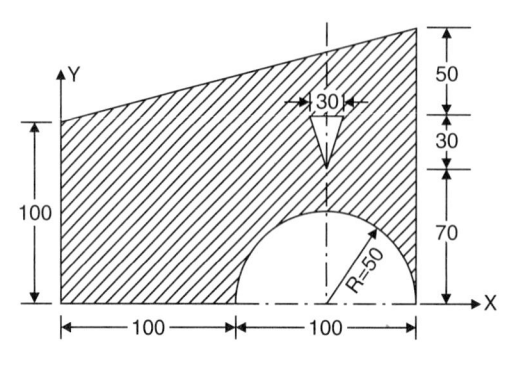

Fig. 9.50

9.6. Locate the centroid of the plane area shown in Fig. 9.51.

[**Ans.** $\bar{x}$ = 104.10 mm; $\bar{y}$ = 44.30 mm]

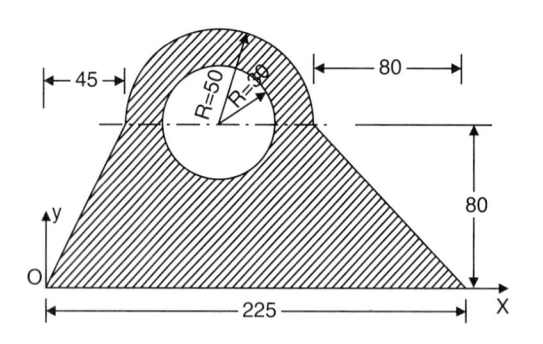

Fig. 9.51

9.7. Determine the coordinates of the centroid of shaded are shown in Fig. 9.52 with respect to the corner point O. Take x = 40 mm [**Ans.** $\bar{x}$ = 2600 mm; $\bar{y}$ = 113.95 mm]

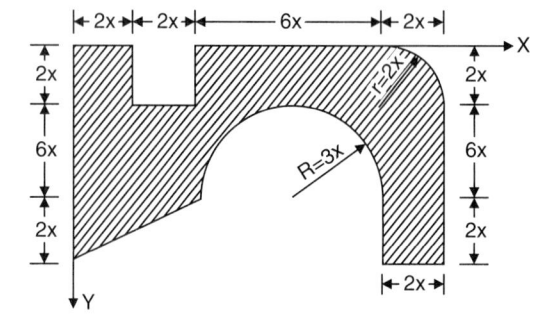

Fig. 9.52

CENTROID AND MOMENT OF INERTIA

9.8. *ABCD* is a square section of sides 100 mm. Determine the ratio of moment of inertia of the section about centroidal axis parallel to a side to that about diagonal *AC*. [**Ans.** 1]

9.9. The cross-section of a rectangular hollow beam is as shown in Fig. 9.53. Determine the polar moment of inertia of the section about centroidal axes.

[**Ans.** I_{xx} = 1,05,38667 mm^4; I_{yy} = 49,06667 mm^4; I_{zz} = 1,54,45334 mm^4]

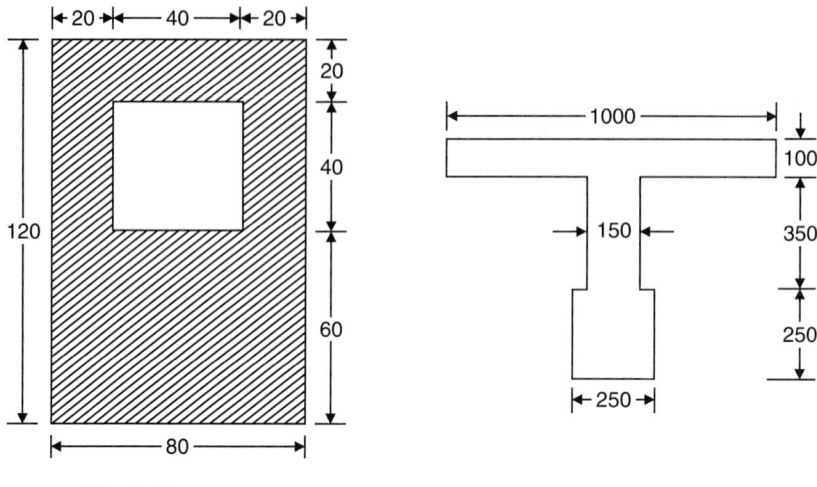

Fig. 9.53

Fig. 9.54

9.10. The cross-section of a prestressed concrete beam is shown in Fig. 9.54. Calculate the moment of inertia of this section about the centroidal axes parallel to and perpendicular to top edge. Also determine the radii of gyration. [**Ans.** I_{xx} = 1.15668 × 10^{10} mm^4; k_{xx} = 231.95 mm;
I_{yy} = 8.75729 × 10^9 mm^4; k_{yy} = 201.82 mm]

9.11. The strength of a 400 mm deep and 200 mm wide *I*-beam of uniform thickness 10 mm, is increased by welding a 250 mm wide and 20 mm thick plate to its upper flanges as shown in Fig. 9.55. Determine the moment of inertia and the radii of gyration of the composite section with respect to cetroidal axes parallel to and perpendicular to the bottom edge *AB*.

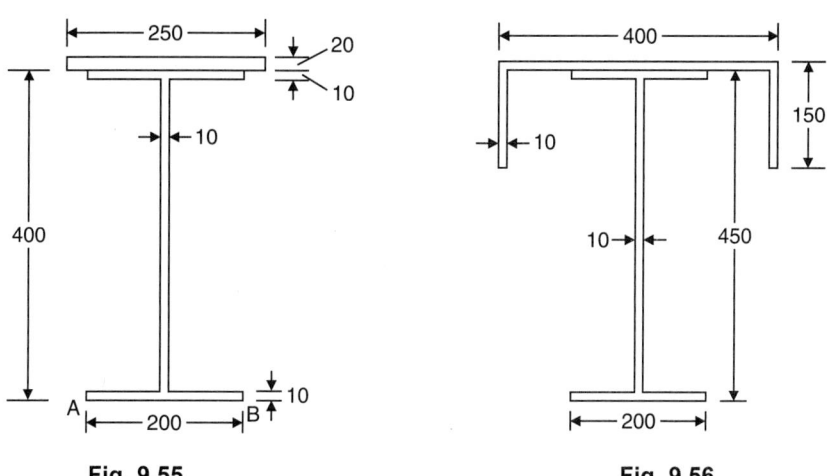

[**Ans.** I_{xx} = 3.32393 × 10^8 mm^4; k_{xx} = 161.15 mm; I_{yy} = 3,94,06667 mm^4; k_{yy} = 55.49 mm]

Fig. 9.55

Fig. 9.56

9.12. The cross-section of a gantry girder is as shown in Fig. 9.56. It is made up of an *I*-section of depth 450 mm, fiange width 200 mm and a channel of size 400 mm × 150 mm. Thickness of all members is 10 mm. Find the moment of inertia of the secion about the horizontal centroid axis.

[**Ans.** $I_{xx} = 4.2198 \times 10^8$ mm^4]

9.13. A plate girder is made up of a web plate of size 400 mm × 10 mm, four angles of size 100 mm × 100 mm × 10 mm and cover plates of size 300 mm × 10 mm as shown in Fig. 9.57. Determine the moment of inertia about horizontal and vertical centroidal axes.

[**Ans.** $I_{xx} = 5.35786 \times 10^8$ mm^4; $I_{yy} = 6,08,50,667$ mm^4]

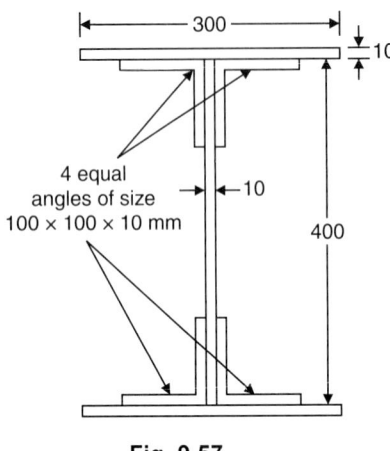

Fig. 9.57

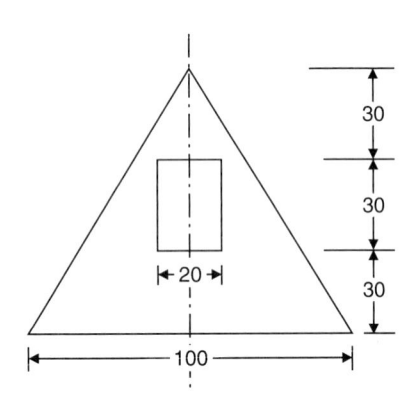

Fig. 9.58

9.14. Determine the moment of inertia and radii of gyration of the area shown in Fig. 9.58 about the base A—B and the centroidal axis parallel to AB.

[**Ans.** $I_{AB} = 48,15000$ mm^4; $I_{xx} = 18,24231$ mm^4]
$k_{AB} = 35.14$ mm $k_{xx} = 21.63$ mm

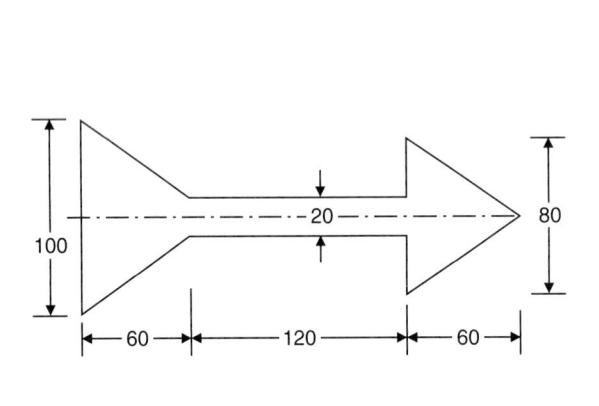

Fig. 9.59

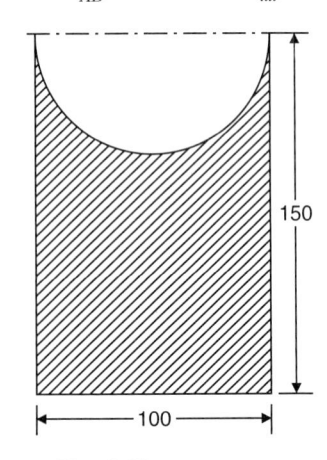

Fig. 9.60

9.15. Determine the moment of inertia of the section shown in Fig. 9.59 about the vertical centroidal axis. [**Ans.** $I_{yy} = 5,03,82857$ mm^4]

9.16. A semi-circular cut is made in rectangular wooden beam as shown in Fig. 9.60. Determine the polar moment of inertia of the section about the centroidal axis.

[**Ans.** $I_{xx} = 12670575$ mm^4; $I_{yy} = 1,00,45631$ mm^4; $I_{zz} = 22716206$ mm^4]

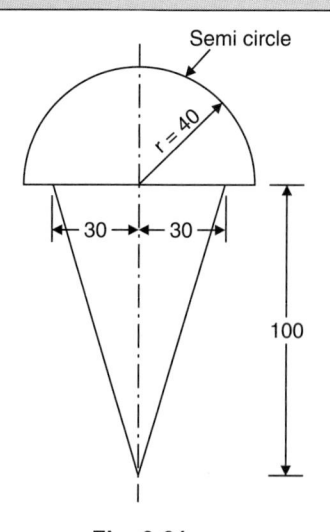

CENTROID AND MOMENT OF INERTIA

Fig. 9.61

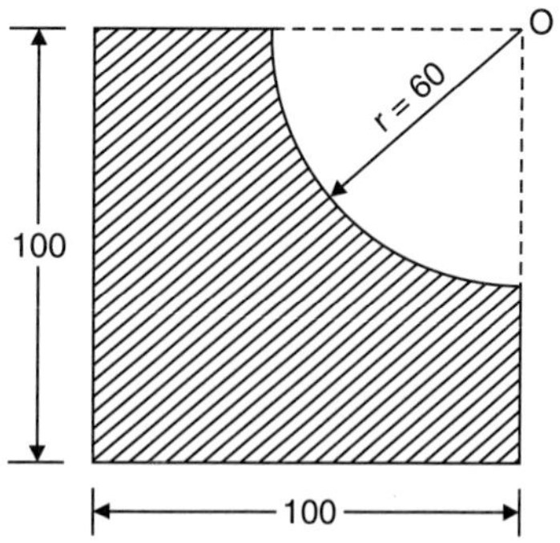

Fig. 9.62

9.17. Determine the moment of inertia of the section shown in Fig. 9.61 about the horizontal centroidal axis. Also find the moment of inertia of the section about the symmetrical axis. Hence find the polar moment of inertia. **[Ans.** I_{xx} = 5409046 mm^4; I_{yy} = 1455310 mm^4; I_{zz} = 6864356 mm^4**]**

9.18. The cross-section of a machine part is as shown in Fig. 9.62. Determine its moment of inertia and radius of gyration about the horizontal centroidal axis.
[Ans. I_{xx} = 5249090.85 mm^4; k_{xx} = 27.05 mm**]**

9.19. The cross-section of a plain concrete culvert is as shown in Fig. 9.63. Determine the moment of inertia about the horizontal centroidal axis. **[Ans.** I_{xx} = 5.45865 × 10^{10} mm^4**]**

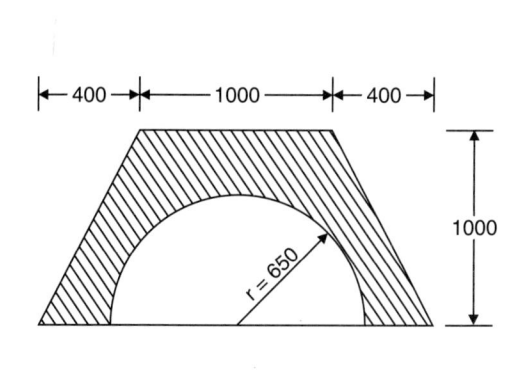

Fig. 9.63

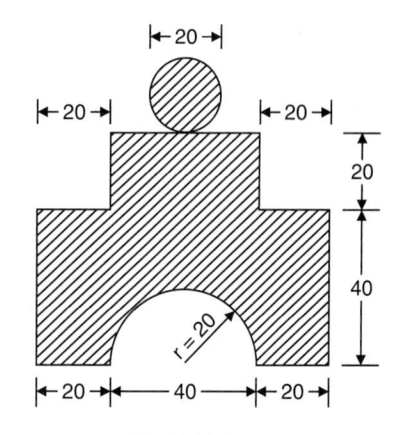

Fig. 9.64

9.20. Determine the centroid of the built-up section shown in Fig. 9.64 and find the moment of inertia and radius of gyration about the horizontal centroidal axis.
[Ans. I_{xx} = 1267942 mm^4; k_{xx} = 18.55 mm**]**

10
Centre of Gravity and Mass Moment of Inertia

Centre of gravity and mass moment of inertia are two important properties of a body required in the study of motion of rigid bodies. In this chapter, the method of finding the centre of gravity of a given body is explained. Centre of gravity is determined from first principles for standard figures and the method of finding it for composite figures is illustrated. The term 'Mass Moment of Inertia', which is similar to moment of inertia of plane figures, is defined and determined for a few standard bodies. The method of finding moment of inertia for composite bodies is illustrated.

10.1. CENTRE OF GRAVITY

Centre of gravity (CG) of a body is the point through which the resultant weight of the body passes through in whichever position the body is kept. Since locating centre of gravity involves locating the line of action of resultant weight, principle of moments (Varignon's theorem) can be used.

Let W_i be weight of ith element of the given body of weight W (Ref. Fig. 10.1). Let the coordinates of this element be (x_i, y_i, z_i) and that of centre of gravity be (x, y, z). Now,

$$W = W_1 + W_2 + \ldots$$
$$= \Sigma W_i \qquad \ldots(10.1)$$

Since the resultant of all elemental weights passes through center of gravity (from definition), we have

$$W\overline{x} = W_1 x_1 + W_2 x_2 + \ldots$$
$$= \Sigma W_i x_i$$

or $\qquad \overline{x} = \dfrac{\Sigma W_i x_i}{W} \qquad \ldots(10.2)$

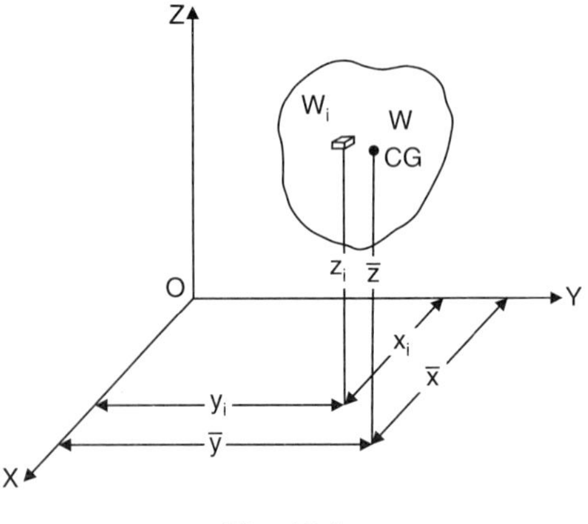

and $\qquad W\overline{y} = \Sigma W_i y_i \qquad \ldots(10.3)$

Fig. 10.1

When reference axis and the body are rotated by 90° so that z axis is horizontal, Varignon's theorem of moments gives,

$$W = \Sigma W_i z_i \qquad \ldots(10.4)$$

Thus, the centre of gravity of a body is given by the expressions:

$$W\overline{x} = \Sigma W_i x_i$$
$$W\overline{y} = \Sigma W_i y_i \qquad \ldots(10.5)$$

and
$$W\bar{z} = \Sigma W_i z_i$$

If M is the mass of the body and m_i the mass of the element, substituting $M = \dfrac{W}{g}$ and $m_i = \dfrac{W_i}{g}$, we get

$$M\bar{x} = \Sigma m_i x_i$$
$$M\bar{y} = \Sigma m_i y_i \qquad \qquad \ldots(10.6)$$
$$M\bar{z} = \Sigma m_i z_i$$

also
$$M = \Sigma m_i \qquad \qquad \ldots(10.7)$$

If the body is made up of uniform material of unit weight γ, then we know, $W_i = V_i\,\gamma$ and hence equation (10.5) reduces to

$$\left.\begin{aligned} V\bar{x} &= \Sigma V_i x_i \\ V\bar{y} &= \Sigma V_i y_i \\ V\bar{z} &= \Sigma V_i z_i \end{aligned}\right\} \qquad \ldots(10.8)$$

where, V is total volume of the body, V_i is the volume of i-th element.

For a body made up of uniform material, the term centre of gravity may be looked upon as centroid of volume.

10.2. USE OF SYMMETRY

As in the case of centroid, the principle of symmetry can be used in locating centre of gravity also. Centroid always lie on the plane of symmetry since in Eqn. (10.5), for every i-th element at a particular distance, we can find another identical element but at negative distance. If there are two planes of symmetry, then centre of gravity lies on the line of intersection of those two symmetric planes. If there are three planes of symmetry, then the point of intersection of planes of symmetry is the centre of gravity.

10.3. CENTRE OF GRAVITY FROM FIRST PRINCIPLES

Selecting the expressions for W_i and x_i for an element in general form, the Equations (10.5) and (10.1) can be written as

$$\left.\begin{aligned} W\bar{x} &= \int x\,dw \\ W\bar{y} &= \int y\,dw \\ W\bar{z} &= \int z\,dw \end{aligned}\right\} \qquad \ldots(10.9)$$

and
$$W = \int dw$$

In the above expressions, the limits of integration are to be suitably selected such that the entire body weight is covered by integration.

Since, weight = mass × gravitational acceleration, equation (10.9) may be written as:

$$M\bar{x} = \int x\,dm$$

$$M\bar{y} = \int y\,dm$$

$$M\bar{z} = \int z\,dm$$

and
$$M = \int dm$$

...(10.10)

where M = mass of the body and dm = mass of elemental body.

If the body is made up of uniform material, equation (10.7) takes the form:

$$V\bar{x} = \int x\,dv$$

$$V\bar{y} = \int y\,dv$$

$$V\bar{z} = \int z\,dv$$

and
$$V = \int dv$$

...(10.11)

where V is the volume of the body and dv is the volume of elemental body.

Using equation (10.9) or (10.10) or (10.11), one can locate the centre of gravity of bodies of simple shape. The method is illustrated with examples (10.1) and (10.2).

Example 10.1. *Locate the centre of gravity of the right circular cone of base radius r and height h shown in Fig. 10.2.*

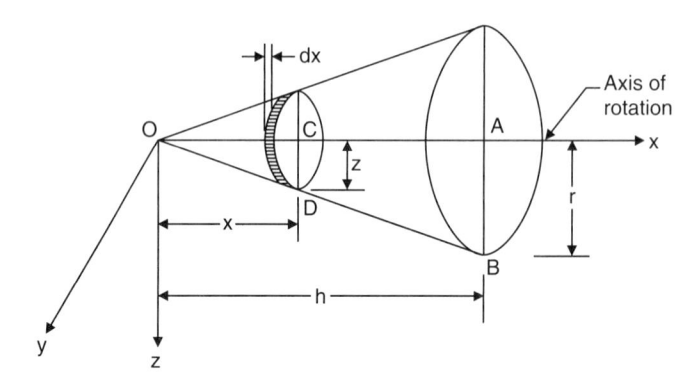

Fig. 10.2

Solution. Taking origin at the vertex of the cone and selecting the axes as shown in Fig. 10.2, it can be observed that due to symmetry the coordinates of centre of gravity $\bar{y}$ and $\bar{z}$ are equal to zero, *i.e.* the centre of gravity lies on the axis of rotation of the cone. To find its distance $\bar{x}$ from the vertex, consider an elemental plate at a distance x. Let the thickness of the elemental plate be dx. From the similar triangles OAB and OCD, the radius of elemental plate z is given by

$$z = \frac{x}{h} r$$

CENTRE OF GRAVITY AND MASS MOMENT OF INERTIA

∴ Volume of the elemental plate dv

$$dv = \pi z^2 \, dx = \pi x^2 \frac{r^2}{h^2} \, dx$$

If γ is the unit weight of the material of the cone, then weight of the elemental plate is given by:

$$dW = \gamma \pi x^2 \frac{r^2}{h^2} \, dx \qquad \ldots(1)$$

∴
$$W = \int_0^h \gamma \frac{\pi r^2}{h^2} x^2 \, dx$$

$$= \gamma \frac{\pi r^2}{h^2} \left[\frac{x^3}{3} \right]_0^h = \gamma \pi \frac{r^2 h}{3} \qquad \ldots(2)$$

$$\left[\text{Note: } \frac{\pi r^2 h}{3} \text{ is volume of cone} \right]$$

Now, substituting the value of dw in (1), above we get:

$$\int x \cdot dw = \int_0^h \gamma \frac{\pi r^2}{h^2} x^2 \cdot x \cdot dx$$

$$= \gamma \frac{\pi r^2}{h^2} \left[\frac{x^4}{4} \right]_0^h = \gamma \frac{\pi r^2 h^2}{4} \qquad \ldots(3)$$

From eqn. (10.9),

$$W \bar{x} = \int x \, dw$$

i.e.
$$\frac{\pi r^3 h}{3} \bar{x} = \frac{\gamma \pi r^2 h^2}{4}$$

∴
$$\bar{x} = \frac{3}{4} h \qquad \ldots(10.12)$$

Thus, in a right circular cone, centre of gravity lies at a distance $\frac{3}{4}h$ from vertex along the axis of rotation i.e., at a distance $\frac{h}{4}$ from the base.

Example 10.2. *Determine the centre of gravity of solid hemisphere of radius r from its diametral axis.*

Solution. Due to symmetry, centre of gravity lies on the axis of rotation. To find its distance $\bar{x}$ from the base along the axis of rotation, consider an elemental plate at a distance x as shown in Fig. 10.3.

Now,
$$x^2 + z^2 = r^2$$
$$z^2 = r^2 - x^2 \qquad \ldots(1)$$

Volume of elemental plate
$$dv = \pi z^2\, dx = \pi(r^2 - x^2)dx \qquad \ldots(2)$$
∴ Weight of elemental plate
$$dw = \gamma dv = \gamma\pi(r^2 - x^2)dx \qquad \ldots(3)$$
∴ Weight of hemisphere
$$W = \int dw = \int_0^r \gamma\pi(r^2 - x^2)dx$$
$$= \gamma\pi\left[r^2 x - \frac{x^3}{3}\right]_0^r$$
$$= \frac{2\gamma\pi r^3}{3} \qquad \ldots(4)$$

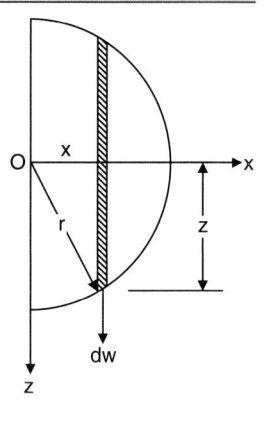

Fig. 10.3

Moment of weight about z axis
$$= \int_0^r x\, dw = \int_0^r x\ \pi(r^2 - x^2)dx$$
$$= \pi\left[r^2\frac{x^2}{2} - \frac{x^4}{4}\right]_0^r = \frac{\pi r^4}{4} \qquad \ldots(5)$$

∴ $\bar{x}$, the distance of centre of gravity from base is given by:
$$W\bar{x} = \int_0^r x\, dw$$

i.e. From (4) and (5) above we get;
$$\frac{2\gamma\pi r^3}{3}\bar{x} = \frac{\gamma\pi r^4}{4}$$

∴ $$\bar{x} = \frac{3}{8}r \qquad \ldots(10.13)$$

Thus, the centre of gravity of a solid hemisphere of radius r is at a distance $\frac{3}{8}r$ from its diametral axis.

10.4. CENTRE OF GRAVITY OF COMPOSITE BODIES

To locate the centre of gravity of a composite body, the composite body will be split into a number of simple bodies, whose weight and centre of gravity can be found using standard expressions. Then, the centre of gravity of the composite body is found using Eqn. (10.5). The method is illustrated with examples 10.3 and 10.4.

Examples 10.3. *Determine the maximum height h of the cylindrical portion of the body with hemispherical base shown in Fig. 10.4 so that it is in stable equilibrium on its base.*

Solution. The body will be stable on its base as long as its centre of gravity is in hemispherical base. The limiting case is when it is on the plane x–x shown in the figure. Centroid lies on the axis of rotation.

Mass of cylindrical portion
$$m_1 = \pi r^2 h \rho \text{ where } \rho \text{ is unit mass of material.}$$

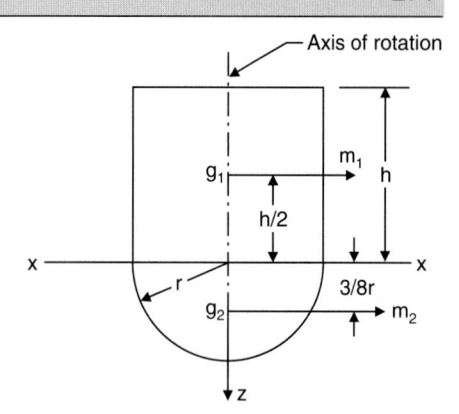

CENTRE OF GRAVITY AND MASS MOMENT OF INERTIA

Its centre of gravity g_1 is at a height

$$z_1 = \frac{h}{2} \text{ from } x \text{ axis.}$$

Mass of hemispherical portion

$$m_2 = \rho \frac{2\pi r^3}{3}$$

and its CG is at a distance

$$z_2 = \frac{3r}{8} \text{ from } x\text{–}x \text{ plane.}$$

Since centroid is to be on x-x plane $\bar{z} = 0$.

i.e. $\Sigma m_i z_i = 0$

∴ $\dfrac{m_1 h}{2} - m_2 \dfrac{3}{8} r = 0$

$$\pi r^2 h \rho \frac{h}{2} = \rho \frac{2\pi r^3}{3} \cdot \frac{3}{8} r$$

∴ $h^2 = \dfrac{1}{2} r^2$ or $h = \dfrac{r}{\sqrt{2}} = 0.707\, r$ **Ans.**

Fig. 10.4

Example 10.4. *A concrete block of size 0.60 m × 0.75 m × 0.5 is cast with a hole of diameter 0.2 m and depth 0.3 m as shown in Fig. 10.5. The hole is completely filled with steel balls weighing 2500 N. Locate the centre of gravity of the body. Take the weight of concrete = 25000 N/m³.*

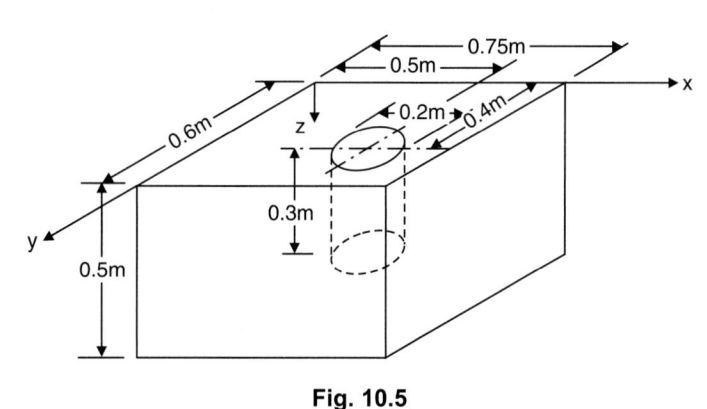

Fig. 10.5

Solution. Weight of solid concrete block:

$W_1 = 0.6 \times 0.75 \times 0.5 \times 25000 = 5625$ N

Weight of concrete (W_2) removed for making hole:

$$W_2 = \frac{\pi}{4} \times 0.2^2 \times 0.3 \times 25000 = 235.62 \text{ N}$$

Taking origin as shown in the figure, the centre of gravity of solid block is (0.375, 0.3, 0.25) and that of hollow portion is (0.5, 0.4, 0.15). The following Table (10.1) may be prepared now.

Table 10.1

	Simple body	W_i	x_i	$W_i x_i$	y_i	$W_i y_i$	z_i	$W_i z_i$
1.	Solid block	5625	0.375	2109.38	0.3	1687.5	0.25	1406.25
2.	Hole in concrete block	−235.62	0.5	−117.81	0.4	−94.25	0.15	−35.34
3.	Steel balls	2500	0.5	1250.0	0.4	1000.0	0.15	375.0

$\Sigma W_i = 7889.38 \qquad \Sigma W_i x_i = 3241.57 \qquad \Sigma W_i y_i = 2593.25 \qquad \Sigma W_i z_i = 1745.91$

$\therefore \qquad \bar{x} = \dfrac{\Sigma W_i x_i}{W} = \dfrac{\Sigma W_i x_i}{\Sigma W_i} \qquad \bar{x} = \dfrac{3241.57}{7889.38} = \mathbf{0.411\ m}$ **Ans.**

Similarly, $\qquad \bar{y} = \dfrac{2593.25}{7887.38} = \mathbf{0.329\ m}$ **Ans.**

$\bar{z} = \dfrac{1745.91}{7889.38} = \mathbf{0.221\ m}$ **Ans.**

Example 10.5. *Locate the centroid of the wire shown in Fig. 10.6. Portion AB is in x–z plane, BC in y–z plane and CD in x–y plane. AB and BC are semicircular in shape.*

Solution. Let w be the weight of wire per mm length. Referring to Fig. 10.6(*b*) for a semicircular wire, $\bar{x} = 0$, due to symmetry and $\bar{z}$ is given by

$$W\bar{z} = \int z\ dw$$

$$\pi r w \bar{z} = \int_0^\pi r \sin\theta\ w\ r d\theta$$

$$= r^2 w\ [-\cos\theta]_0^\pi$$

$$= 2 r^2 w$$

$\therefore \qquad \bar{z} = \dfrac{2r}{\pi}$

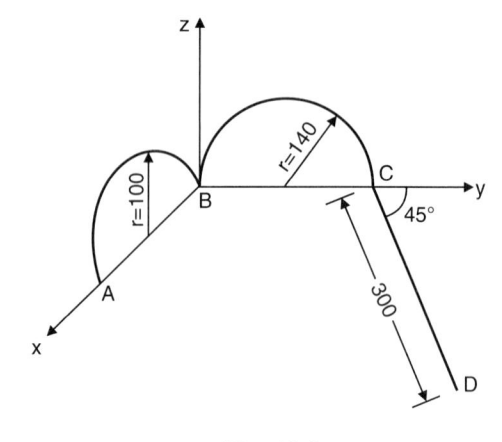

Fig. 10.6

Noting this result, the weight and centre of gravity of portions *AB, BC* and *CD* are tabulated (Table 10.2) and the calculations are carried out.

Table 10.2

Portion	W_i	x_i	$W_i x_i$	y_i	$W_i y_i$	z_i	$W_i z_i$
AB	$100\pi w$	100	$31415.93w$	0	0.0	63.66	$20000w$
BC	$140\pi w$	0	0.0	140	$61575.22w$	89.13	$3920w$
CD	$300w$	$\dfrac{300}{\sqrt{2}}$	$63639.61w$	$\dfrac{300}{\sqrt{2}}$	$63639.61w$	0	0.0

$\Sigma W_i = 1053.98w \qquad \Sigma W_i x_i = 95055.54w \qquad \Sigma W_i y_i = 125214.83w$
$\Sigma W_i z_i = 59200w$

CENTRE OF GRAVITY AND MASS MOMENT OF INERTIA 273

$$\therefore \quad \bar{x} = \frac{\Sigma W_i x_i}{w} = \frac{95055.54}{1053.98} = \mathbf{90.19\ mm}$$

$$\bar{y} = \frac{125214.83}{1053.98} = \mathbf{118.80\ mm} \quad \mathbf{Ans.}$$

and
$$\bar{z} = \frac{59055.54}{1053.98} = \mathbf{56.17\ mm}$$

10.5. THEOREMS OF PAPPUS-GULDINUS

There are two important theorems, first proposed by Greek scientist Pappus (about 340 AD) and then restated by Swiss mathematician Paul Guldinus (1640) for determining the surface area and volumes generated by rotating a curve and a plane area about a non-intersecting axis, some of which are shown in Fig. 10.7. These theorems are known as Pappus-Guldinus theorems.

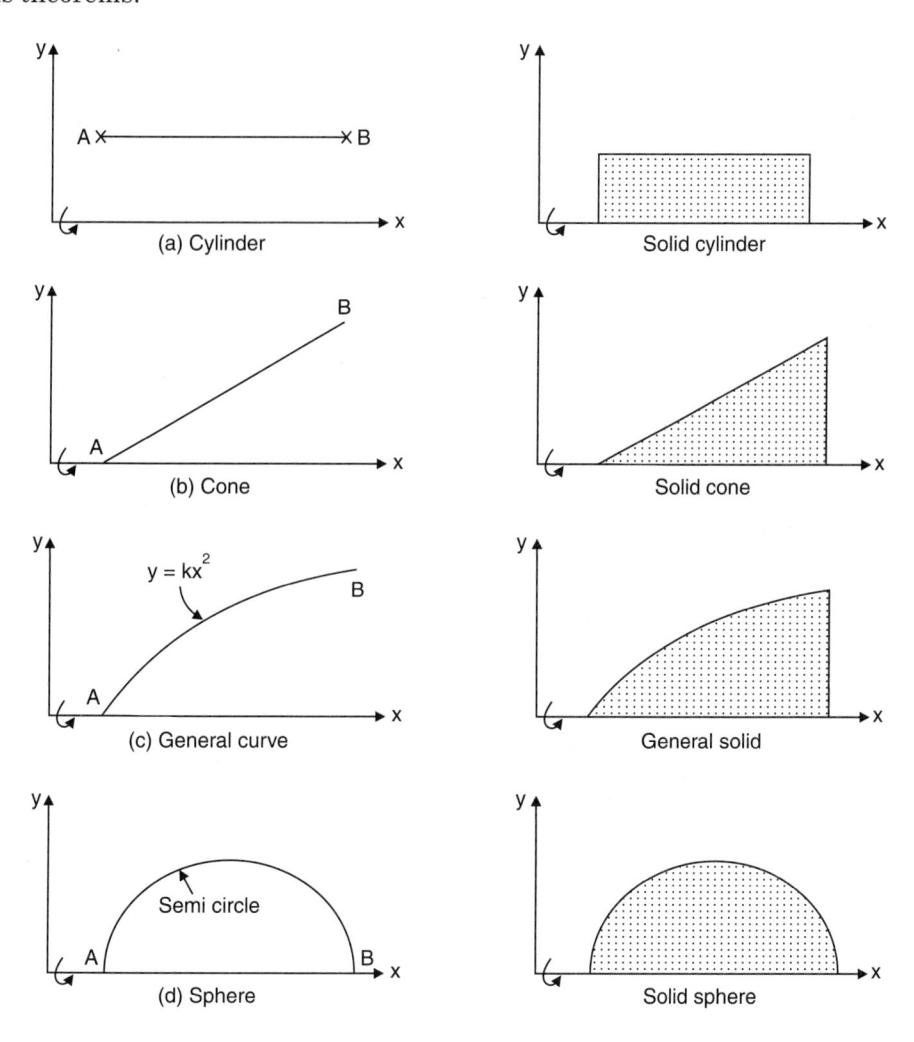

(a) Cylinder

Solid cylinder

(b) Cone

Solid cone

(c) General curve ($y = kx^2$)

General solid

(d) Sphere (Semi circle)

Solid sphere

(e) Torus
Generating surface of revolution

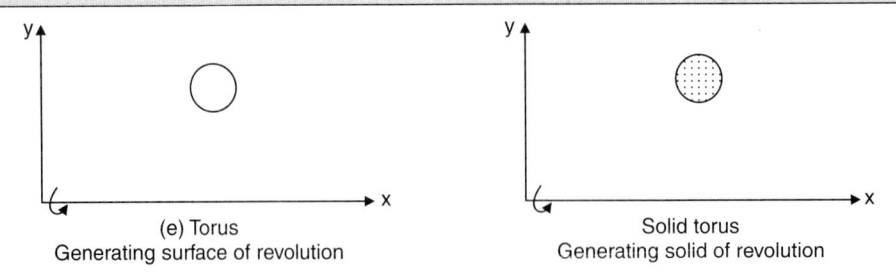

Solid torus
Generating solid of revolution

Fig. 10.7

Theorem I. The area of surface generated by revolving a plane curve about a non-intersecting axis in the plane of the curve is equal to the length of the generating curve times the distance travelled by the centroid of the curve in the revolution.

Proof. Fig. 10.8 shows the isometric view of the plane curve rotated about x-axis by angle θ. We are interested in finding the surface area generated by rotating the curve AB. Let dL be the elemental length on the curve at D. Its coordinate be y. Then the elemental surface area generated by this element at D

$$dA = dL\,(y\theta)$$

∴
$$A = \int dL\,(y\theta) = \theta \int y\,dL$$
$$= \theta\,L y_c$$
$$= L\,(y_c \theta), \text{ where } y_c \text{ is the } y\text{-coordinate or centroid}$$

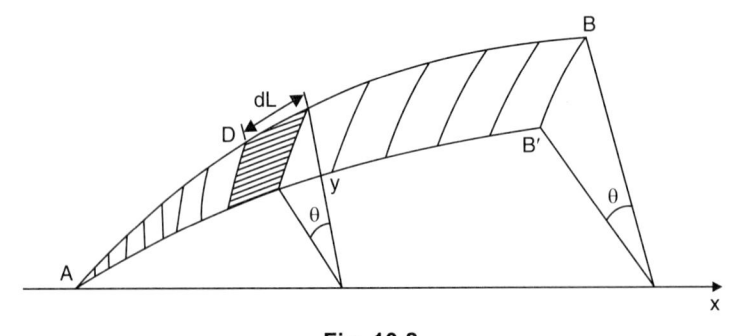

Fig. 10.8

Thus we get area of the surface generated as length of the generating curve times the distance travelled by the centroid.

Theorem II. The volume of the solid generated by revolving a plane area about a non-intersecting axis in the plane is equal to the area of the generating plane times the distance travelled by the centroid of the plane area during the rotation.

Proof. Consider the plane area ABC, which is rotated through an angle θ about x-axis as shown in Fig. 10.9.

Let dA be the elemental area at distance y from x-axis. Then the volume generated by this area during rotation is given by

$$dV = dA\,y\theta$$

∴
$$V = \int dA\,y\theta = \theta \int y\,dA$$
$$= \theta A\,y_c = A(y_c \theta),$$

where y_c is the y-coordinate of centroid of the volume.

CENTRE OF GRAVITY AND MASS MOMENT OF INERTIA

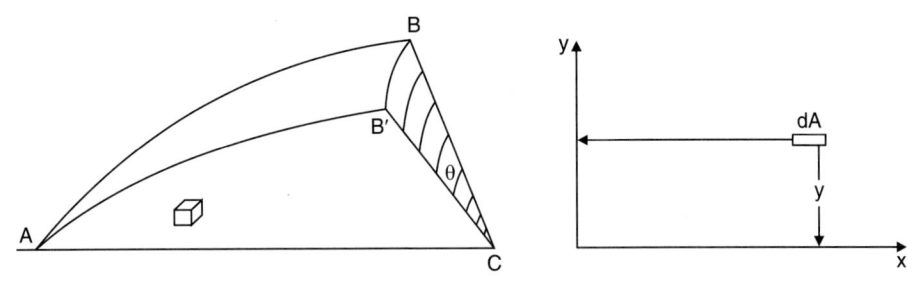

Fig. 10.9

Thus the volume of the solid generated is area times the distance travelled by its centroid during the rotation. Using Pappus-Guldinus theorems surface area and volumes of cones and spheres can be calculated as shown below:

(i) **Surface area of a cone.** Referring to Fig. 10.9(a),

Length of the line generating cone = L

Distance of centroid of the line from the axis of rotation $= y = \dfrac{R}{2}$

In one revolution centroid moves by distance = $2\pi y = \pi R$

∴ Surface area = $L \times (\pi R) = \pi R L$

(ii) **Volume of a cone**

Referring to Fig. 10.10(b),

Area generating solid cone $= \dfrac{1}{2} h R$

Centroid G is at a distance

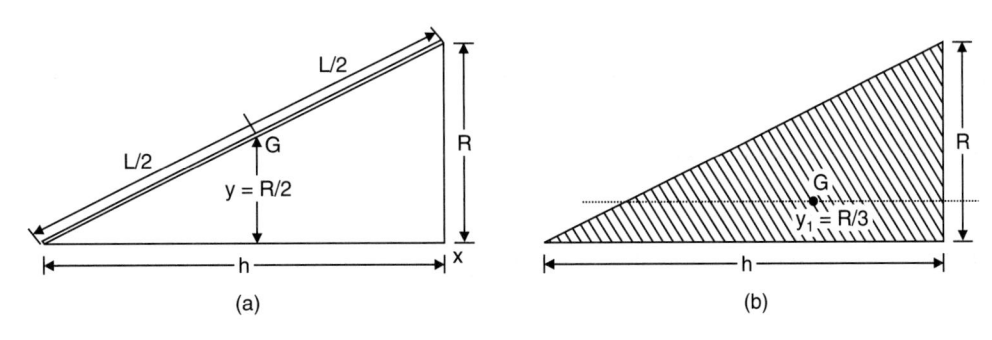

$$y = \dfrac{R}{3}$$

Fig. 10.10

∴ The distance moved by the centroid in one revolution $= 2\pi y_1 = 2\pi \dfrac{R}{3}$

∴ Volume of solid cone $= \dfrac{1}{2} h R \times \dfrac{2\pi R}{3} = \dfrac{\pi R^2 h}{3}$

(iii) **Surface area of sphere.** Sphere of radius R is obtained by rotating a semi circular arc of radius R about its diametral axis. Referring to Fig. 10.11(a),

Length of the arc = πR

Centroid of the area is at $y = \dfrac{2R}{\pi}$ from the diametral axis (*i.e.*, axis of rotation).

∴ Distance travelled by centroid of the arc in one revolution

$$= 2\pi y = 2\pi \dfrac{2R}{\pi} = 4R$$

∴ Surface area of sphere = $\pi R \times 4R = 4\pi R^2$

(*iv*) **Volume of sphere.** Solid sphere of radius R is obtained by rotating a semi-circular area about its diametral axis. Referring to Fig. 6.10(*b*).

Area of semi-circle $= \dfrac{\pi R^2}{2}$

Distance of centroid of semi-circular area from its centroidal axis

$$= y = \dfrac{4R}{3\pi}$$

∴ The distance travelled by the centroid in one revolution

$$= 2\pi y = 2\pi \dfrac{4R}{3\pi} = \dfrac{8\pi R}{3\pi}$$

∴ Volume of sphere $= \dfrac{\pi R^2}{2} \times \dfrac{8\pi R}{3\pi} = \dfrac{4\pi R^3}{3}$

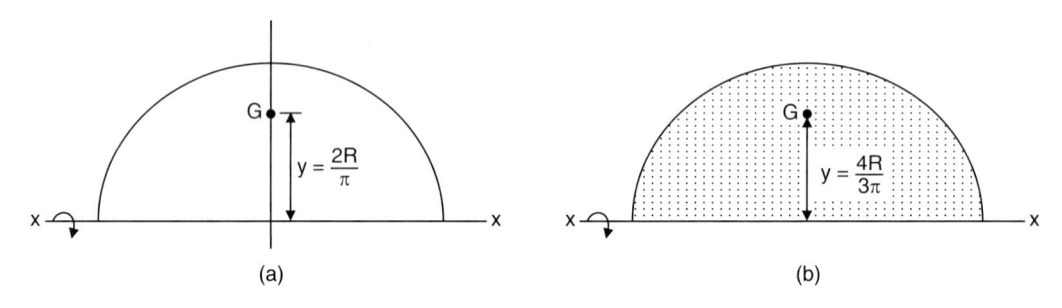

Fig. 10.11

10.6. MASS MOMENT OF INERTIA

Mass moment of inertia of a body about an axis is defined as the *sum total of product of its elemental* masses and square of their distance from the axis.

Thus, the mass moment of the body shown in Fig. 10.12 about axis AB is given by

$$I_{AB} = \Sigma dm\ r^2 = \int r^2\ dm \qquad \ldots(10.14)$$

where r is the distance of element of mass dm from AB.

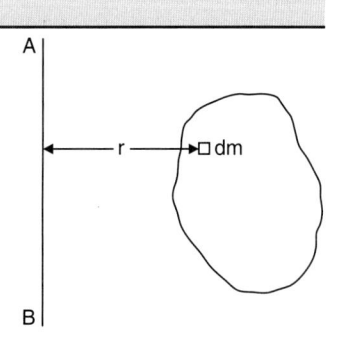

Fig. 10.12

CENTRE OF GRAVITY AND MASS MOMENT OF INERTIA

This term has no physical meaning. It is only a mathematical term which will be very useful in studying rotation of rigid bodies. For mass moment of inertia also, letter I is used, but in a problem where there is a need to differentiate between mass moment of inertia and moment of inertia of area, subscript m may be used with letter I. Since mass moment of inertia is product of mass and square of distance, its unit can be derived as given below.

$$mL^2 = \left(\frac{W}{g}\right)L^2$$

dimensionally, $\dfrac{N}{m/\sec^2}\ m^2$

i.e. N-m-sec^2.

No name has been assigned to this unit. Hence, it may be simply called as unit.

10.7. RADIUS OF GYRATION

Radius of gyration *is that distance which when squared and multiplied with total mass of the body gives the mass moment of inertia of the body.* Thus, if I is moment of inertia of a body of mass M about an axis, then its radius of gyration k about that axis is given by the relation:

$$I = Mk^2 \quad \text{or} \quad k = \sqrt{\frac{I}{M}} \qquad \ldots(10.15)$$

A physical meaning may be assigned to this term. Radius of gyration is the distance at which the entire mass can be assumed to be concentrated such that the moment of inertia of the actual body and the concentrated mass is the same.

10.8. DETERMINATION OF MASS MOMENT OF INERTIA FROM FIRST PRINCIPLES

Mass moment of inertia of simple bodies can be determined from its definition. The following steps may be followed:

1. Take a general element.
2. Write the expression for mass of the element, dm, and its distance, r, from the axis.
3. Integrate the term $r^2 dm$ between suitable limits such that the entire mass of the body is covered.

The procedure is illustrated with the examples given below:

Example 10.6. *Determine the mass moment of inertia of a uniform rod of length L about axes normal to it at its (a) centroid and (b) end.*

Solution. (*a*) *About Centroidal Axis Normal to Rod*

Consider an elemental length dx at a distance x from centroidal axis y–y as shown in Fig. 10.13. Let the mass of rod be m per unit length. Then mass of the element $dm = m\ dx$

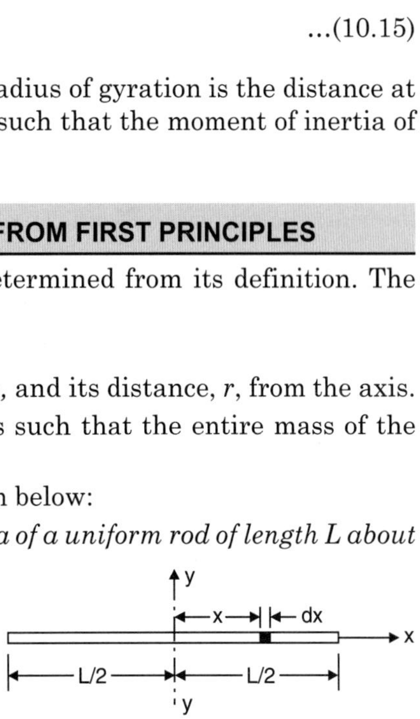

Fig. 10.13

$$I = \int_{-L/2}^{L/2} x^2 \times dm = \int_{-L/2}^{L/2} x^2 \times m\,dx$$

$$= m\left[\frac{x^3}{3}\right]_{-L/2}^{L/2}$$

$$= \frac{mL^3}{12}$$

i.e. $$I = \frac{ML^2}{12} \qquad \ldots(10.16)$$

where $M = mL$ is total mass of the rod.

(b) *About Axis at the end of the rod and normal to it:*

Consider an element of length dx at a distance x from end as shown in Fig. 10.14
Moment of inertia of the rod about y axis:

$$I = \int_0^L x^2 \, m dx$$

$$= m\left[\frac{x^3}{3}\right]_0^L$$

$$= \frac{mL^3}{3}$$

Fig. 10.14

$$I = \frac{ML^2}{3} \qquad \ldots(10.17)$$

Example 10.7. *Determine the moment of inertia of a rectangular plate of size $a \times b$ and thickness t about its centroidal axes.*

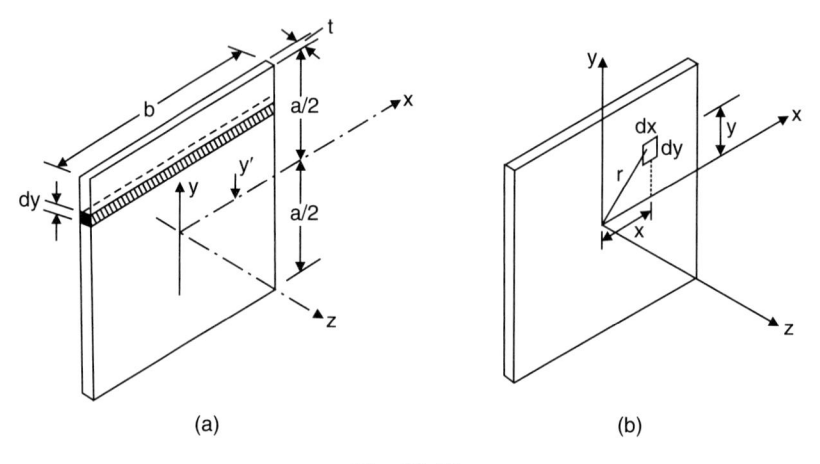

Fig. 10.15

Solution. *To find I_{xx}*

Consider an elemental strip of width dy at a distance y from x axis as shown in Fig. 8.10(a). Mass of the element:

$$dm = \rho b \times t \times dy \qquad (\rho - \text{unit mass of the material})$$

CENTRE OF GRAVITY AND MASS MOMENT OF INERTIA

$$\therefore \quad I_{xx} = \int_{-a/2}^{a/2} y^2 \, dm = \int_{-a/2}^{a/2} y^2 \, \rho bt \, dy$$

$$= \rho bt \left[\frac{y^3}{3} \right]_{-a/2}^{a/2}$$

$$= \frac{\rho bt \, a^3}{12}$$

But mass of the plate $M = \rho \, bta$

$$\therefore \quad I_{xx} = \frac{Ma^2}{12} \qquad \ldots(10.18)$$

To find I_{yy}
Taking an elemental strip parallel to y axis, it can be easily shown that:

$$I_{yy} = \frac{Mb^2}{12} \qquad \ldots(10.19)$$

To find I_{zz}
Consider an element of size $dx \, dy$ and thickness t as shown in Fig. 10.15(b). Now
$$r^2 = x^2 + y^2$$

$$I_{zz} = \int r^2 \, dm = \int (x^2 + y^2) \, dm$$

$$= \int x^2 \, dm + \int y^2 \, dm = I_{xx} + I_{yy}$$

$$= \frac{Ma^2}{12} + \frac{Mb^2}{12}$$

$$I_{zz} = \frac{1}{12} M(a^2 + b^2) \qquad \ldots(10.20)$$

Example 10.8. *Find the moment of inertia of circular plate of radius R and thickness t about its centroidal axis.*

Solution. Consider an elemental area $r \, d\theta \, dr$ and thickness t as shown in Fig. 10.16

dm = mass of the element
$= \rho r \, d\theta \, dr \, t = \rho t \, r d\theta \, dr$

Its distance from x axis = $r \sin \theta$

$$\therefore \quad I_{xx} = \oint (r \sin \theta)^2 \, dm$$

$$= \int_0^R \int_0^{2\pi} r^2 \sin^2 \theta \, \rho t r \, d\theta \, dr$$

$$= \rho t \int_0^R \int_0^{2\pi} r^3 \left(\frac{1 - \cos 2\theta}{2} \right) dr d\theta$$

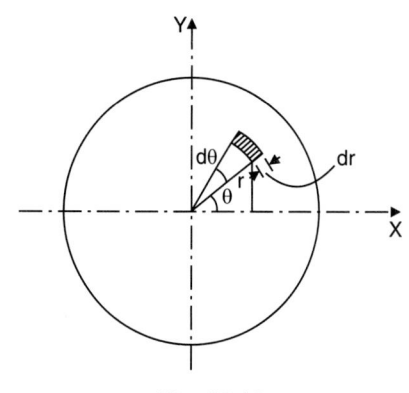

Fig. 10.16

$$= \rho t \int_0^R \frac{r^3}{2} \left[\theta - \frac{\sin 2\theta}{2}\right]_0^{2\pi} dr$$

$$= \rho t \int_0^R \frac{r^3}{2} \times 2\pi \, dr$$

$$= \rho t \, \pi \left[\frac{r^4}{4}\right]_0^R = \rho t \, \frac{\pi R^4}{4}$$

Mass of the plate $M = \rho \times \pi R^2 t$

∴ $$I_{xx} = \frac{MR^2}{4} \qquad \ldots(10.21)$$

Similarly, $$I_{yy} = \frac{MR^2}{4}$$

Actually $I = \dfrac{MR^2}{4}$ is moment of inertia of circular plate about any diametral axis in the plate.

To find I_{zz}, consider the same element.

$$I_{zz} = \oint r^2 \, dm = \int_0^R \int_0^{2\pi} r^2 \rho \, t \, r \, dr \, d\theta$$

$$= \rho \, t \int_0^R r^3 [\theta]_0^{2\pi} \, dr = \rho t \int_0^R 2\pi r^3 \, dr$$

$$= \rho t 2\pi \left[\frac{r^4}{4}\right]_0^R = \rho t \, 2\pi \frac{R^4}{4} = \rho t \frac{\pi R^4}{2}$$

But total mass $M = \rho t \pi R^2$

∴ $$I_{zz} = \frac{MR^2}{2} \qquad \ldots(10.22)$$

Example 10.9. *Determine the mass moment of inertia of a circular ring of uniform cross-section.*

Solution. Consider uniform ring of radius R as shown in Fig. 10.17. Let its mass per unit length be m.

Hence, total mass $M = 2\pi R \, m$

Consider an elemental length $ds = R \, d\theta$ at an angle θ to the diametral axis x–x. The distance of the element from x axis is $R \sin \theta$ and mass of element is $m \, r d \, \theta$.

∴ $$I = \int_0^{2\pi} (R \sin \theta)^2 \, mRd\theta$$

$$= R^3 m \int_0^{2\pi} \sin^2 \theta \, d\theta$$

$$= mR^3 \int_0^{2\pi} \left(\frac{1 - \cos 2\theta}{2}\right) d\theta$$

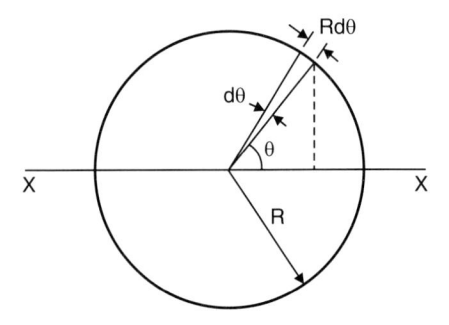

Fig. 10.17

$$= \frac{mR^3}{2}\left[\theta - \frac{\sin 2\theta}{2}\right]_0^{2x} = mR^3\pi$$

But $\qquad M = 2\pi R m$

$\therefore \qquad I = \dfrac{MR^2}{2}$ ...(10.23)

Example 10.10. *Find the mass moment of inertia of the solid cone of height h and base radius R about:*

1. its axis of rotation and

2. an axis through vertex normal to the axis of rotation.

Solution. Moment of Inertia about its Axis of Rotation:

Consider an elemental plate at distance x. Let its radius be r and thickness dx.

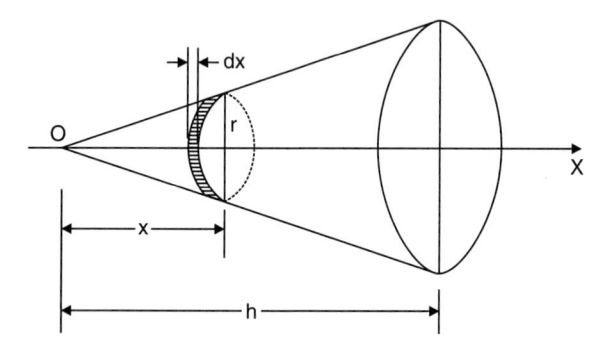

Fig. 10.18

Mass of the elemental plate = $\rho \pi r^2 \, dx$

But from Eqn. (10.22), the moment of inertia of circular plate about normal axis through its centre is

$$= \frac{1}{2} \times \text{mass} \times \text{square of radius}$$

$$= \frac{1}{2} \times \rho \pi r^2 \, dx \cdot r^2$$

$$= \rho \frac{\pi r^4 \, dx}{2}$$

But now, $\qquad r = \left(\dfrac{x}{h}\right) R$

$\therefore \qquad$ Moment of inertia of the elemental plate about x axis

$$= \rho \frac{\pi}{2} \, R^4 \frac{x^4}{h^4} \, dx \qquad \text{...(2)}$$

$\therefore \qquad$ Moment of inertia of the cone about x axis,

$$I_{xx} = \int_0^h \frac{\rho \pi}{2} \, R^4 \frac{x^4}{h^4} \, dx = \frac{\rho \pi}{2} \frac{R^4}{h^4} \left[\frac{x^5}{5}\right]_0^h$$

$$I_{xx} = \frac{\rho \pi R^4 h}{10} \qquad \qquad ...(3)$$

But mass of the cone

$$M = \int_0^h \rho \pi r^2 \, dx = \int_0^h \rho \pi R^2 \frac{x^2}{h^2} \, dx$$

$$= \frac{\rho \pi R^2}{h^2} \left[\frac{x^3}{3} \right]_0^h$$

$$= \frac{\rho \pi R^2 h}{3}$$

From (3) and (4) we get,

$$I_{xx} = \frac{3}{10} MR^2 \qquad \qquad ...(10.24)$$

Moment of Inertia about an Axis through Vertex and Normal to the Axis of Rotation :

Consider an element of size $rd\theta \times dr \times dx$ as shown in Fig. 10.19. Let its coordinates be x, y, z and z_1 be the radius of the plate at distance x from vertex. Now, the mass of this element

$$dm = \rho r d\theta \, dr \, dx$$

and its distance from y axis is given by :

$$l = \sqrt{x^2 + z^2}$$

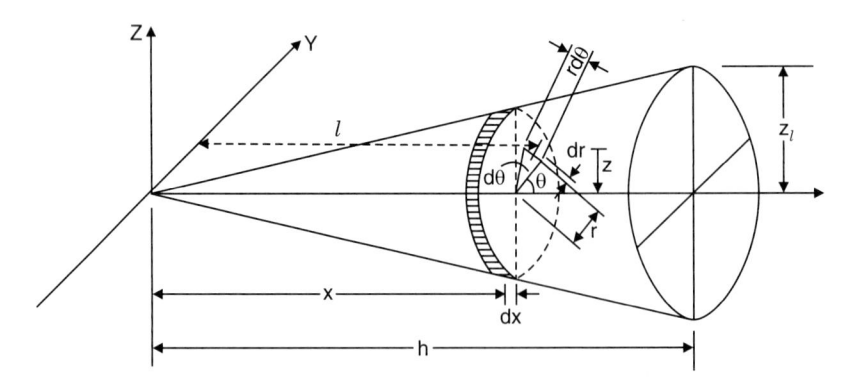

Fig. 10.19

∴ Moment of inertia of the cone about y axis :

$$I_y = \int l^2 \, dm = \int_0^h \int_0^{z_l} \int_0^{2\pi} (x^2 + z^2) \rho \, rd\theta \, dr \, dx$$

But $z = r \sin \theta$

$$I_{yy} = \int_0^h \int_0^{z_l} \int_0^{2\pi} (x^2 + r^2 \sin^2 \theta) \rho \, rd\theta \, dr \, dx$$

CENTRE OF GRAVITY AND MASS MOMENT OF INERTIA

$$= \int_0^h \int_0^{z_l} \int_0^{2\pi} \rho \left(x^2 r + r^3 \frac{1-\cos 2\theta}{2} \right) d\theta\, dr\, dx$$

$$= \int_0^h \int_0^{z_l} \rho \left[x^2 r \theta + \frac{r^3}{2}\left(\theta - \frac{\sin 2\theta}{2}\right) \right]_0^{2\pi} dr\, dx$$

$$= \int_0^h \int_0^{z_l} \rho \left[2\pi x^2 r + \frac{r^3}{2} 2\pi \right] dr\, dx$$

$$= \int_0^h \rho \left[\pi x^2 r^2 + \pi \frac{r^4}{4} \right]_0^{z_l} dx = \int_0^h \rho \left[\pi x^2 z_l^2 + \frac{\pi z_l^4}{4} \right] dx$$

But $z_l = \dfrac{x}{h} R$

∴ $$I_{yy} = \int_0^h \rho \left[\pi x^2 \frac{x^2}{h^2} R^2 + \frac{\pi}{4} \frac{x^4}{h^4} R^4 \right] dx$$

$$= \rho \left[\pi \frac{x^5}{5h^2} R^2 + \frac{\pi x^5}{20 h^4} R^4 \right]_0^h$$

$$= \rho \left[\frac{\pi}{5} h^3 R^2 + \frac{\pi}{20} h R^4 \right] = \frac{\pi \rho R^2 h}{5}\left(h^2 + \frac{R^2}{4} \right)$$

But mass of cone $M = \dfrac{1}{3} \rho \pi R^2 h$

∴ $$I_{yy} = \frac{3M}{5}\left(h^2 + \frac{R^2}{4} \right) \qquad \text{...(10.25)}$$

Example 10.11. *Determine the moment of inertia of a solid plate of radius R about its diametral axis.*

Solution. Consider an elemental plate of thickness dy at distance y from the diametral axis as shown in Fig. 10.20. Radius of this elemental circular plate x is given by the relation :

$$x^2 = R^2 - y^2 \qquad \text{...(1)}$$

∴ Mass of the elemental plate

$dm = \rho \pi x^2\, dy$

$= \rho \pi (R^2 - y^2)\, dy \qquad \text{...(2)}$

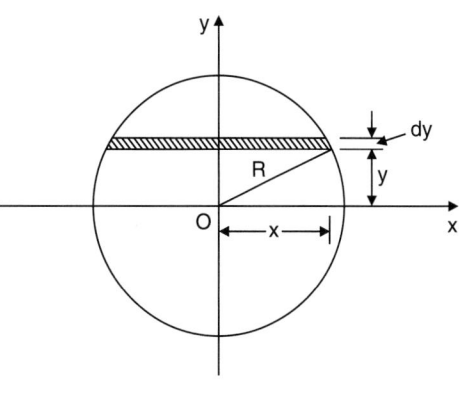

Fig. 10.20

Moment of inertia of this circular plate element about y axis is given by Eqn. (10.22) as:

$$= \frac{1}{2} \times \text{mass} \times \text{square of radius}$$

$$= \frac{1}{2} \times \rho \pi x^2 \, dy \times x^2 = \rho \frac{\pi}{2} x^4 \, dy$$

$$= \rho \frac{\pi}{2} (R^2 - y^2)^2 \, dy$$

$$= \rho \frac{\pi}{2} (R^4 - 2R^2 y^2 + y^4) \, dy$$

$$\therefore \quad I_{yy} = 2 \int_0^R \rho \frac{\pi}{2} (R^4 - 2R^2 y^2 + y^4) \, dy$$

$$= \rho \pi \left[R^4 y - \frac{2R^2 y^3}{3} + \frac{y^5}{5} \right]_0^R$$

$$= \rho \pi R^5 \left[1 - \frac{2}{3} + \frac{1}{5} \right] = \frac{8}{15} \rho \pi R^5 \qquad \ldots(3)$$

But mass of plate, $M = 2 \int_0^R dm = 2 \int_0^R \rho \pi x^2 \, dy$

$$= 2 \int_0^R \rho \pi (R^2 - y^2) \, dy = 2\rho \pi \left(R^2 y - \frac{y^3}{3} \right)_0^R$$

$$= 2\rho \pi \left[R^3 - \frac{R^3}{3} \right]$$

$$M = \frac{4 \pi R^3}{3} \qquad \ldots(4)$$

From (3) and (4) above we get :

$$I_{yy} = \frac{2}{5} MR^2 \qquad \ldots(10.26)$$

Example 10.12. *Using the moment of inertia expression for plates, find the expressions for moment of inertia of*

(a) *Parallelepiped and*

(b) *Circular cylinder about z axis as shown in Fig. 10.21.*

CENTRE OF GRAVITY AND MASS MOMENT OF INERTIA

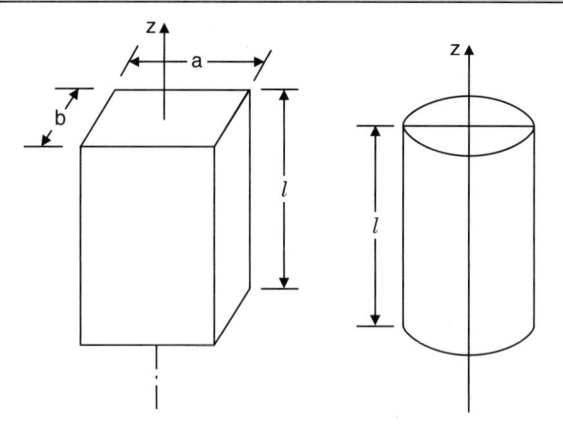

Fig. 10.21

Solution. Parallelepiped can be looked upon as a rectangular plate of thickness $t = l$, and similarly a solid cylinder as a circular plate of thickness l. Hence the expressions for moment of inertia are:

$$\frac{1}{12}M(a^2 + b^2) \text{ for parallelepiped}$$

$$\frac{MR^2}{2} \text{ for cylinder}$$

where

Mass of parallelepiped $= abl\,\rho$
and that of cylinder $= \pi R^2\,l\,\rho$

10.9. PARALLEL AXIS THEOREM/TRANSFER FORMULA

Parallel axis theorem : *The moment of inertia of a body about an axis at a distance d and parallel to a centroidal axis is equal to sum of moment of inertia about centroidal axis and product of mass and square of distance of parallel axis.* Thus, if I_g is the moment of inertia of a body of mass M about a centroidal axis and I_A is the moment of inertia about a parallel axis through A which is at a distance d from centroidal axis, then

$$I_A = I_g + Md^2$$

Proof: Let dm be an element at a distance r from centroidal axis z through centre of gravity as shown in Fig. 10.22.

$$I_g = \int r^2\, dm$$

$$= \int (x^2 + y^2)\, dm \qquad \ldots(1)$$

Fig. 10.22

Now, moment of inertia about z axis through A is,

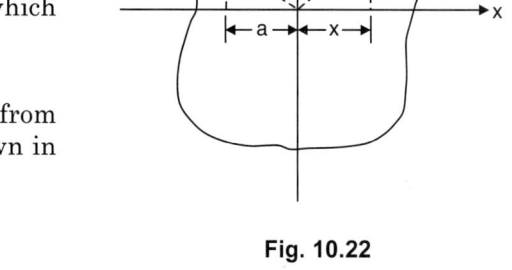

$$I_A = \int r'^2\, dm \text{ where } r' \text{ is the distance of element from } A.$$

$$= \int [(x+a)^2 + (y-b)^2]\, dm$$

$$= \int (x^2 + 2ax + a^2 + y^2 - 2yb + b^2)\, dm$$

$$= \int (x^2 + y^2)\, dm + \int (a^2 + b^2)\, dm + \int 2ax\, dm - \int 2yb\, dm$$

$$= \int r^2\, dm + \int d^2\, dm + \int 2ax\, dm - \int 2b\, y\, dm$$

Since $\quad x^2 + y^2 = r^2 \quad$ and $\quad a^2 + b^2 = d^2$

But, $\quad \int r^2\, dm = I_g$

$$\int d^2\, dm = d^2 \int dm = Md^2$$

$$\int 2ax\, dm = 2a \int x\, dm = 2a\, M\, \bar{x}$$

and $\quad \int 2b\, y\, dm = 2b \int y\, dm = 2b\, M\bar{y}$

where $\bar{x}$ and $\bar{y}$ are distance of centre of gravity from the reference axis. In this case $\bar{x}$ and $\bar{y}$ are zero since reference axis contains centroid. Thus $\int 2a\, x\, dm = \int 2b\, y\, dm = 0$.

Hence $\quad I_A = I_g + Md^2 \quad$ ...(10.27)

10.10. MOMENT OF INERTIA OF COMPOSITE BODIES

In order to determine the moment of inertia the composite body is divided into a set of simple bodies. The centre of gravity and moment of inertia expressions for such simple bodies are known. Moment of inertia of simple bodies about their centroidal axis are calculated and then using parallel axis theorem, moment of inertia of each simple body is found about the required axis. Summing up of the moment of inertia of each simple body about the required axis, gives the moment of inertia of the composite body. The procedure is illustrated with the examples 10.13 and 10.14.

Example 10.13. *Determine the radius of gyration of the body shown in Fig. 10.23 about the centroidal x axis. The grooves are semicircular with radius 40 mm.*

All dimensions shown are in mm.

Solution. The composite body may be divided into
(1) A solid block of size 80 × 120 × 100 mm. and,
(2) Two semicircular grooves each of radius 40 mm and length 80 mm.

Mass of solid block,
$$M_1 = 80 \times 120 \times 100\, \rho$$
$$= 960000\, \rho$$
where, ρ is mass of 1 mm^3 of material.

Its moment of inertia about x axis

$$Ix_1 = M_1 \frac{(100^2 + 120^2)}{12}$$

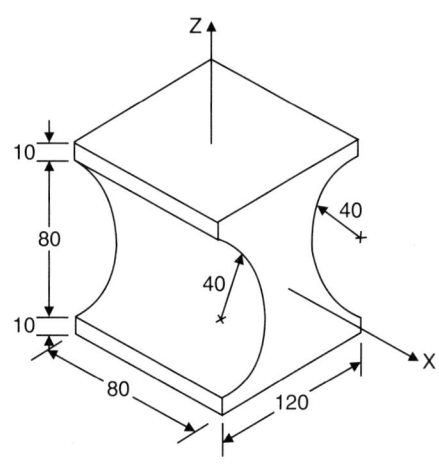

Fig. 10.23

CENTRE OF GRAVITY AND MASS MOMENT OF INERTIA

$$= 960 \times 10^3 \, \rho \, \frac{(100^2 + 120^2)}{12} = 1.952 \times 10^9 \, \rho$$

Semicircular groove :

Mass, $\qquad M_2 = \dfrac{1}{2} \pi r^2 \, l \, \rho$

$$= \frac{1}{2} \pi \, 40^2 \times 80 \, \rho = 201061.93 \, \rho$$

Moment of inertia about the axis parallel to x axis through diametral axis of semicircle:

$$= \frac{1}{2} \text{ of that of cylinder}$$

$$= \frac{1}{2} \times 2M_2 \times \frac{r^2}{2} = \frac{M_2 r^2}{2}$$

Centre of gravity of semicircular groove from this axes is at a distance $d = \dfrac{4r}{3\pi}$

$$= \frac{4 \times 40}{3\pi} = 16.9765 \text{ mm}$$

∴ Moment of inertia about the axis through centre of gravity I_g, is given by:

$$\frac{M_2 r^2}{2} = I_g + M_2 \, d^2$$

∴ $\qquad I_g = 201061.93 \, \rho \left(\dfrac{40^2}{2} - 16.9765^2 \right)$

$$= 1.029 \times 10^8 \, \rho$$

The distance of this centroid from x axis is

$$d' = 60 - 16.9765 = 43.0235 \text{ mm}$$

∴ $\qquad Ix_2 = I_g + M_2 \, d'^2$

$$= 1.029 \times 10^8 \, \rho + 201061.93 \, \rho \times 43.0235^2$$

$$= 4.7507 \times 10^8 \, \rho$$

∴ I_{XX} of composite body

$$= Ix_1 - 2 Ix_2$$

(since there are two semicircular grooves placed symmetrically with respect to x-x axis)

$$I_{xx} = 1.952 \times 10^9 \, \rho - 2 \times 4.7507 \times 10^8 \, \rho$$

$$= 10.0816 \times 10^8 \, \rho$$

Total mass $\qquad M = M_1 - 2M_2 = 960000 \, \rho - 2 \times 201061.93 \, \rho$

$$= 557876.14 \, \rho \text{ units}$$

$$k = \sqrt{\frac{I}{M}} = \sqrt{\frac{10.0816 \times 10^8}{552876.14}}$$

i.e. $\qquad \mathbf{k = 42.57 \text{ mm}}$

Example 10.14. *A cast iron fly wheel has the following dimensions:*

Diameter = 1.5 m
Rim width = 300 mm
Thickness of rim = 50 mm
Hub length = 200 mm
Outer diameter of hub = 250 mm
Inner diameter of hub = 100 mm
Arms : 6 equally spaced uniform slender rods of length 0.575 m
Cross-sectional area of each arm = 8000 mm²

Determine the moment of inertia of the wheel about the axis of rotation. Take mass of cast iron as 7200 kg/m³.

Fig. 10.24

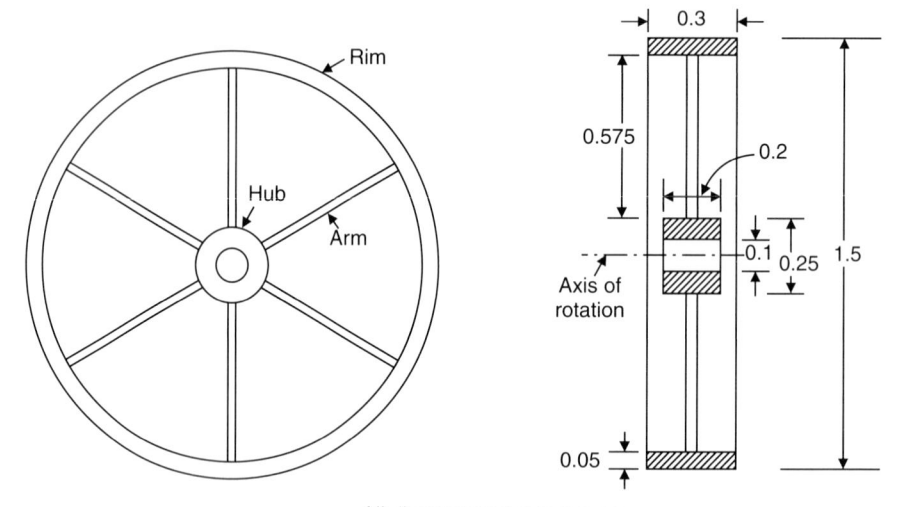

All dimensions are in metres

Solution. *Moment of inertia of rim:*

Outer diameter = 1.5 m

thickness = 0.05 m

∴   Inner diameter = 1.5 – 2 × 0.05 = 1.4 m

Treating it as a solid circular plate of 1.5 m diameter in which a circular plate of diameter 1.4 m is cut and noting that for a plate moment of inertia about the axis is:

$$\frac{\text{Mass} \times \text{Square of radius}}{2}$$

The moment of inertia of rim is

$$I_1 = \frac{M_0 \, R_0^2 - M_i \, R_i^2}{2}$$

$$= \frac{1}{2} [(\pi R_0^2 \, t \, \rho \, R_0^2 - \pi R_i^2 \, t \, \rho \, R_i^2)]$$

where R_0 = outer radius = $\frac{1.5}{2}$ = 0.75

R_i = inner radius = $\frac{1.4}{2}$ = 0.7

$M_0 = M_i$ are the masses of circular plates of radii R_0 and R_i, respectively.

∴ t = width = 0.30.

ρ = mass per cubic meter = 7200 kg/m³

i.e. $I_1 = \left(\frac{1}{2}\right) \pi \, t \, \rho \, [R_0^4 - R_i^4]$

∴ $I_1 = \left(\frac{1}{2}\right) \pi \times 0.3 \times 7200 \, [0.75^4 - 0.7^4]$

= 258.9010 units.

Moment of inertia of hub :

This hollow cylinder may be considered as a circular plate with circular cut out and thickness equal to length of cylinder.

In this case, outer radius $R_0 = \frac{0.25}{2}$ = 0.125 m, and

inner radius $R_i = \frac{0.1}{2}$ = 0.05 m

length of cylinder t = 0.2 m

As in the above case, the moment of inertia can be calculated as the moment of inertia of solid plate minus the moment of inertia of hollow plate

$$I_2 = \frac{1}{2} \times \pi \times 0.2 \times 7200 \; (0.125^4 - 0.05^4)$$

= 0.5381 units.

Moment of Inertia of Arms :

Moment of inertia of arm about its centre of gravity is = $\dfrac{Ml^2}{12}$

and when it is shifted to axis of rotation it will be equal to $\dfrac{Ml^2}{12} + Md^2$

Now, A = 8000 mm² = 8000 × 10⁻⁹ m²

l = 0.575 m

$d = \dfrac{0.575}{2} + 0.125 = 0.4125$ m

$M = l \, A \, \rho = 0.575 \times 8000 \times 10^{-9} \times 7200$

= 0.03312 kg

As there are six such arms,

$$I_3 = 6 \times \left(\frac{Ml^2}{2}\right) \times d^2$$

$$= 6 \times 0.03312 \left(\frac{0.575^2}{12}\right) \times 0.4125^2$$

$$= 0.0393 \text{ units}$$

∴ Moment of inertia of flywheel,
$$I = I_1 + I_2 + I_3$$
$$= 258.90 + 0.5381 + 0.0393$$
$$= 259.4774 \text{ units.}$$

Note: Moment of inertia is contributed mainly by rim.

Important Definitions

1. **Centre of gravity** of a body is the point through which the resultant weight of the body passes through in whichever position the body is kept.
2. **Pappus first theorem** states that the area of surface generated by revolving a plane curve about a non-intersecting axis in the plane of the curve is equal to the length of the generating curve times the distance travelled by the centroid of the curve in the revolution.
3. **Pappus second theorem** states that the volume of the solid generated by revolving a plane area about a non-intersecting axis in the plane is equal to the area of the generating plane times the distance travelled by the centroid of the plane area during the rotation.
4. **Mass moment of inertia** of a body about an axis is defined as the sum total of product of its elemental masses and square of their distances from the axis.
5. **Radius of gyration** is that distance which when squared and multiplied with total mass of the body gives the mass moment of inertia of the body.

    A physical meaning may be assigned to the term radius of gyration. It is the distance at which entire mass can be assumed to be concentrated such that the moment of inertia of the actual body and that of the concentrated mass are the same.
6. **Parallel axis theorem** (transfer formula) states that the moment of inertia of a body about an axis at a distance and parallel to a centroidal axis is equal to sum of moment of inertia about the centroidal axis and product of mass and square of the distance of parallel axis from the centroid.

Important Formulae

1. $M\bar{x} = \Sigma m_i x_i \quad W\bar{x} = \Sigma w_i x_i$
   $M\bar{y} = \Sigma m_i y_i \quad W\bar{y} = \Sigma w_i y_i$
   $M\bar{z} = \Sigma m_i z_i \quad W\bar{z} = \Sigma w_i z_i.$
2. If the body is made up of uniform material,
   $$V\bar{x} = \Sigma V_i x_i, \quad V\bar{y} = \Sigma V_i y_i \quad \text{and} \quad V\bar{z} = \Sigma V_i z_i.$$

CENTRE OF GRAVITY AND MASS MOMENT OF INERTIA

3. In a right circular cone, centre of gravity lies at a distance $\frac{3}{4}h$ from vertex, *i.e.* at a distance $\frac{h}{4}$ from the base, along the axis of rotation.

4. The centre of gravity of a solid hemisphere of radius r is at a distance $\frac{3}{8}r$ from the diametral axis.

5. $I_{AB} = \Sigma\, m_i r_i^2 = \int r^2\, dm$

6. $I = Mk^2$ or $k = \sqrt{\dfrac{I}{M}}$.

7. $I_A = I_g + Md^2$.

PROBLEMS FOR EXERCISE

10.1 Determine the centre of gravity of the pyramid shown in Fig. 10.25. $\left[\text{Ans. } x = \dfrac{3}{4}h\right]$

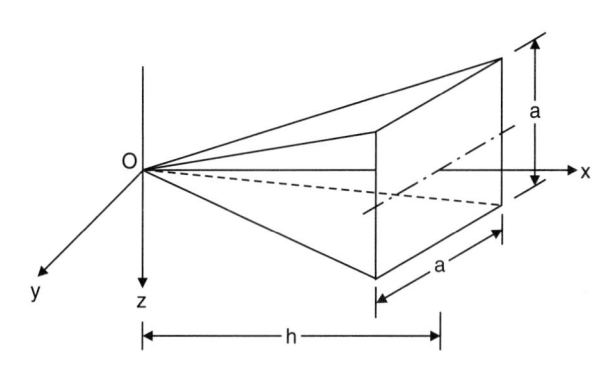

Fig. 10.25

10.2 A steel ball of diameter 150 mm rests centrally over a concrete cube of size 150 mm. Determine the centre of gravity of the system, taking weight of concrete = 25000 N/m³ and that of steel 80000 N/m³. [**Ans.** 168.94 mm from base]

10.3 Locate the centre of gravity of the wire shown in Fig. 10.26. Portion BC is in x-y plane and semicircle CD is parallel to x-z plane.

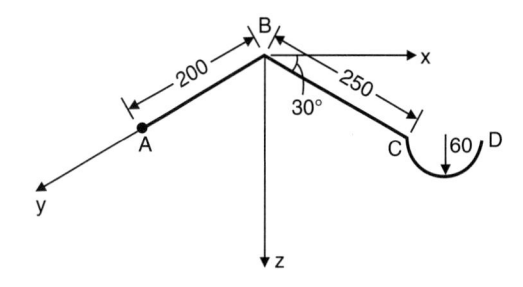

[**Ans.** $\bar{x} = 124.02$ mm ;
$\bar{y} = 110.41$ mm ;
$\bar{z} = 11.28$ mm]

Fig. 10.26

10.4 Determine the moment of inertia of the link shown in Fig. 10.27 about x axis.

[**Ans.** 17.23×10^8 ρ units]

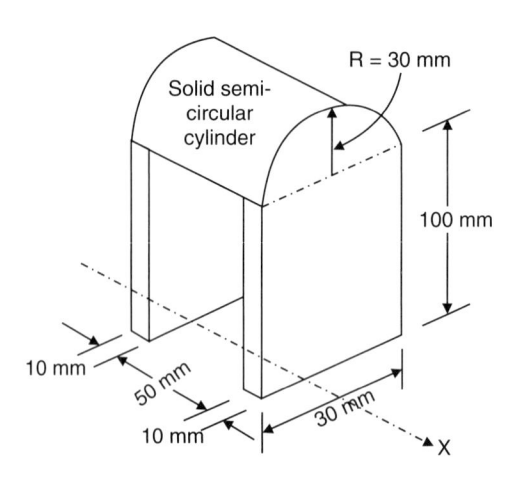

Fig. 10.27

PART—II

11
Introduction to Dynamics

As explained in Chapter I, dynamics is the branch of mechanics dealing with the bodies in motion and it branches into two streams—kinematics and kinetics. **Kinematics** is the branch of dynamics which deals with the motion of bodies without referring to the forces causing the motion of the body and **kinetics** is the branch of dynamics which deals with the motion of bodies referring to the forces causing the motion of the body. In this chapter, the basic terms in dynamics are defined and general principles in dynamics are presented. Different types of motion are explained and outline of part II is presented.

11.1 BASIC TERMS

Motion. *A body is said to be in motion if it is changing its position with respect to a reference point.* A person on a scooter is in motion when referred to the road but is at rest when referred to the scooter itself. In most problems, the reference point is implied. For all engineering problems, any fixed point on the earth is an implied reference point. Other examples of implied reference points are centre of the earth for the study of satellite motion, centre of the sun for the study of motion of the solar system and mass centre of the solar system for the study of interplanetary motion. In this book only engineering problems have been taken up and hence any fixed point on the earth is an implied reference point while considering motion of the body.

Distance and Displacement. Let a body move along the path shown by hatched line in Fig. 11.1 and the tme taken by the body to move from *A* to *B* be *t*, then the *distance* moved by the body in time *t* is the distance measured along the hatched line. Distance is a scalar quantity since it has only magnitude.

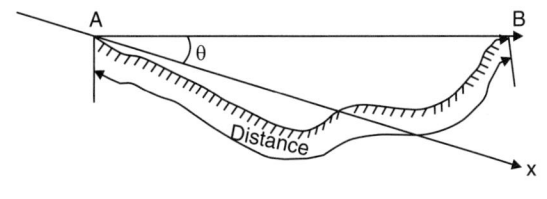

Fig. 11.1

Displacement in this time interval is *the linear distance AB which makes angle* θ *with the x-axis. Displacement of the body in a time interval may be defined as the linear distance between the two positions of the body in the beginning and at the end of the time interval.* Since, displacement has the magnitude and direction, it is a vector quantity.

Speed and Velocity. *The rate of change of distance with respect to time is defined as speed whereas the rate of change of displacement with respect to time is called velocity.* Speed has the magnitude only, whereas velocity has both magnitude and direction. Hence speed is a scalar quantity while velocity is a sector quantity.

If s is the displacement in interval t, the average velocity v is given by $\frac{s}{t}$.

Velocity of a particle at a given instant is called instantaneous velocity and is given by the limiting value of the ratio $\frac{s}{t}$ at time t when both s and t are very small. Let δs be the small displacement in a small time δt. The instantaneous velocity is given by the expression

$$v = \lim_{\delta t \to 0} \frac{\delta s}{\delta t}$$

i.e.
$$= \frac{ds}{dt} \qquad \ldots(11.1)$$

In SI units, metre is the unit for displacement and second is the unit of time. Hence from equation 11.1 it may be observed that unit of velocity is m/sec. Many a times kilometer per hour (km/h or kmph) is also used as unit of velocity. The relation between the two units is given by:

$$1 \text{ kmph} = \frac{1 \times 1000}{60 \times 60} \text{ m/sec}$$

$$= \frac{5}{18} \text{ m/sec.}$$

Acceleration and Retardation. *Rate of change of velocity with respect to time is called acceleration.* Hence mathematically acceleration a is given by

$$a = \frac{dv}{dt} \qquad \ldots(11.2)$$

The acceleration may be positive or negative. The positive acceleration is simply referred as acceleration and the negative acceleration is called as retardation or deceleration.

From equations (11.1) and (11.3) we get,

$$a = \frac{dv}{dt} = \frac{d}{dt}\left(\frac{ds}{dt}\right)$$

i.e.
$$a = \frac{d^2 s}{dt^2} \qquad \ldots(11.3)$$

The unit of velocity in S.I. Units is m/sec. The unit of acceleration which is rate of change of velocity with respect to time (Eqn. 11.2) is m/sec².

11.2. GENERAL PRINCIPLES IN DYNAMICS

Study of dynamics needs not only the devices to measure the force and length but also time. Lack of availability of accurate time measuring devices delayed the developement of dynamics as compared to that of statics. Galileo (1564 –1642) made useful observations which led to development of dynamics. Sir Isaac Newton (1642 – 1727) generalized Galileo's

INTRODUCTION TO DYNAMICS

observations and came out with experimental proof for what are now known as Newton's first law and second law.

Newton's first law has been stated in Chapter I as *Every body continues in its state of rest or of uniform motion, unless it is acted by some external agency.* In other words, if a body is subjected to a balanced system of forces, it will continue to be in the state of rest or of uniform motion. Thus,

A body acted by a balanced system of forces has no acceleration.

In chapter 1, **Newton's second law** has been stated as *the rate of change of momentum is directly proportional to the impressed force, and takes in the direction, in which the force acts.* This statement and the definition of unit of a force lead to the famous equation.

$$\text{Force} = \text{Mass acceleration}$$

i.e. $$F = ma \qquad \ldots(11.4)$$

Hence many a times engineers state *Newton's second law as when an unbalanced system forces act on a particle, the particle moves with an acceleration proportional to the resultant force and it takes place in the direction of the resultant force.*

It may be observed that the Newton's first law is only a particular case of the second law. That is when the resultant force is zero, the acceleration also will be zero.

Newton's third law states that *for every action there is an equal and opposite reaction.* This has been frequently used in statics for drawing free body diagrams and will be used in dynamics also while treating kinetics problems.

Newton's law of gravitation is another important principle in dynamics. This law states that *any two particles of masses m_1 and m_2 separated by distance r, attract each other with a force directly proportional to their masses and inversely proportional to square of their distance.* Thus the force F in Fig. 11.2 is given by

$$F \propto \frac{m_1 m_2}{r^2}$$

$$= G \frac{m_1 m_2}{r^2} \qquad \ldots(11.5)$$

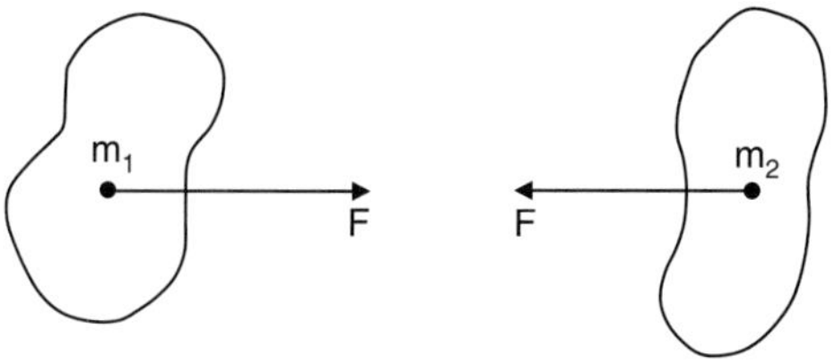

Fig. 11.2

where, G is constant of proportionality and is called universal constant of gravitation.

The force of attraction on a body by the earth is called weight of the body. Let m be the mass of the body and W its weight. Then substituting in equation (11.5),

$$F = W$$
$$m_1 = m$$
$$m_2 = M$$
$$r = R$$

where, M is mass of the earth and R is its radius,

we get
$$W = \frac{mMG}{R^2} = mg \qquad ...(11.6)$$

where,
$$g = \frac{MG}{R^2} \qquad ...(11.7)$$

The radius of earth along polar axis is 63,56400 m and along equitorial axis is 63,28375 m. Hence weight of a body varies from place to place. It is slightly more when the body is near the pole compared to when it is at equator. However, the variation is small. Hence the earth radius R is considered to be the same in all directions. The variation in distances of the bodies considered in engineering problems is very small compared to radius of the earth. Hence it is sufficiently accurate in most engineering computations to assume value of g, and hence that of weight of a body, as constant. Comparing equation (11.6) with eqution (11.4), we can call g as acceleration due to gravity and its value is found to be 9.81 m/sec^2. The weight W is the force always directed towards centre of the earth, *i.e.* in vertically downward direction.

11.3. TYPES OF MOTION

A body may move in any direction in space. In this book, motion in a single plane only is considered. This type of motion is called plane motion. Plane motion may be classified as under:

Translation: A motion is said to be *translation*, if a straight line drawn on the moving body remains parallel to its original position at any time. During translation if the path traced by a point is a straight line, it is called *rectilinear translation* (Fig. 11.3) and if the path is a curve one it is called *curvilinear translation* (Fig. 11.4).

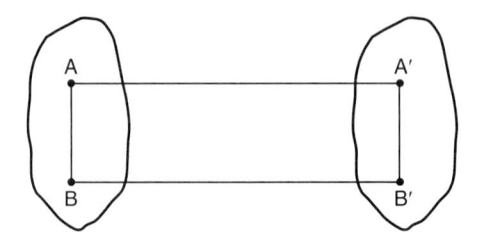

Fig. 11.3. Rectilinear Translation

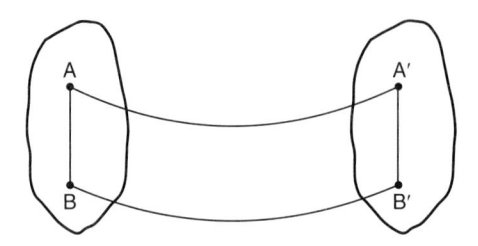

Fig. 11.4. Curvilinear Translation

In the study of the motion of a particles, rectilinear translation and curvilinear translation are usually referred as *linear motion* and *curvilinear motion*.

Rotation: A motion is said to be rotation if all particles of a rigid body move in a concentric circle (Fig. 11.5).

INTRODUCTION TO DYNAMICS

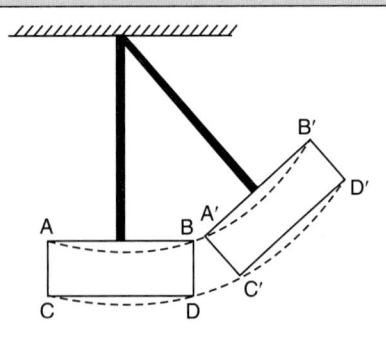

Fig. 11.5

General Plane Motion: The general plane motion is a combination of both translation and rotation. Common examples of such motion are points on wheels of moving vehicles, a ladder sliding down from its position against wall etc.

11.4. OUTLINE OF THE SECTION

The kinematics of a particle is treated under the headings linear motion, projectile and relative motion in chapters 12 to 14. The kinetic problems of a body with plane motion may be analysed by using any one of the following three methods.

(a) D' Alemberts Principle

(b) Work Energy Equation

(c) Impulse Momentum Equation.

These three approaches for the solution of kinetic problems in linear motion are illustrated in Chapters 15 to 17. Impact of elastic bodies is considered in Chapter 18. Circular motion, rotation and general plane motion of rigid bodies are treated in Chapters 19 to 22.

Important Definitions

1. **Kinematics** is the branch of dynamics which deals with the motion of the bodies without referring to the forces causing the motion.
2. **Kinetics** is the branch of dynamics which deals with the motion of bodies referring to the forces causing the motion of the body.
3. **Displacement** of body in a time interval may be defined as the linear distance between two positions of the body in the beginning and at the end of the time interval.
4. Rate of change of displacement with respect to time is called **velocity**.
5. Rate of change of velocity with time is called **acceleration** and negative acceleration is called as **retardation**.

Important Equations

1. $v = \dfrac{ds}{dt}$

2. $a = \dfrac{dv}{dt} = \dfrac{d^2s}{dt^2}$

3. $F = ma$

4. $W = mg$

12
Linear Motion

A particle is said to be in linear motion, if the path traced by it is a straight line. Many kinematic problems in linear motion can be solved just by using the definition of speed velocity and acceleration. Motion curves which will be useful in solving kinematic problems are explained and many problems with uniform velocity, uniform acceleration and varying acceleration are dealt in this chapter.

12.1. MOTION CURVES

Motion curves are the graphical representation of the displacement, velocity and acceleration with time.

Displacement-Time Curve (s – t curve). Displacement-Time curve is a curve with time as abscissa and displacement as ordinate (Fig. 12.1). At any instant of time t, velocity v is given by

$$v = \frac{ds}{dt}$$

If a body is having non-uniform motion, its displacement at various time interval may be observed and $s - t$ curve plotted. Velocity at any time may be found from the slope of $s - t$ curve.

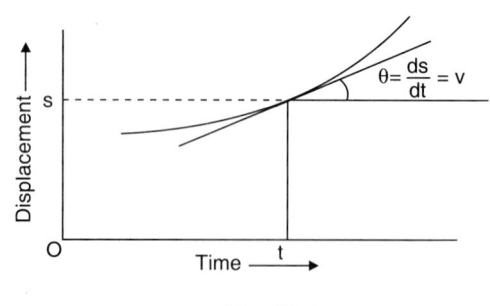

Fig. 12.1

Velocity-Time Curve (v-t curve). In Velocity-Time curve diagram, the abscissa represents time and ordinate, the velocity of the motion. Such a curve is shown in Fig. 12.2. Acceleration 'a' is given by the slope of the $v - t$ curve.

i.e.,
$$a = \frac{dv}{dt} = \theta$$

Thus, acceleration at any time is the slope of $v - t$ curve at the time, as shown in Fig. 12.2.

Now,
$$\frac{ds}{dt} = v$$

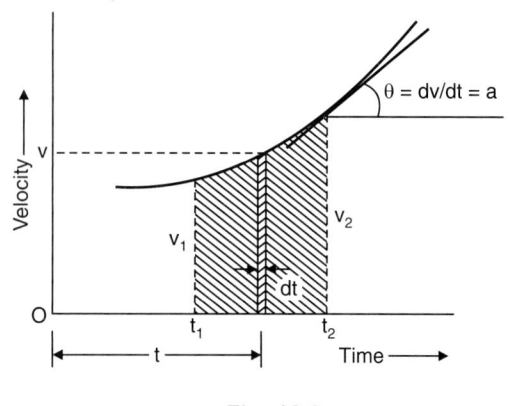

∴
$$ds = v dt$$

Fig. 12.2

LINEAR MOTION

or
$$s = \int v\,dt$$

Referring to Fig. 12.2, $v\,dt$ is the elemental area under the curve at time t in the interval dt. Hence the shaded area under the curve between t_1 and t_2 shown in Fig. 12.2 represents displacement s of the moving body in the time interval between t_1 and t_2. Thus in $v - t$ curve;

(1) Slope of the curve represents acceleration, and

(2) Area under the curve represents displacement.

Accleration-Time Curve (a-t curve). If a body is moving with varying acceleration, its motion can be studied more conveniently by drawing a curve with time as abscissa and acceleration as ordinate. Such a curve is called acceleration-time curve.

Now,
$$\frac{dv}{dt} = a \quad \text{or} \quad dv = a\,dt$$

or
$$v = \int a\,dt$$

Hence the area under the curve represents velocity.

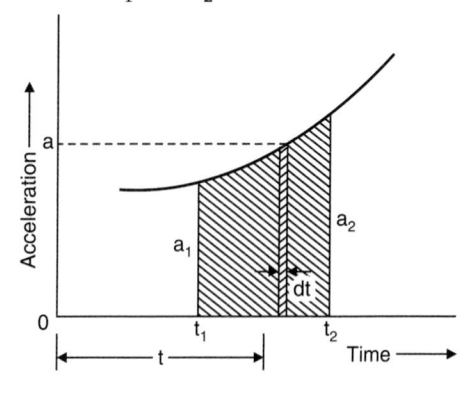

Fig. 12.3

12.2. MOTION WITH UNIFORM VELOCITY

Consider the motion of a body moving with uniform velocity v. Now,

$$\frac{ds}{dt} = v$$

or
$$s = \int v\,dt$$
$$= vt \text{ since } v \text{ is constant} \qquad ...(12.1)$$

$v - t$ curve for such a motion is shown in Fig. 12.4. It can be easily seen that the distance travelled s, from starting point in time t is given by the shaded area, which is a rectangle

or $\quad s = vt$, which is same as equation 12.1

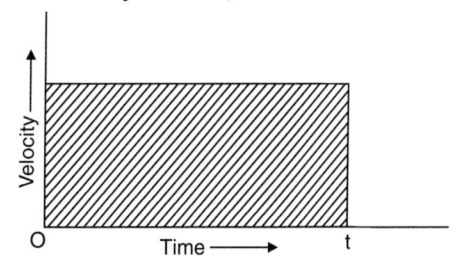

Fig. 12.4

12.3. MOTION WITH UNIFORM ACCELERATION

Consider the motion of a body with uniform acceleration a.

Let u – initial velocity
         v – final velocity
and t – time taken for change of velocity from u to v.

Acceleration is defined as the rate of change of velocity. Since it is uniform, we can write

$$a = \frac{v - u}{t}$$

or
$$v = u + at \qquad \qquad ...(12.2)$$

Displacement s is given by,
$$s = \text{average velocity} \times \text{time}$$
$$= \frac{u+v}{2}t \qquad \ldots(12.3)$$

Substituting the value of v from Eqn. (12.2) into Eqn. (12.3), we get
$$s = \frac{u+u+at}{2}t = ut + \frac{1}{2}at^2 \qquad \ldots(12.4)$$

From (12.2), $\quad t = \dfrac{v-u}{a}$

Substituting it into Eqn. (12.3)
$$s = \frac{u+v}{2}\cdot\frac{v-u}{a} = \frac{v^2 - u^2}{2a}$$

i.e.,
$$v^2 - u^2 = 2as \qquad \ldots(12.5)$$

Thus equations of motion of a body moving with constant acceleration are
$$\left.\begin{array}{ll} v = u + at & (a) \\ s = ut + 1/2\, at^2 & (b) \\ v^2 - u^2 = 2as & (c) \end{array}\right\} \qquad \ldots(12.6)$$
and

Eqn. (12.6) can be derived by integration technique also as given below:

From definition of acceleration
$$\frac{dv}{dt} = a$$
$$dv = a\,dt$$

Since 'a' is constant,
$$v = at + C_1 \qquad \ldots(1)$$
where C_1 is constant of integration

When $t = 0$, velocity = initial velocity, u

Substituting these values in (1), we get
$$\therefore \qquad u = 0 + C_1 \quad \text{or} \quad C_1 = u$$
Thus, $\qquad v = u + at \qquad \ldots(a)$

From the definition of velocity,
$$\frac{ds}{dt} = v = u + at \qquad \text{[from } (a)\text{]}$$
or $\qquad ds = (u + at)dt$

$\therefore \qquad s = ut + \dfrac{1}{2}at^2 + C_2$

where, C_2 is constant of integration.

When $t = 0$, $s = 0$

$\therefore \qquad C_2 = 0 \quad \text{and} \quad \text{hence } s = ut + \dfrac{1}{2}at^2 \qquad \ldots(b)$

From definition of acceleration,
$$a = \frac{dv}{dt} = \frac{dv}{ds}\cdot\frac{ds}{dt}$$
$$= \frac{dv}{ds}v \quad \text{[since } ds/dt = v\text{]} \qquad \ldots(12.7)$$

LINEAR MOTION

$$a\,ds = v\,dv$$

By integrating,

$$a\int_{}^{} ds = \int_u^v v\,dv$$

$$as = [v^2/2]_u^v = \frac{v^2}{2} - \frac{u^2}{2}$$

or $\qquad v^2 - u^2 = 2as \qquad\qquad ...(c)$

The equations of linear motions can be found conveniently referring to $v - t$ diagram. Since acceleration is uniform, the slope of the curve is constant, *i.e.* it is a straight line, as shown in Fig. 12.5.

Now, $\quad a$ = slope of the diagram

$\qquad\quad$ = $\tan\theta$

$$= \frac{BD}{AD} = \frac{BC - CD}{OC} = \frac{BC - OA}{OC}$$

$$= \frac{v - u}{t}$$

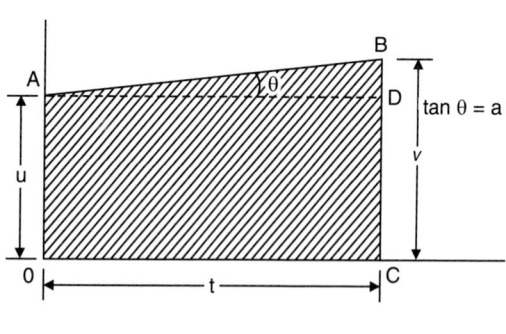

Fig. 12.5

$\therefore \qquad\qquad v = u + at \qquad\qquad\qquad ...(a)$

$\qquad\qquad s$ = Area $AOCB$

$\qquad\qquad\quad$ = Area of rectangle $AOCD$ + Area of $\triangle ABD$

$$= AO \times OC + \frac{1}{2} \times AD \times BD$$

$$= ut + \frac{1}{2} AD \times AD \tan\theta$$

$$= ut + \frac{1}{2} \times t \times t \times a$$

$$s = ut + \frac{1}{2} at^2 \qquad\qquad ...(b)$$

We can also write, $\quad s$ = Area of parallelogram $AOCB$

$$= \frac{1}{2}(AO + BC)\,OC = \frac{1}{2}(u + v)\,t$$

Substituting $\qquad t = \dfrac{v - u}{a}$ from Eqn. (a) we get

$$s = \frac{1}{2}(u + v)\frac{(v - u)}{a}$$

i.e., $\qquad\qquad 2as = v^2 - u^2$

12.4. ACCELERATION DUE TO GRAVITY

In chapter X, it has been shown that the acceleration due to gravity is constant for all practical purposes when we treat the motion of the bodies near earth's surface. Its value is found to be 9.81 m/sec² and is always directed towards centre of the earth, i.e., vertically downwards. Hence, if vertically downward motion of a body is considered, the value of acceleration a in Eqn. (12.6) is 9.81 m/sec² and if vertically upward motion is considered, then,
$$a = -g = -9.81 \text{ m/sec}^2.$$

Example 12.1. *A particle is projected vertically upwards from the ground with an initial velocity of u m/sec.*

Find

(i) the time taken to reach the maximum height;

(ii) the maximum height reached;

(iii) time required for descending; and

(iv) velocity when it strikes the ground.

Solution. Consider the upward motion of the particle.

Initial velocity = u

Since vertically upward motion is considered as positive
$$a = -g = -9.81 \text{ m/sec}^2$$

When maximum height is reached, final velocity $v = 0$.

Form equation of motion,
$$v = u + at$$
we get, $\qquad 0 = u - gt$

$\therefore \qquad\qquad t = \dfrac{u}{g} \qquad\qquad$ ...(1)

Let the maximum height reached (displacement) s be h

From equation of motion, $v^2 - u^2 = 2as$, we get,
$$0 - u^2 = -2gh$$
or $\qquad\qquad h = \dfrac{u^2}{2g} \qquad\qquad$ ...(2)

Now, consider the downward motion of the particle. It starts with zero velocity ($u_2 = 0$) from a height h ($= s$). Let it strike the ground with final velocity v_2. Acceleration due to gravity, $g = 9.81$ m/sec².

From the equation of motion, $v^2 - u^2 = 2as$, we get
$$v_2^2 - 0 = 2gh$$
$$v_2^2 = 2g \dfrac{u^2}{2g} \qquad \left[\text{from eqn. 2, } h = \dfrac{u^2}{2g}\right] \qquad ...(3)$$
$$= u^2$$
$$v_2 = u$$

When a particle is freely projected, the magnitude of its velocity at any given elevation is the same during both upward and downward motion.

From the relation, $\quad v = u + at$

we get, $\qquad\qquad v_2 = 0 + gt$

LINEAR MOTION

$$t = \frac{v_2}{g} = \frac{u}{g} \qquad ...(4)$$

Thus, the time taken for the upward motion is same as the time taken for the downward motion.

Example 12.2. *A small steel ball is shot vertically upwards from the top of a building 25 m above the ground with an initial velocity of 18 m/sec.*

(a) *In what time, it will reach the maximum height?*
(b) *How high above the building will the ball rise?*
(c) *Compute the velocity with which it will strike the ground and the total time it is in motion.*

Solution. For upward motion:

$$u = 18 \text{ m/sec}$$
$$v = 0$$
$$a = -9.81 \text{ m/sec}^2$$
and $$s = h$$

Let t_1 be time taken to reach the maximum height.
From equation of motion,

$$v = u + at$$
we get, $$0 = 18 - 9.81 \, t_1$$
$$t_1 = \mathbf{1.83 \text{ sec}} \quad \text{Ans.}$$

From the relation, $v^2 - u^2 = 2as$ we get,
$$0 - 18^2 = 2(-9.81)h$$

$$\therefore \qquad h = \frac{18^2}{2 \times 9.81} = \mathbf{16.51 \, m} \quad \text{Ans.}$$

$\therefore$ Total height from the ground
$$= 25 + h = 25 + 16.51 = 41.51 \text{ m}$$

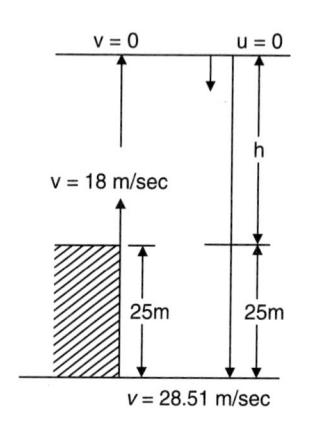

Fig. 12.6

Downward motion:
With usual notations,
$$u = 0, \, v = v_2, \, s = 41.51 \text{ m}, \, a = +9.81 \text{ m/sec}^2$$
$$t = t_2$$
From the relation $v^2 - u^2 = 2as$, we get
$$v_2^2 - 0 = 2 \times 9.81 \times 41.51$$
$$v_2 = \mathbf{28.54 \text{ m/sec}} \quad \text{Ans.}$$
From the relation $v = u + at$ we get,
$$28.54 = 0 + 9.81 \, t_2$$
$\therefore \qquad t_2 = 2.91 \text{ m/sec}$

$\therefore$ **Total time during which the body is in motion**
$$= t_1 + t_2 = 1.83 + 2.91 = \mathbf{4.74 \text{ sec}} \quad \text{Ans.}$$

Example 12.3. *If a stone falls past a window of 2.45 m height in half a second, find the height from which the stone fell.*

Solution. Let the stone be dropped from A (Fig. 12.7) at a height h above the window and BC represent the height of window.
Let t be the time to fall from A to B.

Then for motion from A to B,
$$u = 0, s = h \text{ and } a = +g = 9.81 \text{ m/sec}^2.$$

$$\therefore \quad h = 0 \times t + \frac{1}{2} gt^2$$

$$h = \frac{1}{2} gt^2 \qquad \ldots(1)$$

For motion from A to C:
$$u = 0, s = h + 2.45 \ a = g = 9.81 \text{ m/sec}^3$$
$$\text{time} = t + 0.5$$

$$\therefore \quad h + 2.45 = 0 \times t + \frac{1}{2} g (t + 0.5)^2$$

$$= \frac{1}{2} g (t^2 + t + 0.25) \qquad \ldots(2)$$

Fig. 12.7

Subtracting eqn. (1) from (2), we get
$$2.45 = \frac{1}{2} g (t + 0.25)$$

$$= \frac{1}{2} \times 9.81 (t + 0.25)$$

$$\therefore \quad t = 0.2495 \text{ sec}$$

$$h = \frac{1}{2} gt^2$$

$$= \frac{1}{2} \times 9.81 \times (0.2495)^2$$

$$\mathbf{h = 0.305 \text{ m} \quad Ans.}$$

Example 12.4. *A ball is dropped from the top of a tower 30 m high. At the same instant a second ball is thrown upward from the ground with an initial velocity of 15 m/sec. When and where do they cross and with what relative velocity?*

Solution. Let the two balls cross each other at a height h from the ground (Fig. 12.8) after t seconds.

For the motion of first ball,
$$u = 0, s = 30 - h \text{ and } a = 9.81 \text{ m/sec}^2.$$

Using the equation $s = ut + \frac{1}{2} at^2$, we get

$$30 - h = 0 \times t + \frac{1}{2} \times 9.81 \times t^2 \qquad \ldots(1)$$

For the motion of second ball,
$$u = 15 \text{ m/sec}, s = h \text{ and } a = -9.81 \text{ m/sec}^2$$

$$\therefore \quad h = 15 t - \frac{1}{2} \times 9.81 \ t^2 \qquad \ldots(2)$$

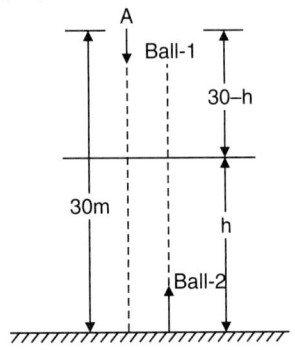

Fig. 12.8

Adding (1) and (2) we get,
$$30 = 15 t$$
or $$t = \mathbf{2 \text{ sec} \quad Ans.}$$

LINEAR MOTION

$$\therefore \quad h = 15 \times 2 - \frac{1}{2} \times 9.81 \times 2^2$$

i.e. $\quad\quad\quad\quad\quad\quad\quad\quad\quad \boldsymbol{h = 10.38 \text{ m} \quad \text{Ans.}}$

At $t = 2$ seconds

(i) Downward velocity of first ball,

$$v_1 = 0 + 9.81 \times 2 = 19.62 \text{ m/sec}$$

(ii) Upward velocity of second ball

$$v_2 = 15 - 9.81 \times 2$$
$$= -4.62 \text{ m/sec}$$
$$v_2 = 4.62 \text{ m/sec downward}$$

$\therefore$ **Relative velocity** $= 19.62 - 4.62$

$$= 15 \text{ m/sec} \quad \text{Ans.}$$

Example 12.5. *A stone dropped into a well is heard to strike the water in 4 seconds. Find the depth of the well, assuming the velocity of sound to be 335 m/sec.*

Solution. Let $\quad h$ – depth of the well

$\quad\quad\quad\quad\quad\quad t_1$ – time taken by stone to strike water

$\quad\quad\quad\quad\quad\quad t_2$ – time taken by sound to travel the height h.

Then, $\quad\quad\quad\quad t_1 + t_2 = 4$ $\quad\quad\quad\quad\quad\quad\quad\quad\quad\quad\quad\quad\quad\quad$ ...(1)

For the downward motion of the stone,

$$h = 0 \times t_1 + \frac{1}{2} g t_1^2$$

i.e. $\quad\quad\quad\quad\quad\quad h = \frac{1}{2} g t_1^2$ $\quad\quad\quad\quad\quad\quad\quad\quad\quad\quad\quad\quad\quad\quad$ ...(2)

Since sound moves with uniform velocity, for the upward motion of sound, we have

$$h = 335 \, t_2 \quad\quad\quad\quad\quad\quad\quad\quad\quad\quad\quad\quad\quad\quad ...(3)$$

From eqns. (2) and (3), we have

$$\frac{1}{2} g t_1^2 = 335 \, t_2$$

But from eqn. (1), $\quad t_2 = 4 - t_1$

Hence $\quad\quad\quad \frac{1}{2} g t_1^2 = 335 (4 - t_1)$

Substituting the value of $g = 9.81$ m/sec^2, we get

$$\frac{9.81}{2} t_1^2 = 335 (4 - t_1)$$

$$t_1^2 + 68.30 t_1 - 273.19 = 0$$

$$t_1 = \frac{-68.3 + \sqrt{68.3^2 - 4 \times 1 \times (-273.19)}}{2} = 3.79 \text{ sec}$$

$\therefore \quad\quad\quad\quad\quad\quad h = \frac{1}{2} g t_1^2 = \frac{1}{2} \times 9.81 \times (3.79)^2$

i.e. $\quad\quad\quad\quad\quad\quad \boldsymbol{h = 70.44 \text{ m} \quad \text{Ans.}}$

Example 12.6. *A particle under a constant declaration is moving in a straight line and covers a distance of 20 m in first two seconds and 40 m in the next 5 seconds. Calculate the distance it covers in the subsequent 3 seconds and the total distance covered, before it comes to rest.*

Solution. Let the particle start from A and come to halt at E as shown in Fig. 12.9. Let the acceleration of the particle be a.

Note that if the particle is having deceleration the value of a will be negative. Let initial velocity be u m/sec.

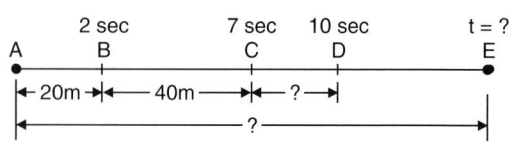

Fig. 12.9

Considering the motion between A and B, we get

$$20 = u \times 2 + \frac{1}{2} a \times 2^2$$

$$20 = 2u + 2a$$

or $\qquad 10 = u + a \qquad$ ...(1)

Considering the motion between A and C, we get

$$60 = 7u + \frac{1}{2} \times a \times 7^2$$

$$60 = 7u + 24.5a \qquad \text{...(2)}$$

From (1), $\qquad 70 = 7u + 7a \qquad$ ...(3)

∴ Subtracting (2) from (3), we get

$$10 = -17.5a$$

or $\qquad a = -.571 \text{ m/sec}^2 \qquad$ (*i.e.*, the body is decelerating)

From (1), $\qquad u = 10 - a = 10 - (-0.571)$

$$= 10.571 \text{ m/sec}$$

Let distance AD be equal to s_1. Then

$$s_1 = 10.571 \times 10 + \frac{1}{2}(-0.571) \times 10^2$$

$$= 77.16 \text{ m}$$

∴ **Distance covered in the interval, 7 seconds to 10 seconds is,**

$$CD = 77.16 - 60 = \mathbf{17.16 \text{ m}} \quad \text{Ans.}$$

Let the particle come to rest at a distance s from the starting point *i.e.* $AE = s$. Then, using the relation $v^2 - u^2 = 2as$, we get

$$0 - (-10.571)^2 = 2(-0.571)s$$

∴ $\qquad s = \dfrac{10.571^2}{2 \times .571} = \mathbf{97.85 \text{ m.}} \quad \text{Ans.}$

Example 12.7. *A motorist is travelling at 80 kmph, when he observes a traffic light 200 m ahead of him turns red. The traffic light is timed to stay red for 10 seconds. If the motorist wishes to pass the light without stopping, just as it turns green, determine: (a) the required uniform deceleration of the motor, and (b) the speed of the motor as it passes the light.*

Solution. Initial velocity $\quad u = 80$ kmph

$$= \frac{80 \times 1000}{60 \times 60} = 22.22 \text{ m/sec}$$

LINEAR MOTION

$t = 10$ seconds, $s = 200$ m

Let 'a' be acceleration.

Using the equation of motion, $s = ut + 1/2\, at^2$

we get,
$$200 = 22.22 \times 10 + \frac{1}{2} a \times 10^2$$

$\therefore$ **$a = -\,0.444$ m/sec^2. Ans.**

(*i.e.* it is decelerating).

Final velocity (speed when it passes the signal)
$$v = u + at = 22.22 - 0.444 \times 10$$
$$v = 17.78 \text{ m/sec}$$
$$= 17.78 \times \frac{60 \times 60}{1000} \text{ kmph}$$

i.e. **$v = 64$ kmph. Ans.**

Example 12.8. *The greatest possible acceleration or deceleration that a train may have is a and its maximum speed is v. Find the minimum time in which the train can get from one station to the next, if the total distance is s.*

Solution. To have the minimum travel time, the train must accelerate at maximum acceleration to reach maximum velocity, then run at that velocity and finally decelerate at maximum rate so that this time is kept the least. Velocity-time diagram for this motion is shown in Fig. 12.10.

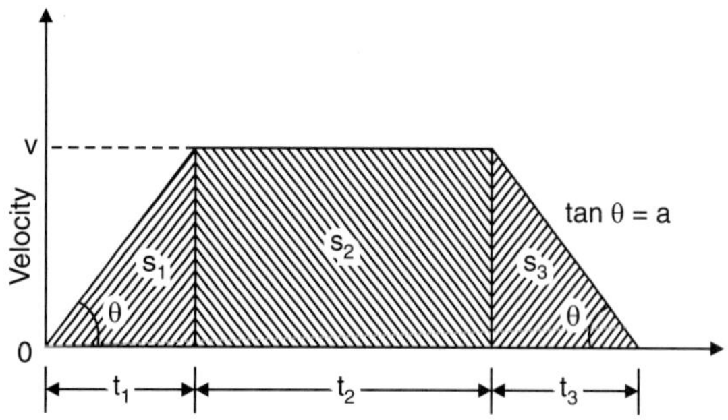

Let t_1 – time for acceleration
   t_2 – time of uniform motion
   t_3 – time for declaration.

$\therefore$ Total time of travel, $t = t_1 + t_2 + t_3$ ...(1)

Fig. 12.10

Now $v = t_1 \tan \theta$
   $= at_1$ [Since $\tan \theta$ = acceleration = a m/sec^2]

and also $v = t_3 \tan \theta = at_3$ ...(2)

$\therefore$ $t_1 = t_3$ ...(3)

Let s_1 – distance travelled while accelerating
   s_2 – distance travelled with uniform velocity
and s_3 – distance travelled while decelerating.

Then
$$s = s_1 + s_2 + s_3 = \frac{1}{2} vt_1 + vt_2 + \frac{1}{2} vt_3$$
$$= vt_1 + vt_2 \qquad \text{(since } t_1 = t_3\text{)}$$

or $$\frac{s}{v} = t_1 + t_2 \qquad ...(4)$$

From (1), (4) and (2) we have
$$t = (t_1 + t_2) + t_3$$
$$t = \frac{s}{v} + \frac{v}{a} \quad \text{Ans.}$$

Example 12.9. *Two stations P and Q are 5.2 km apart. An automobile starts from rest from the station P and accelerates uniformly to attain a speed of 48 kmph in 30 seconds. This speed is maintained until the brakes are applied. The automobile comes to rest at the station Q with a uniform retardation of one metre per second. Determine the total time required to cover the distance between these two stations.*

Solution. Now,
$$v = 48 \text{ kmph}$$
$$= \frac{48 \times 1000}{60 \times 60} = 13.33 \text{ m/sec}$$

and
$$t_1 = 30 \text{ sec}$$

The time-velocity diagram is shown in the Fig. 12.11.

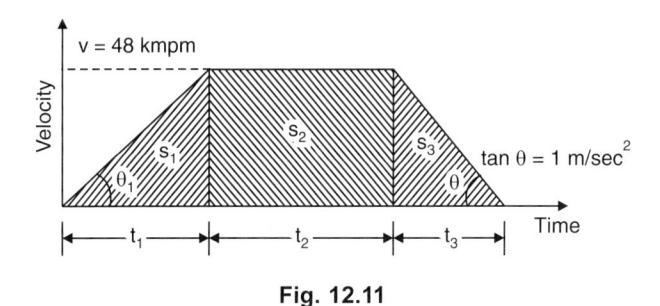

Fig. 12.11

When brakes are applied the automobile retards from a velocity of 13.33 m/sec to zero in t_3 seconds at the rate 1 m/sec². Then
$$t_3 \tan \theta = 13.33$$
$$t_3 \times 1 = 13.33$$
$$t_3 = 13.33 \text{ sec}$$

Let t_2 be the time during which the automobile travels at uniform velocity.

Now,
$$s = s_1 + s_2 + s_3$$

where
$$s = \text{total distance covered} = 5200 \text{ m}$$
$$s_1 = \text{distance covered while accelerating}$$
$$s_2 = \text{distance covered with uniform velocity}$$

and
$$s_3 = \text{distance covered while retarding.}$$

∴
$$5200 = \frac{1}{2} \times 30 \times 13.33 + 13.33 \times t_2 + \frac{1}{2} \times 13.33 \times 13.33$$

∴
$$t_2 = 368.33 \text{ sec}$$

∴ **Total time to cover the distance between the two stations**
$$= 30 + 368.33 + 13.333$$
$$= \mathbf{411.66 \text{ sec}} \quad \textbf{Ans.}$$

Example 12.10. *A cage descends a mine shaft with an acceleration of 0.6 m/sec². After the cage has travelled 30 m a stone is dropped from the top of the shaft.*

Determine: (1) the time taken by the stone to hit the cage, and (2) distance travelled by the cage before impact.

Solution. Let t be the time in seconds during which stone is in motion. Let s be the distance from the top of shaft where impact takes place.

From motion of stone, we get

$$s = \frac{1}{2} gt^2 = \frac{1}{2} \times 9.81 t^2 \qquad \ldots(1)$$

Consider the motion of the cage:
Let t_1 be the time taken to travel first 30 m

$$a = 0.6 \text{ m/sec}^2;\ u = 0$$

∴ $$30 = 0 + \frac{1}{2} \times .6\, t_1^2$$

$$t_1 = 10 \text{ sec}$$

∴ When the stone strikes, the cage has moved for $(t + 10)$ seconds

∴ $$s = 0 + \frac{1}{2} \times 0.6 \times (t+10)^2 \qquad \ldots(2)$$

From (1) and (2), we have

$$\frac{1}{2} \times 9.81\, t^2 = \frac{1}{2} \times 0.6\, (t+10)^2$$

Solving the quadratic equation in t, we get

$t = 3.286$ sec Ans.

∴ $$s = \frac{1}{2} gt^2 = \frac{1}{2} \times 9.81 \times (3.286)^2$$

i.e. **$s = 52.96$ m Ans.**

Example 12.11. *Two trains A and B leave the same station in parallel lines. Train A starts with a uniform acceleration of 0.15 m/sec² and attains a speed of 27 kmph when the steam is reduced to keep speed constant. Train B leaves 40 seconds later with uniform acceleration of 0.3 m/sec² to attain a maximum speed of 54 kmph. When and where will B overtake A?*

Solution. Constant speed of A = 27 kmph

$$= 27 \times \frac{1000}{60 \times 60}$$

$$= 7.5 \text{ m/sec}$$

Constant speed of B = 54 kmph

$$= 15 \text{ m/sec}$$

The velocity-time diagram for the trains A and B are shown in Fig. 12.12(a) and (b) respectively.

Let the train B overtake train A after t seconds at a distance s from the station.

Motion of train A:

Time taken to attain uniform acceleration be t_1
From the relation $v = u + at$, we get

$$7.5 = 0 + 0.15\, t_1$$

∴ $$t_1 = 50 \text{ sec}$$

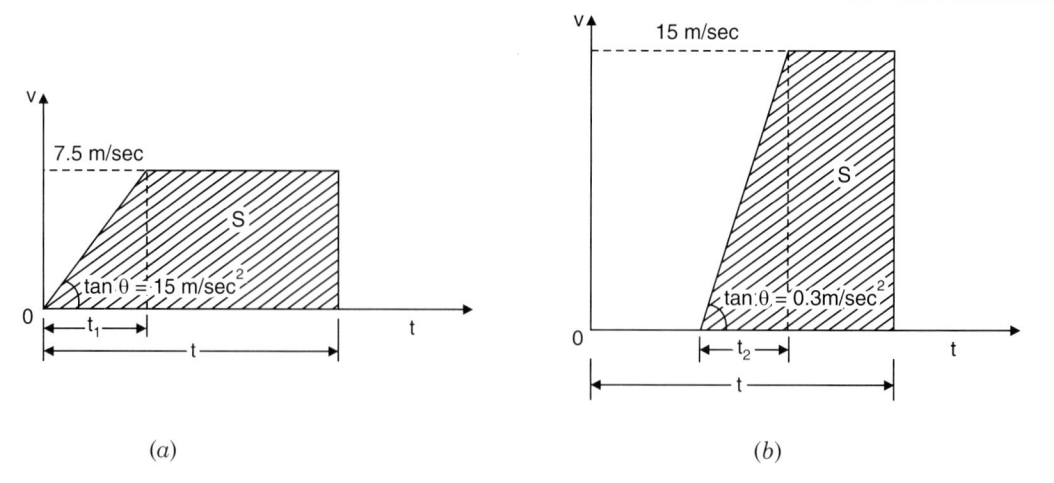

Fig. 12.12

Distance travelled in time t

$$s = \text{shaded area in Fig. 12.12.}(a)$$

$$= \frac{1}{2} t_1 \times 7.5 + 7.5 \, (t - t_1)$$

$$= \frac{1}{2} \times 50 \times 7.5 + 7.5 \, (t - 50)$$

$$= 7.5t - 187.5 \qquad \ldots(1)$$

Motion of train B:

From the relation $v = u + at$, we get

$$15 = 0 + 0.3 \, t_2$$

$$\therefore \qquad t_2 = 50 \text{ sec}$$

Distance travelled in 't' seconds after train A started:

$$s = \frac{1}{2} \times t_2 \times 15 + 15 \, (t - t_2 - 40)$$

$$= \frac{1}{2} \times 50 \times 15 + 15 \, (t - 50 - 40)$$

$$= 15t - 975 \qquad \ldots(2)$$

From (1) and (2), we get

$$7.5t - 187.5 = 15t - 975$$

or

$$t = 105 \text{ sec} \quad \textbf{Ans.}$$

Train B overtakes train A 105 seconds after train A has started.

Distance 's' is given by

$$s = 15t - 975 = 15 \times 105 - 975 = \textbf{600 m} \quad \textbf{Ans.}$$

LINEAR MOTION

Example 12.12. *Two cars are travelling towards each other on a single lane road at the velocities 12 m/sec and 9 m/sec. respectively. When 100 m apart, both drivers realise the situation and apply their brakes. They succeed in stopping simultaneously and just short of colliding. Assume constant deceleration for each case and determine: (a) time required for cars to stop (b) deceleration of each car, and (c) the distance travelled by each car while slowing down.*

Solution. Let '*A*' be the position of one car and '*B*' that of other car when drivers see each other and apply brakes (Fig. 12.13).

After applying brakes, let them meet at *C*. Let $AC = x$ and time taken be t seconds.

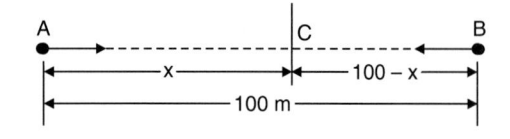

Fig. 12.13

Consider the motion of the first car:

$$u = 12 \text{ m/sec}, \; v = 0, \; s = x.$$

Let a_1 be acceleration (a_1 as it will automatically come out negative since it is retardation case).

From the equation of motion $v = u + at$

We get, $\qquad 0 = 12 + a_1 t$

i.e. $\qquad\qquad a_1 = -\dfrac{12}{t}$ ...(1)

From the equation $v^2 - u^2 = 2as$, we get,

$$0 - 12^2 = 2 a_1 x$$

$$-144 = 2 \times \left(-\dfrac{12}{t}\right) x \quad \text{or} \quad x = 6t \qquad ...(2)$$

Consider the motion of second car:

$$u = 9 \text{ m/sec}, \; v = 0, \; a = a_2, \; \text{time} = t \text{ and } s = 100 - x$$

From the equation of motion $v = u + at$, we get

$$0 = 9 + a_2 t$$

$\therefore \qquad\qquad a_2 = -9/t$ ...(3)

and from the equation of motion, $v^2 - u^2 = 2as$ we get,

$$0 - 9^2 = 2 a_2 (100 - x)$$

$$-81 = 2\left(-\dfrac{9}{t}\right)(100 - x)$$

$$100 - x = 4.5 \, t \qquad ...(4)$$

Adding (2) and (4), we get,

$$100 = 10.5 \, t \quad \text{or} \quad t = 9.524 \text{ sec} \quad \textbf{Ans.}$$

From (1) $\qquad a_1 = -12/t = -1.26 \text{ m/sec}^2 \quad \textbf{Ans.}$

From (3) $\qquad a_2 = -9/t = -0.945 \text{ m/sec}^2 \quad \textbf{Ans.}$

From (2) $\qquad x = 57.14 \text{ m} \quad \textbf{Ans.}$

Distance travelled by the second car is

$$= 100 - x = 42.86 \text{ m} \quad \textbf{Ans.}$$

Example 12.13. *A car and a truck are both travelling at the constant speed of 45 kmph. The car is 10 m behind the truck. The truck driver suddenly applies his brakes, causing the truck to decelerate at the constant rate of 2 m/sec². Two seconds later the driver of the car applies his brakes and just manages to avoid a rear end collision. Determine the constant rate at which the car decelerated.*

Solution. Velocity of the truck and car = 45 kmph

i.e. $\qquad u = 12.5$ m/sec

Taking A as the reference point (Fig. 12.14), the distance moved by the truck S_T is given by

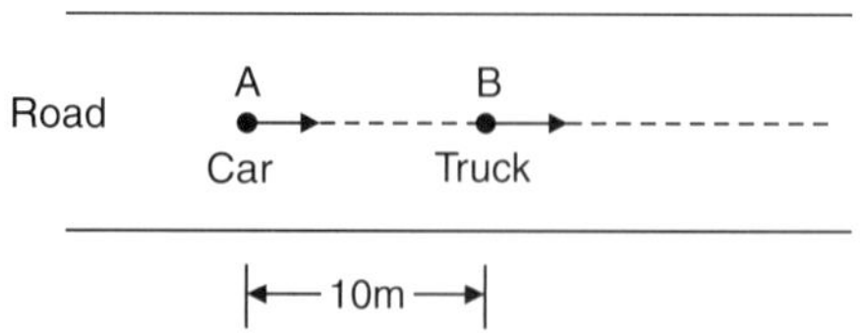

```
                  A         B
Road     •────────•─────────────
         Car      Truck
         |←── 10m ──→|
```

Fig. 12.14

$$s_T = 10 + ut + \frac{1}{2} a_T t^2$$

i.e. $\qquad s_T = 10 + 12.5\, t + \frac{1}{2} a_T t^2$

where, a_T is the acceleration of truck = -2 m/sec²
and t is the time at any instant after the brakes are applied:

∴ $\qquad s_T = 10 + 12.5\, t - t^2 \qquad$ ...(1)

The distance moved by the car s_C during this time is given by

$$s_C = u \times 2 + u\,(t - 2) + \frac{1}{2} a_C (t - 2)^2$$

where, a_c is acceleration of the car.

[Note that $u \times 2$ gives the distance moved by the car before applying the brakes to it.]

∴ $\qquad s_C = 12.5 \times 2 + 12.5\,(t - 2) + \frac{1}{2} a_C (t - 2)^2$

$\qquad\qquad = 12.5t + \frac{1}{2} a_C (t - 2)^2 \qquad$ ...(2)

The $s - t$ curve for the motion of car and truck is given in Fig. 12.15.

Since there is no possibility of car jumping over the truck, there is only one point of contact between the two vehicles which is shown by the point C in Fig. 12.15. At this time, the two vehicles need not have zero velocity. They will have same velocity which will be equal to slope of the curves at C. Beyond point C, the car will be having velocity lesser than that of the truck. Finally, the car will stop at distance s_1 and truck will stop at distance s_2 as shown in the figure.

At point C, $\qquad s_T = s_C$

substituting the values of s_T and s_C from (1) and (2), we get

LINEAR MOTION

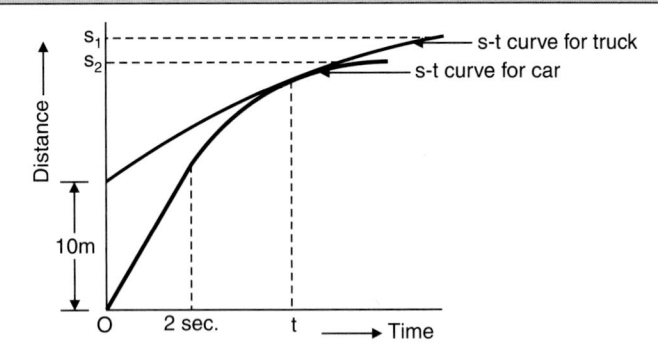

Fig. 12.15

$$10 + 12.5t - t^2 = 12.5t + \frac{1}{2}a_C(t-2)^2$$

$$= 12.5t + \frac{1}{2}a_C t^2 - 2a_C t + 2a_C$$

i.e.
$$t^2\left(\frac{1}{2}a_C + 1\right) - 2a_C t + (2a_C - 10) = 0$$

This is a quadratic equation in t.

∴
$$t = \frac{2a_c}{2(1/2a_C + 1)} \pm \frac{\sqrt{(2a_C)^2 - 4 \times \left(\frac{1}{2}a_C + 1\right)(2a_C - 10)}}{2\left(\frac{1}{2}a_C + 1\right)} \qquad ...(3)$$

Since there is only one contact point c, there should be only one value for t. Hence in (3) the term under the root should be zero, i.e.

$$(2a_C)^2 - 4(1/2 a_C + 1)(2a_C - 10) = 0$$

i.e.
$$4a_C^2 - (2a_C + 4)(2a_C - 10) = 0$$
$$12 a_C + 40 = 0$$

or
$$a_C = -\frac{10}{3}$$

i.e. **the deceleration of the car is** $\frac{10}{3}$ **m/sec^2** **Ans.**

12.5. MOTION WITH VARYING ACCELERATION

A vehicle is normally not accelerated uniformly. Initially it starts with zero acceleration, then the rate of acceleration is increased and when the desired speed is nearing, the rate of acceleration is reduced. By the time desired speed is picked up acceleration is brought to zero. Thus there are situations with varying acceleration. If the variation of acceleration or velocity or displacement with respect to time is known, such problems can be solved using the differential equations:

$$v = \frac{ds}{dt}$$

$$a = \frac{dv}{dt} = \frac{d^2s}{dt^2} = v\frac{dv}{ds}$$

Example 12.14. *The motion of a particle moving in a straight line is given by the expression*

$$s = t^3 - 3t^2 + 2t + 5$$

where, s is the displacement in metres and t is the time in seconds. Determine: (1) velocity and acceleration after 4 seconds; (2) maximum or minimum velocity and corresponding displacement; (3) time at which velocity is zero.

Solution. $\quad s = t^3 - 3t^2 + 2t + 5$ ...(1)

$\therefore \quad v = \dfrac{ds}{dt} = 3t^2 - 6t + 2$ ...(2)

and $\quad a = \dfrac{d^2s}{dt^2} = 6t - 6$ ...(3)

Hence, after 4 seconds,

$v = 3 \times 4^2 - 6 \times 4 + 2 = $ **26 m/sec Ans.**

and $\quad a = 6 \times 4 - 6 = $ **18 m/sec² Ans.**

The velocity is maximum or minimum when $\dfrac{dv}{dt} = a = 0$. From (3) we get

$0 = 6t - 6 \quad$ or $\quad t = 1$

$\therefore \quad$ Corresponding velocity $= 3 \times 1^2 - 6 \times 1 + 2 = -1$ m/sec

Hence it should be minimum velocity.

$\left(\text{or } \dfrac{d^2v}{dt^2} \text{ is + ve, hence it is minimum velocity}\right)$.

Minimum velocity = – 1 m/sec Ans.

Let at time t the velocity be zero. Then

$0 = 3t^2 - 6t + 2$

$\therefore \quad t = \dfrac{6 \pm \sqrt{36 - 4 \times 3 \times 2}}{2 \times 3} = 1.577 \quad \text{and} \quad 0.423$ sec

Velocity is zero at $t = 0.423$ sec and 1.577 sec Ans.

Example 12.15. *The velocity of a particle moving in a straight line is given by the expression*

$$v = t^3 - t^2 - 2t + 2$$

The particle is found to be at a distance 4 m from station A after 2 seconds. Determine: (1) acceleration and displacement after 4 seconds; and (2) maximum/minimum acceleration.

Solution.

Now, $\quad v = t^3 - t^2 - 2t + 2$ ...(1)

$\therefore \quad a = \dfrac{dv}{dt} = 3t^2 - 2t - 2$ ...(2)

LINEAR MOTION

Hence, **acceleration after 4 seconds**
$$= 3 \times 4^2 - 2 \times 4 - 2 = 38 \text{ m/sec}^2 \quad \textbf{Ans.}$$

Now, $\quad \dfrac{ds}{dt} = v = t^3 - t^2 - 2t + 2$

$\therefore \quad s = \dfrac{t^4}{4} - \dfrac{t^3}{3} - t^2 + 2t + C$

where C is constant of integration.

From the given condition, $s = 4$ m when $t = 2$ sec, we get

$$4 = \dfrac{2^4}{4} - \dfrac{2^3}{3} - 2^2 + 2 \times 2 + C$$

i.e. $\quad C = \dfrac{4}{3}$

$\therefore \quad s = \dfrac{t^4}{4} - \dfrac{t^3}{3} - t^2 + 2t + \dfrac{4}{3}$

Hence when $t = 4$ sec,

$$s = \dfrac{4^4}{4} - \dfrac{4^3}{3} - 4^2 + 2 \times 4 + \dfrac{4}{3} = \textbf{36 m} \quad \textbf{Ans.}$$

Acceleration a is maximum or minimum, when

$$\dfrac{da}{dt} = 0$$

i.e. $\quad 6t - 2 = 0$

or $\quad t = \dfrac{1}{3}$ sec

Since $\dfrac{d^2 a}{dt^2}$ is a positive quantity, the above condition is for the minimum value.

$\therefore \quad$ **minimum value of acceleration** $= 3 \times \left(\dfrac{1}{3}\right)^2 - 2 \times \dfrac{1}{3} - 2$

$$= \textbf{– 2.333 m/sec}^2 \quad \textbf{Ans.}$$

Example 12.16. *A body moves along a straight line and its acceleration 'a' which varies with time is given by $a = 2 - 3t$. Five seconds after start of the observations, its velocity is found to be 20 m/sec. Ten seconds after start of the observation, the body is at 85 m from the origin. Determine: (1) its acceleration, velocity and distance from the origin, and (2) the time in which the velocity becomes zero and the corresponding distance from the origin. Describe the motion diagrammatically.*

Solution. In this problem,

$$a = 2 - 3t \quad i.e. \quad \dfrac{dv}{dt} = 2 - 3t$$

$\therefore \quad v = 2t - \dfrac{3t^2}{2} + C_1 \qquad \ldots(1)$

where C_1 is constant of integration.

Now, $v = 20$ m/sec. when $t = 5$ sec

Hence from (1), $20 = 2 \times 5 - \dfrac{3}{2} \times 5^2 + C_1$

∴ $C_1 = 47.5$

Hence, (1) becomes
$$v = 47.5 + 2t - 1.5\, t^2 \qquad \ldots(2)$$

i.e. $\dfrac{ds}{dt} = 47.5 + 2t - 1.5t^2$

∴ $s = 47.5\, t + t^2 - 0.5t^3 + C_2 \qquad \ldots(3)$

where, C_2 is constant of integration.

It is given that,
$$s = 85\ m \text{ when } t = 10 \text{ sec}$$

Hence we get,
$$85 = 47.5 \times 10 + 10^2 - 0.5 \times 10^3 + C_2$$

∴ $C_2 = 10$

Hence, $s = 10 + 47.5\, t + t^2 - 0.5\, t^3 \qquad \ldots(4)$

(1) when $t = 0$,

$$\left.\begin{array}{l} a = 2 - 3 \times 0 = \mathbf{2\ m/sec^2} \\ v = 47.5 + 0 - 0 = \mathbf{47.5\ m/sec} \\ s = 10 - 0 + 0 + 0 - 0 = \mathbf{10\ m.} \end{array}\right\}$$

and

(2) Let the time when velocity becomes zero be t.

Then from (2), we have
$$0 = 47.5 + 2t - 1.5t^2$$

∴ $t = \mathbf{6.33\ sec}$ **Ans.**

Corresponding distance from origin
$$s = 10 + 47.5 \times 6.33 + 6.33^2 - 0.5 \times 6.33^3$$
$$= \mathbf{223.926}\ \ \textbf{Ans.}$$

(3) The values of displacement, velocity and acceleration for various values of t are tabulated below and then $s - t$, $v - t$ and $a - t$ diagrams are drawn (Fig. 12.16).

t	S	v	a
0	10.0	47.5	2
1	58.0	47.0	− 1.0
2	105.0	45.5	− 4.0
3	148.0	40.0	− 7.0
4	184.0	31.5	− 10.0
5	210.0	20.0	− 13.0
6	223.0	5.5	− 16.0
7	220.0	− 12.0	− 19.0
8	198.0	− 32.5	− 22.0

LINEAR MOTION   319

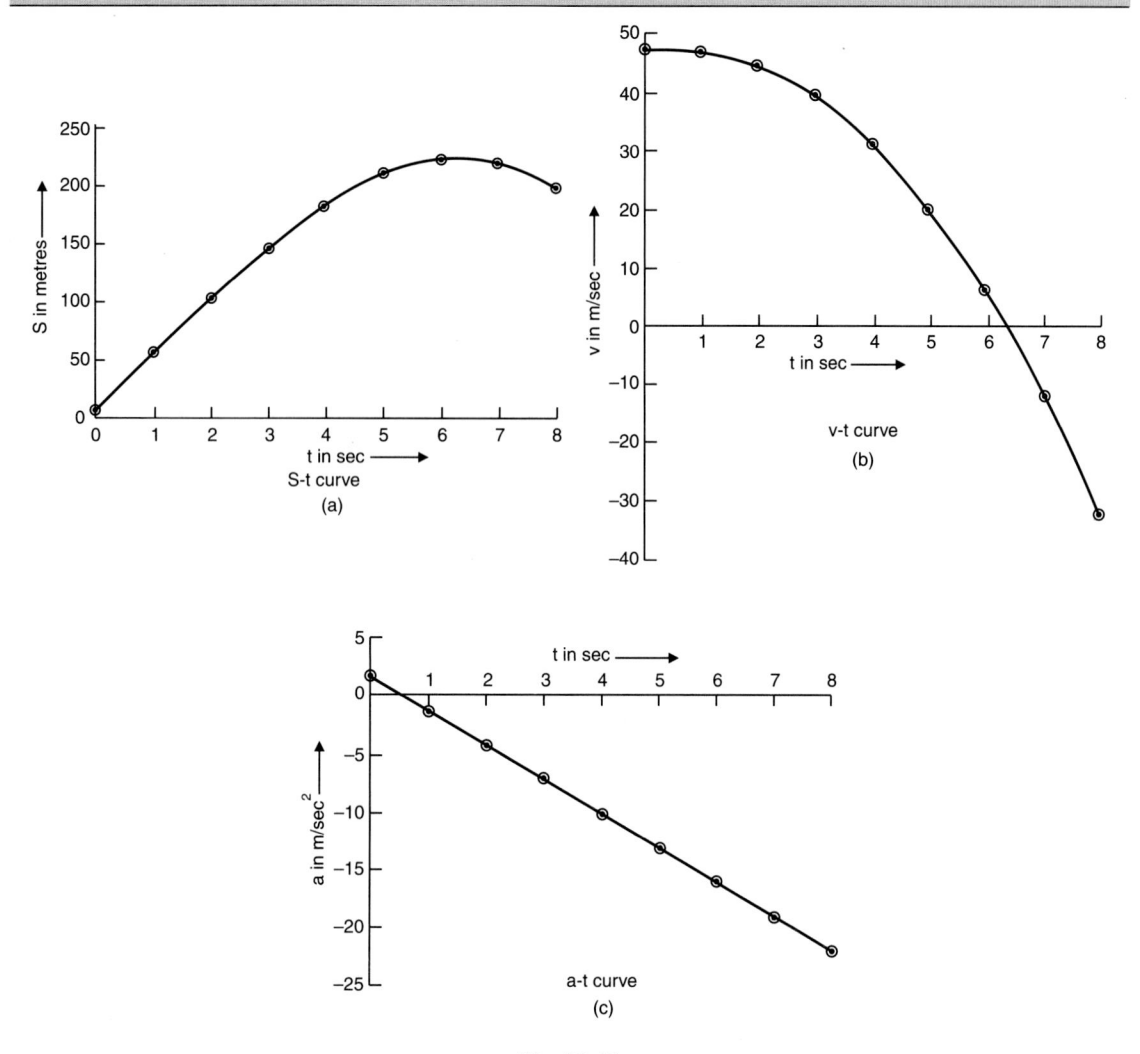

Fig 12.16

Example 12.17. *A particle starts with an initial velocity of 8 m/sec and moves along a straight line. Its acceleration at any time t after start, is given by the expression $\lambda - \mu t$ where λ and μ are constants. Determine the equation for displacement, if the particle covers a distance of 40 m in 5 seconds and stops.*

Solution. $\quad a = \lambda - \mu t \quad i.e. \quad \dfrac{dv}{dt} = \lambda - \mu t$

$\therefore \quad v = \lambda t - \dfrac{\mu t^2}{2} + C_1$

when, $t = 0$, $v = 8$ m/sec

$\therefore \quad 8 = 0 - 0 + C_1 \quad$ or $\quad C_1 = 8$

$$\therefore \qquad v = 8 + \lambda t - \frac{\mu t^2}{2}$$

i.e. $$\frac{ds}{dt} = 8 + \lambda t - \frac{\mu t^2}{2}$$

$$\therefore \qquad s = C_2 + 8t + \frac{\lambda t^2}{2} - \frac{\mu t^3}{6}$$

When, $t = 0$, $s = 0$, $\therefore C_2 = 0$ or $s = 8t + \frac{\lambda t^2}{2} - \frac{\mu t^3}{6}$

After 5 seconds, $s = 40$ m,

$$\therefore \qquad 40 = 8 \times 5 + \frac{25}{2}\lambda - \frac{\mu}{6} \times 125$$

$$\therefore \qquad \mu = 0.6\,\lambda \qquad \qquad \qquad \qquad \qquad \qquad \qquad ...(1)$$

When $t = 5$ sec, $v = 0$

$$\therefore \qquad 0 = 8 + 5\lambda - \mu\frac{25}{2} \qquad \qquad \qquad \qquad ...(2)$$

From (1) and (2), $0 = 8 + 5\lambda - 0.6\,\lambda\frac{25}{2}$

$$\therefore \qquad \lambda = 3.2$$

Hence $\mu = 0.6 \times 3.2 = 1.92$

$\therefore$ Equation of displacement is

$$s = 8t + \frac{3.2 t^2}{2} - \frac{1.92 \times t^3}{6}$$

i.e. $\qquad s = 8t + 1.6t^2 - 0.32t^3$ **Ans.**

Example 12.18. *A car is moving with a velocity of 72 kmph. After seeing a child on the road the brakes are applied and the vehicle is stopped in a distance of 15 m. If the retardation produced is proportional to distance from the point where brakes are applied, find the expression for retardation.*

Solution. Let the expression for retardation be

$$a = -ks, \quad \text{where } s = \text{distance travelled}$$

and $\qquad k = $ constant

$$\therefore \qquad v\frac{dv}{ds} = -ks$$

$$vdv = -ks\,ds$$

$$\therefore \qquad \frac{v^2}{2} = -\frac{ks^2}{2} + C_1$$

At the time the brakes are applied, $s = 0$ and $v = 72$ kmph

i.e. $$v = \frac{72 \times 1000}{36 \times 36} = 20 \text{ m/sec}$$

LINEAR MOTION

$\therefore \qquad \dfrac{20 \times 20}{2} = -0 + C_1 \quad \text{or} \quad C_1 = 200$

$\therefore \qquad \dfrac{v^2}{2} = \dfrac{-ks^2}{2} + 200$

when vehicle stops, $\quad v = 0 \quad$ and $\quad s = 15\ m$

$\therefore \qquad 0 = -k \times \dfrac{15^2}{2} + 200$

$\therefore \qquad k = 400/225 = 1.778$

Hence the expression for retardation is

$\qquad a = -1.778\ s \quad$ **Ans.**

Important Definitions

1. Velocity at any time may be found from the slope of $s - t$ curve.
2. Acceleration 'a' is given by the slope of the $v - t$ curve.
3. Area under $v - t$ curve represents displacement.
4. Area under $a - t$ curve represents velocity.

Important Formulae

1. In case of motion with uniform velocity $s = vt$.
2. In case of motion with uniform acceleration

   $v = u + at$

   $s = ut + \dfrac{1}{2} at^2$

   $v^2 - u^2 = 2as.$

3. Acceleration due to gravity $g = 9.81$ m/sec^2, vertically downward.

PROBLEMS FOR EXERCISE

12.1 A ball is thrown vertically upwards with an initial velocity of 36 m/sec. After 2 seconds, another ball is thrown vertically upwards. What should be its initial velocity so that it crosses the first ball at a height of 30 m? **[Ans.** 28.34 m/sec**]**

12.2 A stone is dropped from the top of a tower. During the last second of its flight, it is found to fall 1/4th of the whole height of tower. Find the height of the tower. What is the velocity with which the stone hits the ground? **[Ans.** $h = 273.27$ m; $v = 73.22$ m/sec**]**

12.3 A stone is dropped into a well without initial velocity. Its splash is heard after 3.5 seconds. Another stone is dropped with some initial velocity and its splash is heard after 3 seconds. Determine the initial velocity of the second stone if velocity of sound is 335 m/sec.

[Ans. $u = 5.34$ m/sec**]**

12.4 A ship, while being launched, slips down the skid with uniform acceleration. If 10 seconds are required to traverse the first 5 metres, what time will be required to slide the total distance of 120 m? With what velocity the ship strikes the water? **[Ans.** 48.99 sec ; 4.899 m/sec**]**

12.5 A train covers a distance of 1.6 km between two stations A and B in 2 minutes, starting from rest. In the first minute of its motion, it accelerates uniformly and in the last 30 seconds it retards uniformly and comes to rest. It moves with uniform velocity during the rest of the period. Find:

(a) Its acceleration in the first minute;

(b) Its retardation in the last 30 second, and

(c) Constant velocity reached by the train.

[**Ans.** (a) 0.3555 m/sec^2 ; (b) 0.711 m/sec^2; (c) 21.3333 m/sec]

12.6 The elevator in an office building starts from ground floor with an acceleration of 0.6 m/sec^2 for 4 seconds. During the next 8 seconds, it travel with uniform velocity. Then suddenly power fails and elevator stops after 3 seconds. If floors are 3.5 m apart, find the floor near which the elevator stops. Assume the retardation is uniform.

[**Ans.** near eighth floor; $h = 27.6$ m]

12.7 A train, starting from rest, is uniformly accelerated during the first 250 m of its run and runs next 750 m at uniform speed. It is then brought to rest in 50 seconds under uniform retardation. If the time taken for the entire journey is 5 minutes, find the acceleration with which the train started. [**Ans.** 0.05 m/sec^2]

12.8 Three marks A, B, C, spaced at a distance of 100 m are made along a straight road. A car starting from rest and accelerating uniformly passes the mark A and takes 10 seconds to reach the mark B and further 8 seconds to reach mark C. Calculate:

(a) the magnitude of the acceleration of the car;

(b) the velocity of the car at A;

(c) the velocity of the car at B; and

(d) the distance of the mark A from the starting point.

[**Ans.** (a) 0.2778 m/sec^2; (b) 8.61 m/sec; (c) 10.833 m/sec; and (d) 133.47 m]

12.9 Two automobiles travelling in the same direction in adjacent lanes are stopped at a highway traffic signal. As the signal turns green, automobile A accelerates at a constant rate of 1.0 m/sec^2. Two seconds later automobile B starts and accelerates at a constant rate of 1.3 m/sec^2. Determine,

(a) when and where B will overtake A and

(b) the speed of each automobile at that time.

[**Ans.** (a) $t = 16.268$ sec; $s = 132.321$ m (b) $v_1 = 16.268$ m/sec; $v_2 = 18.548$ m/sec]

12.10 A particle moves along a straight line. Its motion is represented by the equation

$$s = 16t + 4t^2 - 3t^3$$

where, s is in metres and t, in seconds. Determine

(a) displacement, velocity and acceleration 2 seconds after start;

(b) displacement and acceleration when velocity is zero; and

(c) displacement and velocity when acceleration is zero.

[**Ans.** (a) 24 m, – 4 m/sec, – 28 m/sec^2 ; (b) 24.3 m, – 25.3 m/sec^2 ; (c) 7.64 m, 17.78 m/sec]

12.11 The acceleration of a body starting from rest is given by the equation :

$$a = 12 - 0.1s$$

where, a is the acceleration in m/sec^2 and s is the displacement in metres. Determine the velocity of the body when a distance of 100 m is covered and the distance at which velocity will be zero again. [**Ans.** $s = 100, v = 37.42 ; v = 0, s = 240$]

12.12 A particle moving in a straight line with an initial velocity u m/sec is subjected at any instant to a retardation of λv^3 where λ is a constant and v is the velocity at that instant in m/sec. Calculate the velocity when the distance travelled is s metres and the time for doing so.

$$\left[\textbf{Ans. } v = \frac{u}{\lambda u s + 1} ; \ t = s\left(\frac{\lambda s}{2} + \frac{1}{u_1}\right) \right]$$

13
Projectiles

In the last chapter we considered the motion of a particle along a straight line. But we observe that a particle moves along a curved path if it is freely projected in the air in the direction other than vertical. These freely projected particles which are having the combined effect of a vertical and a horizontal motion are called projectiles. The motion of a projectile has a vertical component and a horizontal component. The vertical component of the motion is subjected to gravitational acceleration/retardation while horizontal component remains constant, if air resistance is neglected. The motion of a projectile can be analysed independently in vertical and horizontal directions and then combined suitably to get the total effect. In this chapter, after defining important terms to be used, horizontal projection, inclined projection on both horizontal and inclined plane are analysed neglecting air resistance.

13.1. DEFINITIONS

The definitions of the terms used in this chapter are given with reference to Fig. 13.1.

Velocity of projection : The velocity with which the particle is projected is called as velocity of projection (u m/sec).

Angle of projection : The angle between the direction of projection and horizontal direction is called as angle of projection (α).

Trajectory : The path traced by the projectile is called as its trajectory.

Horizontal range : The horizontal distance through which the projectile travles in its flight is called the horizontal range or simply range of the projectile.

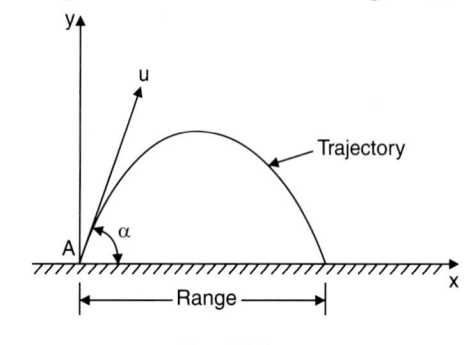

Fig. 13.1

Time of flight : The time interval during which the projectile is in motion is called the time of flight.

13.2. MOTION OF BODY PROJECTED HORIZONTALLY

Consider a particle thrown horizontally from point A with a velocity u m/sec as shown in Fig. 13.2. At any instant the particle is subjected to:

(1) Horizontal motion with constant velocity u m/sec.

(2) Vertical motion with initial velocity zero and moving with acceleration due to gravity g.

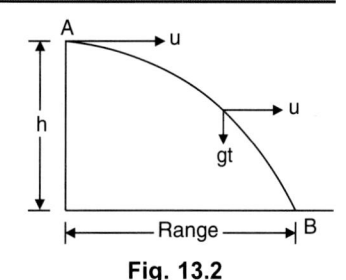

Fig. 13.2

Let h be the height of A from the ground.
Considering vertical motion,

$$h = 0 \times t + \frac{1}{2} gt^2$$

$$= \frac{1}{2} gt^2 \qquad \qquad ...(13.1)$$

This expression gives the time of flight. During this period, the particle moves horizontally with uniform velocity, u m/sec.

∴ Range = ut ...(13.2)

Example 13.1. *A pilot fying his bomber at a height of 2000 m with a uniform horizontal velocity of 600 kmph wants to strike a target (Ref : Fig. 13.3). At what distance from the target, he should release the bomb ?*

Solution. $h = 2000$ m

$u = 600$ kmph

$$= \frac{600 \times 1000}{60 \times 60}$$

$= 166.67$ m/sec

Initial velocity in vertical direction = 0
and gravitational acceleration = 9.81 m/sec².

If t is the time of flight, considering vertical motion, we get

$$2000 = 0 \times t + \frac{1}{2} \times 9.81\ t^2$$

∴ $t = 20.19$ sec

During this period horizontal distance travelled by the bomb

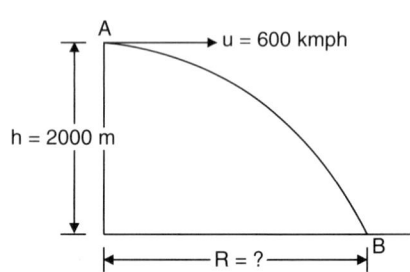

$= ut$
$= 166.67 \times 20.19$
$= 3365.46$ m

Fig. 13.3

Bomb should be released at 3365.46 m from the target. Ans.

Example 13.2 *A person wants to jump over a ditch as shown in Fig. 13.4. Find the minimum velocity with which he should jump.*

Solution. $h = 2$ m
and Range = 3 m

Let t be the time of flight and u the minimum horizontal velocity required. Considering the vertical motion :

$$h = \frac{1}{2} gt^2$$

$$2 = \frac{1}{2}\ 9.81\ t^2$$

∴ $t = 0.6386$ sec

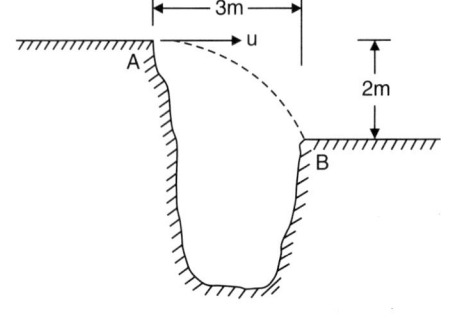

Fig. 13.4

PROJECTILES

Considering horizontal motion of uniform velocity, we get
$$3 = u \times 0.6386$$
∴ **u = 4.698 m/sec Ans.**

Example 13.3. *A pressure tank issues water at A with a horizontal velocity u as shown in Fig. 13.5. For what range of values of u, water will enter the opening BC.*

Solution.

Required velocity to enter at B :
$$h = 1 \text{ m}$$

If t_1 is the time of flight, considering vertical motion,
$$1 = \frac{1}{2} \times 9.81 \, t_1^2$$

or
$$t_1 = 0.4515 \text{ sec}$$

Considering horizontal motion,
$$u_1 t_1 = 3$$

∴
$$u_1 = \frac{3}{0.4515} = 6.44 \text{ m/sec}$$

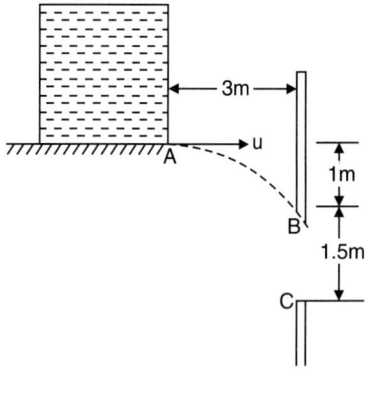

Fig. 13.5

Required velocity to enter at C :

Let t_2 be time required for the flight from A to C.
$$h = 2.5 \text{ m} \, ; \text{Range} = 3 \text{ m}.$$

Considering vertical motion,
$$2.5 = \frac{1}{2} \times 9.81 \, t_2^2$$
$$t_2 = 0.714 \text{ sec}$$

Considering the horizontal motion,
$$u_2 \, t_2 = 3$$

∴
$$u_2 = \frac{3}{0.714} = 4.202 \text{ m/sec}$$

Therefore the range of velocity for which the jet can enter the opening BC is 4.202 m/sec to 6.44 m/sec. Ans.

Example 13.4. *A rocket is released from a jet fighter flying horizontally at 1200 kmph at an altitude of 3000 m above its target. The rocket thrust gives it a constant horizontal acceleration of 6 m/sec². At what angle below the horizontal should pilot see the target at the instant of releasing the rocket in order to score a hit ?*

Solution. Referring to Fig. 13.6, $h = 3000$ m.

In vertical direction, the rocket has initial velocity equal to zero and moves with gravitational acceleration 9.81 m/sec².

Hence if t is the time of flight,
$$3000 = 0 + \frac{1}{2} \times 9.81 \, t^2$$

i.e.
$$t = 24.73 \text{ sec}$$

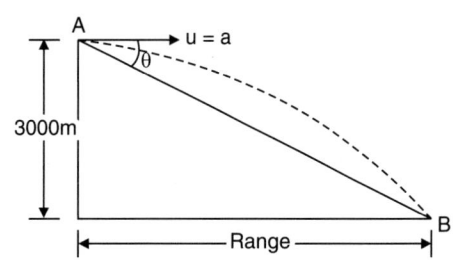

Fig. 13.6

In the horizontal direction, rocket has initial velocity = 1200 kmph and acceleration = 6 m/sec².

Now, $\quad u = 1200$ kmph

$$= \frac{1200 \times 1000}{60 \times 60} = 333.33 \text{ m/sec}$$

∴ Horizontal distance covered during the time of flight = range

$$= ut + \frac{1}{2} at^2$$

$$= 333.33 \times 24.73 + \frac{1}{2} \times 6 \times 24.73^2$$

$$= 10{,}078.5 \text{ m}$$

∴ The angle θ below the horizontal at which the pilot must see the target while releasing the rocket, is given by

$$\tan \theta = \frac{3000}{10{,}078.5}$$

∴ $\quad \theta = \mathbf{16.576°}$ **Ans.**

13.3. INCLINED PROJECTION ON LEVEL GROUND

Consider the motion of a projectile, projected from point A with velocity of projection u and angle of projection α, as shown in Fig. 13.7. Let the ground be a horizontal surface.

The particle has motion in vertical as well as horizontal directions.

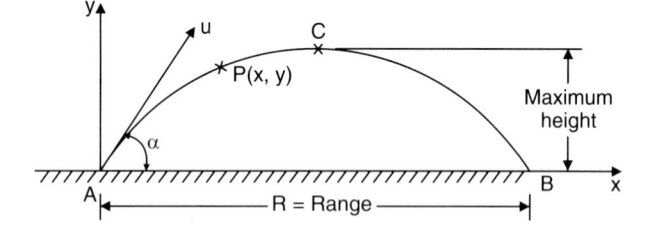

Vertical Motion

Initial velocity = $u \sin \alpha$ upward

Gravitational acceleration = g = 9.81 m/sec² downward

i.e. $\quad\quad\quad a = -g = -9.81 \text{ m/sec}^2$

Hence initially the particle moves upward with velocity $u \sin \alpha$ and retardation 9.81 m/sec².

Velocity becomes zero after some time (at C) and then the particle starts moving downward with gravitational acceleration.

Horizontal Motion

Horizontal component of velocity = $u \cos \alpha$.

Neglecting air resistance, we can say that the projectile is having uniform velocity $u \cos \alpha$ during its entire flight.

Equation of the Trajectory

Let $P(x, y)$ represent the position of projectile after t seconds.

Considering the vertical motion,

$$y = (u \sin \alpha) t - \frac{1}{2} gt^2 \quad\quad\quad \text{...(13.3)}$$

PROJECTILES

Considering horizontal motion,
$$x = (u \cos \alpha) t \qquad \ldots(13.4)$$

$$\therefore \quad t = \frac{x}{u \cos \alpha}$$

Substituting this value in Eqn. 13.3, we get

$$y = u \sin \alpha \, \frac{x}{u \cos \alpha} - \frac{1}{2} g \left(\frac{x}{u \cos \alpha} \right)^2$$

i.e.
$$y = x \tan \alpha - \frac{1}{2} \frac{gx^2}{u^2 \cos^2 \alpha} \qquad \ldots(13.5)$$

But $\dfrac{1}{\cos^2 \alpha} = \sec^2 \alpha = (1 + \tan^2 \alpha).$

Hence Eqn. (13.5) reduces to the form

$$y = x \tan \alpha - \frac{1}{2} \frac{x^2}{u^2} (1 + \tan^2 \alpha) \qquad \ldots(13.6)$$

This is an equation of a parabola. Hence the equation of the trajectory is a parabola.

Maximum Height

When the particle reaches maximum height, the vertical component of the velocity will be zero. Considering vertical motion,

  initial velocity = $u \sin \alpha$
  final velocity = 0
  acceleration = $-g$

Using the equation of linear motion $v^2 - u^2 = 2as$, we get
$$0 - (u \sin \alpha)^2 = -2gh$$

$$h = \frac{u^2 \sin^2 \alpha}{2g} \qquad \ldots(13.7)$$

Time Required to Reach Maximum Height

Using first equation of motion ($v = u + at$), when projectile reaches maximum height,
$$0 = u \sin \alpha - gt$$

$$\therefore \quad t = \frac{u \sin \alpha}{g} \qquad \ldots(13.8)$$

Motion of the projectile in verticle direction is given by Eqn. 13.3 as
$$y = (u \sin \alpha) t - \frac{1}{2} gt^2$$

At the end of flight, $y = 0$

$$\therefore \quad 0 = (u \sin \alpha) t - \frac{1}{2} gt^2$$

$$= t \left(u \sin \alpha - \frac{1}{2} gt \right)$$

$$t = 0$$

or $$t = \frac{2u \sin \alpha}{g}$$

$t = 0$, gives initial position '0' of the projectile.
Hence time of flight is given by

$$t = \frac{2u \sin \alpha}{g} \qquad \qquad ...(13.9)$$

Horizontal Range

During the time of the flight, projectile moves in horizontal direction with uniform velocity $u \cos \alpha$. Hence the horizontal distance traced by the projectile in this time is given by:

$$R = (u \cos \alpha)t$$

$$= u \cos \alpha \frac{2u \sin \alpha}{g}$$

$$R = \frac{u^2 \sin 2\alpha}{g} \qquad \qquad ...(13.10)$$

Maximum Range

In the Eqn (13.10), $\sin 2\alpha$ can have maximum value of 1.

Hence the maximum range $= \dfrac{u^2}{g}$ ...(13.11)

and the angle of projection for this is when
$$\sin 2\alpha = 1$$
or $$2\alpha = 90° \quad i.e. \quad \alpha = 45° \qquad \qquad ...(13.12)$$

Angle of Projection for the Required Range

In Eqn. (13.10), we have the expression for range as

$$R = \frac{u^2 \sin 2\alpha}{g}$$

$$\therefore \quad \sin 2\alpha = \frac{gR}{u^2}$$

Since $\sin (2\alpha) = \sin (180 - 2\alpha)$

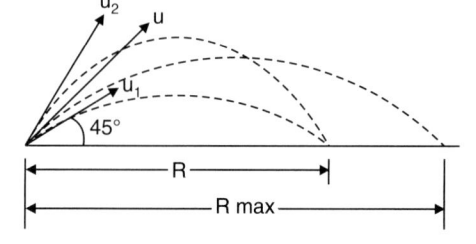

Fig. 13.8

There are two values of α which give the same result.

$$2\alpha_1 = 2\alpha \quad i.e. \quad \alpha_1 = \alpha$$

and another is $2\alpha_2 = 180 - 2\alpha \quad i.e. \quad \alpha_2 = 90° - \alpha$

$$\therefore \quad \alpha_1 + \alpha_2 = 90°$$

Hence if $\alpha_1 = 45° + \theta$
$$\alpha_2 = 45 - \theta$$

Thus there are two angles of projection for the required range as shown in Fig. 13.8.

Example 13.5. *A body is projected at an angle such that its horizontal range is 3 times the maximum height. Find the angle of projection.*

Solution. Let u be velocity of projection and α the angle of projection. Then

$$\text{Maximum height reached} = \frac{u^2 \sin^2 \alpha}{2g}$$

and

$$\text{Range} = \frac{u^2 \sin 2\alpha}{g}$$

∴ In this case

$$\frac{u^2 \sin 2\alpha}{g} = 3 \times \frac{u^2 \sin^2 \alpha}{2g}$$

∴

$$\sin 2\alpha = \frac{3}{2} \sin^2 \alpha$$

i.e.

$$2 \sin \alpha \cos \alpha = \frac{3}{2} \sin^2 \alpha$$

or

$$\tan \alpha = \frac{4}{3}$$

i.e. $\alpha = $ **53.13 degree Ans.**

Example 13.6. *A projectile is aimed at a target on the horizontal plane and falls 12 m short when the angle of projection is 15°, while it overshoots by 24 m when the angle is 45°. Find the, angle of projection to hit the target.*

Solution. Let s be the distance of the target from the point of projection and u be the velocity of projection, as shown in Fig. 13.9.

Range of projection is given by the expression

$$R = \frac{u^2 \sin 2\alpha}{g}$$

Applying it to first case,

$$s - 12 = \frac{u^2}{g} \sin (2 \times 15°)$$

$$= \frac{u^2}{g} \times \frac{1}{2} = \frac{u^2}{2g} \qquad \ldots(1)$$

From the second case

$$s + 24 + = \frac{u^2}{g} \sin (2 \times 45°)$$

$$= \frac{u^2}{g} \qquad \ldots(2)$$

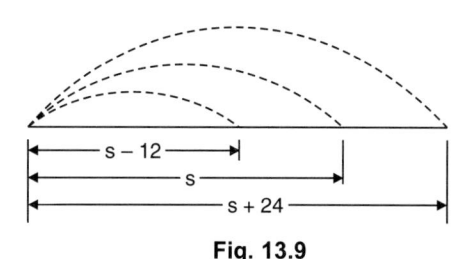

Fig. 13.9

From (1) and (2), we get
$$s + 24 = 2(s - 12)$$
$$\therefore \quad s = 48 \text{ m}$$

Let the correct angle of projection be α. Then
$$48 = \frac{u^2}{g} \sin 2\alpha \qquad \ldots(3)$$

From eqn. (2) $\quad \dfrac{u^2}{g} = s + 24 = 48 + 24$

$$= 72 \text{ m}$$

$\therefore$ From eqn. (3) $\quad 48 = 72 \sin 2\alpha$

$\therefore \quad 2\alpha = 41.81°$

$\quad \alpha = \mathbf{20.905°}$ **Ans.**

Example 13.7. *The horizontal component of the velocity of a projectile is twice its initial vertical component. Find the range on the horizontal plane, if the projectile passes through a point 18 m horizontally and 3 m vertically above the point of projection.*

Solution. Let u be the initial velocity and α its angle of projection.

Vertical component of velocity $= u \sin \alpha$

Horizontal component of velocity $= u \cos \alpha$

In this problem,
$$u \cos \alpha = 2 u \sin \alpha$$

$\therefore \quad \tan \alpha = \dfrac{1}{2}$

$\therefore \quad \alpha = 26.565°$

It is given that when $x = 18$ m, $y = 3$ m. Using the equation of trajectory
$$y = x \tan \alpha - \frac{1}{2} \frac{gx^2}{u^2 \cos^2 \alpha}$$

we get, $\quad 3 = 18 \times \dfrac{1}{2} - \dfrac{1}{2} \times \dfrac{9.81 \times 18^2}{u^2 \cos^2 26.565°}$ [since $\tan \alpha = 1/2$]

i.e. $\quad u^2 = \dfrac{9.81 \times 18^2}{6 \times 2 \times \cos^2 26.565}$

i.e. $\quad u = 18.196$ m/sec

$\therefore$ **Range on the horizontal plane**

$$= \frac{u^2 \sin 2\alpha}{g}$$

$$= \frac{18.196^2 \times \sin(2 \times 26.565°)}{9.81}$$

$$= \mathbf{27.00 \text{ m}} \quad \textbf{Ans.}$$

PROJECTILES

Example 13.8. *Find the least initial velocity with which a projectile is to be projected so that it clears a wall 4 m heigh located at a distance of 5m, and strikes the ground at a distance 4 m beyond the wall as shown in Fig. 13.10. The point of projection is at the same level as the foot of the wall.*

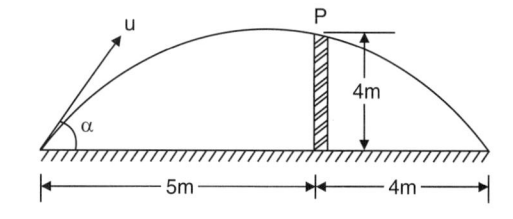

Solution. Let u be the initial velocity required and α the angle of projection.

Fig. 13.10

In this problem, Range = 9 m and at P, the top of wall, $x = 5$ m, $y = 4$ m.

$$\therefore \quad 9 = \frac{u^2 \sin 2\alpha}{g}$$

$$\therefore \quad u^2 = \frac{9g}{\sin 2\alpha} \qquad \ldots(i)$$

From the equation of trajectory,

$$y = x \tan \alpha - \frac{1}{2} \frac{gx^2}{u^2 \cos^2 \alpha}$$

$$4 = 5 \tan \alpha - \frac{1}{2} \frac{g \times 5^2}{\frac{9g}{\sin 2\alpha} \cos^2 \alpha}$$

Substituting $2 \sin \alpha \cos \alpha$ for $\sin 2\alpha$, we get

$$4 = 5 \tan \alpha - \frac{25}{18 \frac{\cos^2 \alpha}{2 \sin \alpha \cos \alpha}}$$

$$= 5 \tan \alpha - \frac{50}{18} \tan \alpha$$

$$= 2.2222 \tan \alpha$$

$\therefore \quad \tan \alpha = 1.8$

$\therefore \quad \alpha = 60.95°$

Form (1), $\quad u^2 = \dfrac{9 \times 9.81}{\sin(2 \times 60.95°)}$

$\therefore \quad \boldsymbol{u = 10.20 \text{ m/sec}}$ **Ans.**

13.4. INCLINED PROJECTION WITH POINT OF PROJECTION AND POINT OF STRIKE AT DIFFERENT LEVELS

Equations (13.8) and (13.9) are to be used only when the point of projection and the point of striking the ground are at the same level. Now let us consider the case, when the point of projection is at a height y_0 above the point of strike as shown in Fig. 13.11.

Any one of the following three methods can be used to analyse such cases :

Method 1

From equation of motion in vertical direction,

$$y = u \sin \alpha \times t - \frac{1}{2} gt^2$$

By putting $y = -y_0$ in the above equation, the time required to reach B (time of flight) is obtained.

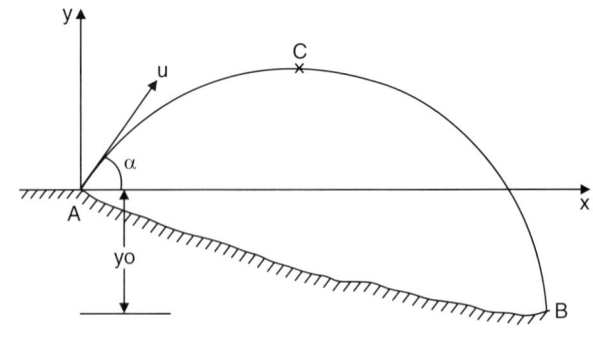

Fig. 13.11

$$-y_0 = u \sin \alpha \times t - \frac{1}{2} gt^2$$

Once the time of flight is known, horizontal range can be found from the relation :

$$R = u \cos \alpha \times t$$

Maximum height above the point of projection and time required to reach it can be found as usual from Eqn. (13.7) and (13.8)

Method 2

Total time of flight can be split into two parts t_1, time required to reach maximum height (point C) and t_2, time required to descend from point C to B.

Considering the vertical motion and noting that vertical component of velocity is zero at C, we get,

$$0 = u \sin \alpha - gt_1$$

$$t_1 = \frac{u \sin \alpha}{g}$$

Height reached in this time interval,

$$h = u \sin \alpha \times t_1 - \frac{1}{2} gt_1^2$$

$$= \frac{u^2 \sin^2 \alpha}{g} - \frac{1}{2} \frac{g u^2 \sin^2 \alpha}{g^2}$$

$$= \frac{u^2 \sin^2 \alpha}{2g}$$

While descending, the vertical distance to be covered $= h + y_0$.

Considering downward motion

$$h + y_0 = 0 \times t_2 + \frac{1}{2} gt_2^2$$

Hence t_2 can be found. Therefore total time of flight $t = t_1 + t_2$.

Horizontal range is given by

$$R = u \cos \alpha \times t$$

PROJECTILES

Method 3

The motion can be split into two parts; AD and DB where D is the point on the trajectory at the same level as A, the point of projection (See Fig. 13.12).

For the portion AD, the Eqn. (13.7) to (13.10) can be used.

For the portion DB :

The vertical component of the velocity of the projectile at point D will be equal to the vertical component of the velocity at A, but downwards, since A and D are at the same elevation. The horizontal component of the velocity remains equal to $u \cos \alpha$ through out.

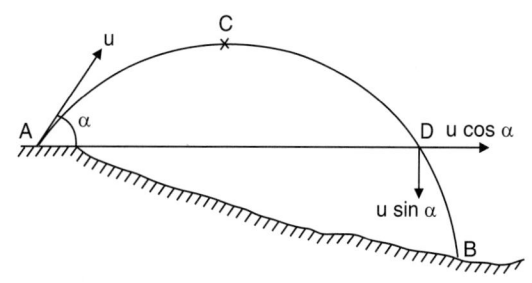

Fig. 13.12

Hence for vertical motion in the portion DB
initial velocity = $u \sin \alpha$, downward
gravitational aceleration = g = 9.81 m/sec² downward.

$$\therefore \quad y_0 = ut_2 \sin \alpha + \frac{1}{2} g t_2^2$$

where, t_2 is the time taken to travel portion DB.

Horizontal distance moved during this time
$$= u \cos \alpha \times t_2$$

Total result may be obtained by combining the motion in the two portions.

Example 13.9. *A bullet is fired from a height of 120 m at a velocity of 360 kmph at an angle of 30° upwards. Neglecting air resistance, find*

(a) *total time of flight,*
(b) *horizontal range of the bullet,*
(c) *maximum height reached by the bullet, and*
(d) *final velocity of the bullet just before touching the ground.*

Solution. Velocity of projection
$$u = 360 \text{ kmph}$$
$$= \frac{360 \times 1000}{60 \times 60} = 100 \text{ m/sec.}$$

(a) Total time of flight

Total time of flight can be found from any one of the three methods. Here all the three methods are illustrated referring to Fig. 13.13.

Method 1
$$y_0 = -120 \text{ m}.$$

Considering vertical motion

$$y = u \sin \alpha \times t - \frac{1}{2} g t^2,$$

$$-120 = 100 \sin 30° \times t - \frac{1}{2} \times 9.81 \, t^2$$

Fig. 13.13

$$t^2 - 10.194\,t - 24.465 = 0$$

or
$$t = \frac{10.194 \pm \sqrt{10.194^2 + 4 \times 1 \times 24.465}}{2}$$

or
$$t = 12.20 \text{ sec} \quad \textbf{Ans.}$$

(The negative value is neglected since it does not give any practical meaning).

Method 2

Let t_1 be the time required for the bullet to reach maximum height and t_2 be the time required by it to touch the ground from this height. When maximum height is reached ;

vertical component of velocity = 0

∴ $$0 = u \sin \alpha - g t_1$$

i.e. $$t_1 = \frac{100 \sin 30°}{9.81} = 5.10 \text{ sec}$$

Maximum height reached in this time is given by

$$h = \frac{u^2 \sin^2 \alpha}{2g} = \frac{100^2 \sin^2 30°}{2 \times 9.81} = 127.42 \text{ m}$$

During downward motion, the distance to be traversed is 127.42 + 120 = 247.42 m. Starting with zero vertical velocity, the particle moves with gravitational acceleration, that is,

$$247.42 = 0 + \frac{1}{2} \times 9.81\, t_2^{\,2}$$

$$t_2 = 7.1 \text{ sec.}$$

∴ $$t = t_1 + t_2 = \textbf{12.20 sec} \quad \textbf{Ans.}$$

Method 3

Time required to travel from A to D be t_1

$$t_1 = \frac{2u \sin \alpha}{g} = \frac{2 \times 100 \sin 30}{9.81}$$

$$t_1 = 10.19 \text{ sec}$$

Let t_2 be the time required to travel from D to B. In the vertical direction, initial velocity (at D) = 100 sin 30° = 50 m/sec

$$g = 9.81 \text{ m/sec}^2$$

distance travelled = 120 m

∴ $$120 = 50\, t_2 + \frac{1}{2} \times 9.81 \times t_2^{\,2}$$

$$t_2^{\,2} + 10.1937\, t_2 - 24.4648 = 0$$

i.e. $$t_2 = 2.01 \text{ sec}$$

∴ **time of flight t = 12.20 sec** **Ans.**

(b) Maximum height reached by the bullet

$$h = \frac{u^2 \sin^2 \alpha}{2g} = \frac{100^2 \sin^2 30°}{2 \times 9.81}$$

= 127.42 m above point A.

i.e. 127.42 + 120 = **247.42 m above the ground.** **Ans.**

PROJECTILES

(c) Horizontal range

$$= u \cos \alpha \times t$$
$$= 100 \cos 30° \times 12.2$$
$$= \mathbf{1056.55 \ m} \quad \textbf{Ans.}$$

(d) **Velocity of the bullet just before striking the ground :**

Vertical component of velocity

$$= u \sin \alpha - gt$$
$$= 100 \sin 30° - 9.81 \times 12.2$$
$$= -69.682 \ m/sec$$
$$= 69.682 \ m/sec \ \text{downward.}$$

Horizontal component of velocity,

$$= 100 \cos \alpha = 86.603 \ m/sec$$

Referring to Fig. 13.14, the velocity at strike

$$v = \sqrt{69.682^2 + 86.603^2}$$

$$v = \mathbf{111.16 \ m/sec} \quad \textbf{Ans.}$$

$$\theta = \tan^{-1} \frac{69.682}{86.603}$$

$$= \mathbf{38.82°} \ \text{as shown in Fig. 13.14} \quad \textbf{Ans.}$$

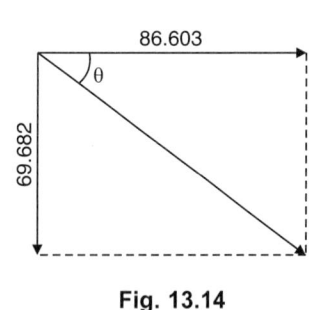

Fig. 13.14

Example 13.10. *A cricket ball thrown by a fielder from a height of 2m, at an angle of 30° to the horizontal, with an initial velocity of 20 m/sec, hits the wickets at a height of 0.5 m from the ground. How far was the fielder from the wickets ?*

Solution. Initial velocity $u = 20$ m/sec.

Angle of projection $\alpha = 30°$

$$y_0 = -(2.0 - 0.5) = -1.5 \ m$$

Time of flight t is given by the expression

$$-1.5 = 20 \sin 30° \times t - \frac{1}{2} \times 9.81 \ t^2$$

$$t^2 - 2.0387 \ t - 0.3058 = 0$$

i.e. $$t = 2.179 \ sec.$$

∴ **The distance of the fielder from the wickets**

$$= \text{Range} = u \cos \alpha \times t$$
$$= 20 \cos 30° \times 2.179$$
$$= \mathbf{37.742 \ m} \quad \textbf{Ans.}$$

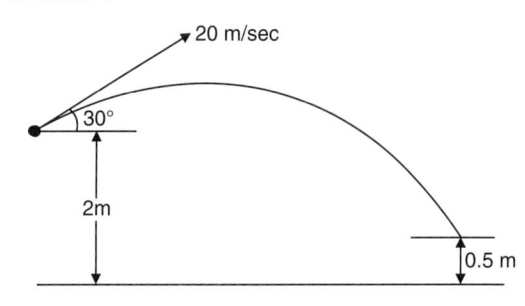

Fig. 13.15

Example 13.11. *Gravel is thrown into a bin from the top of a conveyor (Ref. Fig. 13.16) with a velocity of 5 m/sec. Determine :*

(a) time it takes the gravel to hit the bottom of the bin ;

(b) the horizontal distance from the end of the conveyor to the bin where the gravel strikes the bin, and

(c) the velocity at which the gravel strikes the bin.

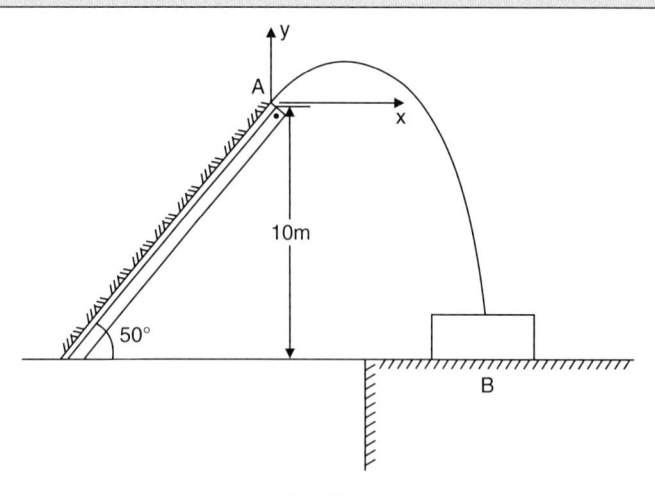

Fig. 13.16

Solution. Taking A as origin of the coordinate system as shown in the figure, vertical motion is represented by :

$$y = u \sin \alpha \times t - \frac{1}{2}gt^2$$

where, $\qquad u = 5$ m/sec
and $\qquad \alpha = 50°$

For the point B; $\quad y = -10$ m

∴ $\qquad -10 = 5 \sin 50° \times t - \frac{1}{2} \, 9.81 \, t^2$

i.e. $\qquad t = 1.871$ sec   **Ans.**

Horizontal distance travelled in this time

$\qquad = u \cos \alpha \times t = 5 \cos 50° \times 1.871$
$\qquad = \mathbf{6.012\ m}$ **Ans.**

Vertical component of velocity of gravel at the time of striking the bin is given by :

$\qquad = u \sin \alpha - gt$
$\qquad = 5 \sin 50° - 9.81 \times 1.871$
$\qquad = -14.524$
$\qquad = 14.524$ m/sec (downward).

Horizontal component of velocity
$\qquad = u \cos \alpha$
$\qquad = 5 \cos 50° = 3.214$ m/sec

Velocity of strike
$\qquad v = \sqrt{14.524^2 + 3.214^2}$
$\qquad = \mathbf{14.875\ m/sec}$ **Ans.**

$\qquad \theta = \tan^{-1} \dfrac{14.524}{3.214}$
$\qquad = \mathbf{77.52°\ to\ horizontal}$ as shown in Fig. 13.17   **Ans.**

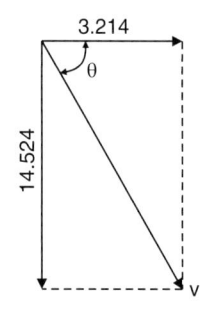

Fig. 13.17

PROJECTILES

Example 13.12. *A soldier fires a bullet with a velocity of 31.32 m/sec at an angle α upwards from the horizontal from his position on a hill to strike a target which is 100 m away and 50 m below his position. Find the angle of projection α. Find also the velocity with which the bullet strikes the object.*

Solution. The equation of the trajectory of bullet is (from Eqn. 13.6)

$$y = x \tan \alpha - \frac{1}{2} \frac{gx^2}{u^2} (\tan^2 \alpha + 1)$$

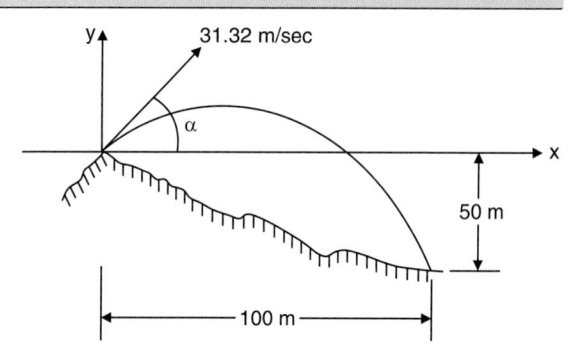

Fig. 13.18

Now, for the point on the ground where bullet strikes,
$$y = -50 \text{ m}; \quad x = 100 \text{ m}$$
and $$u = 31.32 \text{ m/sec}$$

$\therefore \quad -50 = 100 \tan \alpha - \frac{1}{2} \times \frac{9.81 \times 100^2}{31.32^2} (\tan^2 \alpha + 1)$

$= 100 \tan \alpha - 50 (\tan^2 \alpha + 1)$

i.e. $\tan^2 \alpha - 2 \tan \alpha = 0$

$\tan \alpha (\tan \alpha - 2) = 0$

$\therefore \quad \alpha = 0 \quad \text{or} \quad \alpha = \tan^{-1} 2 = \mathbf{63.435°} \quad \mathbf{Ans.}$

When $\alpha = 0$

Horizontal component of velocity
$$v_x = 31.32 \text{ m/sec.}$$

Vertical component of velocity
$$v_y^2 - 0 = 2gh$$
$$v_y = \sqrt{2 \times 9.81 \times 50} = 31.32 \text{ m/sec}$$

∴ **Velocity of strike**
$$v = \sqrt{v_x^2 + v_y^2} = \sqrt{31.32^2 + 31.32^2}$$
$$= \mathbf{44.294 \text{ m/sec}} \quad \mathbf{Ans.}$$

$$\tan \theta = \frac{31.32}{31.32}$$

i.e. $\theta = \mathbf{45°} \quad \mathbf{Ans.}$

When $\alpha = 63.435°$

Horizontal component of velocity
$$v_x = 31.32 \times \cos 63.435°$$
$$= 14.007 \text{ m/sec}$$

Vertical component of velocity of strike v_y will be downward the vertical component of velocity was $31.32 \sin 63.435° = 28.013$ m/sec upward

$$v_y^2 - (-28.013)^2 = 2 \times 9.81 \times 50$$

$$v_y = 42.02 \text{ m/sec}$$

$$v = \sqrt{v_x^2 + v_y^2} = \sqrt{14.007^2 + 42.02^2}$$

$$= \mathbf{44.294 \text{ m/sec}} \quad \mathbf{Ans.}$$

$$\theta = \tan^{-1}\left(\frac{v_y}{v_x}\right) = \frac{42.02}{14.007}$$

$$= \mathbf{71.565° \text{ to horizontal}} \quad \mathbf{Ans.}$$

Example 13.13. *A ball rebounds at A and strikes the inclined plane at point B at a distance s = 76 m as shown in Fig. 13.19. If the ball rises to a maximum height h = 19 m above the point of projection, compute the initial velocity and the angle of projection α.*

Solution. At A the vertical component of velocity = $u \sin \alpha$

When maximum height $h = 19$ m is reached, vertical component of velocity is zero.

∴ $0 - (u \sin \alpha)^2 = 2(-g) 19$

Substituting $g = 9.81$ m/sec^2

$u \sin \alpha = 19.308$ ...(1)

y-co-ordinate of B

$= -AB \times \sin \theta$

$= -76 \times \dfrac{1}{\sqrt{3^2 + 1^2}} = -24.033$ m

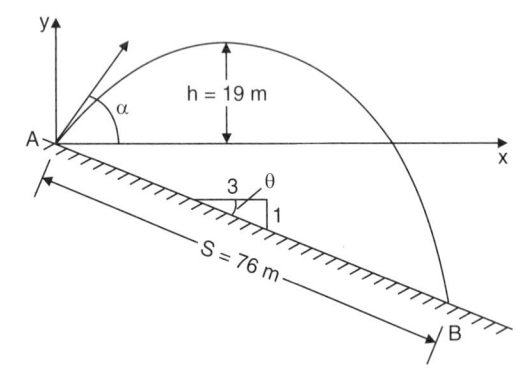

Fig. 13.19

Considering the motion in vertical upward direction,

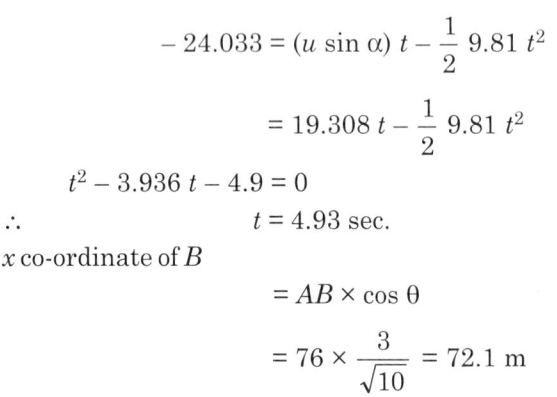

$$-24.033 = (u \sin \alpha) t - \frac{1}{2} 9.81 \, t^2$$

$$= 19.308 \, t - \frac{1}{2} 9.81 \, t^2$$

$$t^2 - 3.936 \, t - 4.9 = 0$$

∴ $t = 4.93$ sec.

x co-ordinate of B

$$= AB \times \cos \theta$$

$$= 76 \times \frac{3}{\sqrt{10}} = 72.1 \text{ m}$$

Considering the horizontal motion of the ball

$u \cos \alpha \times t = 72.1$

$u \cos \alpha \times 4.93 = 72.1$

or $u \cos \alpha = 14.625$ ...(2)

From eqn. (1) and (2)

$$\tan \alpha = \frac{19.308}{14.625} = 1.32$$

∴ α = **52.86°** Ans.

From Eqn. (2),

$$u = \frac{14.625}{\cos 52.86°}$$

i.e. **u = 24.222 m/sec** Ans.

Note : Though this example is about projection on inclined plane, to be discussed in next article, it could be solved now only as *x*, and *y*-coordinates of the point of strike *B* are known.

Example 13.14. *A bomber is flying horizontally at an altitude of 2400 m with the uniform velocity of 1000 kmph to bomb a target. Where should the bomb be released to strike the target?*

At the same time, the target is being defended by an anti-aircraft gun at 60° to the horizontal. The muzzle velocity of the gun is 600 m/sec. When should the shell be fired to hit the bomber ?

Solution. Altitude of the bomber $h = 2400$ m.

Let the time required for bomb to reach ground be t seconds. Then, since initial vertical velocity = 0,

$$2400 = 0 \times t + \frac{1}{2} \times 9.81 \, t^2$$

$$t = 22.12 \text{ sec}$$

Now, $u = 1000 \text{ kmph} = \dfrac{1000 \times 1000}{60 \times 60}$

$$= 277.78 \text{ m/sec.}$$

Horizontal distance moved by bomb

$$= u \times t$$
$$= 277.78 \times 22.12$$
$$= 6144.493 \text{ m.}$$

Bomb should be released when the bomber is 6144.493 m away from the target.
Ans.

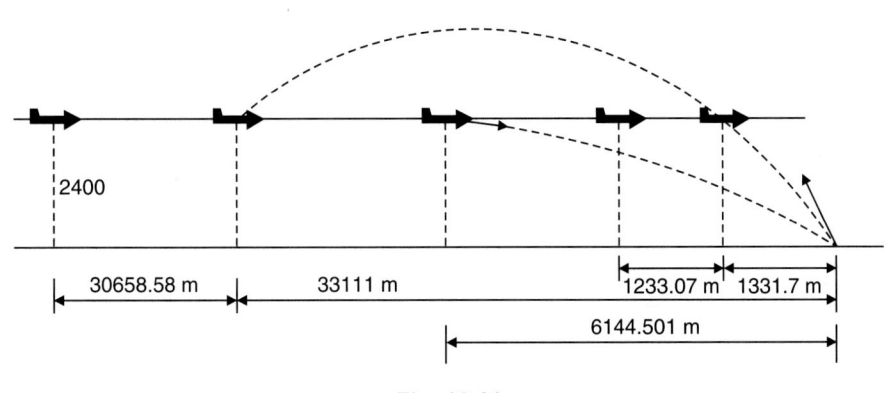

Fig. 13.20

Muzzle velocity of the gun = 600 m/sec

i.e. the velocity of projection u = 600 m/sec

Angle of projection α = 60°

Shell has to hit the bomber at a height,
$$h = 2400 \text{ m}$$
Let the time required for the shell to rise to this height be t_1. Then
$$h = u \sin \alpha \times t_1 - \frac{1}{2} g t_1^2$$
$$2400 = 600 \sin 60° \times t_1 - \frac{1}{2} \times 9.81 \, t_1^2$$
$$t_1^2 - 105.936 \, t_1 + 489.30 = 0$$
i.e. $\quad\quad\quad\quad t_1 = 110.370 \text{ sec} \quad \text{or} \quad 4.433 \text{ sec}$

110.370 sec. is the time required for the shell to strike the bomber while coming down the trajectory and 4.439 sec. is the time required while moving up the trajectory.

When $t_1 = 110.370$ sec

The horizontal distance moved by the shell
$$= u \cos \alpha \times t_1$$
$$= 600 \times \cos 60° \times 110.370$$
$$= 33111 \text{ m}$$

The distance moved by the plane during this period $= ut_1$
$$= 277.78 \times 110.370$$
$$= 30658.58 \text{ m}$$

∴ The gun must fire the shell, when bomber is at a distance
$$= 33111 + 30658.58$$
$$= 63769.58 \text{ m}$$

when $t_1 = 4.839$ sec

The horizontal distance moved by the shell
$$= 600 \times \cos 60° \times 4.439$$
$$= 1331.70 \text{ m}$$

The distance moved by the plane during this period
$$= 277.78 \times 4.439$$
$$= 1233.07 \text{ m}$$

∴ The gun must fire the shell when bomber is at a distance
$$= 1331.70 + 1233.07$$
$$= 2564.77 \text{ m}$$

If the shell is fired when the bomber is at a distance of 63769.58 m, it will hit the plane in its downward motion. If the shell is fired, when the bomber is at a distance of 2795.87 m, the shell will hit the bomber during its upward motion.

13.5. PROJECTION ON INCLINED PLANE

Let AB be a plane inclined at an angle β to the horizontal as shown in the Fig. 13.21. A projectile is fired up the plane from point A with initial velocity u m/sec and an angle α. Now, the range on the inclined plane AB and the time of flight are to be determined.

Let the inclined range AB be denoted by R. AD be the corresponding horizontal range.
∴ $\quad\quad\quad\quad AD = R \cos \beta \quad \text{and} \quad DB = R \sin \beta.$

PROJECTILES

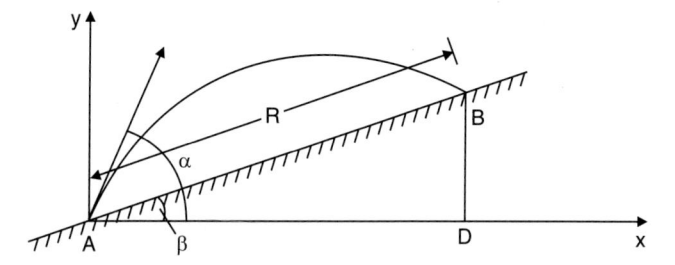

Fig. 13.21

The equation of trajectory of the projectile is given by

$$y = x \tan \alpha - \frac{1}{2} \frac{gx^2}{u^2 \cos^2 \alpha}$$

Applying this equation to point B, we get

$$R \sin \beta = R \cos \beta \tan \alpha - \frac{1}{2} \frac{g R^2 \cos^2 \beta}{u^2 \cos^2 \alpha}$$

i.e.
$$R \left(\frac{1}{2} \frac{g \cos^2 \beta}{u^2 \cos^2 \alpha} \right) = \cos \beta \tan \alpha - \sin \beta$$

$\therefore$
$$R = \frac{2 u^2 \cos^2 \alpha}{g \cos^2 \beta} (\cos \beta \tan \alpha - \sin \beta)$$

$$= \frac{2u^2 \cos \alpha}{g \cos^2 \beta} (\cos \beta \cdot \sin \alpha - \sin \beta \cdot \cos \alpha)$$

$$= \frac{2u^2 \cos \alpha}{g \cos^2 \beta} \sin (\alpha - \beta) \qquad \ldots(13.13(a))$$

i.e.
$$R = \frac{u^2}{g \cos^2 \beta} [\sin (2\alpha - \beta) - \sin \beta] \qquad \ldots(13.13\,(b))$$

[since $2 \cos A \sin B = \sin (A + B) - \sin (A - B)$]

Time of flight :

Let t be the time of flight. The horizontal distance covered during the flight
$$= AD = u \cos \alpha \times t$$

$\therefore$
$$t = \frac{AD}{u \cos \alpha} = \frac{R \cos \beta}{u \cos \alpha}$$

$$= \frac{2u^2 \cos \alpha}{g \cos^2 \beta} \times \frac{\sin (\alpha - \beta)}{u \cos \alpha} \cos \beta$$

$$= \frac{2u \sin (\alpha - \beta)}{g \cos \beta} \qquad \ldots(13.14)$$

For the given values of u and β, the range is maximum when :
$$\sin(2\alpha - \beta) = 1$$

i.e.
$$2\alpha - \beta = \frac{\pi}{2}$$

or
$$\alpha = \frac{\pi}{4} + \frac{\beta}{2} \qquad \ldots(13.15)$$

Referring to Fig. 13.22
$$\theta_1 = \frac{\pi}{4} + \frac{\beta}{2} - \beta$$
$$= \frac{\pi}{4} - \frac{\beta}{2}$$

and
$$\theta_2 = \frac{\pi}{2} - \beta$$

Thus $\theta_2 = 2\theta_1$

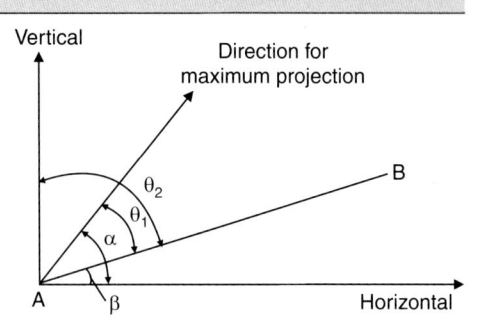

Fig. 13.22

i.e. *the range on the given plane is maximum, when the angle of projection bisects the angle between the vertical and inclined planes.*

If the projection is down the plane, the Fig. 13.13 to 13.15 can be still used, but the value of β should be taken negative.

Example 13.15. *A plane has a slope of 5 in 12. A shot is projected with a velocity of 200 m/sec at an upward angle of 30° to horizontal. Find the range on the plane if :*

(a) the shot is fired up the plane ;

(b) the shot is fired down the plane.

Solution. Initial velocity $u = 200$ m/sec.

Angle of projection $\alpha = 30°$

Inclination of the plane $= \tan^{-1}\left(\dfrac{5}{12}\right) = 22.62°$

(a) When the shot is fired up the plane :
$$\beta = +22.62°$$

$$\text{Range} = \frac{u^2}{g \cos^2 \beta}\,[\sin(2\alpha - \beta) - \sin \beta]$$

$$= \frac{200^2}{9.81 \times \cos^2 22.62}\,[\sin(2 \times 30 - 22.62) - \sin 22.62°]$$

i.e. **Range = 1064.65 m**   **Ans.**

(b) When the shot is fired down the plane :
$$\beta = -22.62°$$

PROJECTILES

$$\text{Range} = \frac{200 \times 200}{9.81 \cos^2(-22.62)} [\sin(2 \times 30 + 22.62) - \sin(-22.62)]$$

$$= \frac{200 \times 200}{9.81 \cos^2 22.62} [\sin 82.62 + \sin 22.62]$$

Range = 6586.27 m   Ans.

Example 13.16. *A p⸱ ⸱n can throw a ball at a maximum velocity of 30 m/sec. If he wants to get maximum range on the plane inclined at 20° to horizontal, at what angle should the ball be projected and what would be the maximum range (a) up the plane (b) down the plane?*

Solution. (a) Up the plane :

For getting maximum range, the direction of projection must bisect the angle between the plane and the vertical. Referring to Fig. 13.23.

$$\theta = \frac{90 - 20}{2} = 35°$$

$$\therefore \quad \alpha = \theta + 20° = 55° \quad \textbf{Ans.}$$

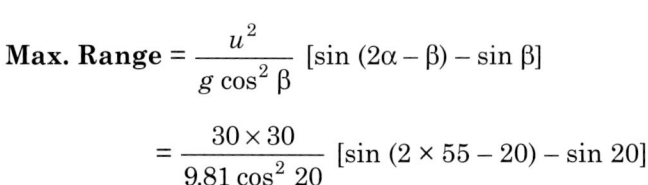

Max. Range $= \dfrac{u^2}{g \cos^2 \beta} [\sin(2\alpha - \beta) - \sin \beta]$

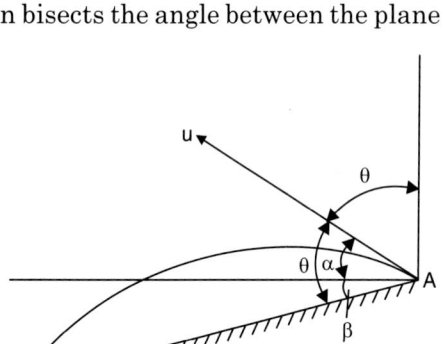

Fig. 13.23

$$= \frac{30 \times 30}{9.81 \cos^2 20} [\sin(2 \times 55 - 20) - \sin 20]$$

$$= \textbf{68.362 m.}$$

(b) Down the plane :

For maximum projection, direction of the projection bisects the angle between the plane and vertical.

Referring to Fig. 13.24

$$\theta = \frac{1}{2}(90 + 20) = 55°$$

$$\therefore \quad \alpha = \theta - \beta = 55 - 20$$

$$= 35°$$

$\therefore$ **Maximum Range down the plane**

$$= \frac{30 \times 30}{9.81 \cos^2(-20)}$$

$$[\sin(2 \times 35 + 20) - \sin(-20)]$$

$$= \textbf{139.432 m   Ans.}$$

Fig. 13.24

Important Definitions

1. Velocity with which the projectile is projected is called as **velocity of projectile.**
2. The angle between the direction of projection and the horizontal direction is called as **angle of projection.**

3. The path traced by the projectile is called as its **trajectory.**
4. The horizontal distance through which the projectile travels in its flight is called as **horizontal range** of projectile.
5. The time interval during which the projectile is in motion is called the **time of flight**.

Important Formulae

1. Motion of body projected horizontally:
$$h = \frac{1}{2} gt^2$$
$$\text{Range} = ut$$

2. Inclined projection on level ground
   (a) Equation of trajectory:
   $$y = x \tan \alpha - \frac{1}{2} \frac{x^2}{u^2} (1 + \tan^2 \alpha)$$
   (b) Maximum height reached
   $$h = \frac{u^2 \sin^2 \alpha}{2g}$$
   (c) Time required to reach maximum height
   $$t = \frac{u \sin \alpha}{g}$$
   (d) Time of flight
   $$t = \frac{2u \sin \alpha}{g}$$
   (e) Horizontal Range:
   $$R = \frac{u^2 \sin 2\alpha}{g}$$
   (f) For maximum range $\alpha = 45°$

3. Projection on inclined plane:
   (a) Range
   $$R = \frac{2u^2 \cos \alpha}{g \cos^2 \beta} \times \sin (\alpha - \beta) = \frac{u^2}{g \cos^2 \beta} [\sin (2\alpha - \beta) - \sin \beta]$$
   (b) Time of flight
   $$t = \frac{2u \sin (\alpha - \beta)}{g \cos \beta}$$
   (c) For maximum range, $\alpha = \frac{\pi}{4} + \frac{\beta}{2}$.

PROBLEMS FOR EXERCISE

13.1 Two particles are dropped simultaneously from a height 100 m above the ground. One of them, in its mid path hits a fixed plane, inclined at an angle as shown in Fig. 13.25. As a result of this impact, the direction of its velocity becomes horizontal. Compute the time taken by the two particles to reach the ground.

[**Ans.** 4.515 sec ; 6.385 sec]

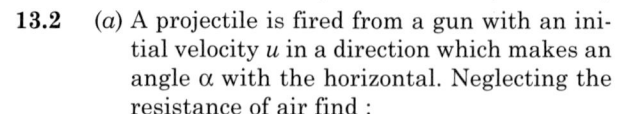

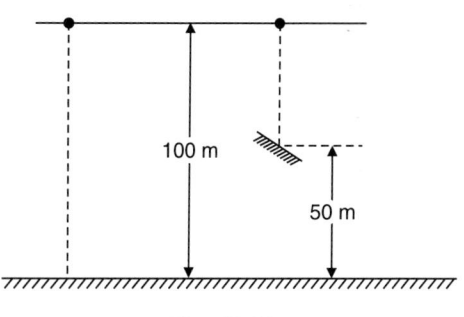

Fig. 13.25

13.2 (a) A projectile is fired from a gun with an initial velocity u in a direction which makes an angle α with the horizontal. Neglecting the resistance of air find :

(1) the time of flight of the projectile to reach the level from which it started.

(2) its range on the horizontal plane, through the point of projection.

(3) the greatest height reached by it.

(4) the equation of its path.

(b) A projectile is fired at a velocity of 800 m/sec at an angle of 40° measured from the horizontal. Determine the time of flight, range, and the greatest height reached.

[**Ans.** $t = 104.838$ sec ; $R = 64248.42$ m ; and $h = 13477.71$ m]

13.3 A stone is projected upwards from the ground with velocity of 16 m/sec at an angle of 60° to the horizontal. With what velocity must another stone be projected at an angle of 45° to the horizontal from the same point in order :

(1) to have the same horizontal range ?

(2) to attain the same maximum height ? [**Ans.** (1) 14.89 m/sec ; (2) 19.6 m/sec]

13.4 A fire brigade man wants to extinguish a fire at a height of 6 m above the nozzle standing at a distance 5 m away from the fire. Find (1) the minimum velocity of the nozzle discharge required ; (2) velocity of discharge if he could extinguish with angle of projection of 60°.

[**Ans.** (1) $u = 11.754$ m/sec ; (2) $u = 13.58$ m/sec]

13.5 Maximum range of a field gun is 2000 m. If a target at a distance of 1200 m is to be hit, what should be the angle of projection ? [**Ans.** $\theta = 18.435°$ or $71.565°$]

13.6 A projectile is fired with a velocity of 300 m/sec at an upward angle of 60° to horizontal. Neglecting air resistance determine : (1) time of flight (2) range (3) its position and (4) its velocity after 35 seconds.

[**Ans.** (1) $t = 52.97$ sec ; (2) $R = 7945.19$ m ; (3) $x = 5250$ m ; $y = 3084.64$ m ;

(4) 171.69 m/sec at $\theta = 29.116°$ to horizontal]

13.7 A rocket is projected vertically upward until it is 50 km above the launching site. At this instant it is turned so that its velocity is directed 3 upward and 4 horizontal and the power is shut off. The velocity, when the power is shut off is 1680 m/sec. The rocket strikes the ground at the same elevation as the launching site. Determine : (1) the horizontal distance travelled by the rocket ; and (2) the velocity with which the rocket strikes the ground.

[**Ans.** (1) 3,31708.4 m ; (2) 1950.23 m/sec ; $\theta = 46.434°$ to horizontal]

13.8 A projectile is fired from a point 'O' with the same velocity as it would be due to a fall of a 100 m from rest. The projectile hits a mark at a depth of 50 m below 'O' and at a horizontal distance of 100 m from the vertical line through 'O'. Determine the two possible directions of projections and show that they are at right angles to each other.

[**Ans.** $\alpha_1 = 76.72°$ above horizontal ; and $\alpha_2 = 13.28°$ below horizontal]

13.9 A stunt man wants to drive his motorbyke across a gap as shown in Fig. 13.26. What should be the minimum velocity of take-off. For smooth landing at this velocity what should be the inclination of the landing ramp ? [**Ans.** $u = 6.56$ m/sec ; $\theta = 51.207°$]

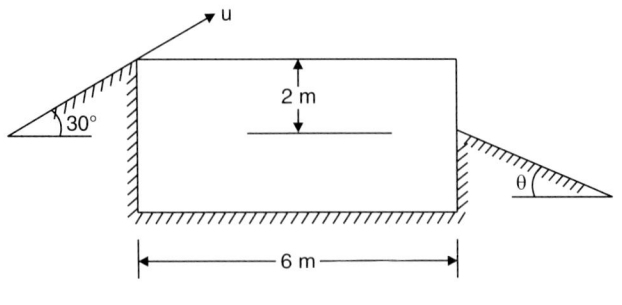

Fig. 13.26

13.10 A gun is fired from the top of a hill 100 m above the sea level. Ten seconds later the shell is found to hit the warship 1000 m away from the position of the gun. Determine the velocity and angle of projection of the shell. With what velocity shell strikes the ship ?
[**Ans.** $u = 107.354$ m/sec ; $\alpha = 21.331°$; $v = 139.868$ m/sec and $\theta = 45°$]

13.11 A missile projected from a rocket travelled 20 km vertically and 10 km horizontally from the launching point, when fuel of the rocket is exhausted. At this stage missile has acquired a velocity of 1600 m/sec at an angle of 35° above the horizontal. Assuming that the rest of the flight is under the influence of the gravity and neglecting air resistance and curvature of the earth, calculate : (1) the total horizontal range from the launching point ; and (2) the time of flight after the fuel has completely burnt. [**Ans.** $R = 271.061$ km ; $t = 206.81$ sec]

13.12 A particle is projected down a plane with an initial velocity of u m/sec. Show that its maximum range is equal to $\dfrac{u^2}{g} \sec\beta\,(\sec\beta + \tan\beta)$ where, β is the inclination of the plane to the horizontal.

13.13 A particle is projected from a point on an inclined plane with a velocity of 30 m/sec. The angle of projection and the angle of the plane are 55° and 20° to the horizontal respectively. Show that the range up the plane is the maximum for the given plane. Find the range and time of flight of the particle. [**Ans.** $R = 68.36$ m ; $t = 3.73$ sec]

14
Relative Velocity

In the previous two chapters, we dealt with the motion of bodies with respect to a stationary point on the earth. In this chapter we shall consider the motion of a body as observed by a person who is also having some motion. When two trains move in parallel lines in the same direction and with the same speed, an observer in one of the train does not feel the motion of the other train. If one train moves a little faster, the observer feels a slight motion. If the two trains are moving in opposite direction, the observer feels that speed is very high. These are some of the common examples of relative motion. The motion of two bodies need not be always in parallel paths. They may be in any direction and in any plane. However, in this book, we shall consider motion in only one plane.

14.1. MOTION ON PARALLEL PATHS IN LIKE DIRECTIONS

Let A be a body moving with a velocity of 60 kmph and B be another body moving on a parallel path, in the same direction as A with the same velocity of 60 kmph. Since both bodies are having the same velocity, the two bodies will be moving side by side and the observer at B feels that A is not at all moving. In other words the relative velocity of A with respect to B is zero. If A is moving with a velocity of 70 kmph and B with 60 kmph, B feels that A is moving with a velocity of 10 kmph only. Thus, the relative velocity of A with respect to B is the vector difference between the velocities of A and B.

Fig. 14.1

14.2. MOTION ON PARALLEL PATH IN OPPOSITE DIRECTION

Let A move with a velocity of 70 kmph in one direction and B with 60 kmph in opposite direction, on a parallel paths, as shown in Fig. 14.2.

In this case relative velocity observed will be 130 kmph *i.e.* vector difference in the two velocities [70 – (– 60) = 130].

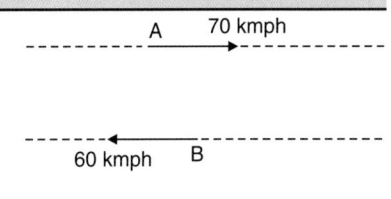

Fig. 14.2

14.3. MOTION IN A PLANE IN ANY DIRECTION

Let A and B be two bodies moving in a plane but in different directions as shown in Fig. 14.3. A has the velocity v_A m/sec and B has the velocity v_B in the directions shown in the Fig. 14.3(a). In this case also relative velocity of A with respect to B ($v_{A/B}$) will be vector difference as shown in Fig. 14.3(b).

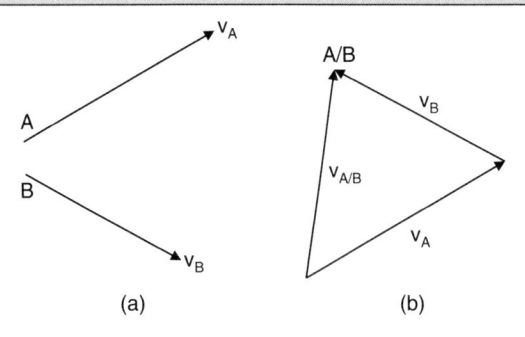

Fig. 14.3

Thus, $$v_{A/B} = v_A - v_B$$
where, $v_{A/B}$ – relative velocity of A w.r.t. B
   v_A – velocity of A
   v_B – velocity of B.

The relative velocity $v_{A/B}$ may be determined graphically as shown in Fig. 14.3(b) or by the analytical method as follows:

(1) find the components of each velocity in any two mutually perpendicular directions ($v_{Ax}, v_{Ay}, v_{Bx}, v_{By}$) (Fig. 14.4).
(2) find the difference in the component velocities in each direction to get the components of relative velocities (v_{rx}, v_{ry});
(3) Combine the components to get the relative velocity.

$$v_{rx} = v_{Ax} - v_{Bx}$$
$$v_{ry} = v_{Ay} - v_{By}$$
$$v_{A/B} = v_r = \sqrt{v_{rx}^2 + v_{ry}^2}$$

$$\tan \alpha = \frac{v_{ry}}{v_{rx}}$$

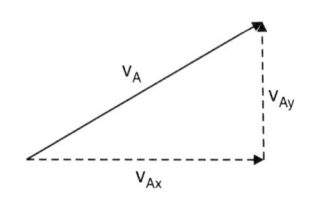

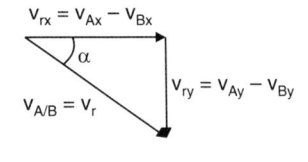

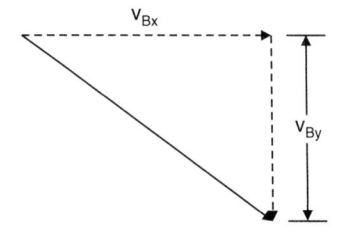

Fig. 14.4

RELATIVE VELOCITY

14.4. RELATIVE DISTANCE

Let A and B be two objects moving with velocities v_A and v_B in a plane. Originally let the object A be at point M and the object B be at N as shown in Fig. 14.5. Now we want to find the relative distance between A and B after a time interval t.

The following procedure may be adopted:

Assume one of the objects stationary, say B is stationary at N.

Move A with relative velocity $v_{A/B} = v_r$. It travels a distance of $v_r \times t$, in time interval t in the direction of relative velocity. This position is shown by the point P in Fig. 14.5. NP represents the relative distance between the objects A and B. Actually B itself has moved by a distance $v_B t$ in the direction of v_B.

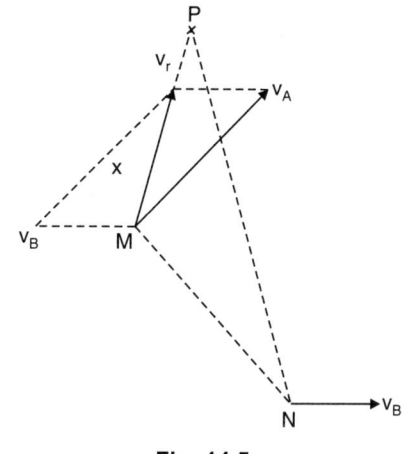

Fig. 14.5

14.5. RELATIVE VELOCITY AND RESULTANT VELOCITY

If a body, moving with a velocity v_1 is acted upon by a force, the effect of which is to cause a velocity v_2 to the body, then the body moves with the resultant velocity v. For example, a motor launch may be started with a velocity v_1 across the river but the river gives it a velocity v_2 in the down stream direction. The combined effect of these two velocities is the resultant velocity as shown in Fig. 14.6. Vector addition will give the resultant velocity. Any one of the methods used in statics (Chapter II) for finding the resultant force can be used here also.

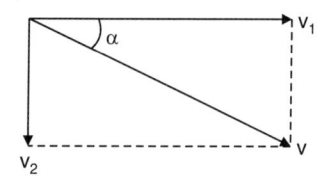

Fig. 14.6

The following differences in relative velocity and resultant velocity may be noted:

(1) Relative velocity is the velocity of an object as observed from another moving object, whereas resultant velocity is the combined effect of two or more forces causing motion of a single body.

(2) Relative velocity of a body with respect to another moving body is obtained as vector difference of the velocities of two bodies, whereas resultant velocity is obtained as the vector addition of the velocities caused by different forces acting on a body.

Example 14.1. *A passenger train 250 m long, moving with a velocity of 72 kmph, overtakes a goods train, moving on a parallel path in the same direction, completely in 45 seconds. If the length of the goods train is 200 m, determine the speed of the goods train.*

Solution. Let v_B = velocity of goods train.

Velocity of passenger train

$$v_A = 72 \text{ kmph}$$

$$= 72 \times \frac{1000}{60 \times 60} = 20 \text{ m/sec}$$

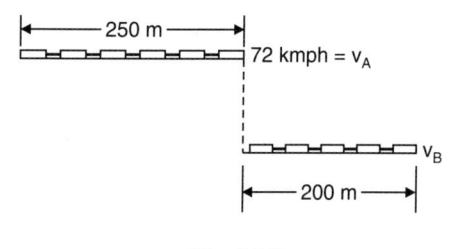

Fig. 14.7

Relative velocity of B with respect to A

$$v_{A/B} = v_r = v_A - v_B = 20 - v_B$$

Relative distance moved to overtake the goods train = 250 + 200 = 450 m

Now, 45 seconds are required to cover this relative distance

$$(20 - v_B)45 = 450$$
$$v_B = 10 \text{ m/sec}$$

i.e. $$v_B = \frac{10 \times 60 \times 60}{1000} = \textbf{36 kmph} \quad \textbf{Ans.}$$

Example 14.2. *A passenger train is 240 m long and is moving with a constant velocity of 72 kmph. At a particular time, its engine approaches last compartment of a goods train moving on a parallel track in the same direction. 25 seconds later its engine starts overtaking the engine of goods train. It took 30 seconds more to completely overtake the goods train. Determine the length and speed of the goods train.*

Solution. Velocity of passenger train

$$v_A = 72 \text{ kmph}$$
$$= 20 \text{ m/sec}$$

Let velocity of goods train be v_B m/sec and its length x metres.

Relative velocity = $20 - v_B$ m/sec

When t = 25 sec, relative distance moved is x metres

∴ $\quad (20 - v_B)25 = x$ ...(1)

In the next t = 30 seconds, relative distance moved = length of passenger train
$$= 240 \text{ m.}$$

∴ $\quad (20 - v_B) 30 = 240$ ...(2)

∴ $\quad v_B = 12 \text{ m/sec}$

$$= \frac{12 \times 60 \times 60}{1000} = \textbf{43.2 km/h} \quad \textbf{Ans.}$$

Substituting the value of v_B in (i)

$$(20 - 12) 25 = x$$

∴ $\quad x = \textbf{200 m} \quad \textbf{Ans.}$

Example 14.3. *Two trains A and B are moving on parallel tracks in opposite direction. Velocity of A is twice that of B. They take 18 seconds to pass each other. Determine their velocities, given, the length of train A is 240 m and that of B is 300 m.*

Solution. Taking the direction of motion of train A as positive,

let velocity of A be v m/sec

∴ Velocity of $B = -v/2$ m/sec

$$\text{Relative velocity} = v - \left(-\frac{v}{2}\right) = 1.5\, v$$

Relative distance moved in 18 seconds
$$= 240 + 300 = 540 \text{ m}$$

∴ $\quad 1.5\, v \times 18 = 540$

$$v = 20 \text{ m/sec} = \textbf{72 kmph} \quad \textbf{Ans.}$$

∴ **Velocity of B** $= -10$ m/sec $= -36$ kmph,

i.e. **36 kmph** **Ans.**

RELATIVE VELOCITY   351

Example 14.4. *Two ships move from a port at the same time. Ship A has velocity of 30 kmph and is moving in N 30° W while ship B is moving in south-west direction with a velocity of 40 kmph. Determine the relative velocity of A with respect to B and the distance between them after half an hour.*

Solution. Taking west direction as x-axis and north direction as y-axis,

$$v_{Ax} = 30 \sin 30° = 15 \text{ kmph}$$
$$v_{Ay} = 30 \cos 30° = 25.98 \text{ kmph}$$
$$v_{Bx} = 40 \sin 45° = 28.284 \text{ kmph}$$
$$v_{By} = -40 \sin 45° = -28.284 \text{ kmph}$$

∴ $v_{rx} = 15 - 28.284 = -13.284 \text{ kmph}$
   $v_{ry} = 25.98 - (-28.284) = 54.264 \text{ kmph}$

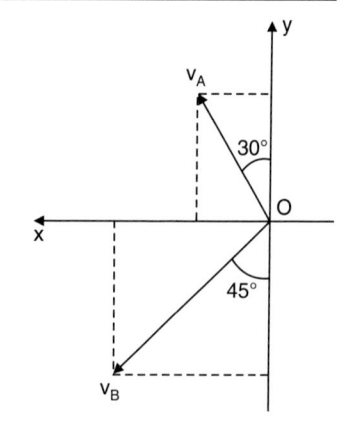

Fig. 14.8

∴ $v_r = \sqrt{(13.284)^2 + (54.264)^2}$
        $= 55.866 \text{ kmph}$

$$\tan \theta = \frac{13.284}{54.264}$$
$$\theta = 13.76°$$

i.e., from B, ship A appears to move with a velocity of **55.866 kmph in N 13.76°E direction. Ans.**

∴ **Relative distance after half an hour**
    = Relative velocity × time
    = 55.866 × 1/2 = **27.933 km Ans.**

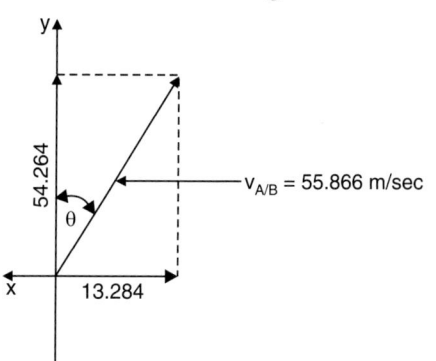

Fig. 14.9

Example 14.5. *An enemy ship was located at a distance of 25 km in north-west direction by a warship. If the enemy ship is moving with a velocity of 18 kmph N 30° E, in which direction the warship must move with a velocity of 36 kmph, to strike at its earliest? Assume the fire range of warship is 5 km. When is the shell to be fired ?*

Solution. Let the enemy ship B be at M and the warship (A) be at N as shown in Fig. 14.10. The required direction of movement of war ship be θ to north direction. Taking north as y direction and west as x direction, it's components of velocity are

$$v_{Ay} = 36 \cos \theta \quad \text{and} \quad v_{Ax} = 36 \sin \theta$$

Components of velocity of enemy ship are

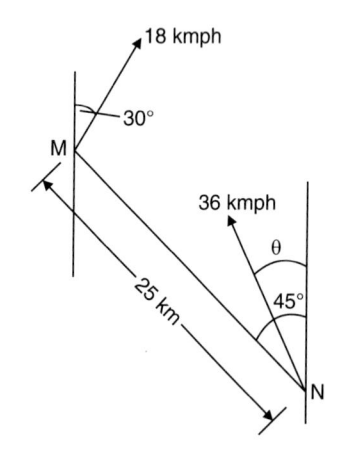

$$v_{By} = 18 \cos 30 \quad \text{and} \quad v_{Bx} = -18 \sin 30°$$
$$= 15.588 \text{ kmph} \qquad = -9 \text{ kmph}$$

Fig. 14.10

The relative velocity of A with respect to B is represented by

$$v_{rx} = 36 \sin \theta - (-9) = 36 \sin \theta + 9$$
$$v_{ry} = 36 \cos \theta - 15.588$$

∴ The direction of relative velocity α with north is given by

$$\tan \alpha = \frac{v_{rx}}{v_{ry}} = \frac{36 \sin \theta + 9}{36 \cos \theta - 15.588}$$

To approach the enemy ship at earliest, the direction of relative velocity α should be MN i.e. 45° to north [Since NM is north west direction]

∴ $$\tan 45° = \frac{36 \sin \theta + 9}{36 \cos \theta - 15.588}$$

or $\quad\quad 36 \cos \theta - 15.588 = 36 \sin \theta + 9 \quad\quad$ (since tan 45° = 1)

$\quad\quad \cos \theta - \sin \theta = 0.683 \quad\quad$ ...(1)

Let $x = \sin \theta$, then $\cos \theta = \sqrt{1 - x^2}$

∴ From (1), $\quad \sqrt{1 - x^2} - x = 0.683$

$$\sqrt{1 - x^2} = 0.683 + x$$
$$1 - x^2 = (0.683 + x)^2$$

i.e. $\quad\quad 2x^2 + 1.366x - 0.5335 = 0$

i.e. $\quad\quad x = 0.2777$

∴ $\quad\quad \theta = 16.12°$

War ship must move in $N\ 16.12\ W$ direction. Ans.

∴ $\quad\quad v_{rx} = 36 \sin \theta + 9 = 19.0$

$\quad\quad v_{ry} = 36 \cos \theta - 15.588 = 19.0$ kmph

∴ Relative velocity $\quad v_{A/B} = v_r = \sqrt{19^2 + 19^2}$

$\quad\quad\quad\quad\quad = 26.870$ kmph

Relative distance to be moved before firing

$\quad\quad\quad\quad = $ Distance between MN – Range of gun

$\quad\quad\quad\quad = 25 - 5 = 20$ km

∴ Time interval t is given by

$$v_r t = 20$$

$$t = \frac{20}{26.870} = 0.744 \text{ hour}$$

i.e. 44 min and 40 sec after sighting the enemy ship the shell is to be fired. Ans.

Example 14.6. *A motor boat has to cross a river 1 km wide and flow of 5 kmph. Assuming the boat moves with uniform velocity of 15 kmph throughout, find the direction in which it should move to reach the other bank in the minimum time. Where and when it will reach the other bank?*

If the boat has to touch exactly the opposite bank, in which direction should it be set and how much time is required for it to reach opposite bank?

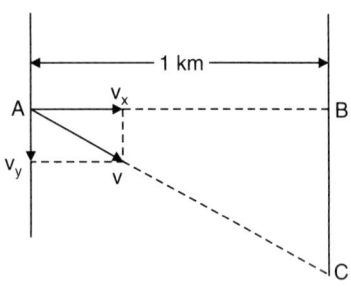

Fig. 14.11

RELATIVE VELOCITY

Solution. Let the motor boat start from point A as shown in Fig. 14.11. The boat moves in the direction of the resultant velocity v and reaches point C on the other bank. This resultant velocity is due to the velocity of boat and velocity of the flow of river.

To reach opposite bank in minimum time, the component of resultant velocity must have the maximum value in x-direction. Only driving force of the boat gives component in x direction. Hence boat should be set in straight across the river.

$$v_{rx} = 15 \text{ kmph}$$

Distance to be moved in x direction = 1 km

∴ **The time interval t required is**

$$15\, t = 1 \quad \text{or} \quad t = \frac{1}{15} \text{ hour}$$

$$= 4 \text{ min} \quad \textbf{Ans.}$$

During this period the boat will move down the stream due to the component of velocity v_y, where

$$v_y = 5 \text{ kmph}$$

∴ **Distance moved in downstream direction**

$$= 5 \times \frac{1}{15} = \frac{1}{3} \text{ km}$$

$$= 333.33 \text{ m} \quad \textbf{Ans.}$$

If the boat has to touch the bank straight across, the resultant velocity of the boat should be directed in x direction, *i.e.* resultant velocity of the boat with respect to stream in y direction = 0. Let the direction of boat be set at θ to x direction as shown in Fig. 14.12.

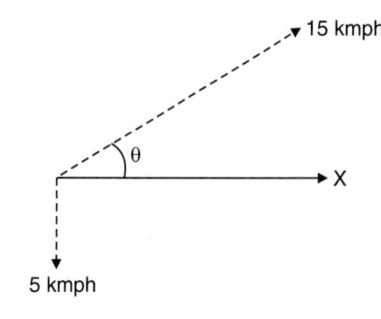

∴ $\qquad 15 \sin \theta = 5$

or $\qquad \sin \theta = \frac{1}{3}$

or $\qquad \theta = 19.47° \quad \textbf{Ans.}$

Fig. 14.12

Example 14.7. *Ship A is approaching a port in due East direction with a velocity of 15 kmph. When this ship was 50 km from port, ship B sails in N 45° W direction with a velocity of 25 kmph from the port. After what time the two ships are at minimum distance and how far each has travelled ?*

Solution. Let west be x and north be y axes

$$v_{Bx} = 25 \sin 45° = 17.678 \text{ kmph}$$
$$v_{By} = 25 \cos 45° = 17.678 \text{ kmph}$$
$$v_{Ax} = -15 \text{ kmph}$$

and $\qquad v_{Ay} = 0.0$

Let v_r be the relative velocity of B with respect to A ($v_{B/A}$).

∴ $\qquad v_{rx} = 17.678 - (-15) = 32.678 \text{ kmph}$

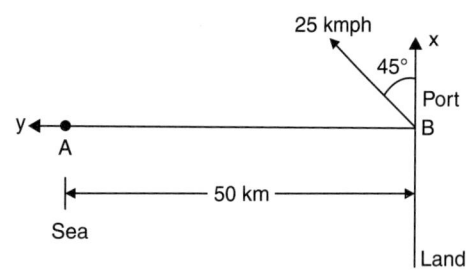

Fig. 14.13

$$v_{ry} = 17.678 \text{ kmph}$$

$$\therefore \quad v_r = \sqrt{32.678^2 + 17.678^2}$$

$$= 37.153 \text{ kmph}$$

$$\tan \alpha = \frac{v_{ry}}{v_{rx}} = \frac{17.678}{32.678}$$

$$\therefore \quad \alpha = 28.41°.$$

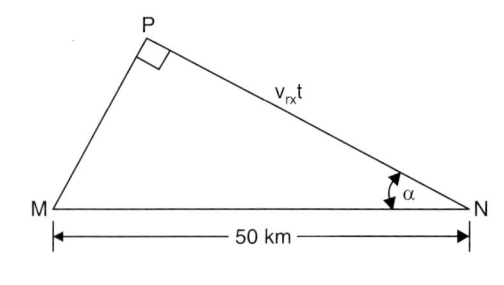

Holding A as stationary at M and allowing B to move with relative velocity $v_{B/A} = v_r$, from N, the two ships will be at minimum distance at point P as shown in Fig. 14.14.

Fig. 14.14

$$\therefore \quad v_r t = 50 \cos \alpha$$

$$37.153 \, t = 50 \cos 28.41°$$

$$\boldsymbol{t = 1.1837 \text{ hours}} \quad \textbf{Ans.}$$

During this period A has moved in due East direction a distance of

$$15 \times 1.1837 = \textbf{17.756 km} \quad \textbf{Ans.}$$

and B has moved in N 45° W a distance of $25 \times 1.1837 = \textbf{29.593 km}$ **Ans.**

Example 14.8. *Ship B is approaching a port at 18 kmph speed moving in East direction. When it was 60 km away, ship A left the port with a speed of 24 kmph in S 60° W. If the two ships can exchange the signal when they are within a range of 25 km, find when and how long they can exchange the signal.*

Solution. Considering west as x axis and south as y axis;

$$v_{Ax} = 24 \cos 30° = 20.785$$
$$v_{Ay} = 24 \sin 30° = 12$$
$$v_{Bx} = -18$$
$$v_{By} = 0$$

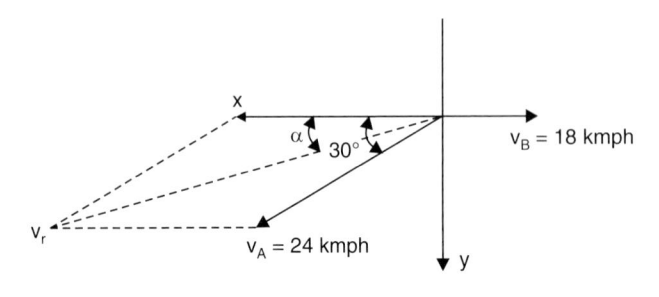

Fig. 14.15

Let relative velocity of A with respect to B ($v_{A/B}$) be v_r at an angle α to western direction.

$$v_{rx} = v_{Ax} - v_{Bx}$$
$$= 38.785 \text{ kmph}$$
$$v_{ry} = v_{Ay} - v_{By} = 12$$
$$v = \sqrt{v_{rx}^2 + v_{ry}^2} = \sqrt{38.785^2 + 12^2}$$
$$= 40.599 \text{ kmph}$$

RELATIVE VELOCITY

$$\tan \alpha = \frac{v_{ry}}{v_{rx}} = \frac{12}{38.785}$$

$$\therefore \quad \alpha = 17.192.$$

Holding B stationary and allowing A to move with relative velocity, the minimum distance between the two ships BC (Fig. 14.16) is given by

$$BC = 60 \sin \alpha = 17.735 \text{ km}$$

The two ships can exchange the signals from point D to E when $BD = BE = 25$ km.

From the triangle BCD,

$$DC = \sqrt{25^2 - 17.735^2}$$
$$= 17.62 \text{ km}$$

Since BC is the foot of isosceles triangle BED,

$$CE = DC = 17.62 \text{ km}$$

From triangle ABC, $\quad AC = 60 \cos \alpha$
$$= 57.319 \text{ km}$$

$\therefore \quad AD = AC - DC = 57.319 - 17.62$
$$= 39.699 \text{ km}$$

and $\quad AE = AC + CE = 57.319 + 17.62$
$$= 74.939 \text{ km}$$

Time taken to reach D, $\quad t_1 = \dfrac{AD}{v_r} = \dfrac{39.699}{40.899}$

$$= 0.9778 \text{ hr}$$
$$= 58 \text{ min } 40 \text{ sec}$$

Time taken to reach E

$$t_2 = \frac{AE}{v_r} = \frac{74.939}{40.599}$$

$$= 1.8458 \text{ hr}$$
$$= 1 \text{ hr, } 50 \text{ min. } 45 \text{ sec}$$

Hence the two ships can start exchanging signals, 58 min 40 sec after ship A leaves the port and continue to do so for 52 min and 15 sec.

Example 14.9. *When a train is moving with a velocity of 36 kmph, a passenger observes rain drops at 30° to vertical. When the velocity of train increases to 54 kmph, he observes rain at 45° to vertical. Determine the true velocity and direction of rain drops. Assume rain drops are in the parallel vertical plane to that of train's vertical plane.*

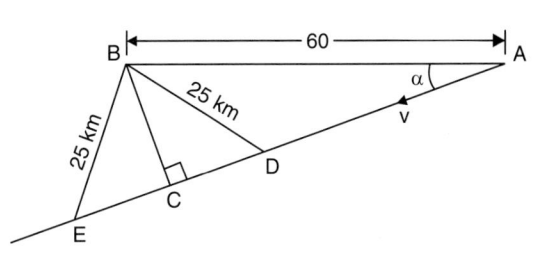

Fig. 14.16

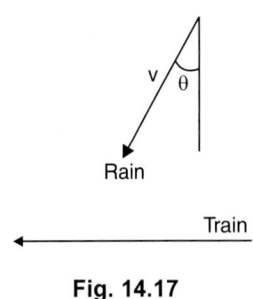

Fig. 14.17

Solution. Let the true velocity of rain be v kmph at a true angle θ with vertical as shown in Fig. 14.17.

Taking the direction of train as x and that of vertical downward as y,
Velocity components of rain are:
$$v_{1x} = v \sin \theta$$
$$v_{1y} = v \cos \theta$$
When the velocity of train was 36 kmph,
$$v_{2x} = 36 \text{ and } v_{2y} = 0$$
∴ The relative velocity components of rain with respect to train are:
$$v_{rx} = v \sin \theta - 36$$
$$v_{ry} = v \cos \theta$$
∴
$$\tan \alpha = \frac{v \sin \theta - 36}{v \cos \theta}$$

Now α the direction of relative velocity is given as 30°.

∴
$$\tan 30° = \frac{v \sin \theta - 36}{v \cos \theta} \qquad \ldots(1)$$

when the velocity of train is 54 kmph, $\alpha = 45°$

∴
$$\tan 45° = \frac{v \sin \theta - 54}{v \cos \theta}$$

$$1 = \frac{v \sin \theta - 54}{v \cos \theta}$$

$$v \cos \theta = v \sin \theta - 54 \qquad \ldots(2)$$

Substituting this value in (1), we get,

$$0.577 = \frac{v \sin \theta - 36}{v \sin \theta - 54}$$

∴ $\qquad v \sin \theta = -11.402 \qquad \ldots(3)$

Substituting it in (2), we get
$$v \cos \theta = -65.402 \qquad \ldots(4)$$
Squaring and adding (3) and (4), we get
$$v^2 = 4407.43$$
$$v = \mathbf{66.388 \text{ kmph}} \quad \textbf{Ans.}$$

$$\sin \theta = -\frac{11.402}{66.388} = -0.1717$$

∴ $\qquad \theta = \mathbf{-9.89°} \quad \textbf{Ans.}$

Example 14.10. *A jet of water is discharged from a tank at A with a velocity of 20 m/sec to strike a moving plate at B as shown in Fig. 14.18. If the plate is moving in vertically downward direction with a velocity of 1 m/sec, determine the relative velocity of water striking the plate.*

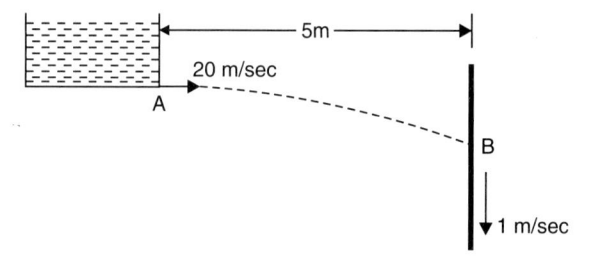

Fig. 14.18

Solution. The water is moving with initial velocity of 20 m/sec horizontally and is subjected to gravitational force, which causes vertical motion also. Horizontal component of the velocity is not altered if we neglect air resistance.

∴ Time taken to move a horizontal distance of 5 m, $t = \dfrac{5}{20} = \dfrac{1}{4}$ seconds.

During this period vertical downward velocity gained by the water

$$= 0 + 9.81 \times \dfrac{1}{4} = 2.453 \text{ m/sec}$$

Velocity of plate:
Horizontal component = 0
Vertical downward component = 1 m/sec.
∴ Relative velocity of water with respect to plate has the components

$$v_{ry} = 2.453 - 1.0 = 1.453 \text{ m/sec}$$
$$v_{rx} = 20 \text{ m/sec}$$

$$v_r = \sqrt{20^2 + 1.4535^2} = \textbf{20.052 m/sec} \quad \textbf{Ans.}$$

$$\tan \alpha = 1.453/20$$
$$\alpha = \textbf{4.16°} \text{ as shown in Fig. 14.19.} \quad \textbf{Ans.}$$

Fig. 14.19

Example 14.11. *A railway carriage 4.8 m long and 1.8 m wide is moving at 96 kmph, when a pistol shot hits the carriage in a direction making an angle of 10° to the direction of the track. The shot enters the carriage in one corner and passes out diagonally at the opposite corner. Find the speed of the shot and the time taken for the shot to traverse the carriage.*

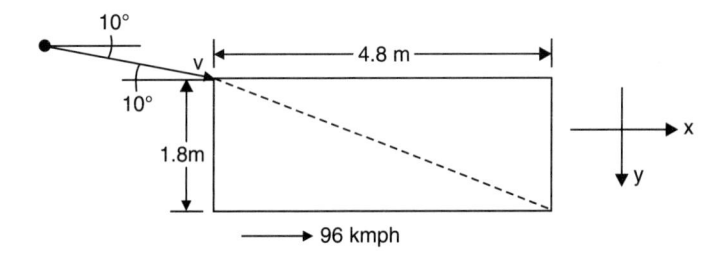

Fig. 14.20

Solution. Let v be the velocity of shot, x be the direction in which train is moving and y be the direction at right angles to it as shown in Fig. 14.20.

Velocity components of train are :
$$v_{1x} = 96 \text{ kmph}$$
$$v_{1y} = 0.$$

Velocity components of bullet are:
$$v_{2x} = v \cos 10° = 0.9848 \, v$$
$$v_{2y} = v \sin 10° = 0.1736 \, v$$

Component of relative velocity of B with respect to A are
$$v_{rx} = 0.9848 \, v - 96$$
and
$$v_{ry} = 0.1736 \, v.$$

Direction of relative velocity α with x axis is given by

$$\tan \alpha = \frac{v_{ry}}{v_{rx}} = \frac{0.1736\, v}{0.9848\, v - 96}$$

It is given that relative velocity of bullet with respect to train makes an angle α such that

$$\tan \alpha = \frac{1.8}{4.8} = 0.375$$

$\therefore \qquad 0.375 = \dfrac{0.1736\, v}{0.9848\, v - 96}$

$$v = 183.96 \text{ kmph} = \mathbf{51.1 \text{ m/sec}} \quad \mathbf{Ans.}$$

Considering the motion in y direction,

$$v_{ry} = 0.1736 \times 51.1 \text{ m/sec}$$

relative distance travelled = 1.8 m

$\therefore \qquad$ **time period** $= t = \dfrac{1.8}{0.1736 \times 51.1} = \mathbf{0.203 \text{ sec}} \quad \mathbf{Ans.}$

Important Definitions

1. **Relative velocity** is the velocity of an object as observed from another moving object.
2. **Resultant velocity** is the combined effect of two or more forces causing motion of the body.

Important Formulae

1. Relative velocity = vector difference of the velocities of two bodies.
2. Resultant velocity = vector sum of the velocities caused by different forces on a body.

PROBLEMS FOR EXERCISE

14.1 Two trains A and B are moving on parallel tracks with velocities 72 kmph and 36 kmph in the same direction. The driver of train A finds that it took 42 seconds to overtake train B while driver of train B finds that train A took 30 seconds to overtake him.
Determine:
(a) length of each train
(b) time taken for complete overtaking.
If the two trains move in opposite direction with the same velocities, how much time is taken for complete crossing? [**Ans.** (a) L_A = 300 m, L_B = 420 m; (b) t = 72 sec, t = 24 sec]

14.2 An aeroplane is flying horizontally at a height of 2000 m at a speed of 800 kmph towards a battle tank. The muzzle velocity of the gun of the tank is 600 metres per second. At what angle the shell should be fired when the plane is at 3000 m from the tank? Neglect air resistance and effect of gravity on the shell. [**Ans.** θ = 81.387°]

14.3 Two ships move simultaneously from a port, one moving in N 45° W and at 25 kmph another at S 60° W, at 15 kmph. Determine the relative velocity of first ship with respect to second ship. When are they going to be 15 km apart? [**Ans.** v_r = 25.61 kmph, t = 35.14 min.]

14.4 Ship A is approaching a port from S 40° W, direction at 20 kmph. When ship A was 20 km from the port, ship B leaves the port at N 60 W with a velocity of 25 kmph. Determine the relative velocity of A with respect to B. When are they at least distance?
[**Ans.** $v_{A/B}$ = 34.62 kmph, θ = S 85°.326 W]

RELATIVE VELOCITY

14.5 Two coastal guard ships are 40 km apart, ship A being due north of B. Ship A is moving S 45° W with a velocity of 18 kmph while ship B is moving west with a velocity of 24 kmph. If the two ships can exchange signals when they are 30 km, apart, when they can begin and how long they can exchange the signals ? [**Ans.** They can begin after 56.18 m and continue for 1 hr. 39 min.]

14.6 To an observer on a ship moving due south with a velocity of 18 kmph, wind appears to blow from due west. After the ship changes the course and moves in due west with the same velocity, wind appears to blow from S 45° W. Assuming the wind has not changed the direction during this period, find the true direction and velocity of the wind. [**Ans.** $v = 40.25$ kmph ; $\theta = 26.565°$]

14.7 A rifleman on a train, moving with a speed of 60 kmph wants to shoot at a stationary object on the ground when it was sighted at 30° to the train. At what angle should he aim if the bullet velocity is 700 kmph ?

If the object on the ground is not stationary, but moves away at a velocity of 40 kmph at right angles to the train, what should be the angle of the rifle at the time of shooting ? Assume object remains in firing range and neglect air resistance and gravitational effect on the bullet.

[**Ans.** (1) $\theta = 32.46°$; (2) $\theta = 33.40°$]

14.8 When a cyclist is riding west at 20 kmph, he feels it is raining at an angle of 45° with the vertical. When he rides at 15 kmph, he feels it is raining at an angle of 30° with the vertical. What is the velocity of the rain ? [**Ans.** $v = 14.377$ kmph]

15
D' Alembert's Principle

In the last three chapters, kinematic problems were considered, *i.e.* the problems which do not involve the cause of motion. The problems which need the consideration of forces causing the motion are treated from this chapter onwards. Such problems are called kinetic problems.

In this chapter analysis of kinetic problems of a body in plane motion using D' Alembert's principle are explained. Dynamic problem can be converted into a static equilibrium problem by applying D' Alembert's principle. This is done by introducing an additional force and then the usual equations of static equilibrium are used to get the solution.

15.1. NEWTON'S SECOND LAW OF MOTION

D' Alembert's principle is in fact an application of Newton's second law to a moving body and looking at it from a different angle. Observing the motion of falling bodies Galileo discovered first two laws of motion which are now commonly known as Newton's laws of motion. However, Newton generalised the laws and demonstrated their validity by astronomical predictions. These laws are presented in the first chapter. According to Newton's second law the rate of change of momentum is directly proportional to the impressed force and takes place in the direction, in which the force acts. The above definition leads to statement that *force is directly proportional to product of mass and acceleration* (Eqn. 1.1) and unit of force is so selected that constant of proportionality reduces to unity (Eqn. 1.2). Hence, finally Newton's second law reduces to the statement, Force = mass × acceleration.

Instead of single force, if a system of forces acts on a particle, the above statement reduces to resultant force is equal to : *the product of mass and acceleration in the direction of the resultant force*. Mathematically

$$R = m\,a \qquad \qquad ...(15.1)$$

where, R is the resultant of forces acting on the particle. Hence many time Newton's second law is stated as : *a particle acted upon by an unbalanced system of forces has an acceleration directly proportional and in line with the resultant force.*

15.2. D' ALEMBERT'S PRINCIPLE

French mathematician Jean le Rond d' Alembert proved in 1743 that the Newton's second law of motion is applicable not only to the motion of a particle but also to the motion of a body and looked at equation (15.1) from different angle. The equation $R = ma$ may be written as

$$R - ma = 0 \qquad \qquad ...(15.2)$$

D' ALEMBERT'S PRINCIPLE

The term '− ma' may be looked as a force of magnitude $m \times a$, applied in the opposite direction of motion and is termed as the inertia force or reverse effective force. D' Alembert looks at equation (15.2) as an equation of equilibrium and states that *the system of forces acting on a body in motion is in dynamic equilibrium with the inertia force of the body.* This is known as D' Alembert's principle.

Let the body shown in Fig. 15.1 be subjected to a system of forces causing the body to move with an acceleration a in the direction of the resultant. Then apply a force equal to '$m\,a$' in the reversed direction of acceleration as shown in Fig. 15.1. Now according to D' Alembert's principle, the equations of equilibrium $\Sigma x = 0$ and $\Sigma y = 0$ may be used for the system of forces shown in Fig. 15.2.

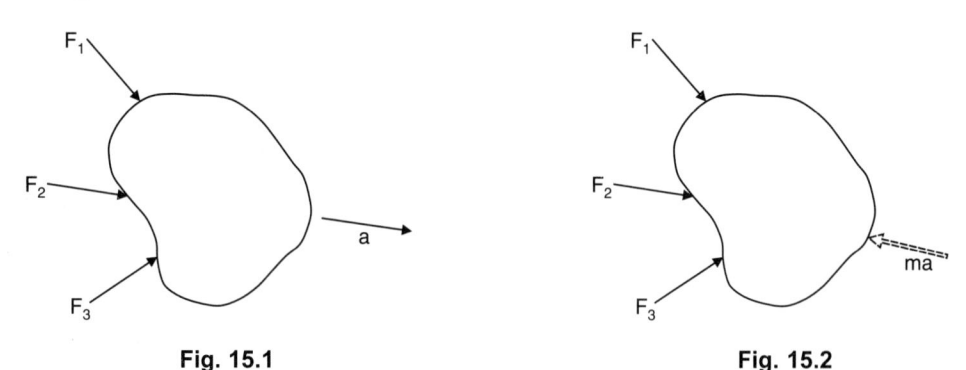

Fig. 15.1

Fig. 15.2

The inertia force $-ma$ has a physical meaning. According to Newton's first law of motion, a body continues to be in the state of rest or of uniform motion in a straight line unless acted by an external force. That means every body has a tendency to continue in its state of rest or of uniform motion. This tendency is called inertia. Hence inertia force is the resistance offered by a body to the change in its state of rest or of uniform motion.

Many scientists are critical of D' Alembert's principle. Usually equilibrium equations are applied to a system of forces acting on a body. Inertia force is not acting on the moving body. Actually this is the force exerted by the moving body to resist the change in its state Hence D' Alembert is criticized for messing-up the concept of equations of equilibrium.

However, many engineers prefer to use D' Alembert's principle, since just by applying a reverse effective force, the moving body can be treated as a body in equilibrium and can be analysed using equations of static equilibrium.

Example 15.1. *A man weighing W Newton entered a lift which moves with an acceleration of a m/sec². Find the force exerted by the man on the floor of lift when*

(a) *lift is moving downward*

(b) *lift is moving upward.*

Solution. (a) **When the lift is moving down-ward,** the inertia force $ma = \dfrac{W}{g}a$, should be applied in upward direction as shown in Fig.(15.3.(a))

Σ Forces in vertical direction = 0 for the lift gives

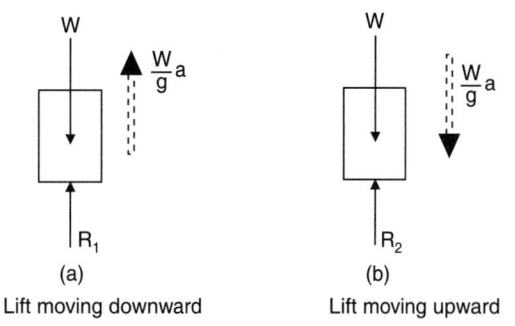

$$R_1 - W + \frac{W}{g}a = 0$$

(a) Lift moving downward

(b) Lift moving upward

Fig. 15.3

$$R_1 = W\left(1 - \frac{a}{g}\right) \quad \ldots(1)$$

(b) **When lift is moving upwards,** inertia force $\frac{W}{g} a$ should be applied in the downward direction as shown in Fig. 15.3 (b). Applying the equation of equilibrium, we get,

$$R_2 - W - \frac{W}{g} a = 0$$

$$R_2 = W\left(1 + \frac{a}{g}\right) \quad \ldots(2)$$

Thus when lift is moving with acceleration downward, the man exerts less force on the floor of the lift and while moving upward he exerts more force. From equations (1) and (2) it may be observed that if the lift moves with uniform velocity then the acceleration a is zero and hence the man exerts force equal to his own weight on the lift.

Example 15.2. *An elevator cage of a mine shaft weighing 8 kN, when empty, is lifted or lowered by means of a wire rope. Once a man weighing 600 N, entered it and lowered with uniform acceleration such that when a distance of 187.5 m was covered, the velocity of the cage was 25 m/sec. Determine the tension in the rope and the force exerted by the man on the floor of the cage.*

Solution. In this problem

initial velocity $u = 0$
final velocity $v = 25$ m/sec
and distance covered, $s = 187.5$ m

Using the equation of motion,
$$v^2 - u^2 = 2\ as$$

we get,
$$25^2 - 0 = 2a\ 187.5$$
or $$a = 1.667 \text{ m/sec}^2$$

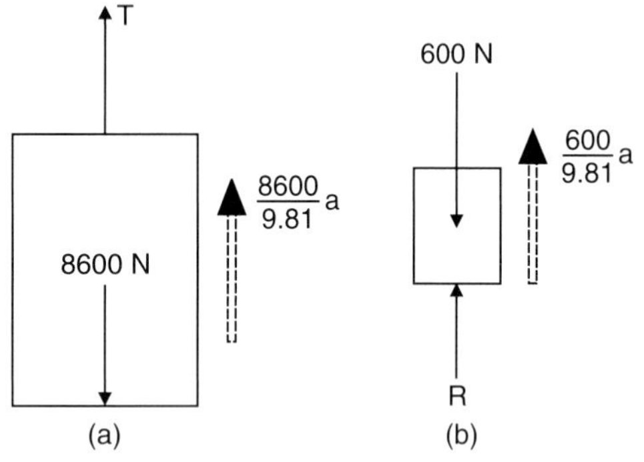

Figure 15.4 (a) shows free body diagram of elevator cage and the man with inertia force $\frac{W}{g} a = \frac{8000 + 600}{9.81} a$ applied in upward direction (since the motion is downward).

Fig. 15.4

Summing up the forces in vertical direction

$$T + \frac{8600}{9.81}\ 1.667 - 8600 = 0$$

$$\therefore \quad T = 7138.90 \text{ N} \quad \textbf{Ans.}$$

Figure 15.4 (b) shows the free body diagram of the man along with inertia force $\frac{600}{9.81} a$ applied in upward direction. Sum of the vertical forces should be zero. That is,

$$\Sigma V = 0 \text{ gives}$$

$$R + \frac{600}{9.81} a - 600 = 0$$

i.e. $$R + \frac{600}{9.81} \times 1.667 - 600 = 0$$

or $$R = 498.06 \text{ N} \quad \textbf{Ans.}$$

D' ALEMBERT'S PRINCIPLE

Example 15.3. *A motorist travelling at a speed of 70 kmph suddenly applies brakes and halts after skidding 50 m. Determine :*

(1) the time required to stop the car ;

(2) the coefficient of friction between the tyres and the road.

Solution. Initial velocity $u = 70$ kmph

$$= \frac{70 \times 1000}{60 \times 60} = 19.44 \text{ m/sec}$$

Final velocity $\quad\quad\quad v = 0$

Displacement $\quad\quad\quad s = 50$ m

Using the equation of linear motion

$$v^2 = u^2 + 2as, \quad \text{we get}$$
$$0 = 19.44^2 + 2a \times 50$$
$$a = -3.78 \text{ m/sec}^2$$

i.e. the retardation is 3.78 m/sec².

Using the relation $v = u + at$, we get

$$0 = 19.44 - 3.78\, t$$

∴ $\quad\quad\quad t = 5.14$ **sec Ans.**

Inertia force must be applied in the opposite direction of acceleration, which means, it should be applied in the direction of motion while retarding. Figure 15.5 shows the free body of the motor along with inertia force.

Σ Forces normal to road = 0

$$N = W$$

From the law of friction, $\quad F = \mu N = \mu W$

Σ Forces in the direction of motion = 0, gives

$$F = \frac{W}{9.81} \, 3.78$$

$$\mu W = \frac{W \times 3.78}{9.81}$$

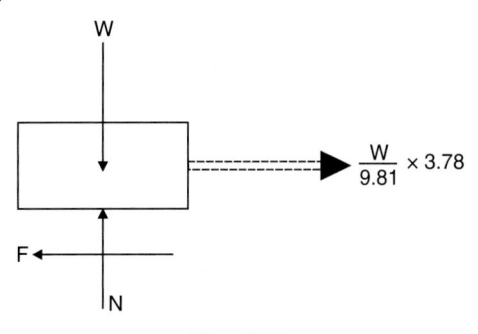

∴ $\quad\quad\quad \mu = \mathbf{0.385}$ **Ans.**

Fig. 15.5

Example 15.4. *A block weighing 1 kN rests on a horizontal plane as shown in Fig. 15.6 (a). Find the magnitude of the force P required to give the block an acceleration of 3 m/sec² to the right. The coefficient of friction between the block and the plane is 0.25.*

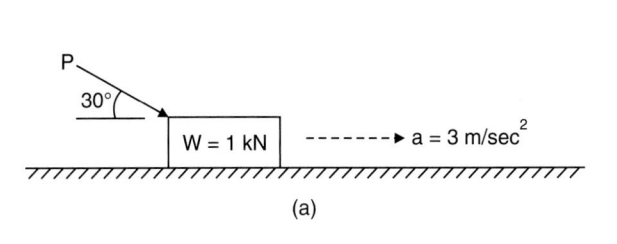

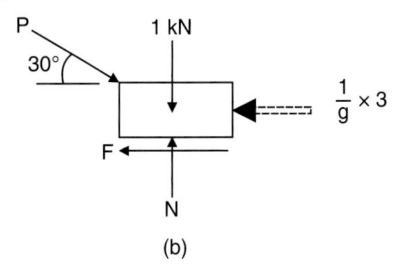

Fig. 15.6

Solution. Free body diagram of the block along with inertia force $ma = \dfrac{1}{g} \times 3 = \dfrac{3}{9.81}$ kN is shown in Fig. 15.6(b). In the figure, N is the normal reaction and F is the frictional force.

$$\Sigma V = 0$$
$$N - 1 - P \sin 30° = 0$$

or
$$N = 1 + \dfrac{P}{2} \qquad \ldots(1)$$

From the law of friction.

$$F = \mu N = 0.25 \left(1 + \dfrac{P}{2}\right) \qquad \ldots(2)$$

$$\Sigma H = 0$$

$$P \cos 30° - F - \dfrac{3}{9.81} = 0$$

$$P \cos 30° - 0.25 \left(1 + \dfrac{P}{2}\right) - \dfrac{3}{9.81} = 0$$

∴ $\qquad P = 0.561$ kN **Ans.**

Example 15.5. *A 750 N crate rests on a 500 N cart. The coefficient of friction between the crate and the cart is 0.3 and between cart and the road is 0.2. If the cart is to be pulled by a force P (Ref. Fig. 15.7 (a)) such that the crate does not slip, determine : (a) the maximum allowable magnitude of P and (b) the corresponding acceleration of the cart.*

Solution. Let the maximum acceleration be a, at which 750 N crate is about to slip. Hence frictional force will have limiting value ($= \mu N$). Consider the free body diagram of crate along with inertia force, shown in Fig. 15.7(b).

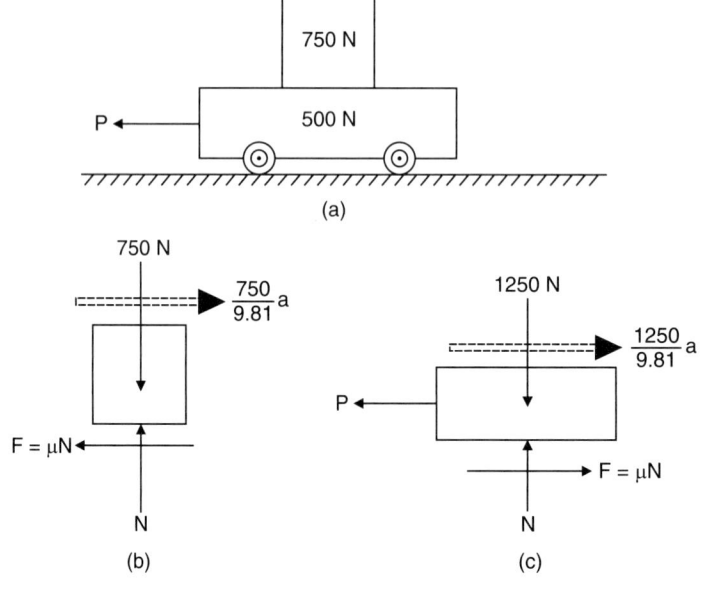

Fig. 15.7

D' ALEMBERT'S PRINCIPLE 365

Frictional force,
$$\Sigma V = 0$$
$$N = W = 750 \text{ Newton}$$
$$F = \mu N = 0.3 \times 750$$
$$= 225 \text{ N}$$
$$\Sigma H = 0$$
$$225 = \frac{750}{9.81} a$$
$$\therefore \quad a = 2.943 \text{ m/sec}^2 \quad \textbf{Ans.}$$

Consider now dynamic equilibrium of crate and the cart shown as a single body in Fig. 15.7(c).

$$\Sigma V = 0$$
$$N = 1250 \text{ Newton}$$
Frictional force $= \mu N = 0.2 \times 1250 = 250$ Newton
$$\Sigma H = 0$$
$$P - 250 - \frac{1250}{9.81} \times 2.943 = 0$$
$$P = 625 \text{ N} \quad \textbf{Ans.}$$

Example 15.6. *A body weighing 1200 N rests on a rough plane inclined at 12° to the horizontal. It is pulled up the plane by means of a light flexible rope running parallel to the plane and passing over a light frictionless pulley at the top of the plane as shown in Fig. 15.8 (a). The portion of the rope beyond the pulley hangs vertically down and carries a weight of 800 N at its end. If the coefficient of friction for the plane and the body is 0.2, find :*

(a) tension in the rope

(b) acceleration with which the body moves up the plane, and

(c) the distance moved by the body in 3 seconds after starting from rest.

Solution. Let a be the acceleration of the system.

Free body diagrams of 1200 N block and 800 N block are shown in Figs. 15.8(b) and 15.8(c) along with inertia forces. According to D' Alembert, we can treat these bodies as in static equilibrium.

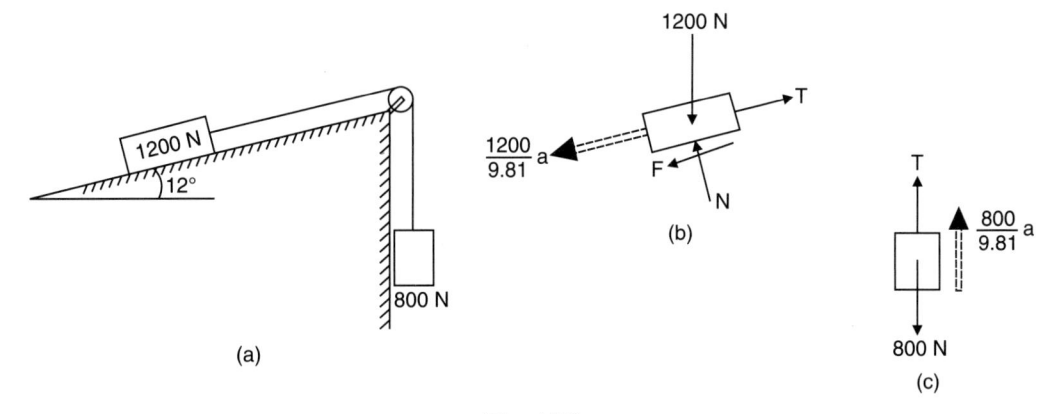

Fig. 15.8

Consider 1200 N block.

Σ Forces normal to the plane $= 0$, gives
$$N - 1200 \cos 12° = 0$$

$$N = 1173.77 \text{ Newton}$$

From the law of friction
$$F = \mu N = 0.2 \times 1173.77 = 234.76 \text{ N}$$

Σ Forces parallel to the inclined plane = 0, gives
$$\frac{1200}{9.81} a + 1200 \sin 12° + F - T = 0$$

i.e.
$$122.32\, a - T = -484.25 \qquad \ldots(1)$$

Consider the free body diagram of 800 N block shown in Fig. 15.8(c).
$$T + \frac{800}{9.81} a = 800 \qquad \ldots(2)$$

Adding equations (1) and (2), we get,
$$\left(122.32 + \frac{800}{9.81}\right) a = 800 - 484.25$$

$\therefore \qquad a = \mathbf{1.549 \text{ m/sec}^2}$ **Ans.**

Substituting it in (2), we get,
$$T = 800 - \frac{800}{9.81} \times 1.549$$

i.e. $\qquad T = \mathbf{673.68 \text{ N}}$ **Ans.**

$$\text{Initial velocity} = 0$$
$$a = 1.549 \text{ m/sec}^2$$
$$t = 3 \text{ sec}$$

Using the equation $s = ut + \frac{1}{2} at^2$, we get **distance moved in 3 seconds** as

$$= 0 \times 3 + \frac{1}{2} \times 1.549 \times 3^2$$
$$= \mathbf{6.971 \text{ m}} \quad \textbf{Ans.}$$

Example. 15.7. *Two weights 800 N and 200 N are connected by a thread and they move along a rough horizontal plane under the action of a force of 400 N applied to the 800 N weight as shown in Fig. 15.9(a). The coefficient of friction between the sliding surface of the weights and the plane is 0.3. Using D' Alembert's principle determine the acceleration of the weight and tension in the thread.*

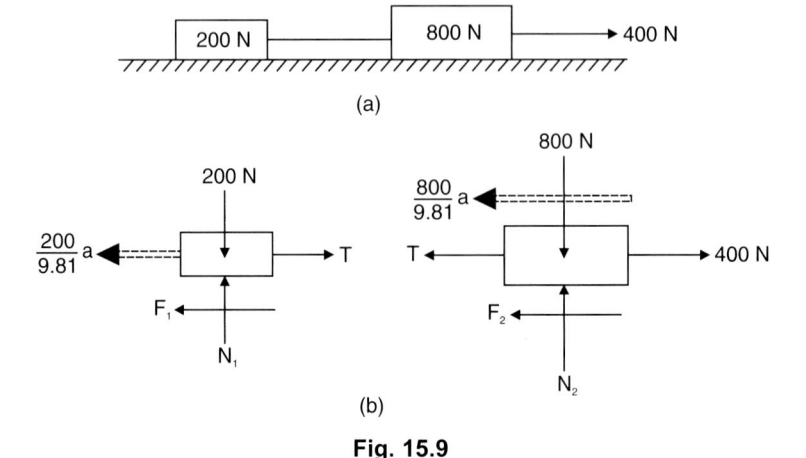

Fig. 15.9

Solution. Free body diagrams of 200 N and 800 N blocks along with inertia forces are shown in Fig. 15.9(b), in which a is acceleration of the system.

Consider the dynamic equilibrium of 200 N weight

$$\Sigma V = 0$$
$$N_1 = 200 \ N. \qquad \ldots(1)$$

From law of friction, $F_1 = \mu \ N_1 = 0.3 \times 200 = 60 \ N \qquad \ldots(2)$

$$\Sigma H = 0$$

$$T - F_1 - \frac{200}{9.81} a = 0$$

i.e., $\qquad T - \dfrac{200}{9.81} a = 60 \qquad \ldots(3)$

since $\qquad F_1 = 60 \ N.$

Consider 800 N body.

$$\Sigma V = 0$$
$$N_2 = 800 \ N \qquad \ldots(4)$$

From law of friction, $\qquad F_2 = \mu N_2 = 0.3 \times 800$
$$= 240 \ N \qquad \ldots(5)$$

$$\Sigma H = 0$$

$$-T - \frac{800}{9.81} a - F_2 + 400 = 0$$

or $\qquad T + \dfrac{800}{9.81} a = 160 \qquad \ldots(6)$

Since $\qquad F_2 = 240 \ N.$

Subtracting equations (3) from (6), we get,

$$\left(\frac{200}{9.81} + \frac{800}{9.81}\right) a = 160 - 60$$

$\therefore \qquad \boldsymbol{a = 0.981 \ m/sec^2}$ **Ans.**

Substituting it in Eqn. (6), we get,

$$T = 160 - \frac{800}{9.81} \times 0.981$$

i.e. $\qquad \boldsymbol{T = 80 \ N}$ **Ans.**

Example 15.8. *Two rough planes inclined at 30° and 60° to horizontal are placed back to back as shown in Fig. 15.10 (a). The blocks of weights 50 N and 100 N are placed on the faces and are connected by a string running parallel to planes and passing over a frictionless pulley. If the coefficient of friction between planes and blocks is* $\dfrac{1}{3}$, *find the resulting acceleration and tension in the string.*

Solution. Let the assembly move down the 60° plane by an acceleration 'a' m/sec². Free body diagrams of 100 N and 50 N blocks along with inertia forces are shown in Figs. 15.10(b) and 15.10(c) respectively.

Consider the block weighing 100 N:

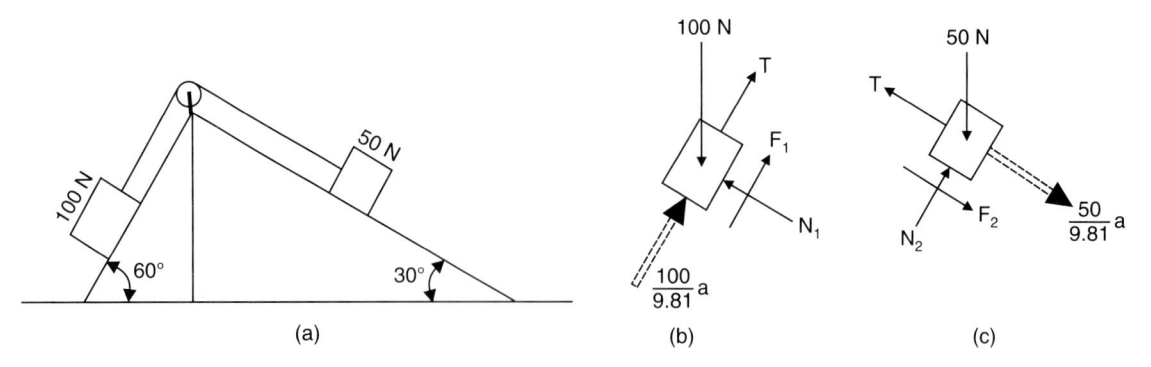

Fig. 15.10

Σ Forces normal to the plane = 0, gives
$$N_1 = 100 \cos 60° = 50 \text{ N} \qquad \ldots(1)$$

From the law of friction,
$$F_1 = \mu N = \frac{1}{3} \times 50 = 16.67 \text{ N} \qquad \ldots(2)$$

Σ Forces parallel to the plane = 0, gives
$$T + \frac{100}{9.81} a - 100 \sin 60° + F_1 = 0$$

$$T + \frac{100}{9.81} a = 69.93 \qquad \ldots(3)$$

since
$$F_1 = 16.67$$

Now consider 50 N block:

Σ Forces normal to plane = 0, gives
$$N_2 = 50 \cos 30° = 43.30 \text{ N} \qquad \ldots(4)$$

From the law of friction, $F_2 = \mu N_2$
$$= \frac{1}{3} \times 43.3 = 14.43 \text{ N} \qquad \ldots(5)$$

Σ Forces parallel to 30° plane = 0

$$\frac{50}{9.81} a + F_2 + 50 \sin 30° - T = 0$$

$$\frac{50}{9.81} a - T = -39.43 \qquad \ldots(6)$$

since
$$F_2 = 14.43$$

Adding equations (3) and (6), we get,
$$\left(\frac{100}{9.81} + \frac{50}{9.81}\right) a = 69.93 - 39.43$$

or
$$a = 1.9947 \text{ m/sec}^2 \quad \text{Ans.}$$

From eqn. (3),
$$T = 69.93 - 100/9.81 \times 1.9947 = \mathbf{49.6 \text{ N}} \quad \text{Ans.}$$

D' ALEMBERT'S PRINCIPLE

Example 15.9. *Two blocks A and B released from rest on a 30° incline, when they are 18 m apart. The coefficient of friction under the upper block A is 0.2 and that under the lower block B is 0.4 (Ref. Fig. 15.11(a)). In what time block A reaches the block B? After they touch and move as a single unit, what will be the contact force between them ? Weights of the block A and B are 100 N and 80 N respectively.*

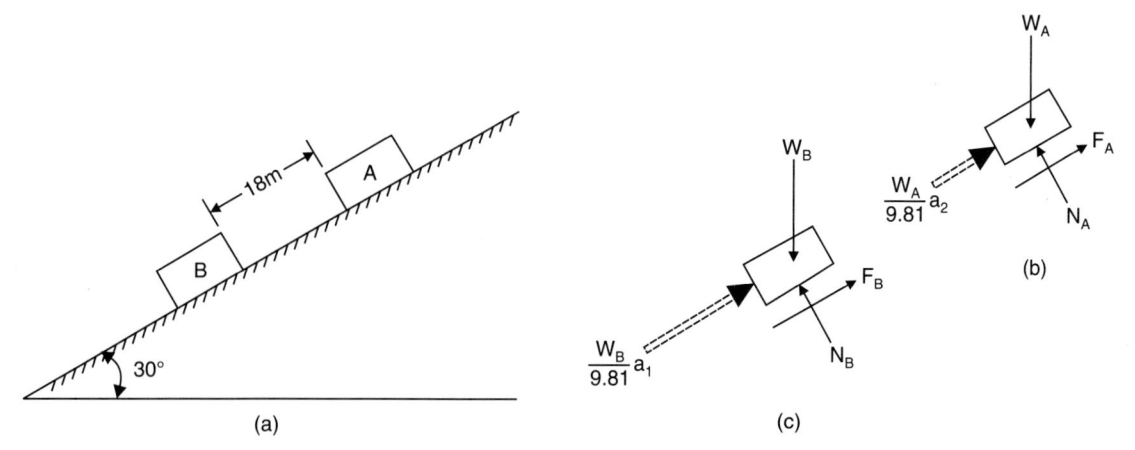

Fig. 15.11 (a, b, c)

Solution. Let block A move with an acceleration a_1 and block B with an acceleration a_2. The free body diagrams of the blocks A and B along with inertia forces are shown in Fig. 15.11(b) and 15.11(c) respectively.

Consider block A.

Σ Forces normal to the plane = 0, gives
$$N_A = W_A \cos\theta = W_A \cos 30° \qquad \ldots(1)$$

From the law of friction $\quad F_A = \mu N_A$
$$= 0.2\, W_A \cos 30° \qquad \ldots(2)$$

Σ Forces parallel to the plane = 0, gives
$$\frac{W_A}{9.81} a_1 + F_A - W_A \sin 30° = 0$$

∴ $\quad \dfrac{W_A}{9.81} a_1 + 0.2\, W_A \cos 30° - W_A \sin 30° = 0$

$$\frac{a_1}{9.81} + 0.2 \cos 30° - \sin 30° = 0$$

$$a_1 = 3.2058 \text{ m/sec}^2 \qquad \ldots(3)$$

Consider block B.

Σ Forces normal to the plane = 0, gives
$$N_B = W_B \cos 30° \qquad \ldots(4)$$

From the law of friction $\quad F_B = \mu N_B$

i.e. $\qquad F_B = 0.4\, W_B \cos 30° \qquad \ldots(5)$

Σ Forces parallel to the plane = 0, gives
$$\frac{W_B}{9.81} a_2 + F_B - W_B \sin 30° = 0$$

$$\frac{W_B}{9.81} a_2 + 0.4 \, W_B \cos 30° - W_B \sin 30° = 0$$

$$\therefore \qquad a_2 = 1.5067 \text{ m/sec}^2$$

Let t be the time elapsed until the blocks touch each other.

Displacement of block A in this period

$$s_1 = u_1 t + \frac{1}{2} a_1 t^2 = \frac{1}{2} \times 3.2058 \, t^2$$

since Initial velocity $u_1 = 0$

Displacement of block B in this time

$$s_2 = u_2 t + \frac{1}{2} a_2 t^2 = \frac{1}{2} \times 1.5067 \, t^2$$

When the two blocks touch each other

$$s_1 = s_2 + 18$$

$$\frac{1}{2} \times 3.2058 \, t^2 = \frac{1}{2} \times 1.5067 \, t^2 + 18$$

$$\therefore \qquad t = 4.60 \text{ sec} \quad \text{Ans.}$$

After the blocks touch each other, let the common acceleration be a. Summing up the forces including inertia forces along the inclined plane.

$$\frac{100}{9.81} a + 0.2 \times 100 \cos 30° - 100 \sin 30 + \frac{80}{9.81} a + 0.4 \times 80 \cos 30° - 80 \sin 30° = 0$$

$$a = 2.45 \text{ m/sec}^2$$

Considering the free body diagram of any one of the blocks, contact force P can be obtained. Free body diagram of block A along with inertia force is shown in Fig. 15.11(d).

Now Σ forces parallel to plane = 0, gives

$$P - 100 \sin 30° + F + \frac{100}{9.81} a = 0$$

$$P - 100 \sin 30° + 0.2 \times 100 \cos 30° + \frac{100}{9.81} \times 2.45$$

$$= 0$$

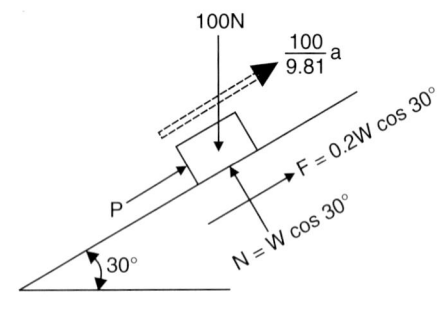

$$P = 7.7 \text{ N} \quad \text{Ans.}$$

Fig. 15.11(d)

Example 15.10. *Two bodies weighing 300 N and 450 N are hung to the ends of a rope passing over an ideal pulley as shown in Fig. 15.12(a). With what acceleration the heavier body comes down ? What is the tension in the string ?*

Solution. Let a be the acceleration with which the system moves and T be the tension in the string.

Free body diagrams of 300 N block and 450 N blocks along with inertia forces are shown in Figs. 15.12(b) and 15.12(c).

D' ALEMBERT'S PRINCIPLE

Consider the body weighing 300 N

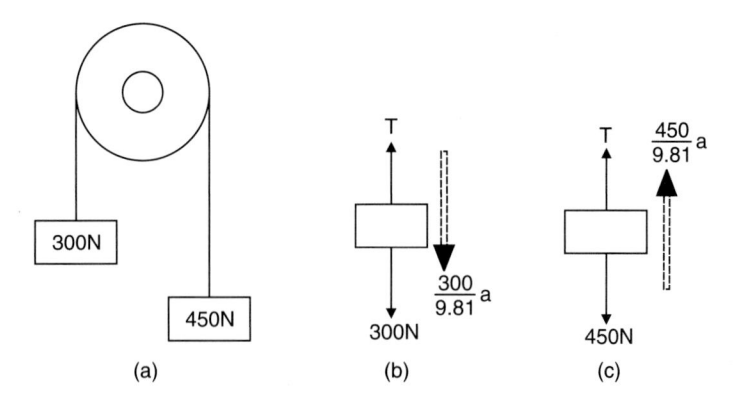

Fig. 15.12

$$\Sigma v = 0$$

$$T - \frac{300}{9.81} a - 300 = 0$$

i.e.
$$T - \frac{300}{9.81} a = 300 \qquad \ldots(1)$$

Considering the body weighing 450 N, we get

$$T + \frac{450}{9.81} a = 450 \qquad \ldots(2)$$

Subtracting eqn. (1) from eqn. (2) we get,

$$\frac{450}{9.81} a + \frac{300}{9.81} a = 450 - 300$$

$$\boldsymbol{a = 1.962 \text{ m/sec}^2} \text{ Ans.}$$

Substituting this value in eqn. (1), we get,

$$T = 300 + \frac{300}{9.81} \times 1.962$$

i.e. $\boldsymbol{T = 360 \text{ N}}$ **Ans.**

Example 15.11. *Determine the tension in the string and accelerations of blocks A and B weighing 1500 N and 500 N connected by an inextensible string as shown in Fig. 15.13(a). Assume pulleys as frictionless and weightless.*

Solution. In this pulley system, it may be observed that if 1500 N block moves downward by distance x, 500 N block moves up by $2x$. Hence if acceleration of 1500 N block is a that of 500 N block is $2a$. The free body diagrams of 1500 N block and 500 N blocks are shown in Fig. 15.13(b) and (c), along with inertia forces. According to D' Alembert's principle the system of forces shown in Fig. 15.13(b) and (c) may be treated to be in equilibrium.

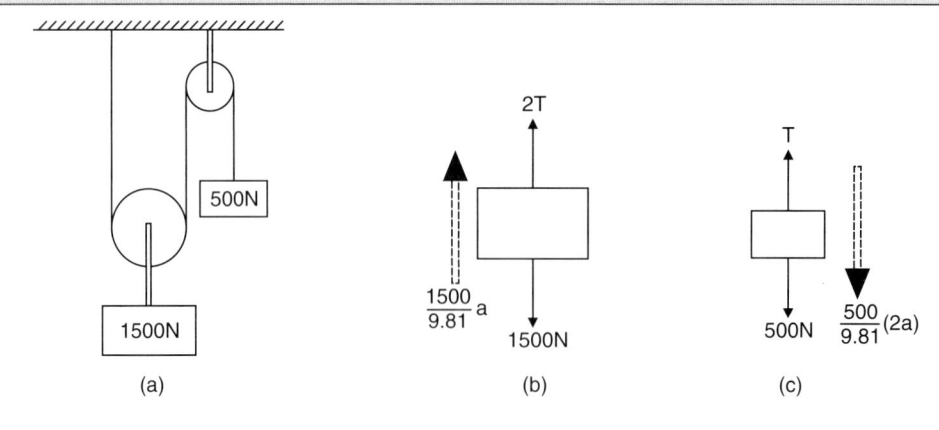

Fig. 15.13

Considering 1500 N block, we get,

$$2T + \frac{1500}{9.81} a = 1500 \qquad \ldots(1)$$

Considering 500 N block, we get,

$$T - \frac{500}{9.81}(2a) = 500 \qquad \ldots(2)$$

From eqns. (1) and (2), we get,

$$\left(\frac{1500}{9.81} + \frac{2000}{9.81}\right) a = 500$$

∴ $a = 1.401$ **m/sec² Ans.**

Substituting it in (1), we get

$$2T = 1500 - \frac{1500}{9.81} \times 1.401$$

$$T = 642.89 \text{ N} \quad \textbf{Ans.}$$

Example 15.12. *An engine of weight 500 kN pulls a carriage weighing 1500 kN up an incline of 1 in 100. The train starts from rest and moves with a constant acceleration against a resistance of 5 N/kN. It attains a maximum speed of 36 kmph in 1 km distance. Determine the tension in the coupling between the carriage and the engine and the tractive force developed by the engine.*

Solution. Initial velocity $u = 0$

Final velocity $v = 36$ kmph $= \dfrac{36 \times 1000}{60 \times 60} = 10$ m/sec²

Displacement $s = 1$ km $= 1000$ m

From the kinematic equation $v^2 = u^2 + 2as$, we get

$$10^2 = 0 + 2 \times a \times 1000$$

$$a = 0.05 \text{ m/sec}^2$$

D' ALEMBERT'S PRINCIPLE

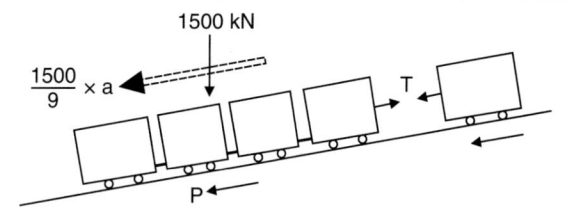

Fig. 15.14

To find the tension in the coupling between engine and train, dynamic equilibrium of the coaches only may be considered. Let T be the tension in coupling. Various forces acting parallel to the track are shown in Fig. 15.14.

Tractive resistance = $5 \times 1500 = 7500$ N = 7.5 kN down the plane.

Component of weight of train = $1500 \times \dfrac{1}{100} = 15$ kN down the plane.

(Note: Since slope of track is very small, $\sin\theta = \tan\theta = \dfrac{1}{100}$.)

Inertia force = $\dfrac{W}{g} a$

$= \dfrac{1500}{9.81} \times 0.05 = 7.645$ kN (down the plane)

Dynamic equilibrium equation is

$T - 7.5 - 15 - 7.645 = 0$

or $\qquad T = 30.145$ **kN Ans.**

To find tractive force developed by the engine, consider dynamic equilibrium of entire train.

Figure 15.15 shows various forces in the direction of track acting on the entire train along with inertia force. Let P be tractive force developed.

Total tractive resistance = $5 \times 2000 = 10,000$ N
$= 10$ kN, down the plane.

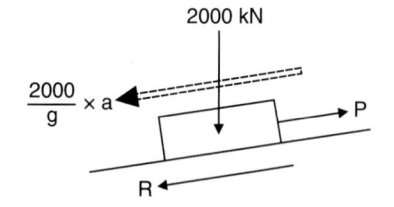

Fig. 15.15

Inertia force = $\dfrac{W}{g} a$

$= \dfrac{2000}{9.81} \times 0.05 = 10.194$ kN, down the plane.

Component of weight down the plane

$= 2000 \times \dfrac{1}{100} = 20$ kN

Dynamic equilibrium equation along the track is

$P - 10 - 10.194 - 20 = 0$

∴ $\qquad P = 40.194$ **kN Ans.**

Important Definition

1. **D' Alembert's principle** states that the system of forces acting on a body in motion is in dynamic equilibrium with the inertia force of the body.

PROBLEMS FOR EXERCISE

15.1 A mine cage weighs 12 kN and can carry a maximum load of 20 kN. The average frictional resistance of the slide guys is 500 N. What constant cable tension is required to give a loaded cage an upward velocity of 3 m/sec, from rest in a distance of 3 m? [**Ans.** 37.393 kN]

15.2 A train weighing 3000 kN is moving up a slope 2 in 100 with an acceleration of 0.04 m/sec². Tractive resistance is 6 N/kN. Determine the acceleration of the train if it moves with the same tractive force:

(a) on a level track;

(b) down the plane inclined at 2 in 100. [**Ans.** 0.2362 m/sec²; 0.4324 m/sec²]

15.3 The coefficient of friction between the crate and the cart shown in Fig. 15.16 is 0.25 and that between cart and the road is 0.15. If the cart is pulled by a force of 500 N determine the acceleration of the cart. Check whether there is slipping of the crate at this stage.

[**Ans.** $a = 1.4138$ m/sec²; No slipping]

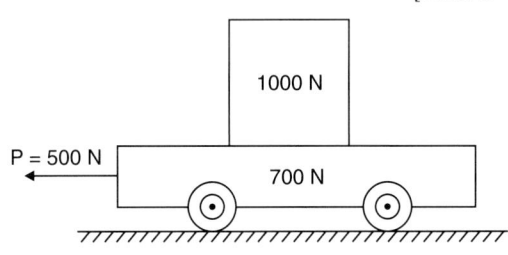

Fig. 15.16

15.4 A block weighing 2500 N rests on a level horizontal plane which has a coefficient of friction 0.20. This block is pulled by a force of 1000 N, which is acting at an angle of 30° to the horizontal. Find the velocity of the block, after it moves 30 m, starting from rest. If the force of 1000 N is then removed, how much further will it move? [**Ans.** $v = 10.4748$ m/sec; $s = 27.961$ m]

15.5 In what distance will body 1 shown in Fig. 15.17 attain a velocity of a 3 m/sec starting from rest? Take coefficient of friction between the blocks and plane as 0.2. Assume pulley is smooth. What is the tension in the chord? [**Ans.** $s = 12.953$ m; $T = 159.08$ N]

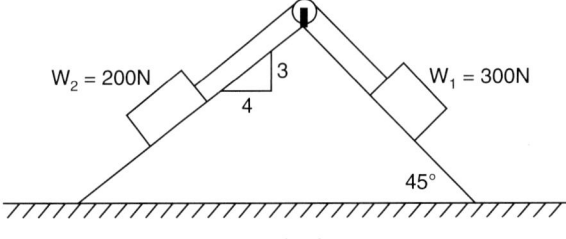

Fig. 15.17

15.6 Determine the constant force P that will give the system of bodies shown in Fig. 15.18 a velocity of 3 m/sec after moving a distance of 4.5 m from the position of rest. Coefficient of friction at all contact points is 0.2. Assume pulleys as frictionless. [**Ans.** 448.39 N]

D' ALEMBERT'S PRINCIPLE

15.7 Two blocks of weight W_1 and W_2 are connected by inextensible wire passing over a smooth pulley as shown in Fig. 15.19. If W_1 is greater than W_2, find the tension in the string and the acceleration of the system.

$$\left[\text{Ans. } a = g\left(\frac{W_1 - W_2}{W_1 + W_2}\right); T = \frac{2 W_1 W_2}{W_1 + W_2}\right]$$

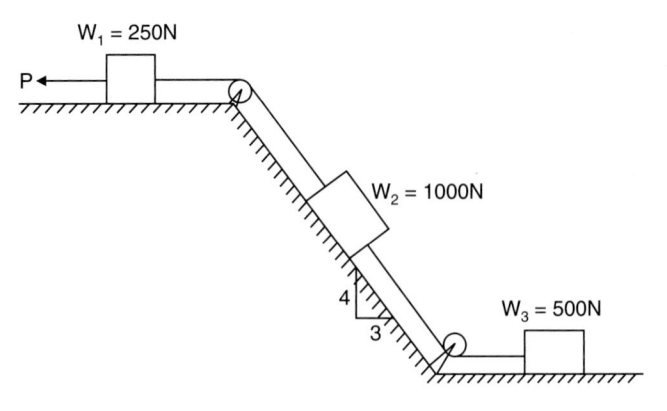

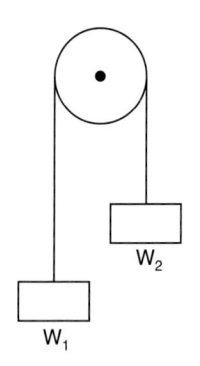

Fig. 15.18

Fig. 15.19

15.8 Find the acceleration of the three weights shown in Fig. 15.20. Assume pulleys are frictionless. Find also the tension in the cables.

$$\left[\text{Ans. Acceleration of 200 N block} = \frac{13}{37} g; \text{ 40 N block} = \frac{23}{37} g; \text{ 40 N block} = \frac{3}{37} g\right.$$

$$\left.\text{Tension in upper cable} = 129.73 \text{ N; Tension in lower cable} = 64.85 \text{ N}\right]$$

15.9 The weights of the three blocks shown in Fig. 15.21 are $W_A = 100$ N, $W_B = 200$ N and $W_C = 200$ N. Coefficient of friction between block A and the floor is 0.2, that between floor and block C is 0.25. Assuming pulleys as weightless and smooth, find the acceleration of each block.

$$\left[\text{Hint: } a_B = \left(\frac{a_A + a_C}{2}\right)\right]$$

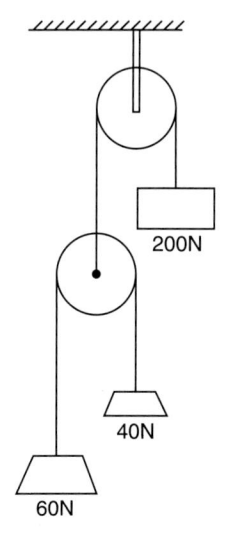

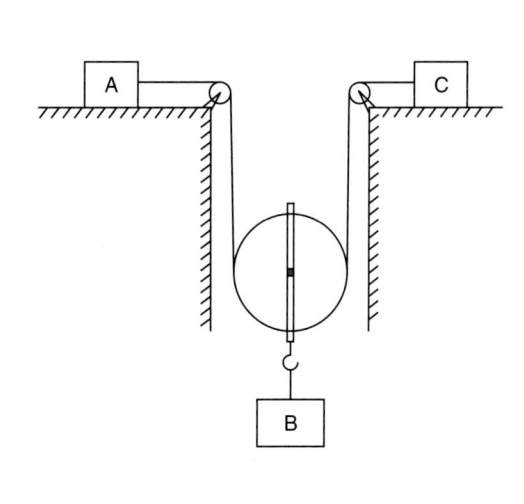

[**Ans.** $a_A = 0.5 g$, $a_B = 0.3 g$, $a_C = 0.1 g$]

Fig. 15.20

Fig. 15.21

16
Work Energy Method

In the last chapter, kinetic problems were solved using D' Alembert's dynamic equilibrium condition. In this chapter, another approach, called Work-Energy approach, is used to solve kinetic problems. This method is advantageous over D' Alembert's method when the problem involves velocities, rather than acceleration. The terms work, energy and power are explained first, then the work energy equation is derived. Using this equation a number of kinetic problems are solved.

16.1. WORK

The *work done* by a force on a moving body is defined as *the product of the force and the distance moved in the direction of the force*. In Fig. 16.1, various forces acting on a particle are shown. If the particle moves a distance s in x direction, from A to B, then the work done by various forces are as given below:

Force	Work done
P_1	$P_1 s$
P_2	$+ P_2 (-s) = -P_2 s$
P_3	$P_3 \times 0 = 0$
P_4	$P_4 s \cos \theta$

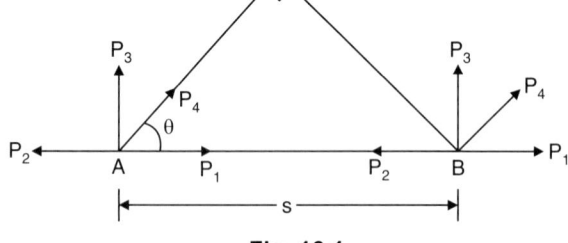

Fig. 16.1

The general expression for work done by a force P is $Ps \cos \theta$, where θ is the angle between the force and the direction of motion. This expression may be rearranged as $P \cos \theta \times s$, giving the new definition of work done. Thus work done by a force may be defined as *the product of component of force in the direction of motion and the distance moved*.

From the definition of work, it is obvious that unit of work is obtained by multiplying unit of force by unit of length. Hence, if newton is unit of force and metre is unit of displacement, unit of work will be N-m. One N-m of work is denoted by the term *Joule (J)*. Hence one Joule may be defined as *the amount of work done by one Newton force when the particle moves 1 metre in the direction of that force*.

The other commonly used units are: *kilo joules kJ* (*i.e.,* kN-m) or *milli Joules mJ* (N-mm) etc.

WORK ENERGY METHOD

16.2. WORK DONE BY A VARYING FORCE

Let the varying force acting at any instance on the particle be P. Now if the particle moves a small distance δs, then the work done by the force is $P \times \delta s$. Work done by the force in moving the body by a distance s is ΣPs. Thus if a force versus displacement curve is drawn (Fig. 16.2), the area under the curve gives the work done by the force. If the variation of P is in a regular fashion, then

$$\Sigma Ps = \int P \, ds \qquad \ldots(16.1)$$

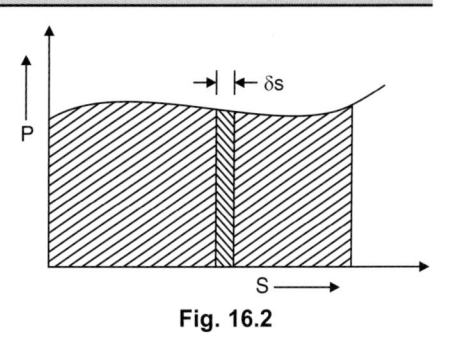

Fig. 16.2

16.3. ENERGY

Energy is defined as the *capacity to do work*. There are many forms of energy like heat energy, mechanical energy, electrical energy and chemical energy. In engineering mechanics, we are interested in mechanical energy. This energy may be classified into potential energy and kinetic energy.

Potential energy is the capacity to do work *due to the position of the body*. A body of weight 'W' held at a height h possesses an energy Wh.

Kinetic energy is *the capacity to do work due to motion of the body*. Consider a car moving with a velocity v m/sec. (Fig. 16.3). If the engine is stopped, it still moves forward, doing work against frictional resistance and stops at a certain distance s. From the kinematic of the motion, we have

$$0 - u^2 = 2as$$

$$a = -\frac{u^2}{2s}$$

from D' Alembert's principle,

$$F + \frac{W}{g} a = 0$$

i.e.,

$$F - \frac{W}{g} \times \frac{u^2}{2s} = 0$$

or

$$F = \frac{Wu^2}{2gs}$$

Fig. 16.3

$\therefore$ Work done $= F \times s = \dfrac{Wu^2}{2g}$ ...(16.2)

This work is done by the energy stored initially in the body.

$\therefore$ Kinetic energy $= \dfrac{1}{2} \times \dfrac{W}{g} v^2$ where v is the velocity of the body. ...(16.3)

Unit of energy is same as that of work, since it is nothing but capacity to do work. It is measured in *Joules* J (N-m) or *kilo Joules* kJ (*i.e.,* kN-m).

16.4. POWER

Power is defined as *time rate of doing work*. Unit of power is *Watt* (*w*) and is defined as one Joule of work done in one second. In practice kilowatt is the commonly used unit which is equal to 1000 watts. Horse power is the unit used in MKS and FPS systems.

$$1 \text{ metric H.P.} = 735.75 \text{ watts}$$
$$1 \text{ British H.P.} = 745.8 \text{ watts}$$

Example 16.1. *A pump lifts 40 m³ of water to a height of 50 m and delivers it with a velocity of 5 m/sec. What is the amount of energy spent during this process? If the job is done in half an hour, what is the input power of the pump which has an overall efficiency of 70%?*

Solution. Output energy of the pump is spent in lifting 40 m³ of water to a height of 50 m and delivering it with the given kinetic energy of delivery.

Work done in lifting 40 m³ of water to a height of 50 m is,

$$= Wh \text{ where } W \text{ is weight of 40 m}^3 \text{ of water}$$
$$= 40 \times 9810 \times 50$$
$$= 1,96,20000 \text{ N-m}$$

[*Note:* 1 m³ of water weighs = 9810 Newtons]

$$\text{Kinetic energy at delivery} = \frac{1}{2} \times \frac{W}{g} v^2$$
$$= \frac{1}{2} \times \frac{40 \times 9810}{9.81} \times 5^2$$
$$= 5,00000 \text{ N-m}$$

Total energy spent = 1,96,20000 + 5,00000

Energy spent = 2,01,20000 N-m Ans.

This energy is spent by the pump in half an hour *i.e.*, in 30 × 60 = 1800 sec.

∴ Output power of pump = output energy spent per second

$$= \frac{2,01,20000}{1800}$$
$$= 11177.8 \text{ watts}$$
$$= 11.1778 \text{ kW}.$$

$$\text{Input power} = \frac{\text{output power}}{\text{efficiency}}$$
$$= \frac{11.1778}{0.7}$$
$$= \mathbf{15.9683 \text{ kW Ans.}}$$

Example 16.2. *A man wishes to move wooden box of 1 metre cube to a distance of 5 m with the least amount of work. If the block weighs 1 kN and the coefficient of friction is 0.3, find whether he should tip it or slide it.*

Solution. *Work done in sliding*

Normal reaction $N = W = 1$ kN

∴ Friction force = $F = \mu N = 0.3$ kN

Applied force $P = F = 0.3$ kN

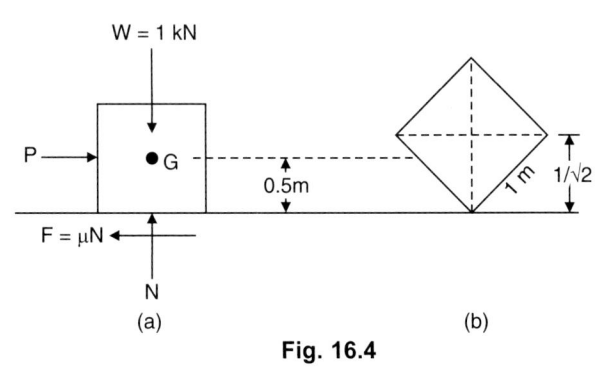

Fig. 16.4

Work to be done in sliding to a distance of 5 m = $P \times 5$
$$= 0.3 \times 5 = 1.5 \text{ kN-m}$$
$$= 1.5 \text{ kJ} \qquad \ldots(1)$$

Work to be done in tipping

In one tipping the centre of gravity of box is to be raised to a height (Ref. Fig. 16.4)

$$= \frac{1}{\sqrt{2}} - 0.5 = 0.207 \text{ m}$$

∴ Work done in one tipping = Wh
$$= 1 \times 0.207$$
$$= 0.207 \text{ kJ}$$

To move a distance of 5 m, five tippings are required. Hence work to be done in moving it by 5 metres, by tipping
$$= 5 \times 0.207 = 1.035 \text{ kJ} \qquad \ldots(2)$$

Since the man needs to spend only 1.035 kJ when tipping and it is less than 1.5 kJ to be spent in sliding, the man should move the box by tipping.

16.5. WORK ENERGY EQUATION FOR TRANSLATION

Consider the body shown in Fig. 16.5 subject to a system of forces F_1, F_2 ... and moving with an acceleration a in x-direction. Let its initial velocity at A be u and final velocity when it moves distance $AB = s$ be v. Then the resultant of system of the forces must be in x direction. Let

$$R = \Sigma F_x \qquad \ldots(16.4)$$

From Newton's second law of motion,

$$R = \frac{W}{g} a$$

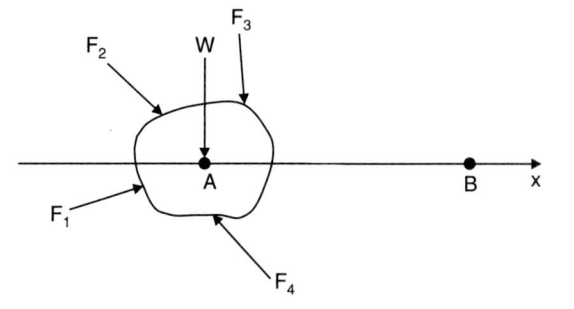

Fig. 16.5

Multiplying both sides by elementary distance ds, we get

$$Rds = \frac{W}{g} a \, ds$$
$$= \frac{W}{g} v \frac{dv}{ds} ds \text{ since } a = v \frac{dv}{ds} \qquad \text{(See eqn. 12.7)}$$
$$= \frac{W}{g} v dv$$

Integrating both sides for the motion from A to B we get,

$$\int_0^s R \, ds = \int_u^v \frac{W}{g} v \, dv$$

$$Rs = \frac{W}{g} [v^2/2]_u^v$$

$$= \frac{W}{2g}(v^2 - u^2) \qquad \ldots(16.5)$$

Now $R.s$ is the work done by the forces acting on the body. $\frac{W}{2g}v^2$ is final kinetic energy and $\frac{Wu^2}{2g}$ is initial kinetic energy. Hence we can say, work done in a motion is equal to change in kinetic energy. That is,

$$\text{Work done} = \text{Final kinetic energy} - \text{Initial kinetic energy} \qquad \ldots(16.6)$$

and it is called *Work Energy Equation*.

This work energy principle may be stated as *the work done by a system of forces acting on a body during a displacement is equal to the change in kinetic energy of the body during the same displacement.*

Using this work energy equation a number of kinetic problems can be solved. This will be found more useful than D' Alembert's principle when we are not interested in finding acceleration in the problem, but mainly interested in velocity and distance.

Example 16.3. *A body weighing 300 N is pushed up a 30° plane by a 400 N force acting parallel to the plane. If the initial velocity of the body is 1.5 m/sec and coefficient of kinetic friction is $\mu = 0.2$, what velocity will the body have after moving 6 m?*

Solution. Consider the free body diagram of the body shown in Fig. 16.6.

Σ Forces normal to plane = 0, gives

$$N = 300 \times \cos 30°$$
$$= 259.81 \text{ Newton}$$

∴ Frictional force $F = \mu N$
$$= 0.2 \times 259.81 = 51.96 \text{ N}$$

Initial velocity $u = 1.5$ m/sec

Displacement $s = 6$ m

Let the final velocity be $= v$ m/sec.

Equating the work done by forces along the plane to change in kinetic energy, we get

$$(400 - F - W \sin \theta)s = \left(\frac{1}{2}\right)\frac{W}{g}(v^2 - u^2)$$

$$\left(400 - 51.96 - 300 \times \frac{1}{2}\right)6 = \frac{1}{2} \times \left(\frac{300}{9.81}\right)(v^2 - 1.5^2)$$

$$77.71 = v^2 - 2.25$$

∴ $v = 8.942$ m/sec **Ans.**

Example 16.4. *Find the power of a locomotive, drawing a train whose weight including that of engine is 420 kN up an incline 1 in 120 at a steady speed of 56 kmph, the frictional resistance being 5 N/kN.*

While the train is ascending the incline, the steam is shut off. Find how far it will move before coming to rest, assuming that the resistance to motion remains the same.

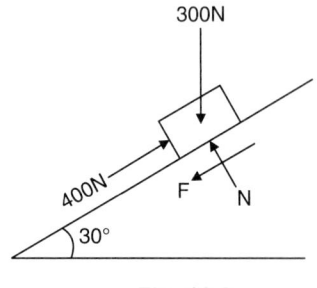

Solution. Fig. 16.7(a) shows the system of forces acting on the locomotive while moving up the incline with steady speed.

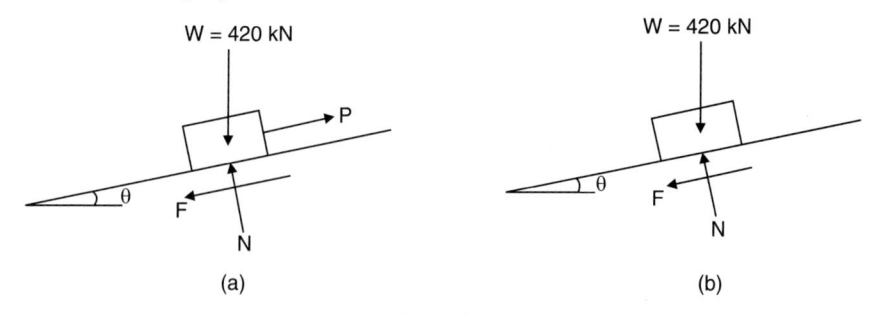

Fig. 16.7

$$v = 56 \text{ kmph}$$
$$= \frac{56 \times 1000}{60 \times 60} = 15.556 \text{ m/sec}$$
$$F = 5 \times 420 = 2100 \text{ Newton}$$
$$= 2.1 \text{ kN}$$
$$P = F + W \sin \theta$$
$$= 2.1 + 420 \times \frac{1}{120}$$
$$= 5.6 \text{ kN}$$

∴ **Power of locomotive** = Work done by P per second
$$= P \times \text{distance moved per second}$$
$$= P \times v$$
$$= 5.6 \times 15.556$$
$$= \mathbf{87.11 \text{ kN}} \textbf{ Ans.}$$

When steam is put off, let it move a distance s before coming to rest
$$\text{initial velocity } u = 15.556 \text{ m/sec}$$
$$\text{Final velocity } v = 0$$

Fig. 16.7(b) shows the system of forces acting in this motion. Resultant force parallel to the plane is
$$= F + W \sin \theta$$
$$= 2.1 + 420 \times \left(\frac{1}{120}\right) = 5.6 \text{ kN (down the plane.)}$$

Writing the work energy equation for motion up the plane, we have,
$$-5.6 \times s = \frac{1}{2} \times \frac{420}{9.81} (0 - 15.556^2)$$
$$s = \mathbf{924.98 \text{ m}} \textbf{ Ans.}$$

Example 16.5. *A tram car weighs 120 kN, the tractive resistance being 5 N/kN. What power will be required to propel the car at a uniform speed of 20 kmph?*

(1) On level surface
(2) up an incline of 1 in 300 and
(3) down an inclination of 1 in 300?
Take efficiency of motor as 80%.

(a)	(b)	(c)
On level track	Up the incline	Down the incline

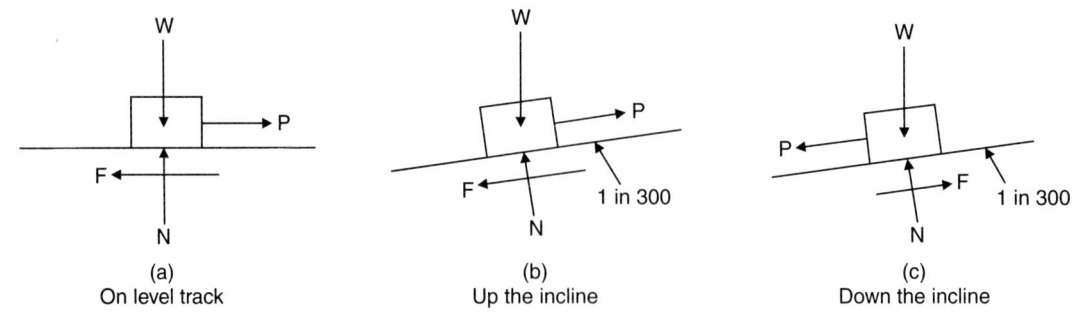

Fig. 16.8

Solution. Fig. 16.8(a), (b) and (c) show the free body diagrams of the locomotive in the three cases given.

In all the cases, frictional resistance

$$F = 5 \text{ N/kN} = 5 \times 120 = 600 \text{ N} = 0.6 \text{ kN}$$

The locomotive is moving with uniform velocity. Hence it is in equilibrium.

Now, $\qquad v = 20 \text{ kmph} = \dfrac{20 \times 1000}{60 \times 60} = 5.556 \text{ m/sec}$

(1) On level track

$$P = F = 0.6 \text{ kN}$$

∴ Output power $= Pv$
$= 0.6 \times 5.556$
$= 3.333 \text{ kW}$

$\eta = 80\% = 0.8$

∴ **Input power** $= \dfrac{\text{Output power}}{\eta}$

$= \dfrac{3.3333}{0.8}$

$= \mathbf{4.167 \text{ kW} \text{ Ans.}}$

(2) Up the plane:

The component of weight $W \sin \theta$ acts down the plane and $\sin \theta \approx \tan \theta = \dfrac{1}{300}$

∴ $\qquad P = F + W \sin \theta = 0.6 + 120 \times \dfrac{1}{300}$

$= 1 \text{ kN}$

Output power required $= P \times v$
$= 1 \times 5.5556$
$= \mathbf{5.5556 \text{ kW}}$

∴ **Input power of engine** $= \dfrac{\text{Output power}}{\eta}$

$= \dfrac{5.5556}{0.8}$

$= \mathbf{6.94 \text{ kW} \text{ Ans.}}$

WORK ENERGY METHOD

(3) Down the incline plane:

Referring to Fig. 16.8(c), we have

$$P = F - W\sin\theta = 0.6 - 120 \times \frac{1}{300} = 0.2$$

∴ Output power $= 0.2 \times 5.5556$
$= 1.1111$ kW

Input power $= \dfrac{1.1111}{0.8} = 1.389$ **kW Ans.**

Example 16.6. *In a police investigation of tyre marks, it was concluded that a car while in motion along a straight level road skidded for a total of 60 metres after the brakes were applied. If the coefficient of friction between the tyres and the pavement is estimated as 0.5, what was the probable speed of the car just before the brakes were applied?*

Solution. Let the probable speed of the car just before brakes were applied be u m/sec. Free body diagram of the car is shown in Fig. 16.9.

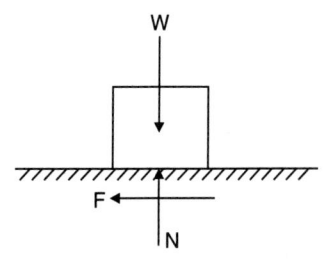

Fig. 16.9

Now, $\Sigma F_v = 0$, gives
$$N = W$$
∴ $$F = \mu N = \mu W = 0.5\,W \qquad \ldots(1)$$

Only force in the direction of motion is F.

Now, final velocity $= 0$
displacement, $s = 60$ m

Applying work energy equation,

$$-F \times s = \frac{W}{2g}(v^2 - u^2)$$

$$-0.5\,W \times 60 = \frac{W}{2 \times 9.81}(0 - u^2)$$

∴ $u = 24.261$ m/sec

$= \dfrac{24.261 \times 60 \times 60}{1000} = $ **87.34 kmph Ans.**

Example 16.7. *A block weighing 2500 N rests on a level horizontal plane for which coefficient of friction is 0.20. This block is pulled by a force of 1000 N acting at an angle of 30° to the horizontal. Find the velocity of the block after it moves 30 m starting from rest. If the force of 1000 N is then removed, how much further will it move? Use work energy method.*

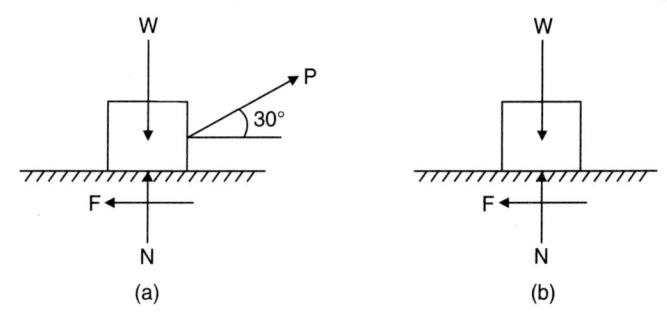

Fig. 16.10

Solution. Free body diagrams of the block for the two cases are shown in Fig. 16.10(a) and (b).

When pull P is acting:
$$N = W - P \sin 30°$$
$$= 2500 - 1000 \sin 30° = 2000 \text{ Newton}$$
$$F = \mu N = 0.2 \times 2000$$
$$= 400 \text{ Newton}$$

Initial velocity = 0

Let   Final velocity be v.

Displacement = s = 30 m

Applying work energy equation for the horizontal motion,

$$(P \cos 30° - F)s = \frac{W}{2g}(v^2 - u^2)$$

$$(0.866 \times 1000 - 400)\,30 = \frac{2500}{2 \times 9.81}(v^2 - 0)$$

$$v = 10.4745 \text{ m/sec Ans.}$$

Now, if the force 1000 N is removed (Ref. Fig. 16.10), let the distance moved be 's' before the body comes to rest.

∴   Initial velocity = 10.4745 m/sec.

Final velocity = 0

Applying work energy equation for the motion in horizontal direction, we get

$$-F \times s = \frac{W}{2g}(v^2 - u^2)$$

$$-400 \times s = \frac{2500}{2 \times 9.81}(0 - 10.4745^2)$$

$$s = 27.96 \text{ m Ans.}$$

Example 16.8. *A small block starts from rest at point A and slides down the inclined plane as shown in Fig. 16.11(a). What distance along the horizontal plane will it travel before coming to rest? The coefficient of kinetic friction between the block of either plane is 0.3. Assume that the initial velocity with which it starts to move along BC is of the same magnitude as that gained in sliding from A to B.*

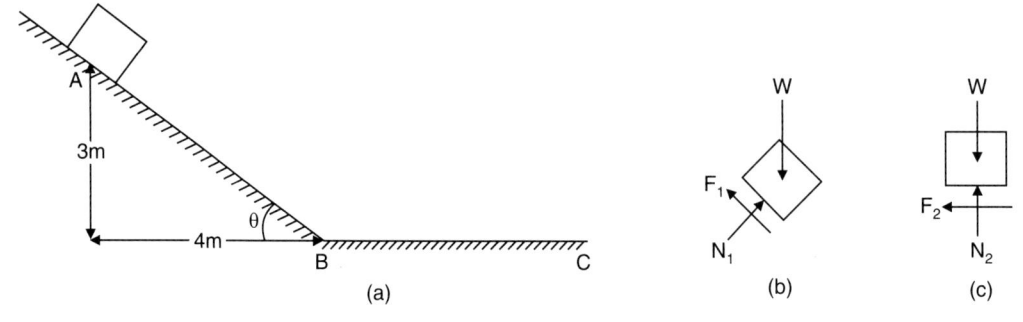

Fig. 16.11

Solution. Length $AB = \sqrt{3^2 + 4^2} = 5$ m

∴ $\sin \theta = 0.6$ and $\cos \theta = \mathbf{0.8}$

Consider the free body diagram of the block on inclined plane at A [Fig. 16.11(b)].
It moves down the plane. Hence

Σ forces normal to plane $= 0$

$$N_1 = W \cos \theta$$
$$= W \times 0.8$$

∴ $F_1 = \mu N_1 = 0.3\, W \times 0.8 = 0.24\, W$

Applying work energy equation for the motion from A to B, velocity at B, v_B is given by,

$$(W \sin \theta - F_1)s = \frac{W}{2g}(v_B^2 - 0)$$

$$(0.6\,W - 0.24\,W) \times 5 = \frac{W}{2 \times 9.81}(v_B^2)$$

$$v_B = 5.943 \text{ m/sec}$$

For the motion on horizontal plane, free body diagram of the block is shown in Fig. 16.11(c).

Initial velocity $= v_B = 5.943$ m/sec

final velocity $= 0$

Writing work energy equation for the motion along BC, we have

$$-F_2 s = \frac{W}{2g}(0 - v_B^2)$$

But $F_2 = \mu N = \mu W = 0.3 W$

∴ $-0.3 Ws = -\dfrac{W}{2 \times 9.81}(5.943)^2$

∴ $s = \mathbf{6}$ **m Ans.**

16.6 MOTION OF CONNECTED BODIES

Work energy equation may be applied to the connected bodies also. There is no need to separate connected bodies and work out for forces in the connecting member. Note the various forces acting on connected bodies. Equate summation of work done by forces acting on bodies to the summation of change in kinetic energy of the bodies. While writing work done note that force components in the direction of motion are to be multiplied by the distance moved.

Consider the two connected bodies shown in Fig. 16.12.

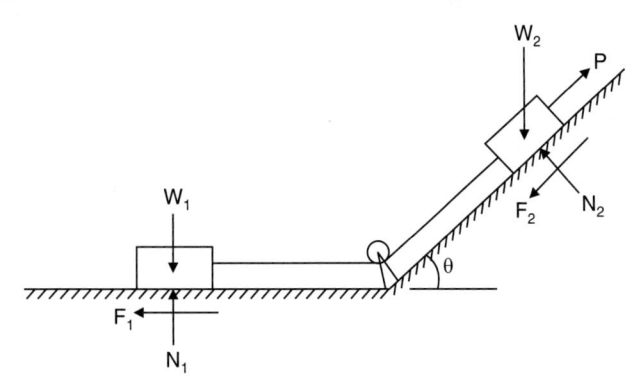

Fig. 16.12

Under the pull P, both bodies move the same distance and with the same velocity. Hence, initial velocity, final velocity and displacement are the same for the two bodies. Out of the three forces W_1, N_1 and F_1 acting on first body only the frictional force F_1 will do the work. Among the various forces acting on the second body, the applied force P, frictional force F_2 and the down the plane component of weight, $W_2 \sin \theta$ will do the work. Hence the work energy equation for that system will be,

$$-F_1 s + (P - F_2 - W_2 \sin \theta)s = \frac{W_1}{2g}(v^2 - u^2) + \frac{W_2}{2g}(v^2 - u^2)$$

i.e., $$(-F_1 + P - F_2 - W_2 \sin \theta)s = \frac{W_1 + W_2}{2g}(v^2 - u^2)$$

Example 16.9. *Determine the constant force P that will give the system of bodies shown in Fig. 16.13(a) a velocity of 3 m/sec after moving 4.5 m from rest. Coefficient of friction between the blocks and the plane is 0.3. Pulleys are smooth.*

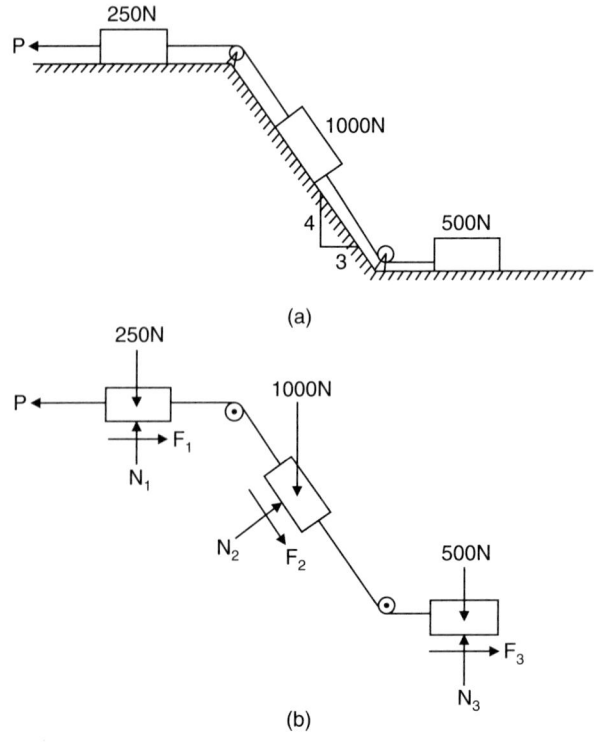

Fig. 16.13(a, b)

Solution. The system of forces acting on connecting bodies is shown in Fig. 16.13(b).

$$N_1 = 250 \text{ N}$$

∴ $$F_1 = \mu N_1 = 0.3 \times 250 = 75 \text{ N}$$

$$N_2 = 1000 \cos \theta = 1000 \times \frac{3}{5} = 600 \text{ N}$$

∴ $$F_2 = 0.3 \times N_2 = 0.3 \times 600$$
$$= 180 \text{ N}$$

WORK ENERGY METHOD

$$N_3 = 500 \text{ N}$$

$$\therefore \quad F_3 = \mu N_3 = 0.3 \times 500 = 150 \text{ N}$$

Let the constant force be *P*. Writing work energy equation,

$$(P - F_1 - F_2 - 1000 \sin\theta - F_3)s = \frac{W_1 + W_2 + W_3}{2g}(v^2 - u^2)$$

$$(P - 75 - 180 - 1000 \times 0.8 - 150)\, 4.5 = \frac{250 + 1000 + 500}{2 \times 9.81}(3^2 - 0)$$

$$P = 1383.39 \text{ N Ans.}$$

[**Note:** Work done is force × distance moved in the direction of force. Hence, N_1, N_2, N_3, 250, 1000 cos θ and 500 forces are not contributing to the work done].

Example 16.10. *In what distance will body A of Fig. 16.14 attain a velocity of 3 m/sec. starting from rest?*

Solution. Take μ = 0.2. Pulleys are frictionless and weightless.

Let θ_1 and θ_2 be the slopes of inclined planes.

Then
$$\sin\theta_1 = \frac{4}{5} = 0.8, \cos\theta_1 = 0.6$$

$$\sin\theta_2 = \frac{3}{5} = 0.6, \cos\theta_2 = 0.8$$

By observing pulley system it may be concluded that if 1500 N body moves a distance '*s*', 2000 N body moves a distance 0.5 *s*, and if velocity of 1500 N block is *v* that of **2000 N block will be 0.5 v. Assuming 2000 N body moves up the plane and 1500 N** body moves down the plane, the forces acting on 1500 N that will do work are:

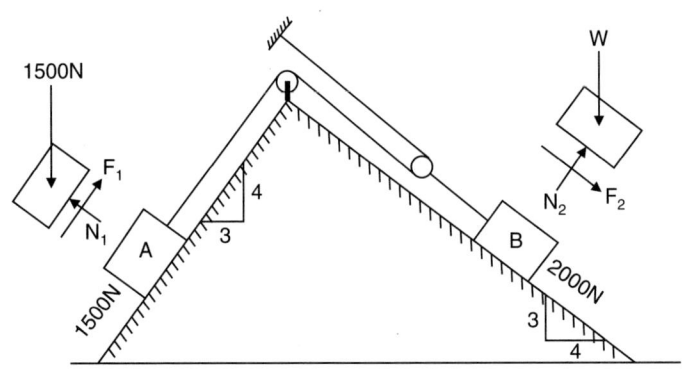

Fig. 16.14

$$1500 \sin\theta_1 = 1200 \text{ N down the plane}$$

and
$$F_1 = \mu \times 1500 \cos\theta_1 = 0.2 \times 1500 \times 0.6 = 180 \text{ N up the plane}$$

The forces acting on 2000 N block that do work when it slides up are:

$$2000 \sin\theta_2 = 2000 \times 0.6 = 1200 \text{ N down the plane}$$

and
$$F_2 = 0.2 \times 2000 \cos\theta_2 = 0.2 \times 2000 \times 0.8$$
$$= 320 \text{ N down the plane.}$$

Equating work done by various forces to change in the kinetic energy of the system, we get

$$(1200 - 180)s - (1200 + 320)\,0.5s = \left\{\frac{1500}{2 \times 9.81}\right\}(v^2 - 0) + \frac{2000}{2 \times 9.81}\{(0.5v)^2 - 0\}$$

In the present case, $v = 3$ m/sec

$$260s = \frac{1500}{2 \times 9.81} \times 3^2 + \frac{2000}{2 \times 9.81} \times 1.5^2$$

$$s = 3.529 \text{ m Ans.}$$

Since s is positive the assumed direction of motion is correct. If it comes out negative note that recalculations are to be made since the frictional force changes the sign, if motion is reversed.

Example 16.11. *Two bodies weighing 300 N and 450 N are hung to the ends of a rope passing over an ideal pulley as shown in Fig. 16.15(a). How much distance the blocks will move in increasing the velocity of system from 2 m/sec to 4 m/sec? How much is the tension in the string? Use work energy method.*

Solution. 450 N block moves down and 300 N block moves up. The arrangement is such that both bodies will be having same velocity and both will move by the same distance. Let 's' be the distance moved. Writing work energy equation for the system, we get,

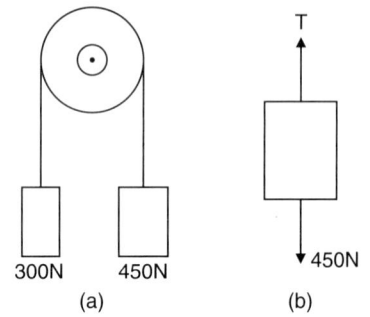

Fig. 16.15

$$450s - 300s = \left[\frac{450}{2 \times 9.81}\right](v^2 - u^2) + \left[\frac{300}{2 \times 9.81}\right](v^2 - u^2)$$

$$150s = \left[\frac{450}{2 \times 9.81}\right] \times (4^2 - 2^2) + \left[\frac{300}{2 \times 9.81}\right](4^2 - 2^2)$$

$$s = 3.058 \text{ m Ans.}$$

Let T be the tension in the string. Consider work energy equation for any one body, say 450 N body as shown in Fig. 16.15(b).

$$450s - Ts = \left[\frac{450}{2 \times 9.81}\right](4^2 - 2^2)$$

$$(450 - T)3.058 = \left[\frac{450}{2 \times 9.81}\right] \times 12$$

$$T = 360 \text{ N Ans.}$$

16.7. WORK DONE BY A SPRING

Consider a body attached to a spring as shown in Fig. 16.16. It is obvious that if the body moves out from its undeformed position, tensile force develops and if it moves towards the supports compressive force develops. In other words the force of a spring is always directed towards its normal position. Experimental results have shown that the magnitude of the force developed in the spring is directly proportional to its displacement from the undeformed position. Thus, if F is the force in the spring due to deformation x from its undeformed position

$F \alpha x$

WORK ENERGY METHOD

i.e., $$F = kx \qquad \ldots(16.7)$$

where the constant of proportionality k is called spring constant and is defined as a force required for unit deformation of the spring. Hence the unit of spring constant is N/m or kN/m.

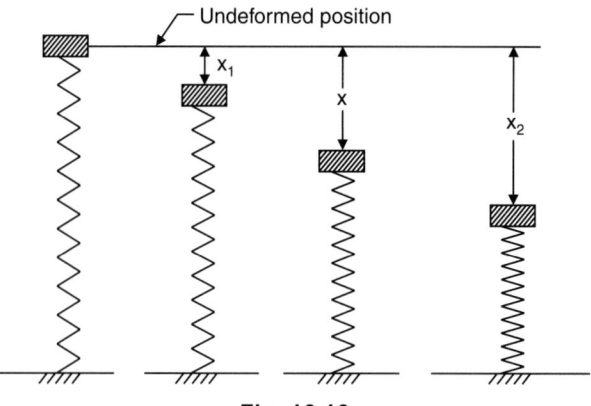

Fig. 16.16

At any instant if the displacement is dx, the work done by spring force dU is given by
$$dU = -Fdx = -kxdx$$

Note: The negative sign is used since the force of spring is in the opposite direction of displacement.

∴ Work done in displacement of the body from x_1 to x_2 is given by

$$U = \int_{x_1}^{x_2} -kxdx = -k\left[\frac{x^2}{2}\right]_{x_1}^{x_2}$$

$$= -\frac{1}{2}k(x_2^2 - x_1^2)$$

If the work done is to be found in moving from undeformed position to displacement x, then

$$U = -\frac{1}{2}k(x^2 - 0^2)$$

$$= -\frac{1}{2}kx^2 \qquad \ldots(16.8)$$

Note: The negative sign is used with the expression for work done by the spring since whenever spring is deformed the force of spring is in the opposite direction of deformation. However, if a deformed spring is allowed to move towards its normal position work done will be positive, since the movement and the force of spring are in the same direction.

Example 16.12. *A 3000 N block starting from rest as shown in Fig. 16.17 slides down a 50° incline. After moving 2 m it strikes a spring whose modulus is 20 N/mm. If the coefficient of friction between the block and the incline is 0.2, determine the maximum deformation of the spring and the maximum velocity of the block.*

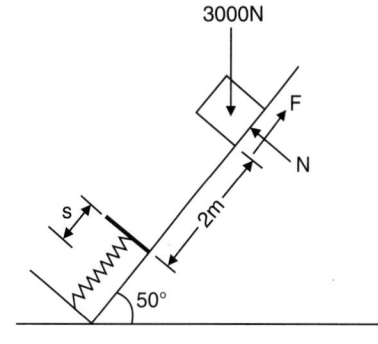

Fig. 16.17

Solution. Normal reaction $N = 3000 \cos 50°$

∴ Frictional force $F = \mu N = 0.2 \times 3000 \cos 50°$
$$= 385.67 \text{ N}.$$

Let the maximum deformation of spring be s mm. The body was at rest and is again at rest, when it moves a distance $(2000 + s)$ milli metres. Applying work energy equation,

$$(3000 \sin 50° - F)(2000 + s) - \frac{1}{2} ks^2 = 0$$

$$(3000 \sin 50° - 385.67)(2000 + s) = \frac{1}{2} \times 20 \times s^2$$

$$191.246 (2000 + s) = s^2$$

Solving the quadratic equation, we get
$$s = 721.43 \text{ mm Ans.}$$

The velocity will be maximum when acceleration $\dfrac{dv}{dt}$ is zero.

Acceleration is zero, when force is zero. Net force acting on the body is zero when spring force developed balances, the force exerted by the body. Let x be the deformation when the net force on the body in the direction of motion is zero.

Then referring to Fig. 16.18,
$$kx = W \sin \theta - F$$
$$20x = 3000 \sin 50 - 385.67$$
$$x = 95.62 \text{ mm}$$

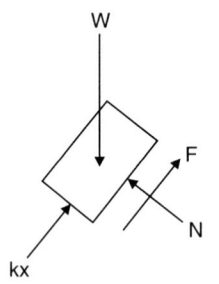

Fig. 16.18

Now applying work energy equations, we have

$$(3000 \sin 50° - 385.67)(2000 + x) - \frac{1}{2} kx^2 = \frac{3000}{2 \times 9810}(v^2 - 0)$$

$$(3000 \sin 50° - 385.67)(2000 + 95.62) - \frac{1}{2} \times 20 \times 95.67^2$$

$$= \left[\frac{3000}{2 \times 9810}\right] v^2$$

$$v = 5060.9 \text{ mm/sec}$$

i.e., $\qquad v = \textbf{5.061 m/sec Ans.}$

Note: Maintain the consistency of units
$$g = 9.81 \text{ m/sec}$$
$$= 9810 \text{ mm/sec}$$

Example 16.13. *A wagon weighing 500 kN starts from rest, runs 30 metre down one percent grade and strikes the bumper post. If the rolling resistance of the track is 5 N/kN, find the velocity of the wagon when it strikes the post.*

If the bumper spring which compresses 1 mm for every 15 kN determine by how much this spring will be compressed.

WORK ENERGY METHOD

Solution. Component of weight down the plane

$$= W \sin \theta = 500 \times \frac{1}{100} = 5 \text{ kN}$$

Track resistance = 5 N/kN

$$= 5 \times 500 = 2500 \text{ N}$$
$$= 2.5 \text{ kN}$$

The wagon starts from rest ($u = 0$), and moves a distance $s = 30$ m before striking the bumper.

Let the velocity of wagon while striking be v m/sec.

Applying work energy equation (Ref. Fig. 16.19), we get

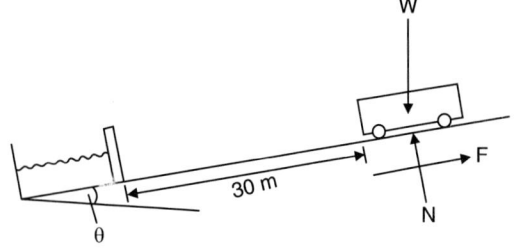

Fig. 16.19

$$(W \sin \theta - F) s = \frac{W}{2g} (v^2 - u^2)$$

$$(5 - 2.5) \times 30 = \frac{500}{2 \times 9.81} (v^2 - 0)$$

$$v = 1.716 \text{ m/sec}$$

Let spring compression be x.

The spring constant is

$$k = 15 \text{ kN/mm}$$
$$= 15000 \text{ kN/m}$$

and velocity is zero.

Applying work energy equation, we get

$$(5.0 - 2.5)(30 + x) - \frac{1}{2} kx^2 = \frac{W}{2g} (0 - 0)$$

$$2.5(30 + x) - \frac{1}{2} \times 15000 \, x^2 = 0$$

$$3000 x^2 - x - 30 = 0$$

$$x = \frac{+1 \pm \sqrt{1 + 4 \times 3000 \times 30}}{2 \times 3000}$$

$$= 0.1002 \text{ m}$$

i.e., $x = $**100.2 mm** **Ans.**

Note: Compared to work done by wagon in moving 30 m, if the work done in moving through x is neglected, then work energy equation reduces to

$$(5 - 2.5)\,30 - \frac{1}{2}\,kx^2 = 0$$

$$x = \sqrt{\frac{2.5 \times 30 \times 2}{15000}} = 0.1 \text{ m}$$

i.e., $\qquad x = 100$ mm.

Important Definitions

1. **Work done** by a force on a moving body is defined as the product of the force and the distance moved in the direction of the force.
2. The amount of work done by one one newton force when the porficle moves 1 meter in the direction of that force is termed as one **Joule** work
3. **Energy** is defined as the capcity to do the work. **Potential energy** is the capcity to do work due to the position of the body and **kinetic energy** is the capacity to do the due to the motion of the body.
4. **Power** is defined as time rate of doing work.

Important Formulae

1. Kinetic energy $= \dfrac{1}{2}\dfrac{W}{g}v^2$
2. Work done = Final kinetic energy – initial kinetic energy.

PROBLEMS FOR EXERCISE

16.1 A 10,000 kN train is accelerated at a constant rate up a 2% grade. The track resistance is constant at 9 N/kN. The velocity increases from 9 m/sec to 18 m/sec in a distance of 600 metres. Determine the maximum power developed by the locomotive.
[**Ans.** $P = 496.422$ kN power = 8935.6 kW]

16.2 Find the power required to pull a train up an incline of 1 in 200 at a speed of 36 kmph, if the weight of the train is 3000 kN and the track resistance is 5 N/kN. Also determine the maximum speed with which the train moves up an incline of 1 in 100 with the same power.
[**Ans.** Power = 300 kW; $v = 24$ kmph]

16.3 On seeing a child on the road, the driver of a car applied brakes instantaneously. The car just ran over the child after skidding a total distance of 28.2 metres in the direction of motion before coming to a stop. The traffic police sued the driver on the ground of overspeeding. The speed limit of the section was 40 kmph. Would you justify the police action if the coefficient of friction between the tyres and the road is 0.5? The weight of the car is 20 kN and it was travelling on a level surface. [**Ans.** $v = 59.877$ kmph, police action is justified]

16.4 A mine cage weighs 12 kN and can carry a maximum load of 20 kN. The average frictional resistance of the side guides is 500 N. What constant cable tension is required to give a loaded cage an upward velocity of 3 m/sec from rest in a distance of 3 m? [**Ans.** $T = 37.47$ kN]

16.5 A 500 N body moves along the two inclines for which the coefficient of friction is 0.2 (Fig. 16.20). If the body starts from rest at A and slides 60 m down the 30° incline, how far will it then move along the other incline? What will be its velocity when it returns to B?

WORK ENERGY METHOD   393

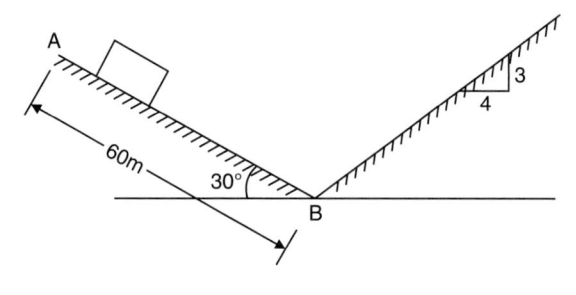

Fig. 16.20

[**Ans.** $s = 25.8$ m; $v = 14.92$ m/sec]

16.6 In what distance will body A shown in Fig. 16.21 attain a velocity of 3 m/sec starting from rest? Take coefficient of friction between the blocks and the plane as 0.2. Assume the pulley is smooth.

[**Ans.** $s = 6.541$ m]

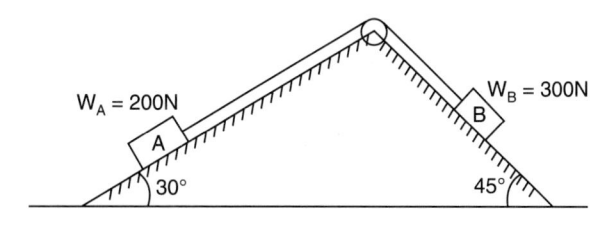

Fig. 16.21

16.7 Two blocks of weights 400 N and 500 N are connected by inextensible flexible wire running around a smooth pulley as shown in Fig. 16.22. Find what will be the velocity of the system if the distance moved by the blocks is 3 m starting from rest.   [**Ans.** $v = 2.557$ m/sec]

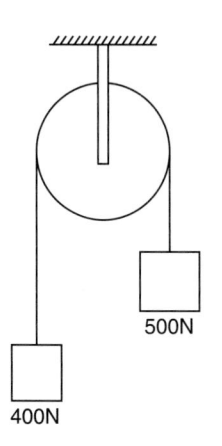

Fig. 16.22

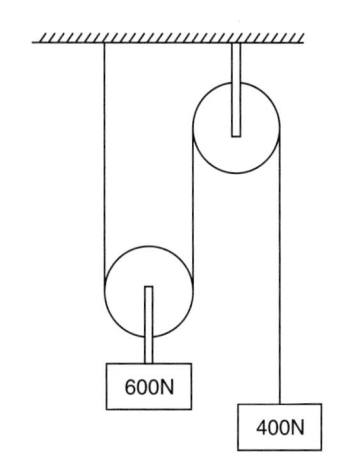

Fig. 16.23

16.8 Two blocks are connected by inextensible wires as shown in Fig. 16.23. Find by how much distance block 400 N will move in increasing its velocity to 5 m/sec from 2 m/sec. Assume pulleys are frictionless and weightless. **[Ans.** $s = 5.89$ m**]**

16.9 By using work energy equation, calculate the velocity and acceleration of the block A shown in Fig. 16.24 after it has moved 6 m from rest. The coefficient of kinetic friction is 0.3 and the pulleys are considered to be frictionless and weightless. Also calculate, the tension in the string attached to A.
[Ans. $v = 3.544$ m/sec; $a = 1.046$ m/sec^2; $T = 189.33$ N**]**

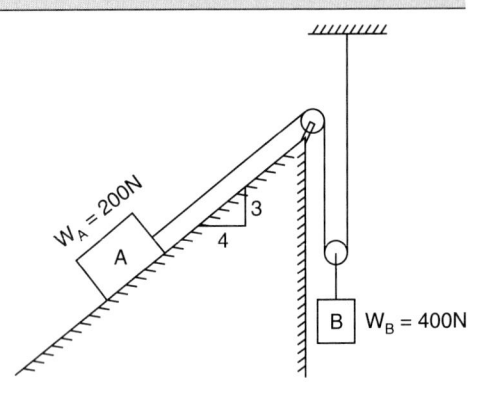

Fig. 16.23

16.10 How far block A shown in Fig. 16.25 will move when its velocity increases from 3 m/sec to 8 m/sec? Assume pulleys are weightless and frictionless. Radius of larger pulley is 0.3 m and that of smaller pulley is 0.2 m. **[Ans.** $s = 15.885$ m**]**

16.11 A wagon weighing 600 kN starts from rest, runs 30 m down a 1% grade and strikes a post. If the rolling resistance of the track is 5 N per kN, find the velocity of the wagon when it strikes the post.

If the impact is to be cushioned by means of one bumper spring, which compresses 1 mm per 25 kN weight, determine how much the bumper spring get compressed. **[Ans.** $v = 1.716$ m/sec; $s = 84.85$ mm**]**

16.12 A spring is used to stop a 1000 N package which is moving down a 20° incline. The spring has a constant $k = 150$ N/mm and is held by the cables so that it is initially compressed by 100 mm. If the velocity of the package is 6 m/sec, when it is 10 m from the spring, determine the maximum additional of the spring in bringing the package to rest. Take coefficient of friction = 0.25. **[Ans.** $s = 195.6$ mm**]**

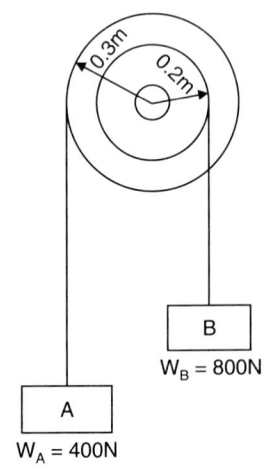

Fig. 16.24

17
Impulse Momentum

It is clear from the discussion of the previous chapters that for solving kinetic problems, involving force and acceleration, D'Alembert's principle is useful and that for the problems involving force, velocity and displacement, the work-energy method is useful. In this chapter, the **Impulse Momentum method** is dealt which is useful for solving the problems involving force, time and velocity.

17.1. LINEAR IMPULSE AND MOMENTUM

If R is the resultant force acting on a body of mass m, then from Newton's second law,
$$R = ma$$
But acceleration $\quad a = dv/dt$
$$\therefore \quad R = m\frac{dv}{dt}$$
i.e., $\quad Rdt = mdv$
$$\therefore \quad \int R\,dt = \int m\,dv$$

If initial velocity is u and after time interval t it becomes v, then
$$\int_0^t R\,dt = m\bigl[v\bigr]_u^v = mv - mu \qquad \ldots(17.1)$$

The term $\int_0^t R \cdot dt$ is called *impulse*. If the resultant force is in newton and time is in second, the unit of impulse will be N-sec.

If R is constant during time interval t, then impulse is equal to $R \times t$.

The term mass × velocity is called *momentum*.

Now, $\quad mv = \dfrac{W}{g} v$

Substituting dimensional equivalence, we get,
$$= \frac{N}{m/sec^2}\, m/sec = N\text{-sec}$$

Thus the momentum has also unit N-sec. The equation (17.1) satisfies the requirement of dimensional homogenity. Equation (17.1) can now be expressed as

Impulse = Final momentum − Initial momentum ...(17.2)

Since the velocity is a vector, impulse is also a vector. The impulse momentum equation (eqn. 17.1 or 17.2) holds good when the directions of R, u and v are the same. Impulse momentum equation can be stated as follows:

The component of the resultant linear impulse along any direction is equal to change in the component of momentum in that direction.

The impulse momentum equation can be applied in any convenient direction and the kinetic problems involving force, velocity and time can be solved easily.

Example 17.1. *A glass marble, whose weight is 0.2 N, falls from a height of 10 m and rebounds to a height of 8 metres. Find the impulse and the average force between the marble and the floor, if the time during which they are in contact is 1/10 of a second.*

Solution. Applying kinematic equations, for the freely falling body (Ref. Fig. 17.1), the velocity with which marble strikes the floor

$$= \sqrt{2gh}$$
$$= \sqrt{2 \times 9.81 \times 10}$$
$$= 14.007 \text{ m/sec (downward)} \quad ...(1)$$

Similarly applying kinematic equations for the marble moving up, we get the velocity of rebound

$$= \sqrt{2g \times 8} = \sqrt{2 \times 9.81 \times 8}$$
$$= 12.528 \text{ m/sec (upward)} \quad ...(2)$$

Taking upward direction as positive and applying impulse momentum equation, we get

$$\text{Impulse} = \frac{W}{g}(v - u)$$
$$= \frac{0.2}{9.81}[12.52 - (-14.007)]$$
$$= \mathbf{0.541 \text{ N-sec}} \quad \textbf{Ans.}$$

If F is the average force, then
$$Ft = 0.541$$
$$F \times 1/10 = 0.541 \text{ N}$$
$$\mathbf{F = 5.41 \text{ N}} \quad \textbf{Ans.}$$

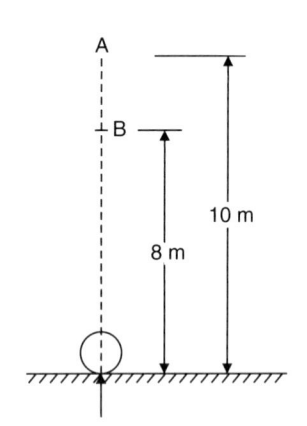

Fig. 17.1

Example 17.2. *A 1 N ball is bowled to a batsman. The velocity of ball was 20 m/sec horizontally just before batsman hit it. After hitting it went away with a velocity of 48 m/sec at an inclination of 30° to horizontal as shown in Fig. 17.2(a). Find the average force exerted on the ball by the bat if the impact lasts for 0.02 sec.*

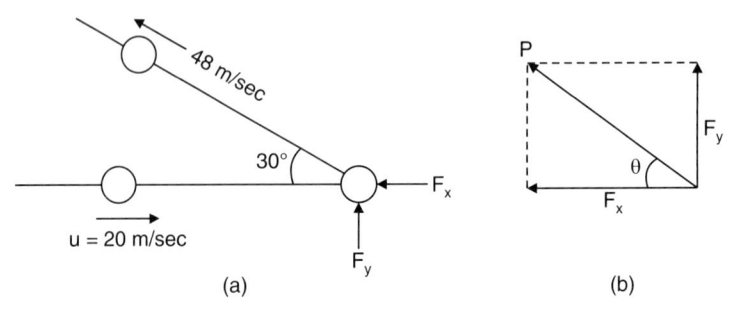

Fig. 17.2(a, b)

IMPULSE MOMENTUM

Solution. Let F_x be the horizontal component of the force and F_y be the vertical component.

Applying impulse moment condition in horizontal direction

$$F_x \times 0.02 = \frac{1}{9.81} [48 \cos 30° - (-20)]$$

$$F_x = 313.81 \text{ N}$$

Applying impulse momentum equation in vertical direction

$$F_y \times 0.02 = \frac{1}{9.81} (48 \sin 30° - 0)$$

$$\therefore \quad F_y = 122.32 \text{ N}$$

$\therefore$ Resultant force

$$F = \sqrt{P_x^2 + P_y^2} = \sqrt{313.81^2 + 122.32^2}$$

$$F = 336.81 \text{ N} \quad \text{Ans.}$$

$$\theta = \tan^{-1}\left(\frac{P_y}{P_x}\right) = \tan^{-1}\left(\frac{122.32}{313.81}\right)$$

i.e., $\quad \theta = 21.30°$ **to horizontal as shown in Fig. 17.2(b)** Ans.

Example 17.3. *A 1500 N block is in contact with a level plane, the coefficient of friction between two contact surfaces being 0.1. If the block is acted upon by a horizontal force of 300 N, what time will elapse before the block reaches a velocity of 16 m/sec starting from rest ? If 300 N force is then removed, how much longer will the block continue to move ? Solve the problem using impulse momentum equation.*

Solution. Consider the FBD of the block shown in Fig. 17.3.

Normal reaction = 1500 N

$\therefore$ Frictional force $F = \mu N = 0.1 \times 1500$
= 150 N

Applying impulse momentum equation in the horizontal direction, we have

$$(300 - 150) t = \left[\frac{1500}{9.81}\right] (v - u)$$

$$= \left[\frac{1500}{9.81}\right] (16 - 0)$$

$$t = 16.31 \text{ sec} \quad \text{Ans.}$$

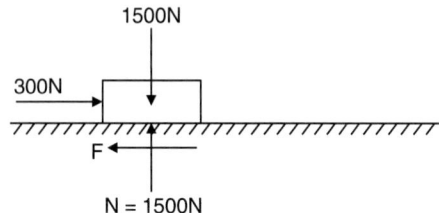

Fig. 17.3

If force is then removed the only horizontal force is $F = 150$ N. Applying impulse momentum equation for the motion towards right, we have

$$-150 t = \frac{1500}{9.81} (0 - 16)$$

$$t = 16.31 \text{ sec}$$

The block takes another 16.31 sec before it comes to rest. Ans.

Example 17.4. *A 20 kN-automobile is moving at a speed of 70 kmph when the brakes are fully applied causing all four wheels to skid. Determine the time required to stop the automobile (a) on concrete road for which $\mu = 0.75$, (b) on ice for which $\mu = 0.08$.*

Solution. Initial velocity of the vehicle

$$u = 70 \text{ kmph} = \frac{70 \times 1000}{60 \times 60}$$

$$= 19.44 \text{ m/sec}$$

Final velocity $v = 0$

Free body diagram is shown in Fig. 17.4

$$F = \mu N = \mu W = 20\mu$$

Applying impulse momentum equation, we have

$$-Ft = \frac{W}{g}(v - u)$$

$$-20\mu t = \left[\frac{20}{9.81}\right](19.44 - 0)$$

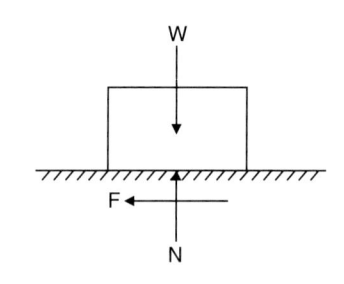

$$\therefore \quad t = \frac{1.982}{\mu}$$

Fig. 17.4

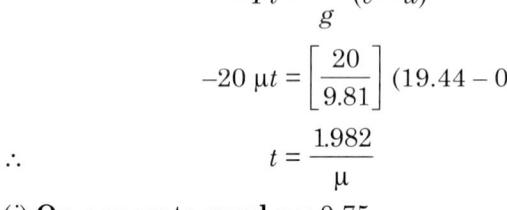

(*i*) **On concrete road** $\mu = 0.75$

$$\therefore \quad t = \frac{1.982}{0.75} = 2.64 \text{ sec} \quad \textbf{Ans.}$$

(*ii*) **On ice,** $\quad \mu = 0.08$

$$\therefore \quad t = \frac{1.982}{0.08} = 24.78 \text{ sec} \quad \textbf{Ans.}$$

Example 17.5. *A block weighing 130 N is on an incline, whose slope is 5 vertical to 12 horizontal. Its initial velocity down the incline is 2.4 m/sec. What will be its velocity 5 sec. later? Take coefficient of friction at contact surface = 0.3.*

Solution. $\tan \theta = \dfrac{5}{12} \Rightarrow \theta = 22.62°$

$$N = W \cos \theta = 130 \cos 22.62° = 120 \text{ newton}$$

$$F = \mu N = 0.3 \times 120 = 36 \text{ newton}$$

Σ Forces down the plane

$$= R = W \sin \theta - F$$

$$= 130 \sin 22.26° - 36$$

$$= 14.0 \text{ N}$$

Initial velocity $\quad u = 2.4$ m/sec.

Let final velocity be v m/sec.

Time interval $\quad t = 5$ sec.

Applying impulse momentum equation,

$$Rt = \frac{W}{g}(v - u), \text{ we get}$$

$$14 \times 5 = \left[\frac{130}{9.81}\right](v - 2.4)$$

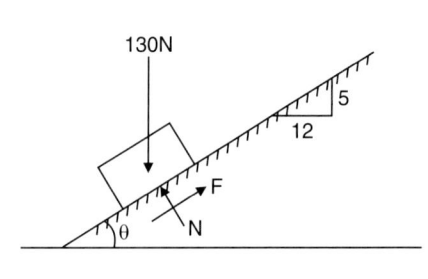

$$\therefore \quad v = 7.68 \text{ m/sec} \quad \textbf{Ans.}$$

Fig. 17.5

17.2. CONNECTED BODIES

The problems involving connected bodies may be solved by any one of the following two methods:

First Method: Free body diagrams of each body is drawn separately. Impulse momentum equation for each body in the direction of its motion is written and then the equations are solved to get the required values.

Second Method: If the connected bodies have same displacement in the same time, the impulse of internal tension in connecting chords will get cancelled. Hence free body diagram of combined bodies may be considered and impulse moment equation applied in the direction of motion of combined bodies. *This method is applicable only if displacement of each body is the same in given time.*

Example 17.6. *Determine the time required for the weights shown in Fig. 17.6(a) to attain a velocity of 9.81 m/sec. What is the tension in the chord ? Take $\mu = 0.2$ for both planes. Assume the pulleys as frictionless.*

Solution. *First Method:* Free body diagram of the two blocks are as shown in Fig. 17.6(b).

For 2000 N block:
$$N_1 = W_1 \cos 30°$$
$$= 2000 \cos 30°$$
$$= 1732.05 \text{ N}$$
$$F_1 = \mu N_1 = 0.2 \times 1732.05$$
$$= 346.41 \text{ N}$$

For 1800 N block:
$$N_2 = W_2 \cos 60°$$
$$= 1800 \cos 60° = 900 \text{ N}$$
$$F_2 = \mu N_2 = 0.2 \times 900$$
$$= 180 \text{ N}.$$

Fig. 17.6

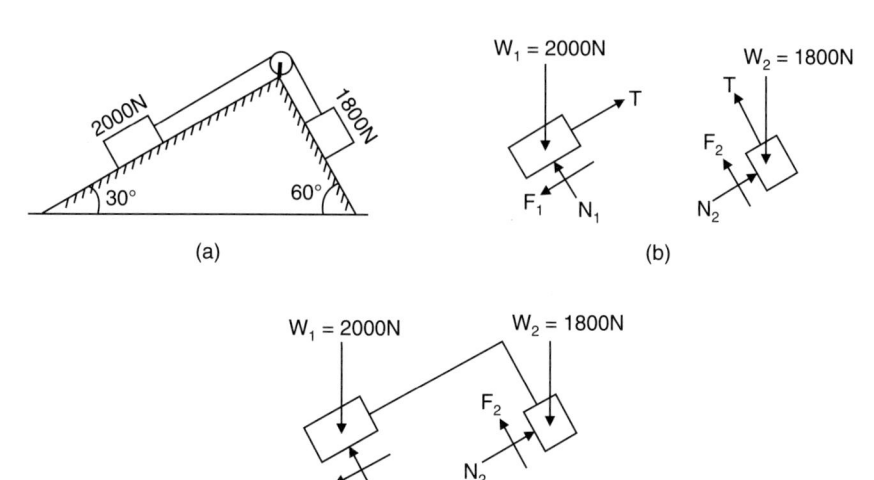

Let T be the tension in the chord.
Initial velocity $u = 0$
Final velocity $v = 9.81$ m/sec.
Applying impulse moment equation for the 2000 N block in upward direction parallel to the plane, we get

$$(T - 2000 \sin 30° - F_1) t = \left[\frac{2000}{9.81}\right] (v - u)$$

$(T - 2000 \sin 30° - 346.41)t = 2000$, Since $v - u = 9.81$

$$(T - 1346.41) t = 2000 \qquad \ldots(1)$$

Applying impulse momentum equation for 1800 N block in the direction parallel to 60° inclined plane, we get

$$(1800 \sin 60° - T - F_2) t = \left[\frac{1800}{9.81}\right] (v - u)$$

$$(1800 \sin 60° - T - 180)t = \left[\frac{1800}{9.81}\right] (9.81 - 0)$$

$$(1378.85 - T) t = 1800 \qquad \ldots(2)$$

Dividing equation (1) by Eqn. (2), we get

$$\frac{(T - 1346.41)}{(1378.85 - T)} = \frac{2000}{1800}$$

∴ $T = 1363.48$ N **Ans.**

Substituting it in Eqn. (2), we get
$(1378.85 - 1363.48) t = 1800$

∴ $t = 117.11$ sec **Ans.**

Second Method: Since the displacement of both bodies are same in given time, consider combined FBD of the blocks as shown in Fig. 17.6(c). Writing impulse momentum equation in the direction of motion, we have

$$(1800 \sin 60° - F_2 - 2000 \sin 30° - F_1)t = \frac{2000 + 1800}{9.81} (v - u)$$

$$(1800 \sin 60° - 180 - 2000 \sin 30° - 346.41)t = \frac{3800}{9.81} \times 9.81$$

∴ $t = 117.11$ sec **Ans.**

To find tension in the chord, consider the impulse momentum equation of any one block, say 2000 N block.

$$(T - 2000 \sin 30° - F_1)t = \frac{2000}{9.81} (v - u)$$

$$(T - 2000 \sin 30° - 346.41) \, 117.11 = \frac{2000}{9.81} \times 9.81$$

∴ $T = 1363.48$ N **Ans.**

Example 17.7. *Determine the tension in the strings and the velocity of 1500 N block shown in Fig. 17.7(a) 5 seconds after starting from*

(a) rest

(b) starting with a downward velocity of 3 m/sec.

Assume pulleys as weightless and frictionless.

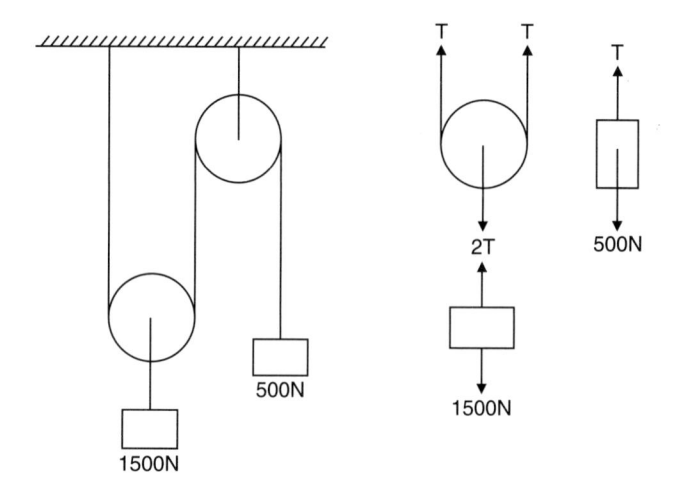

Fig. 17.7

Solution. When 1500 N block moves a distance s, in the same time 500 N block moves a distance $2s$. Hence if velocity of 1500 N block is v m/sec that of 500 N block will be $2v$ m/sec. Let T be the tension in the chord connecting 500 N block. Hence tension in the wire connecting 1500 N block will be $2T$ (see Fig. 17.7(b)). Since the velocities of two blocks are different only first method is to be used.

Case (a): Initial velocity $u = 0$, $t = 5$ sec.

Writing impulse momentum equation for 500 N block, we have

$$(T - 500)t = \frac{500}{9.81}(2v - u)$$

$$(T - 500)5 = \frac{500}{9.81} \times (2v - 0)$$

$$T - 500 = \frac{200\,v}{9.81} \qquad \ldots(1)$$

Applying impulse momentum equation to 1500 N block,

$$(1500 - 2T)\,t = \frac{1500}{9.81}(v - 0)$$

$$(1500 - 2T)\,5 = \frac{1500}{9.81}v$$

$$1500 - 2T = \frac{300\,v}{9.81} \qquad \ldots(2)$$

Adding Eqn. (2) in 2 times Eqn. (1), we get

$$1500 - 2 \times 500 = \frac{400\, v}{9.81} + \frac{300\, v}{9.81} = \frac{700\, v}{9.81}$$

$\therefore \qquad v = 7.007 \text{ m/sec} \quad \textbf{Ans.}$

Substituting it is Eqn. (1), we get

$$T - 500 = \frac{200}{9.81} \times 7.007$$

$\therefore \qquad \boldsymbol{T = 642.86 \text{ N} \quad \textbf{Ans.}}$

Case (b): Initial velocity $u = 3$ m/sec.

Impulse momentum equation for 500 N body will be

$$(T - 500)\, 5 = \frac{500}{9.81}\, (2v - 3)$$

i.e., $\qquad T - 500 = \dfrac{100\,(2v - 3)}{9.81}$ ...(3)

Impulse momentum equation for 1500 N body will be

$$(1500 - 2T)\, 5 = \frac{1500}{9.81}\, (v - 3)$$

i.e., $\qquad 1500 - 2T = \dfrac{300}{9.81}\, (v - 3)$ ...(4)

Adding Eqn. (4) and 2 times Eqn. (1), we get

$$1500 - 1000 = \frac{100}{9.81}\, (7v - 15)$$

$v = 9.15$ **m/sec** **Ans.**

Substituting it in Eqn. (3), we get

$$T - 500 = \frac{100}{9.81}\, (2 \times 9.15 - 3)$$

$\therefore \qquad \boldsymbol{T = 655.96 \text{ N} \quad \textbf{Ans.}}$

Example 17.8. *The system shown in Fig. 17.8(a) has a rightward velocity of 3 m/sec. Determine its velocity after 5 seconds. Take $\mu = 0.2$ for the surfaces in contact. Assume pulleys to be frictionless.*

Solution. Since all bodies have same displacement in given time, consider the combined FBD of the system.

$$N_1 = 500 \text{ N}$$
$$F_1 = 0.2 \times 500 = 100 \text{ N}$$
$$N_2 = 1000 \cos 30° = 866.03$$
$$F_2 = 0.2 \times N_2 = 173.2 \text{ N}$$

IMPULSE MOMENTUM

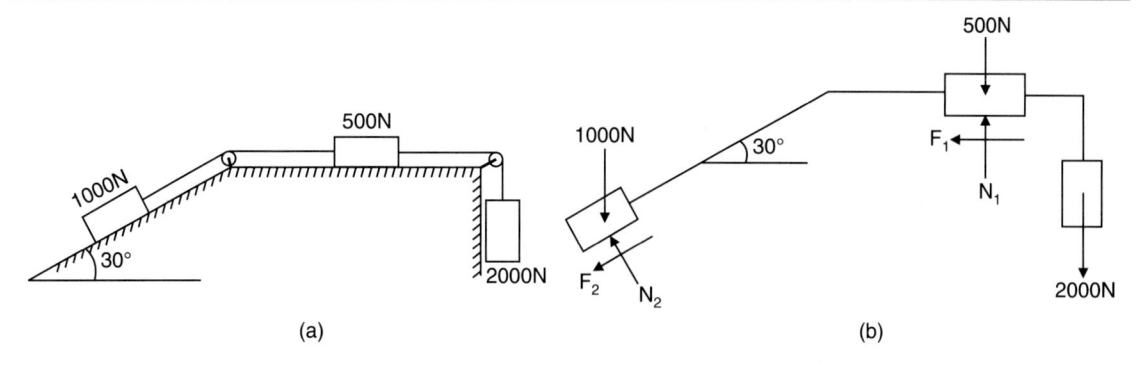

Fig. 17.8

Writing impulse momentum equation for whole system, we get

$$(2000 - F_1 - 1000 \sin 30° - F_2)t = \frac{2000 + 500 + 1000}{9.81}(v - u)$$

$$(2000 - 100 - 1000 \sin 30° - 173.2) \, 5 = \left[\frac{3500}{9.81}\right](v - 3)$$

$$v = 20.19 \text{ m/sec} \quad \textbf{Ans.}$$

Example 17.9. *The system shown in Fig. 17.9(a) has a rightward velocity of 4 m/sec, just before a force P is applied. Determine the value of P that will give a leftward velocity of 6 m/sec in a time interval of 20 sec. Take coefficient of friction = 0.2 and assume ideal pulley.*

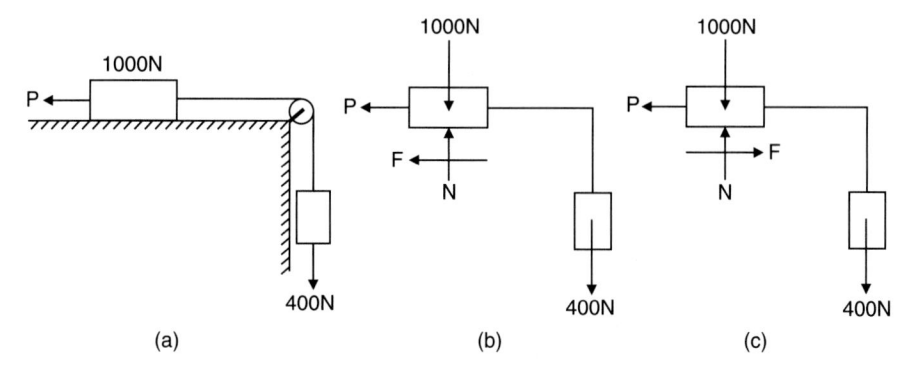

Fig. 17.9

Solution. When the system is moving rightward, frictional force acts leftward as shown in Fig. 17.9(b). Force P first brings the system to stationary position, then starts moving leftward. At this stage, the frictional force acts rightward as shown in Fig. 17.9(c). Let t_1 be the time required to bring the system to the stationary condition.

$$N = 1000 \text{ newton}$$
$$F = 0.2 \times 1000 = 200 \text{ N}.$$

Applying impulse momentum equation for the motion upto stationary condition, we have

$$(400 - 200 - P)\, t_1 = \frac{400 + 1000}{9.81}(0 - 4)$$

$$(P - 200)\, t_1 = \frac{5600}{9.81} \qquad \qquad \ldots(1)$$

Applying impulse momentum equation for the motion from stationary position to leftward motion, after total time of 20 sec, we have

$$(P - F - 400)(20 - t_1) = \frac{1000 + 400}{9.81}(v - 0)$$

$$(P - 600)(20 - t_1) = \frac{1400}{9.81} \times 6 \qquad \qquad ...(2)$$

Simultaneous equations (1) and (2) may be solved to get t_1 and P. Trial and error method may be advantageously used here. Looking at Eqn. (2), the value of P should be more than 600. Let us take a trial value of P as 700. From Eqn. (1)

$$t_1 = 1.142 \text{ sec}$$

Substituting it in Eqn. (2), we get

$$P = 645.41 \text{ N}$$

Substituting this value of P in Eqn. (1), we get

$$t_1 = 1.282 \text{ sec}$$

Substituting it in Eqn. (2), we get

$$P = 645.74 \text{ N}$$

This value is almost same as trial value 646.41 N

Hence, $\quad\quad\quad\quad P = 645.74 \text{ N} \quad \text{Ans.}$

17.3. FORCE OF JET ON A VANE

In hydroelectric generating stations, a jet of water is made to impinge on the vanes of turbines and get deflected by a certain angle. During this process a force is exerted by the jet on the vane and that causes rotation of turbine. This machanical energy is further converted into electric energy. The force exerted by the jet on the vane, moving or stationary, can be determined by applying impulse momentum equations. This is illustrated with examples 17.10 and 17.11.

Example 17.10. *A nozzle issues a jet of water 50 mm in diameter, with a velocity of 30 m/sec which impinges tangentially upon a perfectly smooth and stationary vane, and deflects it through an angle of 30° without any loss of velocity (see Fig. 17.10). What is the total force exerted by the jet upon the vane ?*

Solution. Weight of water whose momentum is changed in t second

$$= (\pi/4)(0.05)^2 \times 30 \times 9810 \times t$$

$$= 577.86 \, t \text{ newtons.}$$

(**Note:** 1 cubic metre of water weighs 9810 newton)

Let P_x and P_y be the components of reactive force of vane.

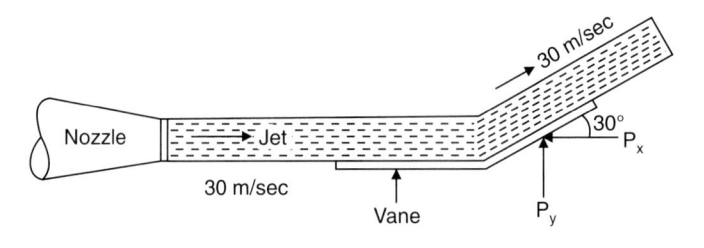

Fig. 17.10

IMPULSE MOMENTUM

Applying impulse momentum equation in x-direction, we get

$$-P_x t = \frac{577.86 t}{9.81} (30 \cos 30° - 30)$$

$$P_x = 236.75 \text{ N}$$

Applying the impulse momentum equation in y-direction

$$P_y t = \frac{577.86 t}{9.81} (30 \sin 30° - 0)$$

$$P_y = 883.58 \text{ N}$$

$$P = \sqrt{P_x^2 + P_y^2} = \sqrt{236.75^2 + 883.58^2}$$

$$= 914.75 \text{ N}$$

Inclination with horizontal,

$$\theta = \tan^{-1}\left(\frac{P_y}{P_x}\right) = \tan^{-1} \frac{883.58}{236.75}$$

$$\theta = 75.0°$$

The force P shown in Figure 17.11(a) is the reactive force of the vane. The force of jet is equal and opposite to this force as shown in Fig. 17.11(b).

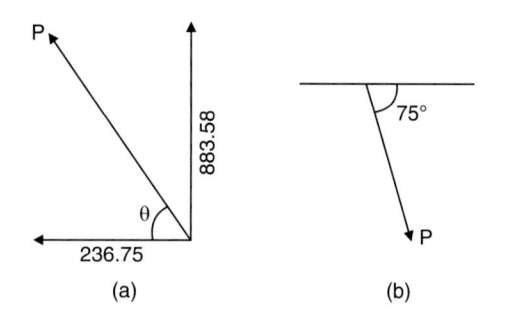

Fig. 17.11

Example 17.11. *In the previous example if the vane is moving with a velocity of 10 m/sec towards right, what will be the pressure exerted by the jet.*

Solution. Velocity of approach

$$= 30 - 10 = 20 \text{ m/sec}$$

Weight of water impinging in t second

$$= (\pi/4) \times (0.05)^2 \times 20 \times t \times 9810$$

$$= 385.24 \, t$$

Velocity of departure = velocity of approach

$$= 30 - 10 = 20 \text{ m/sec}$$

Writing impulse moment equation in x direction

$$-P_x t = \frac{385.24 \, t}{9.81} (20 \cos 30° - 20)$$

$$P_x = 105.22 \text{ N}$$

Applying impulse moment equation in y direction,

$$P_y t = \frac{385.24\, t}{9.81}(20 \sin 30° - 0)$$

$$P_y = 392.70 \text{ N}$$

$$P = \sqrt{P_x^2 + P_y^2} = \sqrt{105.22^2 + 392.70^2}$$

i.e., $\quad P = 406.55 \text{ N} \quad$ Ans.

Its inclination to the horizontal

$$\theta = \tan^{-1}\left(\frac{P_y}{P_x}\right) = \tan^{-1}\left(\frac{392.70}{105.22}\right) = 75°$$

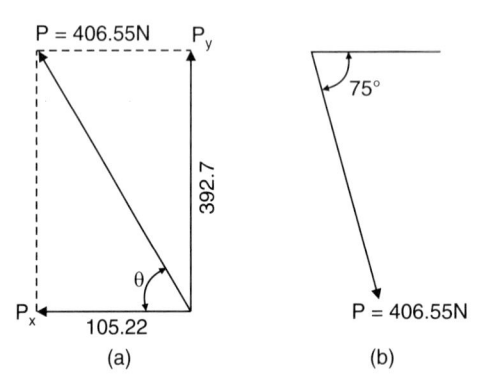

Fig. 17.12

Reaction of vane P is as shown in Fig. 17.12(a). The pressure exerted by the jet is equal and opposite to it and is as shown in Fig. 17.12(b).

17.4. CONSERVATION OF MOMENTUM

In a system, if the resultant force R is zero, the impulse momentum Eqn. 17.2 reduces to final momentum equal to initial momentum. Such situation arises in many cases because the force system consists of only action and reaction on the elements of the system. The resultant force is zero, only when entire system is considered, but not when the free body of each element of the system is considered. When a person jumps off a boat, the action of the person is equal and opposite to the reaction of the boat. Hence, the resultant force is zero in the system. If W_1 is the weight of the person and W_2 that of the boat, v is velocity of the person and the boat before the person jumps out of the boat and v_1, v_2 are the velocities of person and the boat after jumping, according to principle of conservation of momentum,

$$\frac{W_1 + W_2}{g} v = \frac{W_1}{g} v_1 + \frac{W_2}{g} v_2$$

Similar equation holds goods when we consider the system of a gun and shell. The *principle of conservation of momentum* may be stated as, *the momentum is conserved in a system in which resultant force is zero.* In other words, *in a system if the resultant force is zero, initial momentum will remain equal to final momentum.*

It must be noted that conservation of momentum applies to entire system and not to individual elements of the system.

IMPULSE MOMENTUM

Example 17.12. *A 800 N man, moving horizontally with a velocity of 3 m/sec, jumps off the end of a pier into a 3200 N boat. Determine the horizontal velocity of the boat (a) if it had no initial velocity and (b) if it was approaching the pier with an initial velocity of 0.9 m/sec.*

Solution. Weight of man $W_1 = 800$ N

Velocity with which man is running $v = 3$ m/sec.

Weight of the system after man jumps into boat = 800 + 3200 = 4000 N.

(a) Initial velocity of boat = 0

Since the action of the man is equal to the reaction of the boat, the principle of conservation of momentum can be applied to the system consisting of the man and the boat.

Initial momentum = Final momentum

$$\frac{800}{9.81} \times 3 + \frac{3200}{9.81} \times 0 = \frac{4000}{9.81} v$$

$$v = \mathbf{0.6 \text{ m/sec}} \quad \textbf{Ans.}$$

(b) Initial velocity of boat = 0.9 m/sec towards the pier
$$= -0.9 \text{ m/sec}$$

Applying principle of conservation of momentum, we get

$$\frac{800}{9.81} \times 3 + \frac{3200}{9.81} \times (-0.9) = \frac{4000}{9.81} v$$

$$\therefore \quad v = \mathbf{-0.12 \text{ m/sec}}$$

i.e., **velocity of boat and man will be 0.12 m/sec towards the pier. Ans.**

Example 17.13. *A car weighing 11,000 N and running at 10 m/sec holds three men each weighing 700 N. The men jump off from the back end gaining a relative velocity of 5 m/sec with the car. Find the speed of the car if the three men jump off*

(i) *in succession,* (ii) *all together.*

Solution. (i) When three men jump off in succession

initial velocity $u = 10$ m/sec.

Let the velocity when

(a) first man jumps be v_1 m/sec

(b) second man jumps be v_2 m/sec

(c) third man jumps be v_3 m/sec.

Velocity of the first man w.r.t. fixed point when he jumps = $v_1 - 5$.

Applying principle of conservation of momentum when the first man jumps, we get

$$(11{,}000 + 3 \times 700) \, 10 = (11{,}000 + 2 \times 700) \, v_1 + 700 \, (v_1 - 5)$$
$$= (11{,}000 + 3 \times 700) \, v_1 - 700 \times 5$$

$$v_1 = 10 + \frac{700 \times 5}{11{,}000 + 3 \times 700} \qquad \ldots(1)$$

When the second man jumps

$$(11{,}000 + 2 \times 700) v_1 = (11{,}000 + 700) v_2 + 700(v_2 - 5)$$

$$v_2 = v_1 + \frac{700 \times 5}{11{,}000 + 2 \times 700} \qquad \ldots(2)$$

When the third man jumps:
$$(11{,}000 + 700)\, v_2 = 11{,}000 v_3 + 700\,(v_3 - 5)$$
$$v_3 = v_2 + \frac{700 \times 5}{11{,}000 + 700} \qquad \ldots(3)$$

From (1), (2) and (3), we get
$$v_3 = 10 + 700 \times 5 \left(\frac{1}{11{,}000 + 3 \times 700} + \frac{1}{11{,}000 + 2 \times 700} + \frac{1}{11{,}000 + 700} \right)$$
$$= 10.849 \text{ m/sec} \quad \textbf{Ans.}$$

(ii) When three men jump together

Let the velocity of the car be v when three men jump together. Applying principle of conservation of momentum, we get
$$(11{,}000 + 3 \times 700)\, 10 = 11{,}000 v + 3 \times 700 (v - 5)$$
where, $(v - 5)$ is the relative velocity of the men when they jump,
$$v = 10 + \frac{3 \times 700 \times 5}{11{,}000 + 3 \times 700}$$
i.e., $$v = 10.802 \text{ m/sec} \quad \textbf{Ans.}$$

Example 17.14. *A car weighing 50 kN and moving at 54 kmph along the main road collides with a lorry of weight 100 kN which emerges at 18 kmph from a cross road at right angles to main road. If the two vehicles lock after collision, what will be the magnitude and direction of the resulting velocity ?*

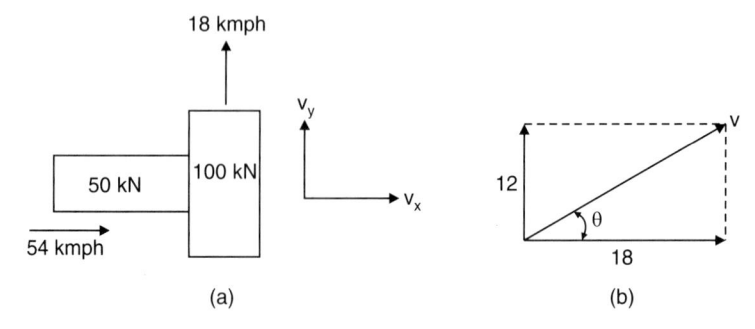

Fig. 17.13

Solution. Let the velocity of the vehicles after collision be v_x in x direction (along main road) and v_y in y direction (along cross road) as shown in Fig. 17.13(a). Applying impulse moment equation along x direction, we get
$$\frac{50 \times 54}{9.81} + 0 = \frac{(50 + 100)}{9.81} v_x$$
∴ $$v_x = 18 \text{ kmph}$$

Applying impulse momentum equation in y direction, we get
$$0 + \frac{100 \times 18}{9.81} = \frac{(50 + 100)}{9.81} v_y$$
$$v_y = 12 \text{ kmph}$$

IMPULSE MOMENTUM

$$\therefore \quad \text{Resultant velocity } v = \sqrt{v_x^2 + v_y^2} = \sqrt{18^2 + 12^2}$$

$$= 21.63 \text{ kmph} \quad \textbf{Ans.}$$

Its inclination to main road

$$\theta = \tan^{-1}\left(\frac{v_y}{v_x}\right) = \tan^{-1} 12/18$$

$$= 33.69° \quad \textbf{Ans.}$$

as shown in Fig. 17.13(b).

Example 17.15. *A gun weighing 300 kN fires a 5 kN projectile with a velocity of 300 m/sec. With what velocity will the gun recoil? If the recoil is overcome by an average force of 600 kN how far will the gun travel? How long will it take?*

Solution. Applying principles of conservation of momentum to the system of gun and the projectile, we get

$$0 = 300 \times v + 5 \times 300$$

$$v = -5 \text{ m/sec}$$

i.e., **Gun will have a velocity of 5 m/sec in the direction opposite to that of bullet.**

Ans.

Let the gun recoil for a distance s.

Using work energy equation, we have

$$-600 \times s = \frac{300}{2 \times 9.81} (0 - 5^2)$$

$$s = 0.637 \text{ m} \quad \textbf{Ans.}$$

Applying impulse momentum equation to gun, we get

$$-600t = (300/9.81)(0 - 5)$$

$$t = 0.255 \text{ sec} \quad \textbf{Ans.}$$

Example 17.16. *A bullet weighing 0.3 N is fired horizontally into a body weighing 100 N which is suspended by a string 0.8 m long. Due to this impact the body swings through an angle of 30°. Find the velocity of the bullet and the loss in the energy of the system.*

Solution. Let the velocity of the block be u immediately after bullet strikes it.

Applying work energy equation for the block, we get

$$-Wh = (W/2g)(v^2 - u^2)$$

$$-100.3 \times 0.8(1 - \cos 30°) = \left[\frac{100.3}{2 \times 9.81}\right](0 - u^2)$$

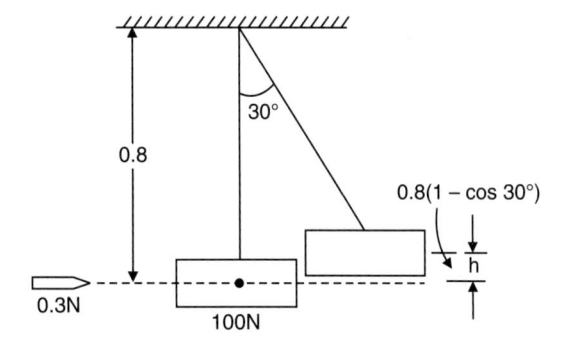

Fig. 17.14

(**Note:** Work done by the weight is negative since weight is a force acting downwards force whereas body has moved upwards).

$$u = 1.025 \text{ m/sec}$$

Let v be the velocity of bullet before striking the block. Applying principle of conservation of momentum to the bullet and block system, we get

$$\frac{0.3}{9.81} v + 0 = \frac{100 + 0.3}{9.81} u$$

$$0.3\, v = 100.3 \times 1.025$$

$$v = 342.69 \text{ m/sec} \quad \textbf{Ans.}$$

Initial energy of bullet $= \dfrac{0.3}{2 \times 9.81} (342.69)^2 = 1795.68$ J

Energy of the block and bullet system

$$= \frac{1}{2} \times \frac{100 + 0.3}{9.81} \, 1.025^2 = 5.37 \text{ J}$$

Loss of energy $= 1795.68 - 5.37$

$$= 1790.31 \text{ J} \quad \textbf{Ans.}$$

Example 17.17. *A bullet weighs 0.5 N and moving with a velocity of 400 m/sec hits centrally a 30 N block of wood moving away at 15 m/sec and gets embedded in it. Find the velocity of the bullet after the impact and the amount of kinetic energy lost.*

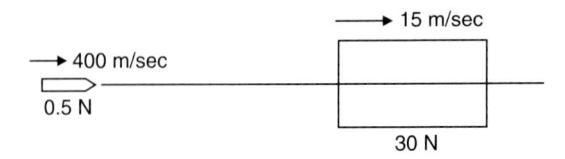

Fig. 17.15

Solution. Initial momentum of the system = Final momentum

$$\frac{0.5}{9.81} \times 400 + \frac{30}{9.81} \times 15 = \frac{(30 + 0.5)}{9.81} v$$

$$v = 21.31 \text{ m/sec}$$

Kinetic Energy lost = Initial K.E. − Final K.E.

$$= \left(\frac{1}{2} \times \frac{0.5}{9.81} \times 400^2 + \frac{1}{2} \times \frac{30}{9.81} \, 15^2 \right) - \frac{1}{2} \times \frac{30.5}{9.81} \times 21.31^2$$

$$= 3715.47 \text{ J} \quad \textbf{Ans.}$$

17.5. PILE AND PILE HAMMER

If the safe bearing capacity of the soil is too less, a set of reinforced concrete or steel poles are driven in the soil. Such poles are known as piles. Over the group of piles concrete cap is cast and on it the structure is built.

IMPULSE MOMENTUM

The piles are driven by pile hammer. It consists of a movable weight called the hammer (see Fig. 17.16). The hammer is raised to a convenient height h and freely dropped. It is guided to fall over the pile. After the hammer strikes the pile the hammer and the pile move downward together. The kinetic energy of the pile and the hammer is utilised in doing the work against resistance of the ground and pile gets driven by a distance s. By repeated hammering the pile is driven to required depth. If the distance moved per blow is known, earth resistance can be calculated. A general equation is derived in Ex. 17.18, and then two specific examples are solved.

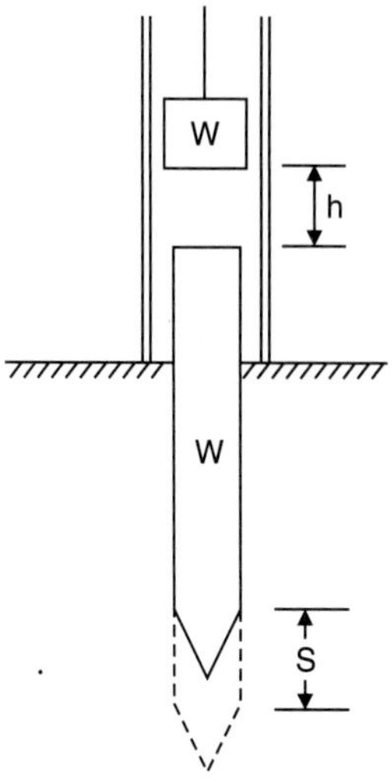

Example 17.18. *A pile of weight W is driven vertically through a distance s when a hammer of weight w is dropped from a height h. Calculate the average resistance of the ground, the loss of kinetic energy during the impact and the time during which the pile is in motion.*

Solution. Initial velocity of hammer = 0
Distance moved before striking pile = h m
Gravitational acceleration = g m/sec^2

∴ Velocity of hammer while striking pile v is given by

$$v^2 - u^2 = 2gh$$
$$v^2 - 0 = 2gh$$
$$v = \sqrt{2gh} \qquad \ldots(1)$$

Fig. 17.16

After striking, the hammer and the pile move together. Hence, the momentum is conserved. Applying the principle of conservation of momentum

$$\frac{wv}{g} = \frac{(w+W)}{g} V$$

where, V is the velocity of hammer and the pile immediately after the strike.

$$V = \frac{w}{w+W} v \qquad \ldots(2)$$

With this initial velocity the pile and hammer start moving downwards and they stop moving after a distance s. During this process, work is done against the resistance R of the earth. Applying work energy equation, we have

$$(w+W)s - R \times s = \frac{1}{2} \frac{w+W}{g} (0 - V^2)$$

$$Rs = (w+W)s + \frac{w+W}{2g} V^2 \qquad \ldots(3)$$

From (2), $$V = \frac{w}{w+W} v$$

∴ $$Rs = (w+W)s + \frac{w+W}{2g} \frac{w^2}{(w+W)^2} v^2$$

i.e., $$Rs = (w + W)s + \frac{w^2}{2g(w + W)} v^2 \qquad ...(4)$$

From (1), $\quad v^2 = 2gh$

$\therefore \qquad Rs = (w + W)s + \dfrac{w^2}{2g(w + W)} 2gh = (w + W)s + \dfrac{w^2}{w + W} h$

or $$R = w + W + \frac{w^2}{w + W} \frac{h}{s} \quad \text{Ans.}$$

Loss of Kinetic Energy during the Impact:

$$= \frac{w}{2g} v^2 - \frac{w + W}{2g} V^2$$

$$= \frac{w}{2g} v^2 - \frac{w + W}{2g} \left(\frac{w}{w + W}\right)^2 v^2 = \frac{v^2}{2g} \left\{ w - \frac{w^2}{w + W} \right\}$$

$$= \frac{2gh}{2g} w \left\{ 1 - \frac{w}{w + W} \right\}$$

$$= wh \frac{W}{w + W}$$

Loss of K.E. $= \dfrac{w W h}{w + W} \quad$ **Ans.**

Time during which the pile is in motion:

Let t be the time during which the pile is in motion. Applying impulse momentum equation

$$[(w + W) - R]t = \frac{(w + W)}{g} (0 - V)$$

$$[R - (w + W)]t = \frac{w + W}{g} V$$

Substituting for R and V, we get

$$\frac{w^2}{w + W} \frac{h}{s} t = \frac{w + W}{g} \frac{w}{w + W} v = \frac{w}{g} \sqrt{2gh}$$

$$t = \frac{w + W}{w} s \sqrt{2gh} \quad \text{Ans.}$$

Example 17.19. *A pile hammer weighing 20 kN drops from a height of 750 mm on a pile of 10 kN. The pile penetrates 100 mm per blow. Assuming that the motion of the pile is resisted by a constant force, find the resistance to penetration of the ground.*

Solution. *Initial velocity of hammer* $u = 0$

Distance moved $= h = 750$ mm $= 0.75$ m

Acceleration $g = 9.81$ m/sec^2

$\therefore \quad$ Velocity at the time of strike $= \sqrt{2gh} = \sqrt{2 \times 9.81 \times 0.75} = 3.836$ m/sec

IMPULSE MOMENTUM

Applying the principle of conservation of momentum of pile and hammer, we get velocity V of the pile and hammer immediately after the impact.

$$\frac{20}{9.81} \times 3.836 = \frac{20 + 10}{9.81} V$$

$$V = \frac{20}{30} \times 3.836 = 2.557 \text{ m/sec}$$

Applying work energy equation to the motion of the hammer and pile, resistance R of the ground can be obtained.

$$(20 + 10 - R)s = \frac{20 + 10}{2g}(0 - V^2)$$

$$(30 - R)\,0.1 = \frac{30}{2 \times 9.81}(-2.557^2)$$

$$R = 130 \text{ kN} \quad \textbf{Ans.}$$

Example 17.20. *A pile hammer, weighing 15 kN drops from a height of 600 mm on a pile of 7.5 kN. How deep does a single blow of hammer drive the pile if the resistance of the ground to pile is 140 kN?*

Assume that ground resistance is constant.

Solution. $h = 600$ mm $= 0.6$ m

Velocity of hammer at the time of strike

$$v = \sqrt{2gh} = \sqrt{2 \times 9.81 \times 0.6} = 3.431 \text{ m/sec}$$

Let V be the velocity of pile and hammer immediately after impact. Applying principle of conservation of momentum to the system of pile and pile hammer, we get

$$\frac{15}{9.81} \times 3.431 + 0 = \frac{15 + 7.5}{9.81} V$$

$$V = \frac{15 \times 3.431}{22.5} = 2.287 \text{ m/sec}$$

Now applying work energy equation to the system, we get

$$(15 + 7.5 - R)s = \frac{15 + 7.5}{2 \times 9.81}(0 - V^2)$$

$$(15 + 7.5 - 140)s = \frac{15 + 7.5}{2 \times 9.81}(-2.287^2)$$

$$s = \frac{22.5}{2 \times 9.81}(+2.287^2)\frac{1}{117.5} = 0.051 \text{ m}$$

$$s = 51 \text{ mm} \quad \textbf{Ans.}$$

Example 17.21. *A hammer weighing 5 N is used to drive a nail of weight 0.2 N with a velocity of 5 m/sec. horizontally into a fixed wooden block. If the nail penetrates by 20 mm per blow, calculate the resistance of the block, which may be assumed uniform.*

Solution. Applying impulse momentum equation to the system of hammer and nail, we get

$$\frac{5}{9.81} \times 5 + 0 = \frac{5 + 0.2}{9.81} V$$

$$V = 4.808 \text{ m/sec}$$

Applying work energy equation to the system shown in Fig. 17.17, we get

$$-Rs = \frac{5.2}{2 \times 9.81}(0 - 4.808^2)$$

$$R \times 0.02 = \frac{5.2}{2 \times 9.81} \times 4.808^2$$

$$R = 306.3 \text{ N} \quad \textbf{Ans.}$$

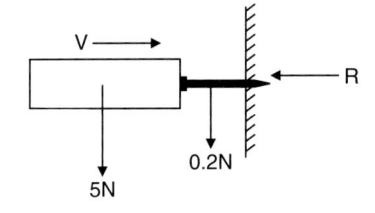

Fig. 17.17

Note: Weights of hammer and nail do not do any work since there is no displacement in the directions of these forces.

Important Definitions

1. The product of mass and velocity is called **momentum**.
2. If R is the resultant force acting on the body for an interval zero to t, then the term $\int_0^t R\,dt$ is called **impulse**.
3. **Impulse momentum equation** can be stated as 'the component of resultant linear impulse along any direction is equal to change in the component of momentum in that direction'.

Important Equations

1. $\int_0^t R\,dt = mv - mu.$
2. If momentum is conserved in a system,

$$\frac{W_1 + W_2}{g} v = \frac{W_1}{g} v_1 + \frac{W_2}{g} v_2.$$

PROBLEMS FOR EXERCISE

17.1 A cricket ball weighing one newton approaches a batsman with a velocity of 18 m/sec in the direction shown in Fig. 17.18. After hit by the bat at B, it moves out with a velocity of 40 m/sec at 45° to horizontal. If the bat and ball were in contact for 0.02 sec, determine the impulsive force exerted by the bat. [**Ans.** 230.37 N]

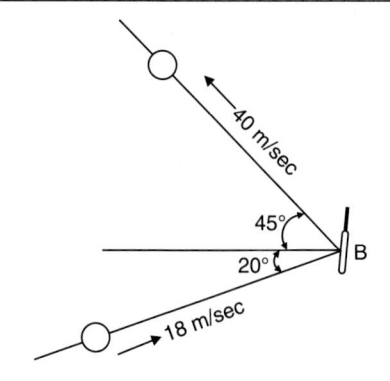

Fig. 17.18

17.2 A block weighing 200 N is pulled up a 30° plane by a force P producing a velocity of 5 m/sec in 5 seconds. If the coefficient of friction is 0.2, determine the magnitude of force P. At this stage if force P is removed, how much more time it will take to come to rest ? [**Ans.** 155.N; 0.757 sec]

17.3 The initial velocity of 500 N block is 6 m/sec towards left. At this stage a weight of 250 N is applied as shown in Fig. 17.19. Determine the time at which the block has (a) no velocity, (b) a velocity of 4 m/sec. to the right. Take coefficient of friction 0.2 and assume pulley as ideal.
[**Ans.** (a) 0.87 sec; (b) 2.23 sec]

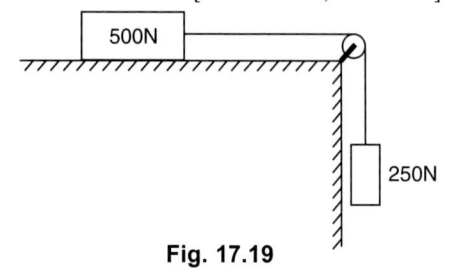

Fig. 17.19

17.4 Determine the tension in the string and the velocity of 2000 N block shown in Fig. 17.20, 6 second after starting with a velocity of 3 m/sec. [**Ans.** v = 25.07 m/sec; T = 1280 N]

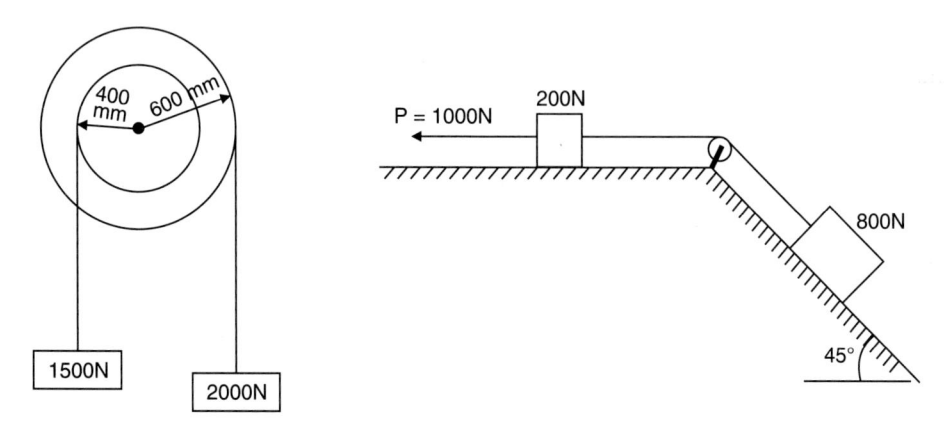

Fig. 17.20   **Fig. 17.21**

17.5 Force P = 1900 N shown in Fig. 17.21 was applied to 200 N block when the block was moving with rightward velocity of 5 m/sec. Determine the time at which the system has (a) no velocity, (b) a velocity of 3 m/sec towards left. Coefficient of friction between blocks and surface = 0.2. Assume pulley to be ideal. [**Ans.** (a) 0.8676 sec; (b) 1.96 sec]

17.6 An engine of weight 500 kN pulls a train weighing 1500 kN up an incline of 1 in 100. The train starts from rest and moves with a constant acceleration against a resistance of 5 N/kN. It attains a speed of 18 kmph in 60 seconds. Determine the tension in the draw bar connecting train and the engine. What will be its speed 90 second after the start? [**Ans.** P = 35.242 kN; v = 27 kmph]

17.7 Determine the force exerted by a 60 mm diameter jet of water flowing at 25 m/sec on (a) a vertical stationary plate, (b) a cup that turns the water through 120°.

[**Ans.** (a) $P = 1767.146$ kN; (b) $P = 3006.787$ kN at $\theta = 30°$ to horizontal]

17.8 A jet of water of 50 mm diameter strikes a series of vanes horizontally at a speed of 36 m/sec and gets deflected through 45°. Determine the force exerted by the jet on vanes, if the vanes are moving away from the jet at a velocity of 6 m/sec.

[**Ans.** $P = 1252.516$ kN and $\theta = 22.5°$ to horizontal]

17.9 A shot is fired horizontally from a gun boat towards a target. The total weight of gun boat including men, gun, shells etc. in it is 15 kN. The weight of shell is 15 N and emerges out at a velocity of 300 m/sec. What will be the velocity of the boat when the shell is fired? If the target weighs 1 kN suspended by a rope of length 2 m, by what angle it will swing if the shell gets embedded in it?

[**Ans.** $v = 0.3$ m/sec; $\theta = 60.06°$]

17.10 A bullet weighing 0.3 N and moving at 600 m/sec penetrates 40 N body shown in Fig. 17.22 and emerges with a velocity of 180 m/sec. How far and how long will the block move, if the coefficient of friction between the body and the horizontal floor is 0.3.   [**Ans.** $s = 1.686$ m; $t = 1.07$ sec]

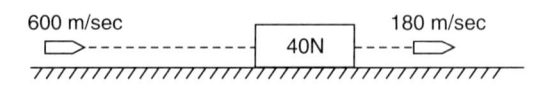

Fig. 17.22

17.11 A cannon weighing 200 kN fires a shell weighing 1 kN with a muzzle velocity of 800 m/sec. Calculate the velocity with which the cannon recoils and the uniform force required to stop it within 400 mm distance. In how much time it will stop?

[**Ans.** $v = 4.0$ m/sec; $t = 0.20$ sec; $F = 407.747$ kN]

17.12 A man weighing 750 N and a boy weighing 500 N jump from a boat to a pier with a horizontal velocity of 5 m/sec relative to the boat. The boat weighs 4000 N and was stationary before they jumped. Determine the velocity of the boat if

(a) they jump together;   (b) boy jump first and the man latter;

(c) man jumps first and the boy latter.   [**Ans.** (a) 1.563 m/sec; (b) 1.464 m/sec (c) 1.458 m/sec]

17.13 A pile hammer weighing 8 kN falls freely from a height of 1.5 m on a pile weighing 5 kN. For each blow, the pile is driven by 80 mm.

Determine:

(a) resistance of the ground;   (b) loss of kinetic energy during the impact;

(c) time during which the pile is in motion for each blow.

[**Ans.** $R = 105.31$ kN; Loss of K.E. $= 4.615$ kN-m, $t = 0.048$ sec]

18

Impact of Elastic Bodies

A collision between two bodies is said to be impact, if the bodies are in contact for a short interval of time and exert very large force on each other during this short period. On impact, the bodies deform first and then recover due to elastic properties and start moving with different velocities. The velocity with which they separate depends not only on their velocity of approach but also on the shape, size, elastic property and the line of impact. In this book, the velocity of the bodies during the short period of impact is not considered. Only the velocities of the colliding bodies before impact and after impact are considered. Some of the new technical terms used in this chapter are first defined, then the cases of direct impact and oblique impact are analysed. The expression for the loss of kinetic energy on impact is derived.

18.1. DEFINITIONS

(*i*) **Line of Impact:** Common normal to the colliding surfaces is known as line of impact.

(*ii*) **Direct Impact:** If the motion of the two colliding bodies is directed along the line of impact, the impact is said to be direct impact.

(*iii*) **Oblique Impact:** If the motion of one or both of the colliding bodies is not directed along the line of impact, the impact is known as oblique impact.

(*iv*) **Central Impact:** If the mass centres of colliding bodies are on the line of impact, the impact is called central impact.

(*v*) **Eccentric Impact:** Even if mass centre of one of the colliding bodies is not on the line of impact, the impact is called eccentric impact.

These terms are illustrated in Fig. 18.1.

18.2. COEFFICIENT OF RESTITUTION

During the collision, the colliding bodies initially undergo a deformation for a small time interval and then recover the deformation in a further small time interval. So the **period of collision** (or time of impact) consists of two time intervals, **Period of Deformation and Period of Restitution.** "Period of Deformation is the time elapse between the instant of the initial contact and the instant of maximum deformation of the bodies." Similarly, "Period of Restitution is the time of elapse between the instant of the maximum deformation condition and the instant of separation of the bodies."

Therefore, Impulse during deformation = $F_D \, dt$

where F_D refers to the force that acts during the period of deformation.

The magnitude of F_D varies from zero at the instant contact to the maximum value at the instant of maximum deformation.

Similarly, impulse during restitution = $F_R \, dt$

where F_R refer to the force that acts during period of restitution.

(a) Direct central impact

(b) Oblique central impact

(c) Direct eccentric impact

(d) Oblique eccentric impact

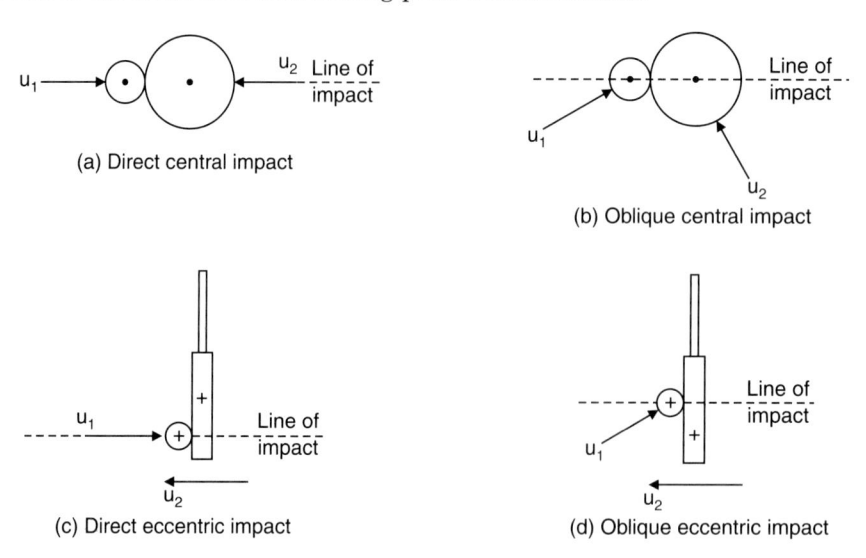

Fig. 18.1

The magnitude of F_R varies from a maximum value at the instant of maximum deformation condition to zero at the instant of just separation of the bodies.

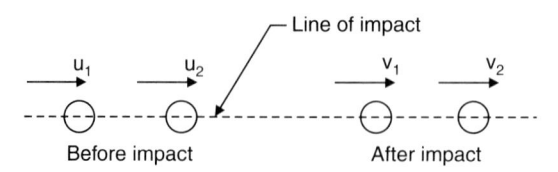

Before impact   After impact

Fig. 18.2

Let m_1—mass of the first body

m_2— mass of the second body

u_1—velocity of the first body before impact

u_2—velocity of the second body before impact

v_1—velocity of the first body after impact

and v_2—velocity of the second body after impact.

At the instant of maximum deformation, the colliding bodies will have same velocity. Let the velocity of the bodies at the instant of maximum deformation be $U_{D\,max}$

Applying Impulse—Momentum principle for the first body

$$F_D dt = m_1 U_{D\,max} - m_1 u_1 \qquad \ldots(1)$$

and
$$F_R dt = m_1 v_1 - m_1 U_{D\,max} \qquad \ldots(2)$$

Now, dividing (1) by (2)

$$\frac{F_R dt}{F_D dt} = \frac{m_1 v_1 - m_1 U_{D\max}}{m_1 U_{D\max} - m_1 u_1}$$

i.e.,
$$\frac{F_R dt}{F_D dt} = \frac{v_1 - U_{D\max}}{U_{D\max} - u_1} \qquad \ldots(3)$$

Similar analysis for the second body gives,

$$\frac{F_R dt}{F_D dt} = \frac{U_{D\max} - v_2}{u_2 - U_{D\max}} \qquad \ldots(4)$$

From (3) and (4)

$$\frac{F_R dt}{F_D dt} = \frac{v_1 - U_{D\max}}{U_{D\max} - u_1} = \frac{U_{D\max} - v_2}{u_2 - U_{D\max}}$$

$$= \frac{v_1 - U_{D\max} + U_{D\max} - v_2}{U_{D\max} - u_1 + v_2 - U_{D\max}} = \frac{v_1 - v_2}{u_2 - u_1} = \frac{v_2 - v_1}{u_1 - u_2}$$

$$= \frac{\text{Relative velocity of separation}}{\text{Relative velocity of approach}} \qquad \ldots(5)$$

Sir Isaac Newton conducted the experiments and observed that when collision of two bodies takes place relative velocity of separation bears a constant ratio to the relative velocity of approach, the relative velocities being measured along the line of impact. This constant ratio is called as the coefficient of restitution and is denoted by letter *e*. Hence from (5), we have

$$e = \frac{F_R dt}{F_D dt} = \frac{v_2 - v_1}{u_1 - u_2} \qquad \ldots (18.1)$$

The coefficient of restitution of two colliding bodies may be taken as the ratio of impulse during restitution period to the impulse during the deformation period. This is also equal to the ratio of relative velocity of separation to the relative velocity of approach of the colliding bodies, the relative velocities being measured in the line of impact.

Notes: (*i*) The relative velocities are to be considered only along the line of impact,

(*ii*) Signs of the velocities are to be considered carefully in eqn. 18.1.

For perfectly elastic bodies, the magnitude of relative velocity after impact will be same as that before impact and hence coefficient of restitution will be 1. Perfectly inelastic bodies cling together and hence the velocity of separation will be zero *i.e.*, coefficient of restitution will be zero. The coefficient of restitution always lies between 0 and 1.

The value of coefficient of restitution depends not only on the material property but it also depends on the shape and size of the body. Hence, the coefficient of restitution is the property of two colliding bodies but not merely of material of the colliding bodies.

18.3. DIRECT CENTRAL IMPACT

Let u_1, u_2 be initial velocities and v_1 and v_2 be the velocities after collision of bodies 1 and 2 respectively.

Let W_1 and W_2 be the weight of the colliding bodies and 'e' be coefficient of restitution. u_1, u_2, W_1, W_2 and e are known quantities. Velocities after collision v_1 and v_2 are the unknowns. To find these two unknowns, we need two equations. One equation is obtained by the principle of conservation of momentum as,

$$\frac{W_1}{g} u_1 + \frac{W_2}{g} u_2 = \frac{W_1}{g} v_1 + \frac{W_2}{g} v_2$$

i.e.,
$$W_1 u_1 + W_2 u_2 = W_1 v_1 + W_2 v_2 \qquad \text{...(18.2)}$$

Another equation based on the definition of coefficient of restitution (Eqn. 18.1), may be written as,

$$e(u_1 - u_2) = v_2 - v_1 \qquad \text{...(18.3)}$$

From equations (18.2) and 18.3 the unknown velocities v_1 and v_2 may be found.

Example 18.1. *Direct central impact occurs between a 300 N body moving to the right with a velocity of 6 m/sec and 150 N body moving to the left with a velocity of 10 m/sec. Find the velocity of each body after impact if the coefficient of restitution is 0.8.*

Fig. 18.3

Solution. Referring to Fig. 18.3, initial velocity of 300 N body

$$u_1 = 6 \text{ m/sec}$$

Initial velocity of 150 N body $u_2 = -10$ m/sec

Let final velocity of 300 N body $= v_1$ m/sec and

Final velocity of 150 N body $= v_2$ m/sec

From principle of conservation of momentum,

$$\frac{300}{g} \times 6 + \frac{150}{g}(-10) = \frac{300}{g} v_1 + \frac{150}{g} v_2$$

i.e.
$$2v_1 + v_2 = 2 \qquad \text{...(1)}$$

From the definition of coeff. of restitution, we have,

$$e(u_1 - u_2) = v_2 - v_1$$
$$0.8(6 + 10) = v_2 - v_1$$

i.e.
$$v_2 - v_1 = 12.8 \qquad \text{...(2)}$$

From (1) and (2), we get,

$$3v_1 = -10.8$$
$$v_1 = -\textbf{3.6 m/sec} \quad \textbf{Ans.}$$

IMPACT OF ELASTIC BODIES

and hence $v_2 = 12.8 - v_1 = 12.8 - 3.6$

$v_2 = $ **9.2 m/sec** Ans.

Example 18.2. *A 80 N body moving to the right at a speed of 3 m/sec strikes a 10 N body that is moving to the left at a speed of 10 m/sec. The final velocity of 10 N body is 4 m/sec to the right. Calculate the coefficient of restitution and the final velocity of the 80 N body.*

Solution. $v_2 = 3$ m/sec, $\quad u_2 = -10$ m/sec

$v_1 = ?$, $\quad v_2 = 4$ m/sec

Applying the principles of conservation of momentum to the colliding bodies, we get

$$\frac{80}{g} \times 3 + \frac{10}{g}(-10) = \frac{80}{g}v_1 + \frac{10}{g} \times 4$$

i.e., $\quad v_1 = \dfrac{80 \times 3 - 100 - 40}{80} = $ **1.25 m/sec** Ans.

From the definition of coefficient of restitution, we get

$e(u_1 - u_2) = v_2 - v_1$

$e(3 + 10) = 4 - 1.25$

$e = $ **0.212** Ans.

Example 18.3. *A golf ball is dropped from a height of 10 m on a fixed steel plate. The coefficient of restitution is 0.894. Find the height to which the ball rebounds on the first, second and third bounces.*

Solution. Initial height $h_0 = 10$ m

Velocity of golf ball before impact $u_1 = \sqrt{2gh_0}$

Velocity of steel plate before impact $u_2 = 0$

Velocity of steel plate after impact $v_2 = 0$

Let velocity of golf ball after impact be v_1.

From the definition of coefficient of restitution, we have

$e(u_1 - u_2) = v_2 - v_1$

$e(\sqrt{2gh_0} - 0) = 0 - v_1$

$v_1 = -e\sqrt{2gh_0} = e\sqrt{2gh_0}$ in upward direction.

From kinematic equation, the height h_1 to which the ball will rise is given by

$v_1^2 - 0 = 2gh_1$

$$h_1 = \frac{v_1^2}{2g} = \frac{e^2 \times 2gh_0}{2g}$$

i.e., $\quad h_1 = e^2 h_0$ ...(i)

Now, $e = 0.894, h_0 = 10$ m

∴ $h_1 = 0.894^2 \times 10 = $ **7.992 m** Ans.

Similarly, after second bounce the height to which the ball will rise is given by

$h_2 = e^2 h_1 = 0.894^2 \times 7.992$

$= $ **6.388 m** Ans.

And, after third bounce, the height
$$h_3 = e^2 h_2 = 0.894^2 \times 6.388$$
$$h_3 = 5.105 \text{ m} \quad \text{Ans.}$$

Example 18.4. *A ball is dropped from a height of 1 m on a smooth floor. The height of first bounce is 0.810 m.*

Determine,

(a) coefficient of the restitution

(b) expected height of second bounce.

Solution. Velocity of ball before striking the floor
$$u_1 = \sqrt{2gh} = \sqrt{2 \times 9.81 \times 1} = 4.429 \text{ m/sec}$$

Velocity of ball after striking the floor
$$v_1 = -\sqrt{2gh_1} = -\sqrt{2 \times 9.81 \times 0.810} = -3.987 \text{ m/sec}$$

There is no movement of the floor, before and after striking *i.e.*, $u_2 = v_2 = 0$

From the definition of coefficient of restitution, we have
$$e(u_1 - u_2) = v_2 - v_1$$
$$e \times 4.429 = 0 + 3.987$$
$$e = \frac{3.987}{4.429} = 0.9 \quad \text{Ans.}$$

Let velocity of the ball after second bounce be v_2

Velocity of strike in this case $v_1 = 3.987$ m/sec downward.

$\therefore \quad e(3.987 - 0) = 0 - v_2$
$$v_2 = -3.586 \text{ m/sec}$$
$$= 3.586 \text{ m/sec} \quad \text{upward.}$$

Expected height h_2 is given by the kinematic equation as
$$h_2 = \frac{v_2^2}{2g} = \frac{3.576^2}{2 \times 9.81}$$
$$h_2 = 0.6561 \text{ m} \quad \text{Ans.}$$

Example 18.5. *A 10 N ball traverses a frictionless tube shown in Fig. 18.4. falling through a height of 2m. It then strikes a 20 N ball hung from a rope 1.2 m long.*

Determine the height to which the hanging ball will rise

(i) if the collision is perfectly elastic

(ii) if the coefficient of restitution is 0.7.

IMPACT OF ELASTIC BODIES

Solution. Velocity of 10 N ball after falling through 2 m height

$$u_1 = \sqrt{2gh} = \sqrt{2 \times 9.81 \times 2} = 6.264 \text{ m/sec}$$

Velocity of 20 N ball before impact

$$u_2 = 0$$

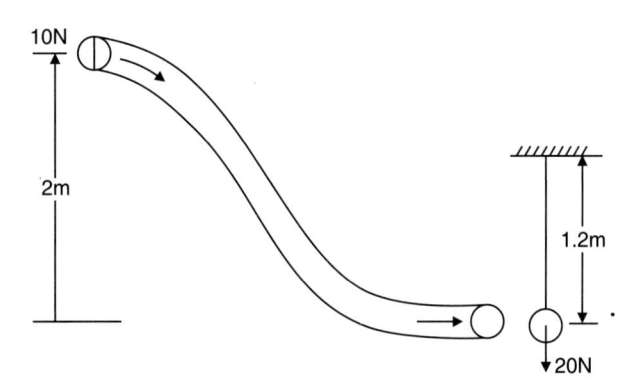

Fig. 18.4

Let the velocities of 10 N and 20 N balls after impact be v_1 and v_2 respectively.

By principle of conservation of momentum

$$m_1 u_1 + m_2 u_2 = m_1 v_1 + m_2 v_2$$

$$\frac{10}{g} 6.264 + 0 = \frac{10}{g} v_1 + \frac{20}{g} v_2$$

i.e.,
$$v_1 + 2v_2 = 6.264 \qquad \ldots(1)$$

From the definition of coefficient of restitution, we have

$$e(u_1 - u_2) = v_2 - v_1 \qquad \ldots(2)$$

Case (i): $e = 1$

∴ $\quad u_1 - u_2 = v_2 - v_1$

i.e.,
$$v_2 - v_1 = 6.264 \qquad \ldots(3)$$

From eqns. (1) and (3),

$$v_2 = \frac{2 \times 6.264}{3} = 4.176 \text{ m/sec}$$

∴ $\quad v_1 = 6.264 - 2 \times 4.176$

$\quad\quad\quad = -2.088$ m/sec.

Let h be the height to which hanging ball will rise after impact. Applying work energy equation

Change in K.E. = work done

$$0 - \frac{20}{2g} v_2^2 = -20 \times h$$

$$h = \frac{v_2^2}{2g} = \frac{4.176^2}{2 \times 9.81}$$

$$h = 0.889 \text{ m} \quad \text{Ans.}$$

Case (ii): $e = 0.7$

From Eqn. (2), we have

$$0.7(6.264) = v_2 - v_1$$

i.e. $\quad v_2 - v_1 = 4.385 \quad \quad \quad \quad \quad \quad \quad \quad \quad \quad \quad \quad \quad \quad \quad \quad \quad ...(4)$

From Eqns. (1) and (4), we get

$$v_2 = \frac{6.264 + 4.385}{3} = 3.55 \text{ m/sec}$$

∴ The height to which ball will rise

$$h_2 = \frac{v_2^2}{2g} = \frac{4.176^2}{2 \times 9.81}$$

i.e. $\quad h_2 = 0.642 \text{ m} \quad \text{Ans.}$

18.4. OBLIQUE IMPACT

Referring to Fig. 18.1(b) note that, the expression for coefficient of restitution, $e(u_1 - u_2) = v_2 - v_1$, holds good only for the component of velocities in the line of impact. The component of velocities in a direction at right angles to line of impact remain unaltered.

Example 18.6. *The magnitude and direction of the two identical smooth balls before central oblique impact are as shown in Fig. 18.5. Assuming coefficient of restitution e = 0.9, determine the magnitude and direction of the velocity of each ball after the impact.*

Solution. Component of velocity of A before impact

(i) Normal to line of impact = 9 sin 30°

$$u_{AY} = 4.5 \text{ m/sec}$$

(ii) In the line of impact = 9 cos 30°

$$u_{AX} = 7.794 \text{ m/sec}$$

Component of velocity of B before impact:

(i) Normal to line of impact = 12 sin 60°

$$u_{BY} = 10.392 \text{ m/sec}$$

(ii) In the line of impact $u_{BX} = -12 \cos 60°$

$$= -6 \text{ m/sec}$$

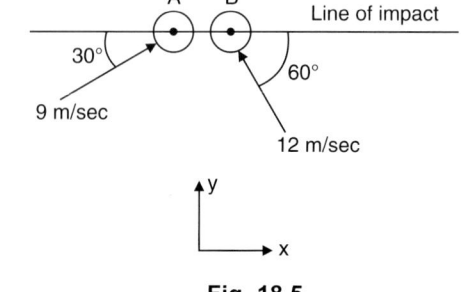

Fig. 18.5

Component of velocities only in the line of impact get affected by impact.

$$v_{AY} = u_{AY} = 4.5 \text{ m/sec}$$

$$v_{BY} = u_{BY} = 12 \sin 60° = 10.392 \text{ m/sec}$$

IMPACT OF ELASTIC BODIES

Let v_{AX} and v_{BX} be the velocities after impact. Applying principles of conservation of momentum in the line of impact, we get

$$mu_{AX} + mu_{BX} = mv_{AX} + mv_{BX}$$
$$7.794 - 6 = v_{AX} + v_{BX}$$

or $\qquad v_{AX} + v_{BX} = 1.794$ ...(1)

From the definition of coefficient of restitution

$$0.9(u_{AX} - u_{BX}) = v_{BX} - v_{AX}$$
$$v_{BX} - v_{AX} = 0.9(7.794 + 6) = 12.415 \qquad ...(2)$$

From Eqns. (1) and (2), we get

$$v_{BX} = \frac{12.415 + 1.794}{2} = 7.104 \text{ m/sec}$$

$\therefore \qquad v_{AX} = 1.794 - 7.104 = -5.310 \text{ m/sec}$

$$v_A = \sqrt{5.31^2 + 4.5^2} = \textbf{6.960 m/sec} \quad \textbf{Ans.}$$

$$\theta_A = \tan^{-1}\frac{4.5}{5.31} = \textbf{40.28°} \text{ as shown in Fig. 16.16(a).} \quad \textbf{Ans.}$$

$$v_B = \sqrt{7.104^2 + 10.392^2} = \textbf{12.588 m/sec} \quad \textbf{Ans.}$$

and $\qquad \theta_B = \tan^{-1}\left(\dfrac{10.392}{7.104}\right) = \textbf{55.643}$ as shown in Fig. 16.6 (b) $\quad$ **Ans.**

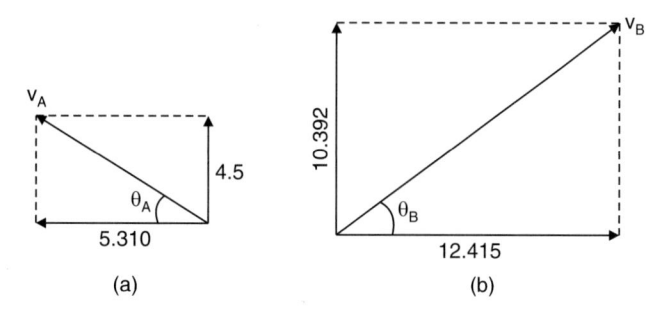

Fig. 18.6

Example 18.7 *A ball is dropped from a height of 3 m upon a 15° incline. If e = 0.8, find the resultant velocity of the ball after impact.*

Solution. The ball falls freely under gravity from a height of $h_0 = 3$ m. Hence, its vertical downward velocity at the instance of striking the incline

$$u_1 = \sqrt{2gh_0} = \sqrt{2 \times 9.81 \times 3}$$
$$= 7.672 \text{ m/sec}$$

Line of impact is normal to the plane *i.e.,* at 15° to vertical. Taking axes x and y as shown in Fig. 18.7, velocity of the ball before striking the plane,

(*i*) normal to line of impact $u_{1x} = u_1 \sin 15° = 1.986$ m/sec

(*ii*) In the line of impact $u_{1y} = -u \cos 15° = -7.411$ m/sec

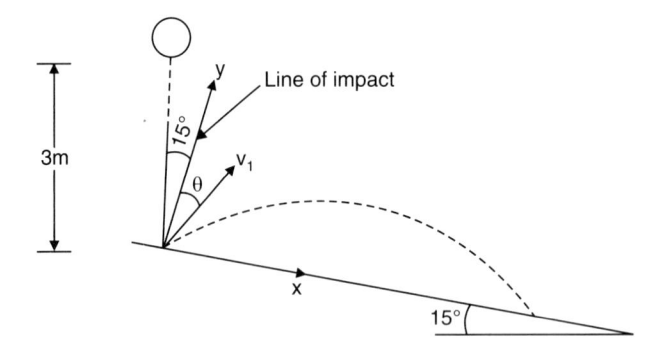

Fig. 18.7

Let the velocity after impact be v_2.

In the direction normal to line of impact, the component of velocity is not affected.

$$v_{1x} = u_{1x} = 1.986 \text{ m/sec}$$

Initial and final velocities of floor = 0

From the definition of coefficient of restitution

$$0.8(-7.412 - 0) = 0 - v_{1y}$$
$$v_{1y} = 5.929 \text{ m/sec}$$

$$v_1 = \sqrt{v_{1x}^2 + v_{1y}^2} = \sqrt{1.986^2 + 5.929^2} = \textbf{6.253 m/sec}$$

$$\theta = \tan^{-1}\left(\frac{v_{1x}}{v_{1y}}\right) = \tan^{-1}\left(\frac{1.986}{5.929}\right)$$

= 18.52° to the line of impact   **Ans.**

∴ **Inclination to the plane**

$$= 90° - 18.52° = \textbf{73.48°} \quad \textbf{Ans.}$$

Example 18.8. *A ball falls vertically for 3 seconds on a plane inclined at 20° to the horizontal axis. If the coefficient of restitution is 0.8, when and where will the ball strike the plane again?*

Solution. Velocity of ball which striking the plane

$$= 3 \times g = 3g \text{ downward}$$

Component of velocity down the plane

$$= 3g \sin 20°$$

IMPACT OF ELASTIC BODIES

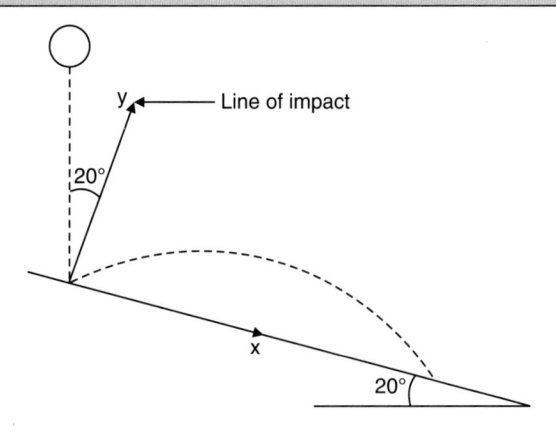

Fig. 18.8

Component of velocity in the line of impact
$$U_y = -3g \cos 20°$$
Velocity after the impact
$$v_y = eu_y = 0.8 \times 3g \cos 20° = 2.4g \cos 20°$$
Acceleration in the line of impact $= -g \cos 20°$

Considering the motion normal to the plane and using, kinematic equation $s = ut + \dfrac{1}{2}at^2$ we get,

$$0 = 2.4g \cos 20° \, t - \left(\dfrac{1}{2}\right) g \cos 20° \, t^2$$

$$t = 4.8 \text{ sec}$$

Component of velocity parallel to plane is not affected by the impact, since it is normal to the line of impact.
$$v_x = 3g \sin 20°$$
Acceleration in this direction $= g \sin 20°$

∴ Distance travelled in 4.8 sec is given by

$$s = 3g \sin 20° \, t - \dfrac{1}{2} g \sin 20° \, t^2$$

$$= g \sin 20° \left(3 \times 4.8 - \dfrac{1}{2} \times 4.8^2\right)$$

$$= 9.81 \sin 20° \left(3 \times 4.8 - \dfrac{1}{2} \times 4.8^2\right)$$

$$s = 86.967 \text{ m} \quad \textbf{Ans.}$$

Example 18.9. *A ball is thrown at an angle θ with the normal to a smooth wall. It rebounds at an angle θ' with the normal. Show that the coefficient of restitution is expressed by*

$$e = \dfrac{\tan \theta}{\tan \theta'}.$$

Solution. Component normal to line of impact is unaltered after impact. If u_1 is velocity of approach and v_1 is the velocity of separation of the ball (Fig. 18.9), we get

$$u_1 \sin \theta = v_1 \sin \theta'$$

or
$$v_1 = u_1 \frac{\sin \theta}{\sin \theta'} \qquad ...(1)$$

The velocities of wall before and after impact i.e., u_2, $v_2 = 0$. Hence from the definition of coefficient of restitution,

$$e(u_1 \cos \theta - 0) = 0 - (-v_1 \cos \theta') = v_1 \cos \theta' \qquad ...(2)$$

From (1) and (2),

$$eu_1 \cos \theta = u_1 \frac{\sin \theta}{\sin \theta'} \cos \theta'$$

or
$$e = \frac{\tan \theta}{\tan \theta'}.$$

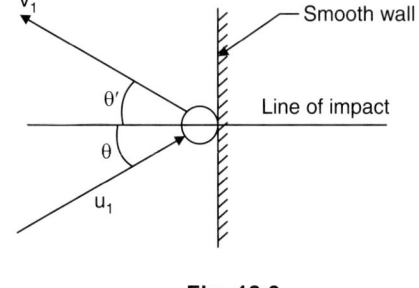

Fig. 18.9

Example 18.10. *A ball is dropped on an inclined plane and is observed to move horizontally after the impact. If coefficient of restitution is 'e', determine the inclination of the plane and the velocity after impact.*

Solution. Let θ be the inclination of the plane to horizontal (Ref. Fig. 18.10). Hence the line of impact is at right angles to the plane. The vertical downward velocity before striking the plane be u and the horizontal velocity after impact be v. During impact component of velocity normal to the line of impact is not altered. Hence

$$u \sin \theta = v \cos \theta$$
$$v = u \tan \theta \qquad ...(1)$$

Considering the velocities of ball and inclined plane (which is zero) the coefficient of restitution e is given by

$$e(u \cos \theta - 0) = [0 - (-v \sin \theta)]$$
$$e u \cos \theta = v \sin \theta$$

Substituting the value of v from (1), we get

$$e u \cos \theta = u \tan \theta \sin \theta$$

or
$$e = \tan^2 \theta$$

or
$$\tan \theta = \sqrt{e}$$

From (1),
$$v = u \tan \theta \quad \textbf{Ans.}$$

i.e.,
$$\mathbf{v = u\sqrt{e}} \quad \textbf{Ans.}$$

Fig. 18.10

IMPACT OF ELASTIC BODIES

Example 18.11. *A ball is dropped from a height $h_0 = 1.2$ m on a smooth floor as shown in Fig. 18.11. Knowing that for the first bounce, $h_1 = 1$ m and $D_1 = 0.4$ m, determine*

(a) the coefficient of restitution

(b) the height and the range of the second bounce.

Solution. Line of impact being vertical, horizontal component of velocity is not affected by impact.

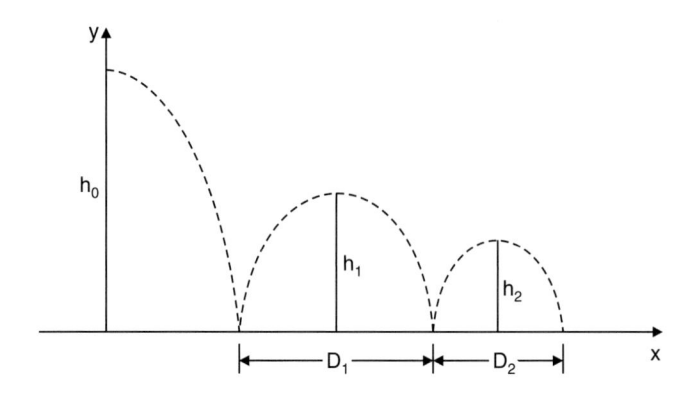

Fig. 18.11

Let this value be u_x.

The ball is dropped from a height of $h_0 = 1.2$ m.

Hence vertical component of velocity before first impact

$$u_y = \sqrt{2gh_0} \text{ downward.}$$

Let the vertical component of velocity after first impact be v_{1y}.

Since after first bounce, the ball has raised to

$$h_1 = 1 \text{ m,}$$

$$v_{1y} = \sqrt{2gh_1} \text{ upward.}$$

$$e = \frac{\text{Rel. velocity of separation}}{\text{Rel. velocity of approach}}$$

$$= \frac{\sqrt{2gh_1}}{\sqrt{2gh_0}} = \sqrt{h_1/h_0} = \sqrt{1/1.2} = \mathbf{0.913} \quad \textbf{Ans.}$$

Time of flight t_1 in first bounce $= \dfrac{2v_{1y}}{g} = \dfrac{2\sqrt{2gh_1}}{g} = \dfrac{2\sqrt{2 \times 9.81 \times 1}}{9.81}$

$$= 0.903 \text{ sec}$$

Range, $D_1 = u_x t$

$$0.4 = u_x \times 0.903$$
$$u_x = 0.443 \text{ m/sec}$$

Vertical component of velocity after second bounce

$$v_{2y} = e\sqrt{2gh_1} = 0.903\sqrt{2 \times 9.81 \times 1} = 4.0 \text{ m/sec}$$

$$\therefore \quad h_2 = \frac{v_{2y}^2}{2g} = \textbf{0.815 m} \quad \textbf{Ans.}$$

Time of flight, $\quad t_2 = 2 \times \dfrac{v_{2y}}{g} = \dfrac{2 \times 4}{9.81} = 0.815 \text{ sec}$

Horizontal range $\quad D_2 = u_x \times t_2 = 0.443 \times 0.815$
$$= \textbf{0.316 m} \quad \textbf{Ans.}$$

18.5. LOSS OF KINETIC ENERGY

During collision the kinetic energy is lost due to imperfect elastic action. Energy is also lost due to

(a) heat generated

(b) sound generated and

(c) vibration of colliding bodies.

The loss of kinetic energy can be found by finding kinetic energy before impact and after impact. Let u_1, u_2 be initial velocities and v_1, v_2 be final velocities of two bodies colliding in the line of impact, their weights being W_1 and W_2. Then

$$\text{Initial K.E.} = \frac{W_1}{2g} u_1^2 + \frac{W_2}{2g} u_2^2$$

$$\text{Final K.E.} = \frac{W_1}{2g} v_1^2 + \frac{W_2}{2g} v_2^2$$

$\therefore \quad$ Loss of K.E.

$$= \text{Initial K.E.} - \text{Final K.E.}$$

$$= \left(\frac{W_1}{2g} u_1^2 + \frac{W_2}{2g} u_2^2 \right) - \left(\frac{W_1}{2g} v_1^2 + \frac{W_2}{2g} v_2^2 \right)$$

$$= \frac{W_1}{2g}(u_1^2 - v_1^2) + \frac{W_2}{2g}(u_2^2 - v_2^2)$$

Example 18.12. *A sphere of weight 10 N moving at 3 m/sec collides with another sphere of weight 50 N moving in the same line at 0.6 m/sec. Find the loss of kinetic energy during impact and show that the direction of motion of the first sphere is reversed after the impact. Assume coefficient of restitution as 0.75.*

IMPACT OF ELASTIC BODIES

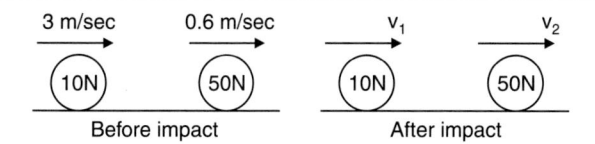

```
    3 m/sec         0.6 m/sec              v₁              v₂
    ──→             ──→                    ──→             ──→
   (10N)           (50N)                  (10N)           (50N)
   ─────────────────────              ─────────────────────
       Before impact                       After impact
```

Fig. 18.12

$$u_1 = 3 \text{ m/sec}, \quad u_2 = 0.6 \text{ m/sec}$$

Solution. Let velocities after impact be v_1 and v_2.

From principles of conservation of momentum,

$$\frac{10}{g}(3) + \frac{50}{g}(0.6) = \frac{10}{g} v_1 + \frac{50}{g} v_2$$

$$v_1 + 5v_2 = 6 \qquad \qquad ...(1)$$

From the definition of coefficient of restitution, we have

$$0.75 (3 - 0.6) = v_2 - v_1 \quad \text{or} \quad v_2 - v_1 = 1.8 \qquad \qquad ...(2)$$

From (1) and (2), $\quad v_2 = 1.3$ m/sec

and $\quad v_1 = 6 - 1.3 \times 5 = -0.5$ m/sec

Thus, **the velocity of first ball is reversed after impact.**

Loss of K.E. = Initial K.E. − Final K.E.

$$= \frac{10}{2g} \times 3^2 + \frac{50}{2g} \times 0.6^2 - \left\{ \frac{10}{2g} \times (0.5)^2 + \frac{50}{2g} \times 1.3^2 \right\}$$

$$= \frac{10}{2 \times 9.81} (9 + 1.8 - 0.25 - 8.45)$$

$$= \textbf{1.07 joules} \quad \textbf{Ans.}$$

Example 18.13. *Find the loss of kinetic energy in example 18.6 if the weight of each ball is 10 N.*

Solution. Component of velocities in *y*-direction is not affected. Hence no change in K.E. due to the components of velocities in *y*-direction. In *x*-direction,

$$u_{AX} = 7.79 \text{ m/sec.}; \quad u_{BX} = -6 \text{ m/sec.};$$

$$v_{AX} = -5.31 \text{ m/sec.}; \quad v_{BX} = 7.104 \text{ m/sec.};$$

Mass of the ball $= \dfrac{10}{g} = \dfrac{10}{9.81}$

$\therefore \quad$ Loss of K.E. $= \dfrac{10}{2 \times 9.81} [u_{AX}^2 + u_{BX}^2 - v_{AX}^2 - v_{BX}^2]$

$$= \frac{10}{2 \times 9.81} [7.79^2 + (-6)^2 - (-5.31)^2 - 7.104^2]$$

$$= \textbf{18.02 joules} \quad \textbf{Ans.}$$

Important Definitions

1. For definitions of the terms 'line of impact', direct impact, oblique impact, central impact, eccentric impact, refer Art 18.1.
2. The **coefficient of restitution** of two colliding bodies may be taken as the ratio of impulse during restitution period to the impulse during the deformation period. This is also equal to the ratio of relative velocity of separation to the relative velocity of approach of the colliding bodies, the relative velocities being measured in the line of impact.

Important Formulae

1. $e = \dfrac{v_2 - v_1}{v_1 - v_2}$, for the component of velocities in the line of impact.
2. For a freely following body from height h_0 and having bounce h_1,
$$e = \sqrt{h_1/h_0}.$$

PROBLEMS FOR EXERCISE

18.1 Two bodies, one of which is 400 N with a velocity of 8 m/sec and the other of 250 N with a velocity of 12 m/sec, move towards each other along a straight line and impinge centrally. Find the velocity of each body after impact if the coefficient of restitution is 0.8.
[**Ans.** $v_1 = -5.846$ m/sec; $v_2 = 10.154$ m/sec]

18.2 A ball is dropped from a height of h_0 metres on a floor. Show that after first bounce it will rise to height h_1 given by $h_1 = e^2 h_0$, where e is coefficient of restitution. Determine to what height the ball will rise after 3 bounces if drooped from a height of 2 m if the coefficient of restitution is 0.8.
[**Ans.** 0.524 m]

18.3 10 N sphere shown in Fig. 18.13 is released from rest when $\theta_A = 90°$. The coefficient of restitution between the sphere and the block is 0.70. If the coefficient of friction between the block and the horizontal surface is 0.3, determine how far the block will move after impact. [**Ans.** $s = 0.934$]

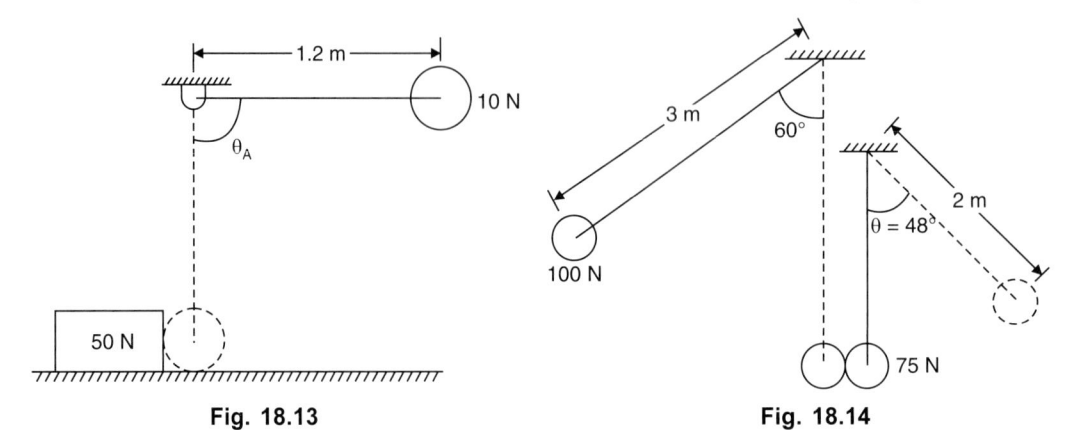

Fig. 18.13

Fig. 18.14

18.4 100 N ball shown in Fig. 18.14 is released from the position shown by continuous line. It strikes a freely suspended 75 N ball. After impact, 75 N ball is raised by an angle $\theta = 48°$. Determine the coefficient of restitution. [**Ans.** $e = 0.162$ m]

[**Hint:** $u_1 = \sqrt{2gh_1}$ where, $h_1 = 3(1 - \cos 60°)$

$v_2 = \sqrt{2gh_2}$ where $h_2 = 2(1 - \cos 48°)$]

IMPACT OF ELASTIC BODIES

18.5 Two identical balls, moving horizontally collide as shown in Fig. 18.15. Determine their velocities after impact, if the coefficient of restitution is 0.75.

[**Ans.** $v_1 = 8.718$ m/sec; $\theta_1 = 23.41°$; $v_2 = 8.718$ m/sec; $\theta_2 = 96.587°$]

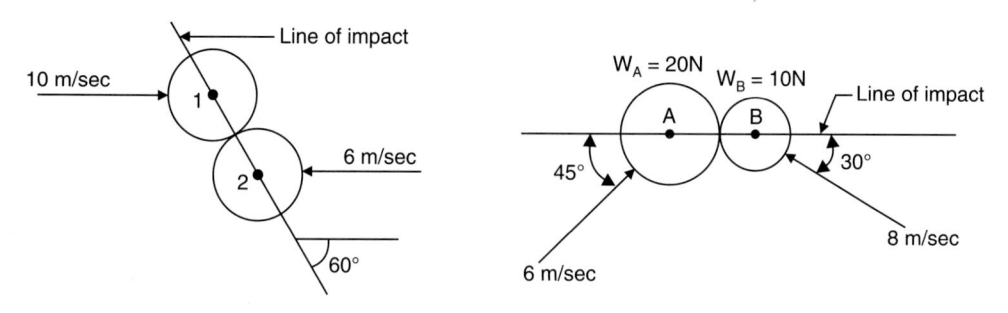

Fig. 18.15

Fig. 18.16

18.6 Find the velocities of the two balls shown in Fig. 18.16 after impact, if the coefficient of restitution = 0.6. [**Ans.** $v_A = 4.576$ m/sec ; $\theta_A = 67.99°$; $v_B = 6.3932$ m/sec ; $\theta_B = 38.73°$]

18.7 Central impact takes place between a 40 N ball and a stationary block of 60 N as shown in Fig. 18.17. The block rests on rollers and move, freely on horizontal surface. Find the velocity of the block and ball after impact. Take coefficient of restitution as 0.8.

[**Ans.** Velocity of block = 5.76 m/sec; velocity of ball = 4.04 m/sec at 67.88° to horizontal]

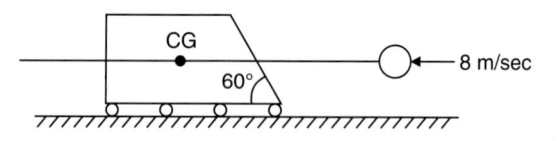

Fig. 18.17

18.8 A ball drops on to a smooth horizontal floor and bounces as shown in Fig. 18.18. Derive expressions for coefficient of restitution in terms of (a) two successive heights, (b) two successive ranges. Determine also time of f lights in nth bounce.

[**Ans.** (a) $e = \sqrt{h_n/(h_{n-1})}$; (b) $e = D_n/D_{n-1}$, $t = 2\sqrt{2h_n/g}$]

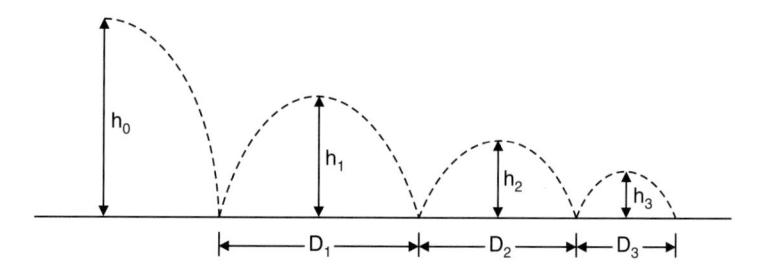

Fig. 18.18

18.9 80 N and 150 N bodies are approaching each other with a velocity of 20 m/sec, and 6 m/sec respectively. What will be the velocity of each body after impact ? How much is the loss of kinetic energy? Take coefficient of restitution = 0.6. Assume 80 N block is moving from left to right.

[**Ans.** For 80 N block $v_1 = \overleftarrow{7.13}$ m/sec ; 150 N block, $v_2 = \overrightarrow{8.47}$ m/sec ;

Loss of K.E. = 1150.43 joules]

19
Circular Motion of Rigid Bodies

Whenever there is change in directions of roads/railways, circular curves are provide to get smooth change over of the direction. It is common experience that passengers in vehicles feel a bit uncomfortable while negotiating the curve. While negotiating curves sometimes vehicles skid and sometime overturn. Though the vehicles move with uniform velocity, there will be acceleration towards centre of the circle and hence inertia force acts in radially outward direction. This force is known as *centrifugal force*. It is this force which causes instability of vehicles. In this chapter, the expression for acceleration towards the centre of circle is derived and stability conditions of the vehicles on plane track and banked track are analysed.

19.1. ACCELERATION DURING CIRCULAR MOTION

Let a body moves with uniform velocity along a curved path of radius r as shown in Fig. 19.1 (a). In time dt, let it moves from A to B. Since the body has uniform velocity, tangential velocity v_A at A is equal to tangential velocity v_B at B in magnitude, say $v_A = v_B = v$. However, there is change in direction equal to $d\theta$. Drawing the vector diagram of velocities as shown in Fig. 19.1 (b), we find that there is change in velocity, dV at right angles to v_A (i.e., in radial inward direction) of magnitude dv.

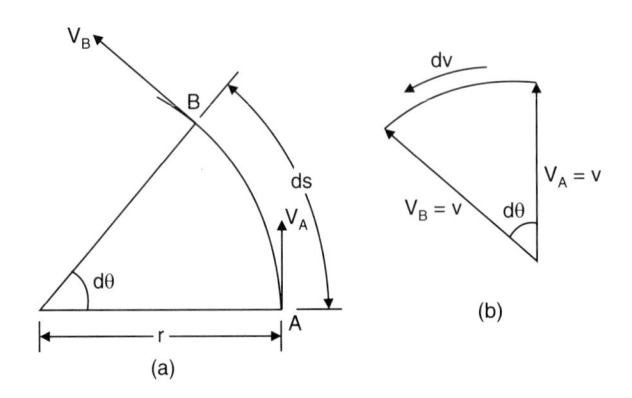

Fig. 19.1

Now, $$dv = v\, d\theta = v\frac{ds}{r} = \frac{v}{r}ds$$

CIRCULAR MOTION OF RIGID BODIES

$$\text{Acceleration} = \frac{dv}{dt} = \frac{v}{r}\frac{ds}{dt} = \frac{v}{r} \cdot v$$

$$a = \frac{v^2}{r} \qquad \qquad \ldots(19.1)$$

Thus when a body moves with uniform velocity v along a curved path of radius r, it has a radial inward acceleration of magnitude v^2/r.

If we apply D'Alembert's principle to get equilibrium condition, an inertia force of magnitude $\dfrac{w}{g}a = \dfrac{w}{g}\dfrac{v^2}{r}$ must be applied in radial outward direction. This force is called as *centrifugal force*.

19.2. MOTION ON LEVEL ROAD

Consider a body moving with uniform velocity on a circular curve of radius r. Let the road be flat. The cross-sectional view is shown in Fig. 19.2. If W is the weight of the body, the inertia force is given by

$$\frac{W}{g}a = \frac{W}{g}\frac{v^2}{r}$$

It acts radially outward. Let pressure on inner wheel be R_1 and on outer wheel be R_2. Total frictional forces be F and centre of gravity be at a height h.

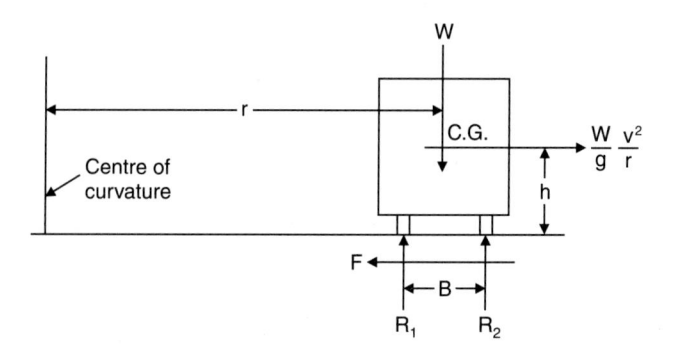

Fig. 19.2

Condition for skidding:

According to D'Alembert's principle the body is in dynamic equilibrium under the action of forces;

W – weight of the vehicle
R_1, R_2 – the reactions
F – frictional force

and $\dfrac{W}{g}\dfrac{v^2}{r}$ – inertia force.

Skidding takes place when frictional force reaches limiting value i.e., when
$$F = \mu W$$
where, μ is the coefficient of friction between road surface and the wheels. When skidding is about to take place,

Forces in horizontal direction = 0

i.e.,
$$\frac{W}{g}\frac{v^2}{r} = F = \mu W$$

or
$$v^2 = \mu g r$$

$$v = \sqrt{\mu g r} \qquad \ldots(19.2)$$

This is the limiting value of velocity from the consideration of skidding.

Condition for overturning:

Let $\quad$ B–distance between inner and outer wheels

h–height of C.G. from ground level

Taking moment about outer wheel, we get,

$$R_1 B + \frac{W}{g}\frac{v^2}{r}h - W\frac{B}{2} = 0$$

or
$$R_1 = \frac{W}{2} - \frac{W}{g}\frac{v^2}{r}\frac{h}{B} \qquad \ldots(19.3)$$

and hence $\quad R_2 = W - R_1$

$$= W - \frac{W}{2} + \frac{W}{g}\frac{v^2}{r}\frac{h}{B} = \frac{W}{2} + \frac{W}{g}\frac{v^2}{r}\frac{h}{B} \qquad \ldots(19.4)$$

From equation (19.3), it may be observed that as velocity of the body increases R_1 decreases. A stage may reach when R_1 becomes zero and after that vehicle overturns on outer wheel. To avoid this situation, the maximum speed should be limited to value of v given by eqn. (19.3). when $R_1 = 0$.

i.e.,
$$0 = \frac{W}{2} - \frac{W}{g}\frac{v^2}{r}\frac{h}{B}$$

or
$$v^2 = \frac{gr}{2}\frac{B}{h}$$

i.e.,
$$v = \sqrt{\frac{gr}{2}\frac{B}{h}}$$
$$\ldots(19.5)$$

In case of rail cars, the required lateral force F is offered by bolts holding the rail to the sleepers as shown in Fig. 19.3. The bolts are designed to resist such force and there is no chance of skidding. However, chance of overturning is there.

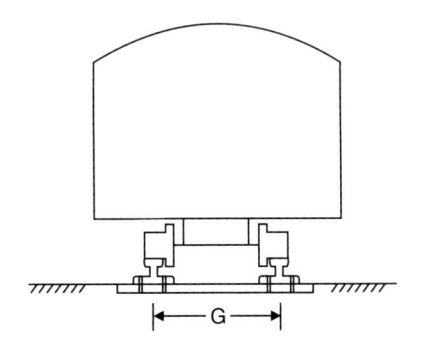

Fig. 19.3

CIRCULAR MOTION OF RIGID BODIES

The distance between inner and outer wheel is equal to gauge of the railway track, which is normally represented by letter G. Hence, eqn. (19.5). takes the form

$$v = \sqrt{\frac{gr}{2}\frac{G}{h}} \qquad \ldots(19.6)$$

19.3. NEED FOR BANKING OF ROADS AND SUPER ELEVATION OF RAILS

When a vehicle moves around a flat curve, a radial outward force $\dfrac{w}{g}\dfrac{v^2}{r}$ is experienced increasing the pressure on the outer wheel and decreasing the pressure on the inner wheel. It creates discomfort to passengers. Apart from this, the speed is to be reduced considerably on curved path to avoid skidding and overturning. If the road is banked (sloping downward towards centre of curve in the cross-section) or the track is given super elevation (raising the outer rail over inner rail) as shown in Fig. 19.4, the following may be achieved:

(i) Skidding and overturning avoided
(ii) Higher speed permitted
(iii) Lateral pressure F eliminated, giving comfort to passenger
(iv) Excess wear and tear of wheels avoided.

Banking of road

Super elevation $\alpha = \tan^{-1} e/G$

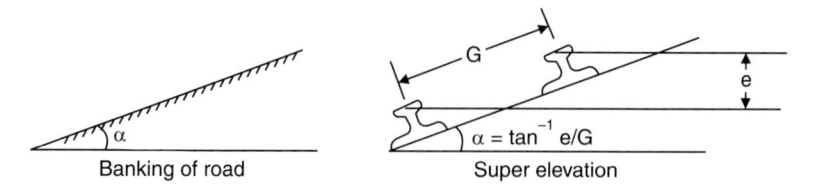

Fig. 19.4

19.4. DESIGNED SPEED

The speed on a banked curved path for which no lateral pressure develops is called the *design speed* on that curve. If vehicle moves with designed speed, there will be equal pressure on inner and outer wheels and hence passengers will not experience any discomfort.

Let α be the angle of banking as shown in Fig. 19.5. Various forces to be considered for dynamic equilibrium at this stage are:

(i) Self weight W acting vertically downward

(ii) Centrifugal force $\dfrac{w}{g}\dfrac{v^2}{r}$ acting radially outward.

(horizontal),

(iii) Reactions R_1 and R_2 acting normal to the road surface.

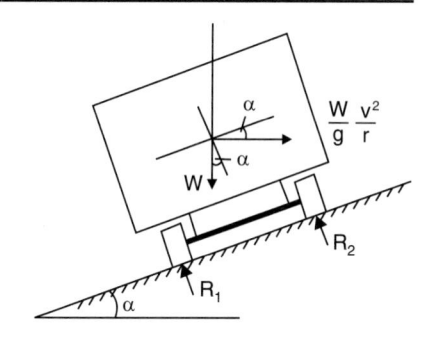

Fig. 19.5

For designed speed there will not be any lateral (frictional) force and wheel reactions are normal to road.

Σ Forces to the inclined plane = 0, gives

$$\frac{W}{g}\frac{v^2}{r}\cos\alpha - W\sin\alpha = 0$$

$$\tan\alpha = \frac{v^2}{gr} \qquad \ldots(19.7)$$

This is the relationship between angle of banking and design speed. If designed speed is higher, angle of banking is also higher.

19.5. SKIDDING AND OVERTURNING ON BANKED ROADS

The chances of skidding and overturning and banked roads is not ruled out, if the vehicles move at speeds much higher than designed speed. However, this speed works out to be much more than what it would be (Eqns. 19.2 and 19.6) on flat roads.

Let the frictional forces developed at inner and outer wheels be F_1 and F_2 respectively. Free body diagram of the vehicle moving on a banked curved path along with inertia force $\frac{W}{g}\frac{v^2}{r}$ is shown in Fig. 19.6.

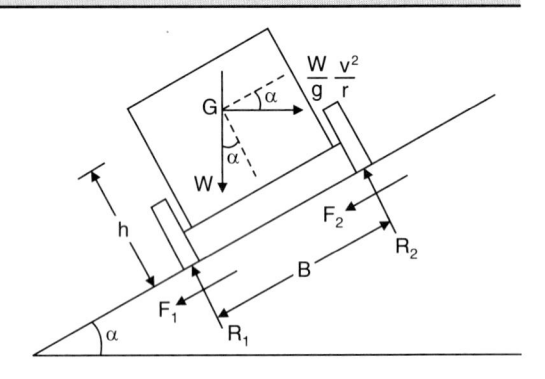

Fig. 19.6

(i) Condition for skidding:

Consider the dynamic equilibrium of the body in lateral direction.

Summing up the forces normal to the inclined surface, we get

$$R_1 + R_2 = W\cos\alpha + \frac{W}{g}\frac{v^2}{r}\sin\alpha \qquad \ldots(1)$$

Summing up forces parallel to the inclined surface, we get,

$$F_1 + F_2 + W\sin\alpha = \frac{W}{g}\frac{v^2}{r}\cos\alpha \qquad \ldots(2)$$

At the time of skidding,
$$F_1 = \mu R_1 \quad \text{and} \quad F_2 = \mu R_2$$

where, μ is coefficient of friction.

∴ Eqn. (2) reduces to

$$\mu(R_1 + R_2) + W\sin\alpha = \frac{W}{g}\frac{v^2}{r}\cos\alpha$$

From Eqn. (1), substituting the value of $R_1 + R_2$, we get

$$\mu \left(W \cos \alpha + \frac{W}{g} \frac{v^2}{r} \sin \alpha \right) + W \sin \alpha = \frac{W}{g} \frac{v^2}{r} \cos \alpha$$

or

$$\mu \cos \alpha + \sin \alpha = \frac{v^2}{gr} (\cos \alpha - \mu \sin \alpha)$$

or

$$\frac{v^2}{gr} = \frac{\mu \cos \alpha + \sin \alpha}{\cos \alpha - \mu \sin \alpha} = \frac{\mu + \tan \alpha}{1 - \mu \tan \alpha}$$

$\therefore$

$$v = \sqrt{gr \frac{\mu + \tan \alpha}{1 - \mu \tan \alpha}} \qquad \ldots (19.8)$$

Substituting $\mu = \tan \phi$
where, ϕ is angle of friction, we get

$$\frac{v^2}{gr} = \frac{\tan \phi + \tan \alpha}{1 - \tan \phi \tan \alpha} = \tan (\alpha + \phi)$$

i.e.,

$$v = \sqrt{\tan (\alpha + \phi) \times gr} \qquad \ldots (19.9)$$

(ii) Condition for overturning:

Let h be the height of centre of gravity of vehicle above road surface and B be the distance between inner and outer wheels. At the time of overturning $R_1 = 0$. Taking moment about contact point of outer wheel with the road

$$W \cos \alpha \frac{B}{2} + W (\sin \alpha) h + \frac{Wv^2}{gr} (\sin \alpha) \frac{B}{2} - \frac{Wv^2}{gr} (\cos \alpha) h = 0$$

$$\frac{B}{2} \cos \alpha + h \sin \alpha = \frac{v^2}{gr} \left[\left(-\frac{B}{2} \sin \alpha + h \cos \alpha \right) \right]$$

i.e.,

$$\frac{v^2}{gr} = \frac{(B/2) \cos \alpha + h \sin \alpha}{h \cos \alpha - (B/2) \sin \alpha} = \frac{B + 2h \tan \alpha}{2h - B \tan \alpha}$$

$$v = \sqrt{gr \frac{B + 2h \tan \alpha}{2h - B \tan \alpha}} \qquad \ldots (19.10)$$

In case of railcars, if e is super elevation and G is the gauge, we have

$$\tan \alpha = \frac{e}{G} \quad \text{and} \quad B = G$$

Hence, limiting speed from the consideration of overturning is

$$v = \sqrt{gr \frac{G + 2h \, e/G}{2h - (e/G) G}} = \sqrt{gr \frac{G + (2h \, e/G)}{2h - e}} \qquad \ldots (19.11)$$

Note: There is no question of skidding in railcars.

Example 19.1. *An automobile weighing 25 kN moves on a road, the longitudinal section of which is shown in Fig. 19.7. If it moves with uniform velocity of 50 kmph, what is the vertical reaction experienced at points A, B, C.*

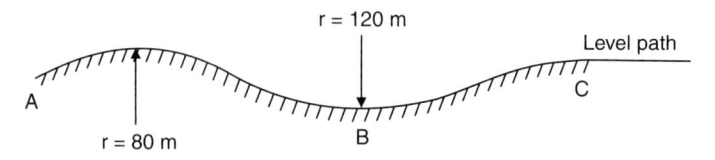

Fig. 19.7

Solution. Velocity $v = 50$ kmph $= 13.889$ m/sec.

(*i*) **When vehicle is at A:** Centrifugal force is vertically upward (radially outward). Its magnitude is

$$= \frac{Wv^2}{gr} = \frac{25}{9.81} \times \frac{13.889^2}{80}$$

$$= 6.146 \text{ kN}$$

∴ Vertical reaction $= W - 6.145$

$= 25 - 6.145 =$ **18.855 kN Ans.**

(*ii*) **When automobile is at B:** Centrifugal force is vertically downward and is equal to

$$= \frac{W}{g} \frac{v^2}{r} = \frac{25}{9.81} \times \frac{13.889^2}{120}$$

$$= 4.097 \text{ kN}$$

∴ Vertical reaction $= W + 4.097$

$= 25 + 4.097 =$ **29.097 kN Ans.**

(*iii*) *On level track at C:*

Vertical reaction $= W =$ **25 kN Ans.**

Example 19.2. *A car weighing 15 kN goes round a flat curve of 50 m radius. The distance between inner and outer wheel is 1.5 m and the C.G. is 0.75 m above the road level. What is the limiting speed of the car on this curve ? Determine the normal reactions developed at the inner and outer wheels if the car negotiates the curve with a speed of 40 kmph. Take coefft. of friction $\mu = 0.4$.*

Solution. Consider the dynamic equilibrium of the car, forces acting on which are as shown in Fig. 19.8.

Limiting speed v from the consideration of avoiding skidding is given by

$$\mu W = \frac{W}{g} \frac{v^2}{r}$$

$$v = \sqrt{\mu g r}$$

$$= \sqrt{0.4 \times 9.81 \times 50}$$

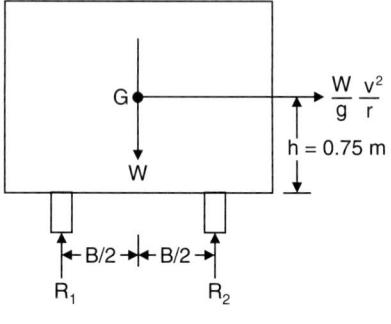

Fig. 19.8

$$= 14.007 \text{ m/sec}$$
$$= 14.007 \times \frac{60 \times 60}{1000} = 50.42 \text{ kmph}$$

Limiting speed from the consideration of preventing overturning:

Taking moment about point of contact of outer wheel with road and noting that $R_1 = 0$ when, the vehicle is about to overturn, we get

$$W \times \frac{B}{2} = \frac{W}{g}\frac{v^2}{r} h$$

$$v = \sqrt{\frac{gr}{2}\frac{B}{h}} = \sqrt{\frac{9.81 \times 50}{2} \times \frac{1.50}{0.75}}$$

$$= 22.147 \text{ m/sec} = 79.73 \text{ kmph},$$

∴ **Limiting speed v = 50.42 kmph   Ans.**

If the vehicle moves with a velocity of 40 kmph,

$$v = 40 \text{ kmph} = 40 \times \frac{1000}{60 \times 60} = 11.111 \text{ m/sec}$$

Taking moment about outer wheel, we get,

$$R_1 \times B + \frac{W}{g}\frac{v^2}{r} h = W \times \frac{B}{2}$$

$$R_1 = \frac{W}{2} - \frac{W}{g}\frac{v^2}{r}\frac{h}{B}$$

$$= \frac{15}{2} - \frac{15}{9.81} \times \frac{11.111^2}{50} \times \frac{0.75}{1.5}$$

$$R_1 = \textbf{5.612 kN   Ans.}$$

∴ $R_2 = W - R_1 = 15 - 5.612$

i.e., $R_2 = \textbf{9.388 kN   Ans.}$

Example 19.3. *Find the angle of banking for a highway curve 200 m radius designed to accommodate cars travelling at 120 kmph, if the coefficient of friction between tyres and the road is 0.6.*

Solution. $v = 120 \text{ kmph} = 33.333 \text{ m/sec}$

If α is the angle of banking, then

$$\tan \alpha = \frac{v^2}{gr} = \frac{33.333^2}{9.81 \times 200} = 0.566$$

$$\alpha = \textbf{29.52°   Ans.}$$

Example 19.4. *Find at what speed a vehicle can move round a curve of 40 m radius without side slip (i) on a level road, (ii) on a road banked to an inclination of 1 in 10.*

At what speed can the vehicle travel on banked road without any lateral frictional force ? Assume the coefficient of friction between the vehicle and the road as 0.4.

Solution. $r = 40$ m, $\mu = 0.4$

(*i*) On level road, limiting speed from the consideration of avoiding skidding is given by

$$v = \sqrt{\mu g r} = \sqrt{0.4 \times 9.81 \times 40}$$

$$= 12.528 \text{ m/sec}$$

i.e., $\qquad v = \textbf{45.102 kmph}\quad\textbf{Ans.}$

(*ii*) On a road banked to an inclination of 1 in 10:

$$\tan \alpha = 1/10 = 0.1$$

$$v = \sqrt{gr \frac{\mu + \tan \alpha}{1 - \mu \tan \alpha}} = \sqrt{9.81 \times 40 \frac{0.4 + 0.1}{1 - 0.4 \times 0.1}}$$

$$= 14.296 \text{ m/sec}$$

i.e., $\qquad v = \textbf{51.466 kmph}\quad\textbf{Ans.}$

If lateral forces are not to be experienced the vehicle should travel with designed speed of curve. It is given by

$$\tan \alpha = \frac{v^2}{gr}$$

$$0.1 = \frac{v^2}{9.81 \times 40}$$

$$v = 6.264 \text{ m/sec}$$

i.e., $\qquad v = \textbf{22.551 kmph}\quad\textbf{Ans.}$

Example 19.5. *A car weighing 20 kN, goes around a curve of 60 m radius banked at an angle of 30°. Find the frictional force acting on tyres and normal reaction on outer and inner wheels, when the car is travelling at 96 kmph. The coefficient of friction between the tyres and the road is 0.6. Take width of wheel base B = 1.6 m and height of C.G. of the vehicle above road level h = 0.8.*

Solution. Free body diagram of the vehicle along with inertia force (centrifugal force) $\frac{w}{g} \cdot \frac{v^2}{r}$ is shown in Fig. 19.9.

Now, the velocity of car $v = 96$ kmph $= 26.667$ m/sec

Consider the dynamic equilibrium of the vehicle.

Σ Force $\parallel$ to the road surface $= 0$, gives

$$F + W \sin 30° = \frac{W}{g} \frac{v^2}{r} \cos 30°$$

CIRCULAR MOTION OF RIGID BODIES

$$F = W[(v^2/gr)\cos 30° - \sin 30°]$$

$$= 20\left[\frac{26.667^2}{9.81 \times 60}\cos 30° - \sin 30°\right]$$

$$= 10.925 \text{ kN} \quad \text{Ans.}$$

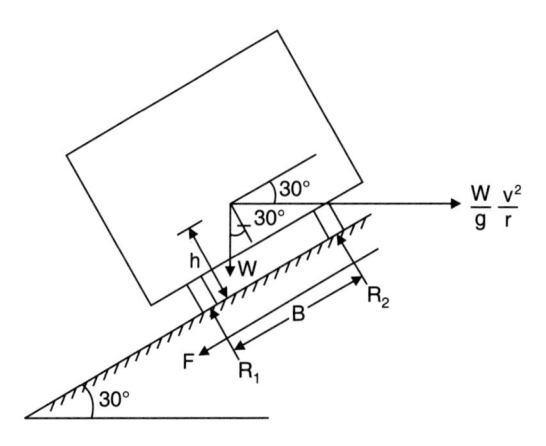

Fig. 19.9

Taking moment about point of contact of outer wheel with road surface, we get

$$R_1 \times B - W\sin 30° \times h - W\cos 30° \times \frac{B}{2} + \frac{W}{g}\frac{v^2}{r}\cos 30° \times h - \frac{W}{g}\frac{v^2}{r}\sin 30° \times \frac{B}{2} = 0$$

$$R_1 = W\left[\frac{h}{B}\sin 30° + \frac{1}{2}\cos 30° + \frac{v^2}{gr}\left(\frac{1}{2}\sin 30° - \frac{h}{B}\cos 30°\right)\right]$$

$$= 20\left[\frac{0.8}{1.6}\sin 30° + \frac{1}{2}\cos 30° + \frac{26.667^2}{9.81 \times 60}\left(\frac{1}{2}\sin 30° - \frac{0.8}{1.6}\cos 30°\right)\right]$$

$$R_1 = 9.238 \text{ kN} \quad \text{Ans.}$$

Taking summation of forces normal to road surface, we get

$$R_1 + R_2 = W\cos 30° + \frac{W}{g}\frac{v^2}{r}\sin 30°$$

$$\therefore \quad R_2 = 20\left[\cos 30° + \frac{26.667^2}{7.81 \times 60}\sin 30°\right] - 9.238$$

i.e., $$R_2 = 20.164 \text{ kN} \quad \text{Ans.}$$

Example 19.6. *Calculate the super elevation of the rail on a curved track for a locomotive running at 60 kmph, gauge and radius of curvature being 1.68 m and 800 m respectively. Find the lateral thrust exerted on the outer rail, if the speed of the locomotive is changed to 80 kmph. Weight of locomotive is 1000 kN.*

Solution. $G = 1.68$ m, $r = 800$ m

(i) $v = 60$ kmph $= (60 \times 1000)/(60 \times 60) = 16.667$ m/sec

If α is the angle of banking required, for comfortable journey (*i.e.*, without experiencing any lateral thrust), α is given by

$$\tan \alpha = \frac{v^2}{gr} = \frac{16.667^2}{9.81 \times 800} = 0.03539$$

∴ Superelevation is given by

$$\frac{e}{G} = \tan \alpha$$

or
$$e = G \tan \alpha = 1.68 \times 0.03539 = 0.0595 \text{ m}$$
$$= \mathbf{59.5 \text{ mm}} \quad \mathbf{Ans.}$$

(*ii*) When speed is 80 kmph:

$$v = 80 \text{ kmph} = \frac{80 \times 1000}{60 \times 60} = 22.222 \text{ m/sec}$$

$$\tan \alpha = \frac{e}{G} = 0.03539$$

$$\alpha = 2.02685°$$
$$\sin \alpha = 0.03537$$
$$\cos \alpha = 0.99937$$

Summing up the forces parallel to the line joining top of rails, we get

$$F + W \sin \alpha = \frac{W}{g} \frac{v^2}{r} \cos \alpha$$

$$F = W \left[\frac{v^2}{gr} \cos \alpha - \sin \alpha \right]$$

$$= 1000 \left[\frac{22.222^2}{9.81 \times 800} \times 0.99937 - 0.03537 \right]$$

i.e., $\qquad \mathbf{F = 27.514 \text{ kN}} \quad \mathbf{Ans.}$

Example 19.7. *What angle of bank the pilot of an aeroplane should maintain, if he wants to fly a horizontal circular path with a radius of 1300 m at a speed of 400 kmph? Calculate the normal force on the aeroplane under the flight condition if the plane is weighing 80 kN.*

Solution. $\qquad r = 1300 \text{ m}, W = 8 \text{ kN}$

$$v = 400 \text{ kmph} = \frac{400 \times 1000}{60 \times 60} = 111.111 \text{ m/sec}$$

Angle of bank required is given by

$$\tan \alpha = \frac{v^2}{gr} = \frac{111.111^2}{9.81 \times 1300} = 0.968$$

i.e., $\qquad \mathbf{\alpha = 44.07°} \quad \mathbf{Ans.}$

CIRCULAR MOTION OF RIGID BODIES

Referring to Fig. 19.10, the magnitude of lift under the flight condition is given by

$$N = W \cos \alpha + \frac{W}{g} \cdot \frac{v^2}{r} \sin \alpha$$

$$= 80 \left[\cos 44.07° + \frac{111.111^2}{9.81 \times 1300} \sin 44.07° \right]$$

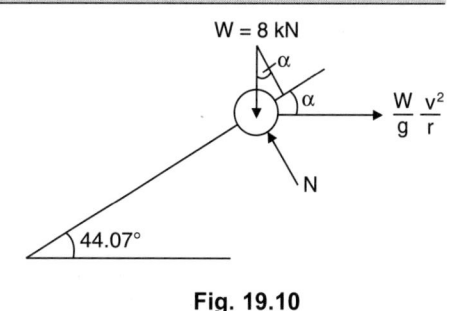

i.e. $\quad N = 111.34$ kN **Ans.**

Fig. 19.10

Important Formulae

1. If v is the uniform velocity of a body along a curved path of radius r, its radial inward acceleration is given by

$$a = \frac{v^2}{r}$$

and its inertia force

$$= \frac{W}{g} \cdot \frac{v^2}{r}.$$

2. If circular motion is on a level road, condition for skidding gives $v = \sqrt{\mu g r}$ and condition for overturning gives,

$$v = \sqrt{\frac{gr}{2} \frac{B}{h}}.$$

3. For a design speed banking required is given by

$$\tan \alpha = \frac{v^2}{gr}.$$

4. For a banked road, condition for skidding is

$$v = \sqrt{gr \tan (\alpha + \phi)} \quad \text{and condition for overturning is } v = \sqrt{gr \frac{G + 2h \tan \alpha}{2h - B \tan \alpha}}.$$

PROBLEMS FOR EXERCISE

19.1 A roller of weight 1 kN starts rolling down from position A on a smooth surface as shown in Fig. 19.11. What will be its velocity at B and the vertical reaction on it from the surface ?

[**Ans.** 17.155 m/sec, 2.5 kN]

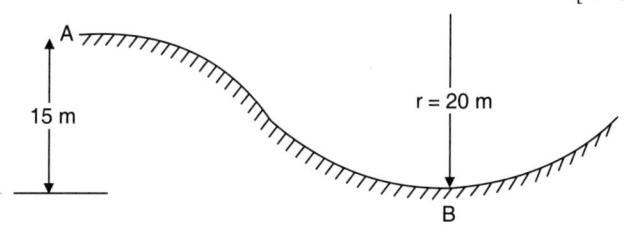

Fig. 19.11

19.2 Find at what maximum speed a vehicle can move round a flat curve of 60 m without side slip. Find also the limiting value of height of C.G. of vehicle, if overturning consideration should not limit speed of the vehicle on this curve. Given, weight of vehicle = 22 kN, base width 1.6 m and coefficient of friction between tyres and road surface = 0.5. [**Ans.** $v = 17.155$ m/sec; $h = 1.6$ m]

19.3 Find the limiting speed of vehicle on a curve of 100 m radius from the consideration of skidding and overturning

(i) if no banking is provided (ii) if 20° banking is provided.

With what speed the vehicle should negotiate the above banked curve, so that passengers do not feel any discomfort. Take weight of vehicle = 15 kN, distance between centres of inner and outer wheels = 1.5 m, height of C.G. above road surface = 0.65 m and coefficient of friction = 0.5.

[**Ans.** With no banking $v = 22.147$ m/sec.; with banking $v = 33.081$ m/sec.; Design velocity $v = 18.896$ m/sec]

19.4 A cyclist is riding in a horizontal circle of radius 20 m at a speed of 18 kmph. What should be the angle to the vertical of the centre line of the bicycle to ensure stability? What is the maximum velocity with which he can negotiate the curve if weight of bicycle and the man is 800 N and the C.G. is 450 mm above the road surface? Take coefficient of friction $\mu = 0.6$.

[**Ans.** $\alpha = 7.26°$; $v = 39.06$ kmph]

19.5 A 20 kN vehicle is going round a circular curve of radius 80 m with a velocity of 72 kmph. If the road has a banking of 25°, find the total frictional force developed and normal reactions at inner and outer wheels. Take base width of vehicle $B = 1.5$ m, height of C.G. above road surface = 0.6 m.

Coefficient of friction $\mu = 0.6$. [**Ans.** $R_{inner} = 10.903$ kN; $R_{outer} = 11.532$ kN; $F = 0.787$ kN]

20

Rotation of Rigid Bodies

So far the discussion was about rigid bodies motion having rectilinear or curvilinear translation. During translation a straight line drawn on the rigid body remains parallel to its original position at any time. In this chapter the discussion is on the motion of body rotating about a fixed axis. In such motion every particles of rigid body move along concentric circles as shown in Fig. 20.1.

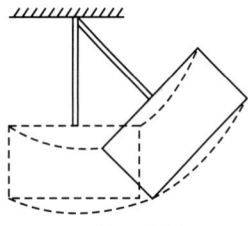

Fig. 20.1

20.1. ANGULAR MOTION

Translation has been treated as linear motion in previous chapters and now we treat rotation as angular motion. The displacement of the body in rotation is measured in terms of angular displacement θ, where θ is in radian. When a particle in a body moves from position A to B, the displacement is θ as shown in Fig. 20.2. This displacement is a vector quantity since it has magnitude as well as direction. The direction is a rotation-either clockwise or counter-clockwise.

The rate of change of angular displacement with time is called angular velocity and is denoted by ω. Thus

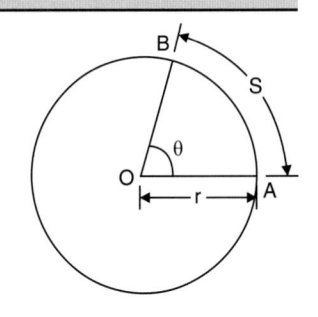

$$\omega = \frac{d\theta}{dt} \qquad \ldots (20.1)$$

Fig. 20.2

The rate of change of angular velocity with time is called angular acceleration and is denoted by α.

Thus $\qquad \alpha = \dfrac{d\omega}{dt} = \dfrac{d^2\theta}{dt^2} \qquad \ldots(20.2)$

The angular acceleration may be expressed in another useful form.

Now $\qquad \alpha = \dfrac{d\omega}{dt} = \dfrac{d\omega}{d\theta} \cdot \dfrac{d\theta}{dt}$

Thus $\qquad \alpha = \omega \dfrac{d\omega}{d\theta} \qquad \ldots(20.3)$

20.2. RELATIONSHIP BETWEEN ANGULAR MOTION AND LINEAR MOTION

When the particle moves from A to B (Ref. Fig. 20.2), the distance travelled by it is s. If r is the distance of the particle from the centre of rotation then

$$s = r\theta \qquad \ldots(20.4)$$

The tangential velocity of the particle is called as linear velocity and is denoted by v. Then

$$v = \frac{ds}{dt} = r\frac{d\theta}{dt} \qquad \ldots(20.5)$$

The linear acceleration of the particle in tangential direction a_t is given by

$$a_t = \frac{dv}{dt} = r\frac{d^2\theta}{dt^2} \qquad \ldots(20.6)$$

While treating the curvilinear motion in previous chapter, it has been shown that, if v is the tangential velocity, then there is radial acceleration v^2/r.

Denoting radial acceleration by a_n, we get

$$a_n = \frac{v^2}{r} = r\omega^2 \qquad \ldots(20.7)$$

20.3. UNIFORM ANGULAR VELOCITY

If the angular velocity is uniform, the angular distance moved in t second by a body having angular velocity ω radian/second is given by,

$$\theta = \omega t \quad \text{radian} \qquad \ldots(20.8)$$

Note that the uniform angular velocity is characterized by zero angular acceleration.

Many a times the angular velocity is given in terms of number of revolution per minute (rpm). Since there are 2π radians in one revolution and 60 seconds in one minute the angular acceleration ω is given by

$$\omega = \frac{2\pi N}{60} \quad \text{rad/sec.} \qquad \ldots(20.9)$$

where, N – is in rpm.

Since angular velocity is uniform, the time taken for one revolution T is given by Eqn. 20.8, as,

$$2\pi = \omega T$$

or $\qquad T = 2\pi/\omega \qquad \ldots(20.10)$

20.4. UNIFORMLY ACCELERATED ROTATION

Let us consider the uniformly accelerated motion with angular acceleration α. From the definition

$$\frac{d\omega}{dt} = \alpha$$

$\therefore \qquad \omega = \alpha t + C_1$

where, C_1 is constant of integration.

If the initial velocity is ω_0, then
$$\omega_0 = \alpha \times 0 + C_1 \quad \text{or} \quad C_1 = \omega_0$$
$\therefore \quad \omega = \omega_0 + \alpha t \qquad \qquad \ldots(1)$

Again from the definition of angular velocity
$$\frac{d\theta}{dt} = \omega = \omega_0 + \alpha t$$
$\therefore \quad \theta = \omega_0 t + \frac{1}{2}\alpha t^2 + C_2$

where, C_2 is constant of integration.

Measuring the angular displacement from the instant of reckoning the interval of time, we get,
$$0 = 0 + 0 + C_2$$
$$\theta = \omega_0 t + 1/2\,\alpha t^2 \qquad \qquad \ldots(2)$$

From eqn. (20.3),
$$\alpha = \omega \frac{d\omega}{d\theta} \quad \text{or} \quad \alpha\, d\theta = \omega\, d\omega$$

Integrating, we get
$$\alpha\theta = \frac{\omega^2}{2} + C_3$$

where, C_3 is constant of integration.

Initially, $\theta = 0$ and $\omega = \omega_0$

Hence we get,
$$\alpha \times 0 = \frac{\omega_0^2}{2} + C_3 \quad \text{or} \quad C_3 = \frac{-\omega_0^2}{2}$$
$$\alpha\theta = \frac{\omega^2}{2} + \frac{-\omega_0^2}{2}$$

or $\quad \omega^2 - \omega_0^2 = 2\alpha\theta \qquad \qquad \ldots(3)$

Thus for uniformly accelerated angular motion

$$\omega = \omega_0 + \alpha t \qquad \ldots(i)$$
$$\theta = \omega_0 t + \frac{1}{2}\alpha t^2 \qquad \ldots(ii) \qquad \ldots(20.11)$$
and $\quad \omega^2 - \omega_0^2 = 2\alpha\theta \qquad \ldots(iii)$

Carefully note similarly of eqn. (20.11) with uniformly accelerated linear motion equations listed below:

$$v = u + at \qquad \ldots(i)$$
$$s = ut + \frac{1}{2}\alpha t^2 \qquad \ldots(ii) \qquad \ldots(20.12)$$
and $\quad v^2 - u^2 = 2as \qquad \ldots(iii)$

The problems of uniform angular retardation can be handled with equation (20.11) by noting that retardation is negative quantity of acceleration, as has been done in linear motion.

Example 20.1. *The rotation of a fly wheel is governed by the equation* $\omega = 3t^2 - 2t + 2$ *where ω is in radian per second and t is in seconds. After one second from the start the angular displacement was 4 radians. Determine the angular displacement, angular velocity and angular acceleration of the fly wheel when t = 3 seconds.*

Solution. Now, $\omega = 3t^2 - 2t + 2$

i.e., $$\frac{d\theta}{dt} = 3t^2 - 2t + 2$$

or $$\theta = t^3 - t^2 + 2t + C$$

where, C is constant of integration.

When $t = 1$, $\theta = 4$,

∴ $4 = 1 - 1 + 2 + C$

i.e., $C = 2$

∴ $\theta = t^3 - t^2 + 2t + 2$

When, $t = 3$,

$\theta = 3^3 - 3^2 + 2 \times 3 + 2 =$ **26 radian** Ans.

$\omega = 3 \times 3^2 - 2 \times 3 + 2 =$ **23 rad/sec** Ans.

Angular acceleration α is given by

$$\alpha = \frac{d\omega}{dt} = 6t - 2$$

∴ When $t = 3$, $\alpha = 6 \times 3 - 2 =$ **16 rad/sec²** Ans.

Example 20.2. *The angular acceleration of a flywheel is given by $\alpha = 12 - t$, where, α is in rad/sec² and t is in seconds. If the angular velocity of the flywheel is 60 rad/sec at the end of 4 seconds, determine the angular velocity at the end of 6 seconds. How many revolutions take place in these 6 seconds?*

Solution. $\alpha = 12 - t$

i.e., $$\frac{d\omega}{dt} = 12 - t$$

or $$\omega = 12t - \frac{t^2}{2} + C$$

where, C is constant of integration.

When $t = 4$ sec, $\omega = 60$ rad/sec

∴ $60 = 12 \times 4 - \frac{4-4}{2} + C_1$

i.e., $C_1 = 20$

∴ $\omega = 12t - \frac{t^2}{2} + 20$

When $t = 6$ sec

$$\omega = 12 \times 6 - \frac{6^2}{2} + 20 = \textbf{74 rad/sec} \quad \textbf{Ans.}$$

ROTATION OF RIGID BODIES

Now, $\dfrac{d\theta}{dt} = \omega = 12t - \dfrac{t^2}{2} + 20$

$\therefore \quad \theta = 6t^2 - t^3/6 + 20t + C_2$

where, C_2 is constant of integration.

When $t = 0$, $\quad \theta_0 = C_2$

When $t = 6$ sec, $\quad \theta_6 = 6 \times 6^2 - 6^3/6 + 20 \times 6 + C_2 = 180 + C_2$

$\therefore$ Angular displacement during 6 seconds

$$= \theta_6 - \theta_0 = 180 + C_2 - C_2 = 180 \text{ rad.}$$

$\therefore$ **Number of revolution** $= \dfrac{180}{2\pi} = $ **28.648   Ans.**

Example 20.3. *A wheel rotating about a fixed axis at 20 revolutions per minute is uniformly accelerated for 70 seconds during which it makes 50 revolutions. Find the (i) angular velocity at the end of this interval and (ii) time required for the velocity to reach 100 revolutions per minute.*

Solution. Initial velocity

$\omega_0 = 20$ rpm

$\quad = \dfrac{20 \times 2\pi}{60}$ rad/sec

$\quad = 2.0944$ rad/sec

$t = 70$ sec

Angular displacement $\theta = 50$ revolutions

$\quad = 50 \times 2\pi = 100\pi$ radian

Using kinematic equation,

$$\theta = \omega_0 t + \dfrac{1}{2} \alpha t^2$$

we get, $\quad 100\pi = 2.0944 \times 70 + \dfrac{1}{2} \alpha \times 70^2$

$\alpha = 0.06839$ rad/sec^2

$\therefore$ Angular velocity at the end of 70 seconds interval

$\omega = \omega_0 + \alpha t = 2.0944 + 0.06839 \times 70$

$\quad = $ **6.8816 rad/sec   Ans.**

Let the time required for the velocity to reach 100 rpm be t.

$\omega = 100$ rpm

$\quad = \dfrac{100 \times 2\pi}{60}$ rad/sec

Using the relation $\omega = \omega_0 + \alpha t$, we get

$$t = \dfrac{\omega - \omega_0}{\alpha} = \left(\dfrac{200\pi}{60} - 2.0944\right) \dfrac{1}{0.06839}$$

i.e., $\quad t = $ **122.497 sec   Ans.**

Example 20.4. *A fly-wheel, which accelerates at uniform velocity is observed to have made 100 revolution to increase its velocity from 120 rpm to 160 rpm. If the flywheel originally started from rest, determine*

(i) *the value of acceleration,*

(ii) *time taken to increase the velocity from 120 rpm to 160 rpm and*

(iii) *revolution made in reaching a velocity of 160 rpm, starting from rest.*

Solution. $\theta = 100$ revolution $= 200\pi$ radian

$$\omega_0 = 120 \text{ rpm} = \frac{120 \times 2\pi}{60} = 4\pi \text{ rad/sec}$$

$$\omega = 160 \text{ rpm} = \frac{160 \times 2\pi}{60} = 16.755 \text{ rad/sec}$$

Using the kinematic relation $\omega^2 - \omega_0^2 = 2\alpha\theta$, we get,

$$16.755^2 - (4\pi)^2 = 2\alpha \times 200\pi$$

$$\alpha = 0.0977 \text{ rad/sec}^2$$

From the kinematic relation $\omega = \omega_0 + \alpha t$,

we get, $16.755 = 4\pi + 0.0977\, t$

∴ **$t = 42.86$ sec Ans.**

Let θ' be the total angular displacement in reaching the velocity of 160 rpm. Then,

$$16.755^2 - 0^2 = 2 \times 0.0977 \times \theta'$$

$$\theta' = 1436.1 \text{ rad}$$

$$\theta' = \frac{1436.1}{2\pi} = \mathbf{228.57 \text{ revolution}} \quad \mathbf{Ans.}$$

Example 20.5. *Power supply was cut off to a power driven wheel when it was rotating at a speed of 900 rpm. It was observed to come to rest after making 360 revolutions. Determine its angular retardation and time it took to come to rest after power supply was cut off.*

Solution. $\omega_0 = 900 \text{ rpm} = \dfrac{900 \times 2\pi}{60} = 30\pi$

$$\omega = 0$$

$$\theta = 360 \text{ revolution} = 360 \times 2\pi = 720\pi \text{ rad.}$$

From the relation $\omega^2 - \omega_0^2 = 2\alpha\theta$,

we get $0 - (30\pi)^2 = 2\alpha \times 720\pi$

$$\alpha = -1.9635 \text{ rad/sec}^2$$

i.e. **retardation is 1.9635 rad/sec² Ans.**

Using the relation $\omega = \omega_0 + \alpha t$, we get

$$0 = 30\pi - 1.9635 t$$

$$t = \frac{30\pi}{1.9635} = \mathbf{48 \text{ sec}} \quad \mathbf{Ans.}$$

Example 20.6. *The step pulley shown in Fig. 20.3 starts from rest and accelerates at 2 rad/sec². How much time is required for block A to move 20 m? Find also the velocity of A and B at that time.*

Solution. When A moves by 20 m, the angular displacement of pulley θ is given by

$$r\theta = s$$

ROTATION OF RIGID BODIES

$$1 \times \theta = 20$$
$$\theta = 20 \text{ radian}$$

Now, $\alpha = 2 \text{ rad/sec}^2$ and $\omega_0 = 0$

From the kinematic relation

$$\theta = \omega_0 t + \frac{1}{2}\alpha t^2,$$

we get

$$20 = 0 + \frac{1}{2} \times 2t^2$$

$$t = 4.472 \text{ sec} \quad \text{Ans.}$$

Velocity of pulley at this time

$$\omega = \omega_0 + \alpha t = 0 + 2 \times 4.472$$
$$= 8.944 \text{ rad/sec}$$

Velocity of A, $v_A = 1 \times 8.944$
$$= \textbf{8.944 m/sec} \quad \textbf{Ans.}$$

and **velocity of B,** $v_B = 0.6 \times 8.944 = \textbf{5.367 m/sec} \quad \textbf{Ans.}$

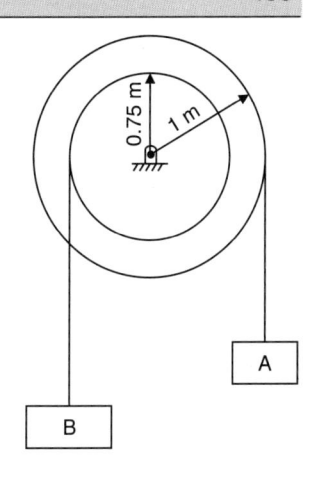

Fig. 20.3

20.5. KINETICS OF RIGID BODY ROTATION

Consider the wheel shown in Fig. 20.4 rotating about its axis in clockwise direction with an acceleration α. Let δm be mass of an element at a distance r from the axis of rotation. If δp be the resulting force on this element,

$\delta p = \delta m \times a$ where, a is tangential acceleration

But $a = r\alpha$ where α is angular acceleration.

$\therefore \quad \delta p = \delta m r \alpha$

Rotational moment δM_t due to this force δp is given by

$$\delta M_t = \delta p \times r$$
$$= \delta m \, r^2 \, \alpha$$
$$M_t = \Sigma \, \delta M_t = \Sigma \, \delta m \, r^2 \alpha$$
$$= \alpha \, \Sigma \, \delta m \, r^2$$
$$= \alpha \, I$$

where I is mass moment of inertia of the rotating body (Eqn. 8.14).

Fig. 20.4

Thus $\quad M_t = I\alpha \quad ...(20.13)$

Note the similarly between the expressions $M_t = I\alpha$ and $F = ma$ used in linear motion.

Force causes linear motion while rotational moment causes angular motion. The force is equal to the product of mass and the linear acceleration whereas rotational moment is the product of mass moment of inertia and the angular acceleration.

The product of the mass moment of inertia and the angular velocity of a rotating body is called *Angular Momentum*. Thus.

Angular Momentum = $I \omega \quad ...(20.14)$

20.6. KINETIC ENERGY OF ROTATING BODIES

Consider the rotating body shown in Fig. 20.5 with angular velocity ω. Let δm be mass of an element which is at a distance r from the axis of rotation. Hence, if v is the linear velocity of the element

$$v = r\omega$$

Now, Kinetic Energy of the elements mass

$$= \frac{1}{2} \delta m \, v^2$$

∴ K.E. of the rotating body

$$= \Sigma \frac{1}{2} \delta m \, v^2$$

$$= \Sigma \frac{1}{2} \delta m \, r^2 \omega^2$$

$$= \frac{1}{2} \omega^2 \Sigma \delta m r^2$$

But from the definition of mass moment of inertia, (Eqn. 20.14),

$$I = \Sigma \delta m r^2$$

∴ $\quad$ K.E. $= \frac{1}{2} I \omega^2 \quad$ ...(20.15)

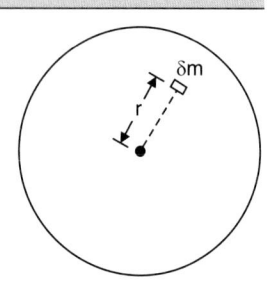

Fig. 20.5

Table 20.1 gives comparison between various terms used in linear motion and rotation.

Table 20.1

Particulars	Linear Motion	Angular Motion
Displacement	s	θ
Initial velocity	u	ω_0
Final Velocity	v	ω
Acceleration	a	α
Formula for final velocity	$v = u + at$	$\omega = \omega_0 + \alpha t$
Formula for displacement	$s = ut + \frac{1}{2}at^2$	$\theta = \omega_0 t + \frac{1}{2}\alpha t^2$
Formual in terms of displacement velocity and acceleration	$v^2 - u^2 = 2as$	$\omega^2 - \omega_0^2 = 2\alpha\theta$
Force causing motion	$F = ma$	$M_t = I\alpha$
Momentum	mv	$I\omega$
Kinetic Energy	$\frac{1}{2}mv^2$	$\frac{1}{2}I\omega^2$

Example 20.7. *A flywheel weighing 50 kN and having radius of gyration 1m loses its speed from 400 rpm to 280 rpm in 2 minutes. Calculate*

(i) *the retarding torque acting on it*

(ii) *change in its kinetic energy during the above period*

(iii) *change in its angular momentum during the same period.*

Solution. $\quad \omega_0 = 400 \text{ rpm} = \dfrac{400 \times 2\pi}{60} = 41.888 \text{ rad/sec}$

$$\omega = 280 \text{ rpm} = \frac{280 \times 2\pi}{60} = 29.322 \text{ rad/sec}$$

ROTATION OF RIGID BODIES

but,
$$t = 2 \text{ min} = 120 \text{ sec}$$
$$\omega = \omega_0 + \alpha t$$

∴
$$\alpha = \frac{\omega - \omega_0}{t} = \frac{29.3224 - 41.888}{120}$$
$$= -0.1047 \text{ rad/sec}^2$$

i.e., retardation is 0.1047 rad/sec^2 ...(1)

Weight of flywheel = 50 kN = 50,000 N
Radius of gyration $k = 1$ m

∴
$$I = mk^2 = \frac{50,000}{9.81} \times 1^2 = 5096.84$$

(i) Retarding Torque Acting on the Flywheel
$$= I\alpha = 5096.84 \times 0.1047$$
$$= \mathbf{533.74 \text{ N·m}} \quad \textbf{Ans.}$$

(ii) Change in Kinetic Energy
$$= \text{Initial K.E.} - \text{Final K.E.}$$
$$= \frac{1}{2} I\omega_0^2 - \frac{1}{2} I\omega^2$$
$$= \frac{1}{2} \times 5096.84 \, (41.888^2 - 27.322^2)$$
$$= \mathbf{22{,}80442.9 \text{ N·m}} \quad \textbf{Ans.}$$

(iii) Change in its Angular Momentum
$$= I\omega_0 - I\omega$$
$$= 5096.84 \, (41.888 - 29.322)$$
$$= \mathbf{64048.93 \text{ N-sec}} \quad \textbf{Ans.}$$

Example 20.8. *A pulley of weight 400 N has a radius of 0.6 m. A block of 600 N is suspended by a tight rope wound round the pulley, the other end being attached to the pulley as shown in Fig. 20.6. Determine the resulting acceleration of the weight and the tension in the rope.*

Solution. Let a be the resulting acceleration and T be the tension in the rope. Hence angular acceleration of pulley.
$$\alpha = a/r = a/0.6 = 1.667a \text{ m/sec}^2 \qquad ...(1)$$

An intertia force of $(600/g)\,a$ may be considered and the dynamic equilibrium condition can be written for the block as:
$$T = \left(600 - \frac{600}{9.81} a\right) \qquad ...(2)$$

From Kinetic equation for pulley, we get,
$$M_t = I\alpha \quad i.e., \quad T(0.6) = I \times 1.667a \qquad ...(3)$$

But
$$I = \frac{Mr^2}{2} = \frac{400}{9.81} \times \frac{0.6^2}{2}$$

From (3), $T = \dfrac{400}{9.81} \times \dfrac{0.6^2}{2} \times \dfrac{1.667\,a}{0.6}$

$= \dfrac{200}{9.81}\, a$ N·m

Substituting it in (2), we get

$$\dfrac{200}{9.81} a = 600 - \dfrac{600}{9.81} a$$

$$a = \dfrac{600 \times 9.81}{800}$$

$$= 7.358 \text{ m/sec}^2 \quad \text{Ans.}$$

∴ $T = \dfrac{200}{9.81} \times 7.358$

$= 150$ N   Ans.

Fig. 20.6

Example 20.9. *The composite pulley shown in Fig. 20.7 weighs 800 N and has a radius of gyration of 0.6 m. The 2000 N and 4000 N blocks are attached to the pulley by inextensible strings as shown in the figure.*

Neglecting weight of the strings, determine the tension in the strings and angular acceleration of the pulley.

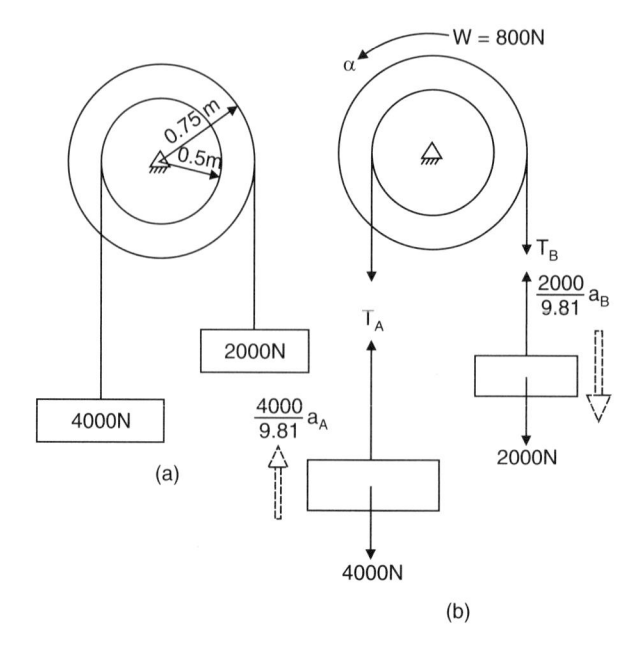

Fig. 20.7

Solution. Since the moment of 4000 N block is more than that of 2000 N block about the axis of rotation, the pulley rotates in anticlockwise direction as shown in Fig. 20.7 (b). Let a_A be acceleration of 4000 N block and a_B that of 2000 N block. Then

$$a_A = 0.5\,\alpha \quad \text{and} \quad a_B = 0.75\alpha$$

where, α – angular acceleration of pulley.

ROTATION OF RIGID BODIES

Writing dynamic equilibrium equation for the two blocks, we get

$$T_A = 4000\left(1 - \frac{a_A}{9.81}\right) = 4000\left(1 - \frac{0.5}{9.81}\alpha\right)$$

$$T_B = 2000\left(1 + \frac{a_B}{9.81}\right) = 2000\left(1 + \frac{0.75\,\alpha}{9.81}\right)$$

From Kinetic equation of pulley, we have
$$M_t = I\alpha$$

i.e., $\quad T_A \times 0.5 - T_B \times 0.75 = \dfrac{800}{9.81} \, 0.6^2 \, \alpha$

$$4000\left(1 - \frac{0.5}{9.81}\alpha\right)0.5 - 2000\left(1 + \frac{0.75}{9.81}\alpha\right)0.75 = \frac{800}{9.81}(0.6)^2\,\alpha$$

$$245.97\,\alpha = 500$$
$$\alpha = \mathbf{2.033\ rad/sec^2}\quad\text{Ans.}$$

$\therefore \quad T_A = 4000\left(1 - \dfrac{0.5}{9.81} \times 2.033\right) = \mathbf{3585.58\ N}\quad \text{Ans.}$

$\quad T_B = 2000\left(1 + \dfrac{0.75}{9.81}\, 2.033\right) = \mathbf{2310.83\ N}\quad \text{Ans.}$

Example 20.10. *A cylinder weighing 500 N is welded to a 1 m long uniform bar of 200 N as shown in Fig. 20.8. Determine the acceleration with which the assembly will rotate about point A, if released from rest in horizontal position. Determine the reaction at A at this instant.*

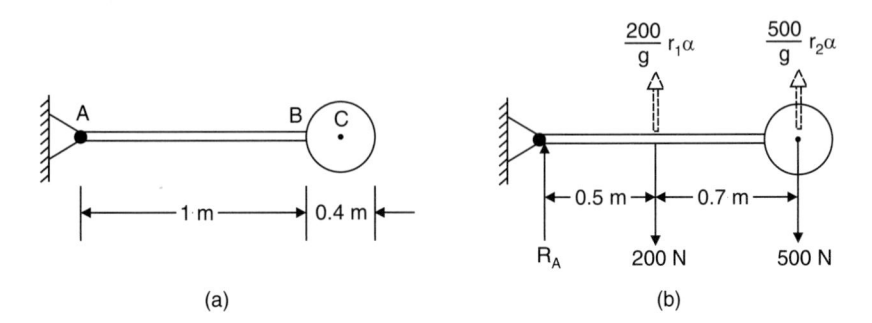

Fig. 20.8

Solution. Let α be the angular acceleration with which the assembly will rotate. Let I be the mass moment of inertia of the assembly about the axis of rotation A. Using the transfer formula $I = I_g + Md^2$ we can assemble mass moment of inertia about the axis of rotation through A.

Mass moment of inertia of the bar about A
$$= \frac{1}{2} \times \frac{200}{9.81} \times 1^2 + \frac{200}{9.81} \times (0.5)^2 = 6.7968$$

Now moment of inertia of the cylinder about A
$$= \frac{1}{2} \times \frac{500}{9.81} \times 0.2^2 + \frac{500}{9.81} \times 1.2^2 = 74.414$$

∴ Mass moment of inertia of the system about A
$$I = 6.7958 + 74.41 = 81.2097$$
Rotational moment about A (Ref. Fig. 20.8(b))
$$M_t = 200 \times 0.5 + 500 \times 1.2 = 700 \text{ N·m}$$
equating it to $I\alpha$, we get,
$$81.2097 \, \alpha = 700 \quad \text{or} \quad \alpha = \mathbf{8.6197 \text{ rad/sec}} \quad \textbf{Ans.}$$

Instantaneous acceleration of rod AB is vertical and its magnitude is given by
$$= r_1 \alpha = 0.5 \times 8.6197$$
$$= 4.310 \text{ m/sec}$$

Similarly the instantaneous acceleration of the cylinder is also vertical and is equal to
$$r_2 \alpha = 1.2 \times 8.6197 = 10.344 \text{ m/sec}$$

Applying D' Alembert's dynamic equilibrium equation to the system of forces shown in Fig. 20.8 (b), we get
$$R_A = 200 + 500 - \frac{200}{9.81} \times 4.3100 - \frac{500}{9.81} \times 10.344$$

i.e., $\mathbf{R_A = 84.934 \text{ N}}$ **Ans.**

Example 20.11. *Rod AB, weighing 200 N is welded to the rod CD weighing 100 N as shown in Fig. 20.9 (a). The assembly is hinged at A and is freely held. Determine the instantaneous vertical and horizontal reactions of A when a horizontal force of 300 N acts at a distance of 0.75 m from A.*

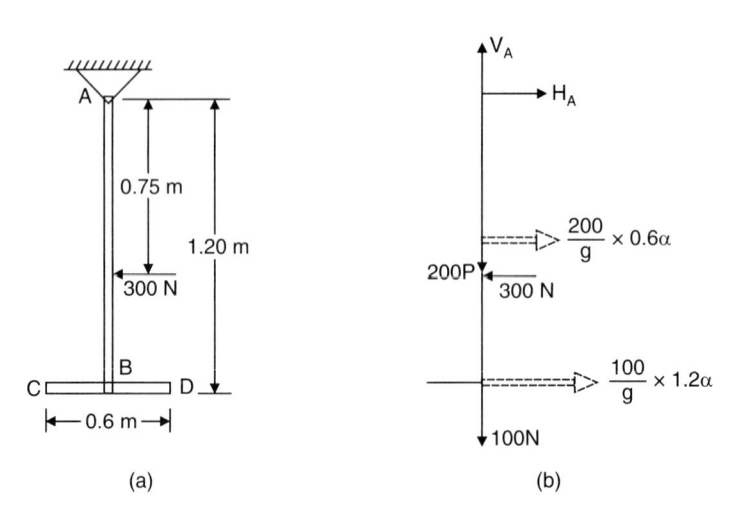

Fig. 20.9

Solution. Mass moment of inertia of AB about axis of rotation
$$A = \frac{1}{12} \times \frac{200}{9.81} \times 1.2^2 + \frac{200}{9.81} \times 0.6^2 = 9.786$$

Mass moment of inertia of rod CD about A
$$= \frac{1}{12} \times \frac{100}{9.81} \times 0.6^2 + \frac{100}{9.81} \times 1.2^2 = 147.0$$

ROTATION OF RIGID BODIES

∴ Total mass moment of the system about A
$$9.786 + 147.0 = 156.78$$

Let α be the instantaneous angular acceleration. Writing the kinetic equation for the rotation about A, we get

$$I\alpha = M_t$$
$$156.786\,\alpha = 300 \times 0.75$$
$$\alpha = 1.4351 \text{ rad/sec}$$

At the instant 300 N force is applied, the linear accelerations of AB of CD are horizontal and are equal to $0.6\,\alpha$ and $1.2\,\alpha$ respectively. Hence the inertia forces are as shown in Fig. 20.9 (b) by dotted arrows. Let V_A and H_A be the vertical and horizontal reactions at A. Writing the dynamic equilibrium conditions, we get,

$$V_A = 200 + 100 = \mathbf{300\ N} \quad \textbf{Ans.}$$

and
$$H_A = 300 - \frac{200}{9.81} \times 0.6\,\alpha - \frac{100}{9.81} \times 1.2\,\alpha$$

Substituting the value of α, we get,
$$H_A = \mathbf{264.891\ N} \quad \textbf{Ans.}$$

Important Formulae : Refer table 20.1

PROBLEMS FOR EXERCISE

20.1 The motion of a cam is defined by the relation $\theta = t^3 - 8t + 15$, where θ is expressed in radians and t in seconds. Determine the angular displacement, angular velocity and angular acceleration after (a) 2 sec. (b) 4 sec.
[**Ans.** $\theta = 7$ rad and 47 rad; $\omega = 4$ rad/sec and 40 rad/sec, $\alpha = 12$ rad/sec² and 24 rad/sec²]

20.2 The angular acceleration of a flywheel is given by the expression $\alpha = 6t - t^2$ where α is in rad/sec² and t is in second. When will the flywheel stop momentarily prior to reversing the direction, given that the flywheel starts from rest. How many revolutions are made during this time.
[**Ans.** $t = 9$ sec, revolution = 29.006]

20.3 A wheel is rotating about its axis with a constant acceleration of 1 rad/sec². If the initial and final velocities are 50 rpm and 100 rpm, determine the time taken and number of revolution made during this period. [**Ans.** $t = 5.236$ sec and revolutioins = 6.545]

20.4 A rotor of an electric motor is uniformly accelerated to a speed of 1800 rpm from rest for 5 seconds and then immediately power is switched off and the rotor decelerate uniformly. If the total time elapsed from start to stop is 12.5 sec, determine the number of revolutions made while (i) acceleration, (ii) deceleration. Also determine the value of deceleration.
[**Ans.** 75 revolution, 175 revolution and 25.1327 rad/sec²]

20.5 A wheel rotates with uniform acceleration. During 3rd and 6th second wheel rotates 8 radians and 12 radians respectively. Determine the initial velocity and acceleration of the wheel.
[**Ans.** $\omega_0 = 4.667$ rad/sec, $\alpha = 1.333$ rad/sec²]

20.6 The pulley B shown in Fig. 20.10 is rotating at the rate of 5 rad/sec, clockwise. If the deceleration is observed to be 1.2 rad/sec², how much distance blocks C and D move before coming to rest?

How many revolutions are made by pulley A and B during this period? Assume no slip between the wire and the pulley.

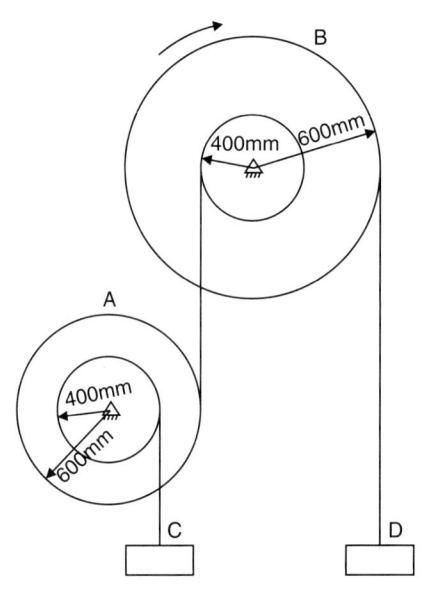

Fig. 20.10

[**Ans.** s_C = 4.166 m, s_D = 8.333 m, 0.829 revolution by A, 1.658 revolution by B]

20.7 A pulley weighs 500 N and has a radius of 0.75 m. A block weighing 400 N is supported by inextensible wire wound round the pulley. Determine the velocity of the block 2 sec after it is released from rest. Assume the motion is under constant acceleration. [**Ans.** 12.074 m/sec]

20.8 The pulley shown in Fig. 20.11 weighs 600 N and has a radius of 0.8 m. A rope passing over this pulley supports 800 N load at one end and 400 N at another end. Determine the tension in the string and the angular acceleration of the pulley if the blocks are allowed to move.

[**Ans.** α = 3.27 rad/sec^2, T_1 = 506.67 N, T_2 = 586.67 N]

Fig. 20.11   Fig. 20.12

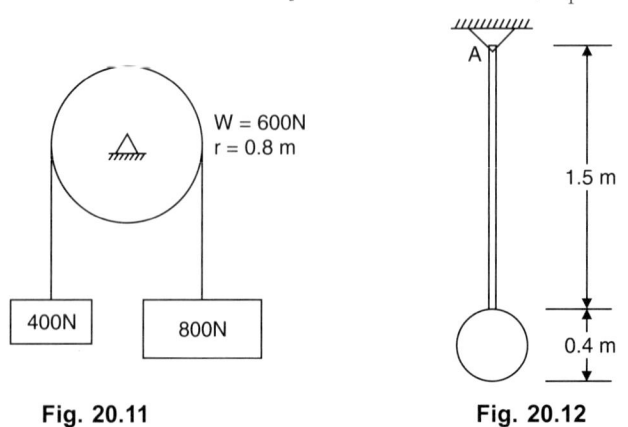

20.9 A solid sphere of radius 0.2 m and weighing 80 N is joined to a bar, 1.5 m long and weighing 100 N, as shown in Fig. 20.12. At what distance a horizontal force of 250 N should be applied so that the system will get an instantaneous acceleration of 8 rad/sec.? What are the reactions produced at the support. [**Ans.** V_A = 110 N, H_A = 17.807 N]

21
Mechanical Vibration

If an elastic structure in equilibrium is disturbed by external forces and then external force is removed the structure tries to come back to its original equilibrium position. However it overshoots the equilibrium position due to energy stored and again tries to reach equilibrium position from the other side. It sets in the oscillation of the structure about *its equilibrium position. This is called vibration.* Common examples of mechanical vibrations are high rise structures vibrating due to wind forces or earthquake forces, vehicles vibrating due to movement on uneven road, pendulums oscillating. Due to air resistance or due to specially designed dampening mechanism the structure comes to static equilibrium position after vibrating for some time.

If the disturbing force do not act during vibration, it is called free vibration. If disturbing force acts at periodical interval on the elastic body, it is called forced vibration.

Consider the vibration of the pendulum shown in Fig. 21.1. Since vibration is a periodic phenomenon, it comes back to some position after a definite time interval. Thus if the pendulum is released from its position B shown in Fig. 21.1, it overshoots its equilibrium position A, reaches position C and comes back. It starts oscillating about the position A. The time interval required to complete one cycle of vibration (say position B to reach position C and coming back to B) is called the *period* of vibration and may be represented by the notation T. Each repeatation of motion is called a cycle. The reciprocal of the period $f = \dfrac{1}{T}$ is called *frequency* and is measured in cycles per second. A cycle per second is called one Hertz. Usual notation for a Hertz is Hz. The term *amplitude* is used to refer to the maximum displacement from the equilibrium position.

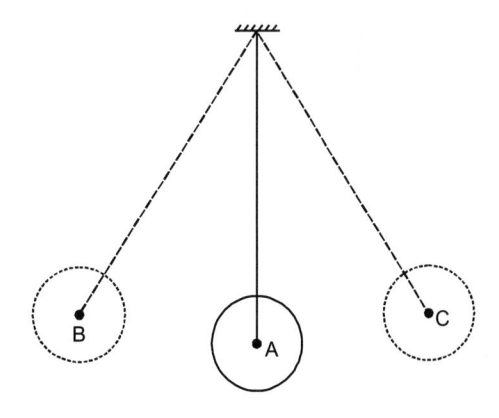

Fig. 21.1

21.1 SIMPLE HARMONIC MOTION

A motion in which acceleration of the body is directly proportional to its displacement and directed towards origin is called simple harmonic motion. Mathematically S.H.M. means

$$a = -ks \qquad \ldots(21.1)$$

where k is constant.

There is similarity between the above expression and the motion of the projection of a particle on a diametral axis, when the particle moves with constant velocity. Referring to Fig. 21.2.

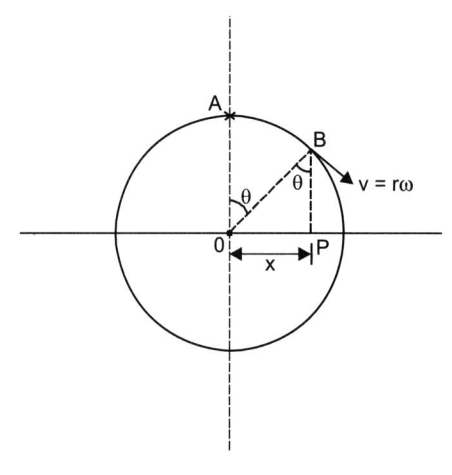

Fig. 21.2

Let a particle move from position A to B with constant velocity v. Let P be its projection on the diametrical axis x. Then, if ω is the angular velocity,

$$v = r\omega \qquad \ldots(1)$$
$$x = r\sin\theta \qquad \ldots(2)$$

$\therefore \qquad \dfrac{dx}{dt} = r\cos\theta \dfrac{d\theta}{dt}$

i.e., $\qquad v_x = r\omega \cos\theta \qquad \ldots(3)$

$\dfrac{dv_x}{dt} = -r\omega \sin\theta \dfrac{d\theta}{dt}$

i.e., $\qquad a_x = -r\omega^2 \sin\theta$

$\qquad\qquad = -\omega^2 x$ since $x = r\sin\theta \qquad \ldots(4)$

Since ω is constant, eqn. (5) is

$$a_x = -kx, \text{ where } k = \omega^2 \qquad \ldots(5)$$

Thus a simple harmonic motion can be imagined as the motion of projection of the position of a particle rotating about a fixed axis with constant velocity. From definition of period T,

we get $\qquad T = \dfrac{2\pi}{\omega} \qquad \ldots (21.2)$

$\qquad f = \dfrac{1}{T} = \dfrac{\omega}{2\pi} \qquad \ldots (21.3)$

$\qquad \text{amplitude} = r \qquad \ldots (21.4)$

Example 21.1. *A simple harmonic motion is defined by the expression $a = -25\,s$. Determine its period and frequency.*

Solution.

$\qquad\qquad a = -25.0\,s$

i.e. $\qquad\qquad \omega^2 = 25$

i.e. $\qquad\qquad \omega = 5$

MECHANICAL VIBRATION

$\therefore$ Period $T = \dfrac{2\pi}{\omega} = \dfrac{2\pi}{5} = 1.257$ sec

and frequency $f = \dfrac{1}{T} = \dfrac{1}{1.257} = 0.796$ osc per second.

Example 21.2. *The amplitude of a particle in simple harmonic motion is 0.75 m and the period is 1.2 sec. Determine the maximum velocity and the maximum acceleration.*

Solution. Amplitude $r = 0.75$ m

Period $T = 1.2$ sec

i.e. $\dfrac{2\pi}{\omega} = 1.2$

or $\omega = \dfrac{2\pi}{1.2} = 5.236$ rad/sec

$\therefore \quad v_x = \dfrac{dx}{dt} = r\cos\theta \dfrac{d\theta}{dt}$

$\qquad = r\omega\cos\theta.$

$\therefore \quad v_x \max = r\omega = 0.75 \times 5.236$

$\qquad = \mathbf{3.927 \text{ m/sec}}$ **Ans.**

$\qquad a_x = -r\omega^2 \sin\theta$

$\therefore \quad a_x, \max = -r\omega^2 = 0.75 \times 5.236^2$

$\qquad = \mathbf{20.56 \text{ m/sec}^2}$ **Ans.**

Example 21.3. *In the Ex. 21.2 determine the displacement, velocity and acceleration after 0.5 seconds.*

Solution.

After 0.5 sec, $\theta = \omega t = 5.236 \times 0.5 = 2.618$ radians $= 150°$

$\therefore$ Displacement $x = r\sin\theta$

$\qquad = 0.75 \sin 150°$

$x = \mathbf{0.375 \text{ m}}$ **Ans.**

Velocity $= r\omega\cos\theta$

$\qquad = 0.75 \times 5.236 \cos 150°$

$v_x = \mathbf{-3.4 \text{ m/sec}}$ **Ans.**

acceleration $a_x = -\omega^2 x$

$\qquad = 5.236^2 \times 0.375$

$a_x = \mathbf{10.28 \text{ m/sec}^2}$ **Ans.**

Example 21.4. *A particle is in simple harmonic motion. It has a velocity of 0.5 m/sec when it is 0.2 m from its static equilibrium position and has a velocity of 0.35 m/sec when it is 0.3 m from the equilibrium position. Determine the maximum velocity, maximum acceleration and the frequency of vibration.*

Solution. Let r be amplitude and θ be angular displacement. Then

$\qquad x = r\sin\theta$

and $\qquad v = r\omega\cos\theta$

Hence in this case

$$0.2 = r \sin \theta_1 \qquad \ldots(1)$$
$$0.5 = r\omega \cos \theta_1 \qquad \ldots(2)$$
$$0.3 = r \sin \theta_2 \qquad \ldots(3)$$
and $$0.35 = r\omega \cos \theta_2 \qquad \ldots(4)$$

From eqns. (1) and (3)

$$\frac{0.2}{0.3} = \frac{\sin \theta_1}{\sin \theta_2} \text{ or } \sin \theta_2 = 1.5 \sin \theta_1, \qquad \ldots(5)$$

From eqns. (2) and (4), we get

$$\frac{0.5}{0.35} = \frac{\cos \theta_1}{\cos \theta_2}$$

or $$\cos \theta_2 = 0.7 \cos \theta_1 \qquad \ldots(6)$$

$$\sqrt{1 - \sin^2 \theta_2} = 0.7\sqrt{1 - \sin^2 \theta_1}$$
$$1 - \sin^2 \theta_2 = 0.7(1 - \sin^2 \theta_1)$$

Using eqn. (5), the equation reduces to
$$1 - 2.25 \sin^2 \theta_1 = 0.7(1 - \sin^2 \theta_1)$$
$$\therefore \qquad 0.3 = 1.55 \sin^2 \theta_1$$
$$\sin \theta_1 = 0.44$$
$$\theta_1 = 2.16°$$

From eqn. (1), $$0.2 = r \sin \theta_1$$
$$= r \sin 26.1°$$
$$\therefore \qquad r = 0.454 \text{ m}$$

and from eqn. (2), $$0.5 = 0.454 \omega \cos 26.1°$$
$$\therefore \qquad \omega = 1.225 \text{ rad/sec}^2$$
$$v_x = r\omega \cos \theta$$
$$\therefore \qquad v_{max} = r\omega = 0.454 \times 1.225 = \mathbf{0.557 \text{ m/sec}} \quad \mathbf{Ans.}$$
$$a = -\omega^2 r \sin\theta$$
$$\therefore \qquad a_{max} = \omega^2 r = -1.225^2 \times 0.454$$
$$\mathbf{a_{max} = -0.682 \text{ m/sec}^2} \quad \mathbf{Ans.}$$

Frequency $$f = \frac{\omega}{2\pi} = \frac{1.225}{2\pi}$$
$$\mathbf{f = 0.195 \text{ osc. per sec}} \quad \mathbf{Ans.}$$

Example 21.5. *A ball weighing W is tightly held by an elastic string as shown in Fig. 21.3. Show that for small displacements, the ball will have a simple harmonic motion. Compute the period and frequency of vibration.*

Solution. Let the force developed in the string the P. The ball has inertia force $\dfrac{W}{g} a$ in the horizontal direction, which is the direction of motion.

$$\Sigma H = 0$$

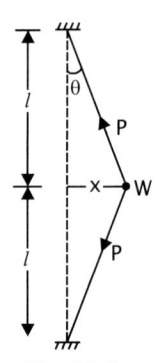

Fig. 21.3

MECHANICAL VIBRATION

$$\Sigma P \sin \theta + \frac{W}{g} a = 0 \qquad \ldots(1)$$

Since θ is small, $\sin \theta = \frac{x}{l}$,

$$\text{gives } \theta = \frac{x}{l}$$

∴ The equilibrium equation (1) reduces to

$$2P \frac{x}{l} + \frac{W}{g} a = 0$$

or

$$a = -\frac{2Pg}{Wl} x$$

$$= -Kx$$

where

$$K = \frac{2Pg}{Wl}$$

Thus it is of the form $a = -ks$, which is a simple harmonic motion.

Since

$$\omega^2 = K$$

We have

$$\omega^2 = \frac{2Pg}{Wl}$$

or

$$\omega = \sqrt{\frac{2Pg}{Wl}}$$

∴

$$T = \frac{2\pi}{\omega} = 2\pi \sqrt{\frac{Wg}{2Pg}} \quad \textbf{Ans.}$$

$$f = \frac{1}{T} = \frac{1}{2\pi} \sqrt{\frac{2Pg}{Wl}} \quad \textbf{Ans.}$$

21.2 SIMPLE HARMONIC MOTION AS A SINE WAVE

In the previous article we defined simple harmonic motion mathematically as $a = -ks$ and showed that it can be viewed as motion of projection of a particle on a diametrical axis, when the particle rotates with uniform velocity about a fixed axis. We saw in S.H.M.

$$s = x = r \sin \theta$$
$$= r \sin \omega t.$$

Hence the above expression can be viewed as propagation of sine wave as shown in Fig. 21.4.

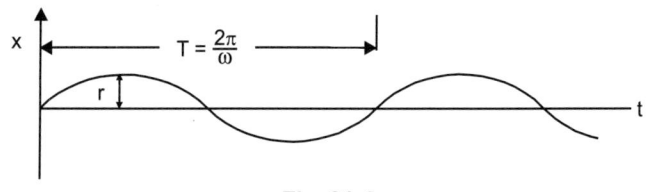

Fig. 21.4

21.3 SIMPLE PENDULUM

A pendulum is a small mass, which is suspended with a string of negligible weight. Referring to Fig. 21.5, if c is the point of suspension and L is the length of pendulum, let us consider its motion when it is release from position B. Let us consider the equilibrium condition when the mass is at position P.

Consider the equation of equilibrium in its tangential position. Let T be tension in the string. Inertia force is in the tangential position and is equal to $\dfrac{W}{g}a$, where a is the acceleration in this position.

Fig. 21.5

Σ Forces in tangential direction $= 0 \rightarrow$

$$W \sin \theta + \frac{W}{g} a = 0$$

If θ is maintained small, $\sin \theta = \theta = \dfrac{s}{L}$ ...(1)

$\therefore \qquad W\theta = -\dfrac{W}{g} a$ ...(2)

$\qquad W\dfrac{s}{L} = -\dfrac{W}{g} a$ ...(3)

i.e. $\qquad a = -\dfrac{g}{L} s$ ...(4)

Thus it is the equation of simple harmonic motion $[a = -ks = -\omega^2 s]$ in which

$$\omega^2 = \frac{g}{L}$$

i.e. $\qquad \omega = \sqrt{\dfrac{g}{L}}$

$\therefore \qquad T = \dfrac{2\pi}{\omega} = 2\pi \sqrt{\dfrac{L}{g}}$ ...(21.5)

and frequency $\qquad f = \dfrac{1}{T} = \dfrac{1}{2\pi} \sqrt{\dfrac{g}{L}}$ ...(21.6)

Thus in simple pendulum, for small angle of vibration, period and frequency depend only on the length of the chord.

Example 21.6. *Find the period of simple pendulum whose length is 1.5 m.*
Solution.

$$T = 2\pi \sqrt{\frac{L}{g}} = 2\pi \sqrt{\frac{1.5}{9.81}}$$

$$T = 2.457 \text{ sec} \quad \text{Ans.}$$

MECHANICAL VIBRATION

Example 21.7. *What length is required if the period of pendulum it is to be kept 1 sec.?*

Solution. From the relation,

$$T = 2\pi\sqrt{\frac{L}{g}}, \text{ we get}$$

$$L = 2\pi\sqrt{\frac{L}{9.81}}$$

$$\therefore \quad L = 0.248 \text{ m} \quad \textbf{Ans.}$$

21.4 COMPOUND PENDULUM

A rigid body pinned at a point and allowed to oscillate under the influence of gravity may be called as compound pendulum, Fig. 21.6 shows some of the compound pendulums. Thus in case of simple pendulum, mass is concentrated at the end of suspended weightless wire where as in case of compound pendulum it is distributed over a large area.

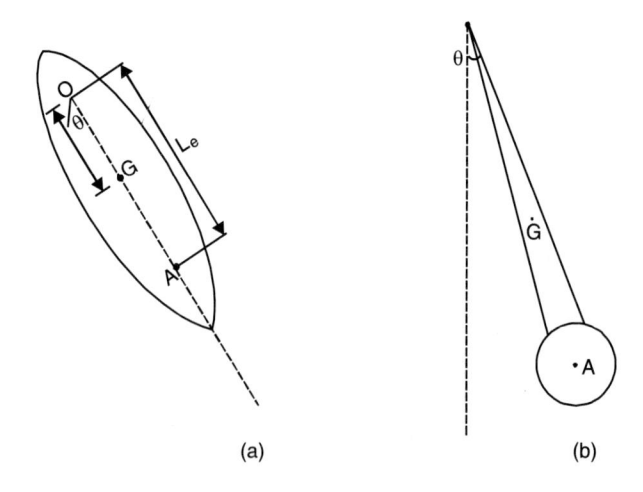

(a)   (b)

Fig. 21.6

Let the mass moment of inertia of the body about centre of suspension 'O' be I_z and radius of gyration be K_z while M be mass of the compound pendulum. Taking a as the angular acceleration when angular displacement is θ, the equilibrium equation in tangential direction gives,

$$-Wr \sin \theta = I_z a \qquad ...(1)$$

where r is the distance of centroid G from 'O'. From definition,

$$-I_z MK_z^2 = \frac{W}{g} K_z^2 \qquad ...(2)$$

$\therefore$ From eqns. (1) and (2) we get

$$-Wr \sin \theta = \frac{W}{g} K_z^2 a$$

since θ is small, $\sin \theta = \theta$

$$\therefore \quad -Wr\theta = \frac{W}{g} K_z^2 a$$

or
$$a = -\frac{gr}{K_z^2}\theta \qquad \ldots(3)$$

The equation (3) is similar to $a = -\omega^2 r$. Hence we can say, the angular velocity of compound pendulum is given by

$$\omega = \sqrt{\frac{gr}{K_z^2}}$$

∴ Period of oscillation

$$T = \frac{2\pi}{\omega} = 2\pi \sqrt{\frac{K_z^2}{gr}} \qquad \ldots(4)$$

For simple pendulum we have seen

$$T = 2\pi \sqrt{\frac{L}{g}} \qquad \ldots(5)$$

Comparing equations (4) and (5) we can say, for compound pendulum.

$$T = 2\pi \sqrt{\frac{L_e}{g}} \qquad \ldots(21.7(a))$$

where
$$L_e = \frac{K_z^2}{r} \qquad \ldots(21.7(b))$$

Let A be the point at distance L_e from the centre of suspension. Then from parallel axis theorem for mass moment of inertia, we know,

$$I_A = I_g + Mr^2$$
$$MK_z^2 = MK^2 + Mr^2$$

or
$$K_z^2 = K^2 + r^2$$

∴
$$L_e = \frac{K_z^2}{r} = \frac{K^2}{r} + r \qquad \ldots(21.8)$$

From eqn. 21.8 we conclude, the period of oscillation is the same for all parallel axis of suspension located at the same distance from the centre of gravity. Point A is called centre of oscillation.

It can be proved that centre of oscillation A and centre of suspension O are interchangeable. When O is centre of suspension, we know,

$$GO \times GA = r(L_e - r) = r\left(\frac{K^2}{r} + r - r\right) = K^2 \qquad \ldots(6)$$

It A is made centre of suspension, let O' be centre of oscillation. Then

$$GO' \times GA = (L_e - r') r' = \left(\frac{K^2}{r'} + r' - r'\right) r' = K^2 \qquad \ldots(7)$$

From eqns. (6) and (7), we conclude O' will concide with O. Thus point O becomes centre of oscillation when A becomes centre of suspension $i.e.$, centre of suspensions and centre of oscillations are interchangeable.

Example 21.8. *A uniform bar shaped pendulum of length L is suspended on knife edge at a distance $\frac{L}{4}$ from its centre of gravity as shown in Fig. 21.7. Determine its period, frequency and equivalent length.*

Solution. For the bar,

$$I_G = \frac{ML^2}{12}$$

$$\therefore \quad I_A = \frac{ML^2}{12} + M\left(\frac{L}{4}\right)^2$$

$$= ML^2\left[\frac{1}{12} + \frac{1}{16}\right] = \frac{7ML^2}{48}$$

$$K_z^2 = \frac{7L^2}{48}$$

$$T = 2\pi\sqrt{\frac{K_z^2}{gr}} = 2\pi\sqrt{\frac{7L^2}{48} \times \frac{1}{9.81} \times \frac{4}{L}}$$

Since $r = L/4$

$r = 1.53\sqrt{L}$ sec **Ans.**

$$f = \frac{1}{T} = \frac{1}{1.53\sqrt{L}} = \frac{0.654}{\sqrt{L}} \quad \textbf{Ans.}$$

$$L_e = \frac{K_z^2}{r} = \frac{7}{48}\frac{L^2}{L/4} = 0.583\,L \quad \textbf{Ans.}$$

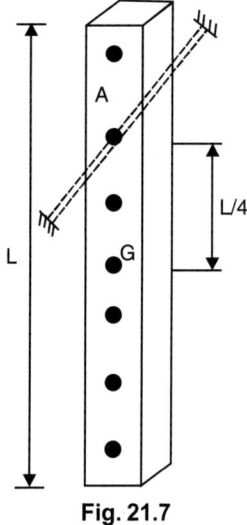

Fig. 21.7

Example 21.9. *The compound pendulum shown in Fig. 21.8 consists of a uniform bar of 600 mm length weighing 25 newton to which is attached a circular disc of 300 mm diameter, weighing 40 newton. Determine the period of oscillation and equivalent length corresponding to simple pendulum of same period.*

Solution.

Length of uniform rod = 600 mm = 0.6 m
Radius of uniform disc = 0.3 m.

$\therefore \quad I_z$ = Mass moment of inertia about centre of suspension

$$= M_1\frac{L_1^2}{12} + M_1\left(\frac{L_1}{2}\right)^2 + M_2\frac{R_z^2}{z} + M_2 d_2^2$$

$$= \frac{25}{9.81} \times \frac{0.6^2}{12} + \frac{25}{9.81}\left(\frac{0.6}{2}\right)^2 + \frac{40}{9.81}\frac{0.15^2}{2} + \frac{40}{9.81}(0.6 + 0.15)^2$$

$$= 2.645 \text{ units}$$

$$M = M_1 + M_2 = \frac{25}{9.81} + \frac{40}{9.81} = \frac{65}{9.81}$$

Fig. 21.8

$$\therefore \quad \frac{65}{9.81} K_{zz}^2 = 2.645$$

$$K_{zz}^2 = 0.3992$$

Distance of centre of gravity of compound pendulum from centre of suspension

$$r = \frac{\text{Moment of mass about centre of suspension}}{\text{Total mass}}$$

$$= \frac{\frac{25}{9.81} \times 0.3 + \frac{40}{9.81} \times 0.75}{\frac{25}{9.81} + \frac{40}{9.81}} = 0.5769 \text{ m}$$

$$\therefore \quad T = 2\pi \sqrt{\frac{K_{zz}^2}{r}} = 2\pi \sqrt{\frac{0.3992}{0.5769}}$$

i.e., $\quad T = 5.23$ sec **Ans.**

Equivalent length $\quad L_e = \dfrac{K_z^2}{r} = \dfrac{0.3992}{0.5769} = \mathbf{0.692 \text{ m}}$ **Ans.**

Example 21.10. *A circular ring of outer radius 1 m and inner radius 0.75 m is supported on a knife edge as shown in Fig. 21.9 (a). Determine the period for small vibration, if it weighs 200 N.*

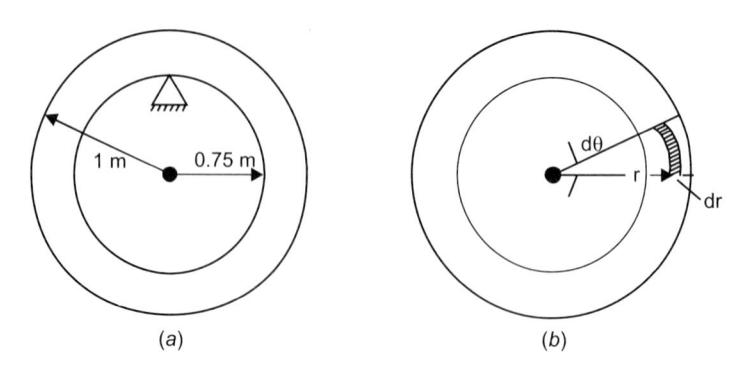

Fig. 21.9

Solution : Considering on elemental mass as shown in Fig. 21.9(b) we know for a ring

$$dm = \rho r\, d\theta dr\, t = \rho tr\, d\theta dr$$

$$\therefore \quad I = \oint r^2 dm = \int_{R_2}^{R_1} \int_0^{2\pi} \rho tr^3\, d\theta dr$$

$$= \int_{R_2}^{R_1} \rho tr^3 [\theta]_0^{2\pi} dr = 2\pi \rho t \left[\frac{r^4}{4}\right]_{R_2}^{R_1}$$

$$= \frac{1}{2} \pi \rho t \left[R_1^4 - R_2^4\right]$$

Mass of the ring, $\quad = \rho t \pi \left(R_1^2 - R_2^2\right)$

$$\therefore \int \rho t \pi \left(R_1^2 - R_2^2 \right) K^2 = \frac{1}{2} \pi \rho t \left[R_1^4 - R_2^4 \right]$$

$$\therefore \qquad K^2 = \frac{R_1^2 + R_2^2}{2}$$

where K is radius of gyration about centre of the ring.

$$\therefore \qquad K_z^2 = \frac{R_1^2 + R_2^2}{2} + d^2 = \frac{R_1^2 + R_2^2}{2} + R_2^2$$

In this problem,

$$K_z^2 = \frac{1^2 + 0.75^2}{2} + 0.75^2 = 1.34375$$

$$\therefore \qquad T = 2\pi \sqrt{\frac{K_z^2}{gr}} = 2\pi \sqrt{\frac{1.34375}{9.81 \times 0.75}}$$

Thus $\qquad T = 2.685$ sec   **Ans.**

Important Definition and Formulae

1. If disturbing force do not act during vibration it is called free vibration. If the disturbing force acts at periodical interval during the vibration of the elastic body, it is called forced vibration.
2. Time interval required to complete one cycle of vibration is called period of vibration (T).
3. Number of cycles per second is called frequency $\left(f = \frac{1}{T} \right)$.
4. The maximum displacement form the equilibrium position is called amplitude.
5. A motion in which acceleration of the body is directly proportional to its displacement and directed towards equilibrium position is called Simple Harmonic Motion. Thus in S.H.M.
$$a = -ks$$
6. Simple Harmonic Motion may be visualised as motion of a pendulum or as a sine wave motion. In such cases.
$$a = -\omega^2 x$$
where $\qquad x = r \sin \theta$
$\therefore \qquad v = r\omega \cos \theta$
7. Centre of suspension and centre of oscillations are interchangeable.
8. In compound pendulum
$$L_e = \frac{K_z^2}{r}$$

$$T = 2\pi \sqrt{\frac{L_e}{g}} = 2\pi \sqrt{\frac{K_z^2}{gr}}$$

PROBLEMS FOR EXERCISE

21.1 A particle has simple harmonic motion defined by $a = -16\,s$ and an amplitude of 0.4 m. Find the velocity and acceleration of the particle when it has $s = 0.15$ m.

[**Ans.** $v = 1.48$ m/sec, $a = -2.4$ m/sec^2]

21.2 A particle has simple harmonic motion. Its maximum velocity was 6 m/sec and the maximum acceleration was found to be 12 m/sec^2. Determine its angular velocity, amplitude. Also determine its velocity and acceleration when displacement is half the amplitude.

[**Ans.** $\omega = 2$ rad/sec, $r = 3$ m, $v = 2.598$ m/sec, $a = 6$ m/sec^2]

21.3 In a place in space it was found that a simple pendulum of length 2 m has period of oscillation is 3 sec. Determine the magnitude of gravitational acceleration.

[**Ans.** $g = 8.773$ m/sec^2]

21.4 A compound pendulum consists of a uniform rod of length 0.5 m weighing 40 N and another uniform rod of 0.4 m width weighing 32 N welded together as shown in Fig. 21.10. Determine its period of vibration.

[**Ans.** $T = 1.4$ sec]

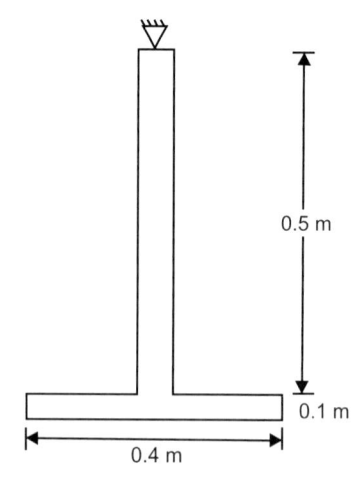

Fig. 21.10

22
General Plane Motion of Rigid Bodies

So far in the previous chapters the motion of bodies possessing either translation or rotation is considered. In this chapter the motion of bodies having both translation and rotation simultaneously is considered. Rolling is a special case of this type of motion. Kinematics and kinetics of rolling bodies as well as bodies possessing general plane motion are considered.

22.1. MEANING OF GENERAL PLANE MOTION

A body is said to have general plane motion if it possesses translation and rotation simultaneously. Common examples of such motion are a wheel rolling on straight line and a rod sliding against wall at one end and on the floor at other end. Figures 22.1 and 22.2 illustrate how these two cases can be split into translation and rotation:

Plane motion = Translation + Rotation

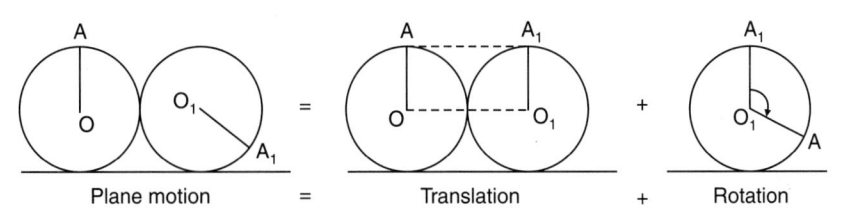

Fig. 22.1

Plane motion = Translation + Rotation

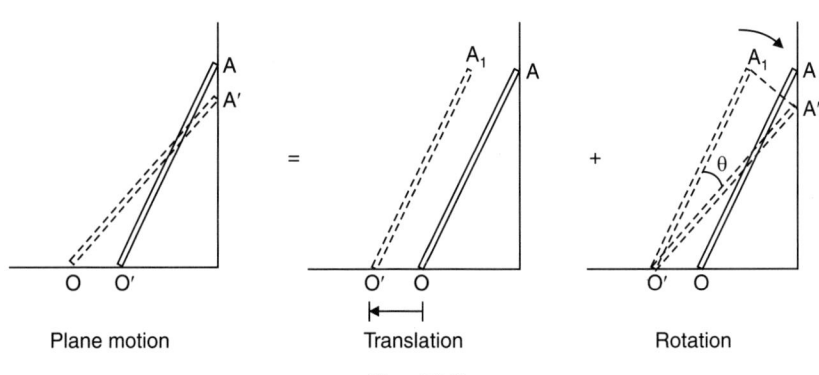

Fig. 22.2

22.2. KINEMATICS OF GENERAL PLANE MOTION

For the analysis of general plane motion, it is convenient to split the motion into translation and pure rotation cases. The analysis for these two cases is carried out separately and then combined to get the final motion. The kinematic analysis of general plane motion is illustrated with the following four problems in this article.

Example 22.1. *Derive the relationship between the linear motion of geometric centre and angular motion of a wheel rolling without slipping.*

Solution. Consider a wheel of radius r rolling with angular velocity ω and angular acceleration α. In Fig. 22.3 the dotted line shows the original position and solid line shows the position after rotating through angular distance θ. Since there is no slip, linear distance $BC = s_A$ must be equal to angular distance $BC = r\theta$, i.e.,

$$s_A = r\theta \qquad ...(1)$$

$$\frac{ds_A}{dt} = \frac{r\,d\theta}{dt} \quad i.e., \quad v_A = r\omega \quad ...(2)$$

and
$$\frac{d^2 s_A}{dt^2} = r\frac{d^2\theta}{dt^2} \quad i.e., \quad a_A = r\alpha \qquad ...(3)$$

Fig. 22.3

Thus when wheel rotates without slip, the relationship between motion of its geometric centre and its angular motion are

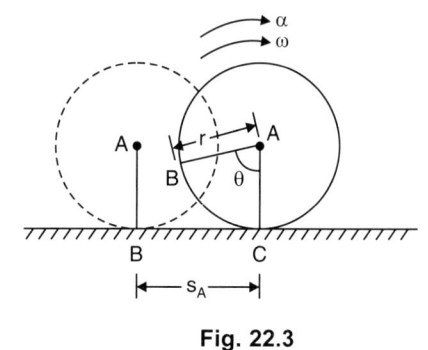

$$\left.\begin{array}{ll} s_A = r\theta & \quad ...(i) \\ v_A = r\omega & \quad ...(ii) \\ a_A = r\alpha & \quad ...(iii) \end{array}\right\} \qquad ...(22.1)$$

and

Example 22.2. *A wheel of radius 1m rolls freely with an angular velocity of 5 rad/sec and with an angular acceleration of 4 rad/sec², both clockwise as shown in Fig. 22.4 (a). Determine the velocity and acceleration of points B and D shown in the figure.*

Solution. The motion of points B and D may be split into translation of geometric centre A and rotation about A. Translation of A is given by

$$v_A = r\omega = 1 \times 5 = 5 \text{ m/sec}$$
$$a_A = r\alpha = 1 \times 4 = 4 \text{ m/sec}^2$$

Now, consider the rotation of B about A. Its relative linear velocity with respect to A is horizontal (normal to AB) and is given by

$$v_{B/A} = AB \times \omega = 1 \times 5 = 5 \text{ m/sec}$$
$$\therefore \quad v_B = v_A + v_{B/A} = 5 + 5 = 10 \text{ m/sec.}$$

Similarly the tangential acceleration of B with respect to A is, $a_{B/A} = 1 \times 4 = 4 \text{ m/sec}^2$. B has also got radial inward acceleration of magnitude,

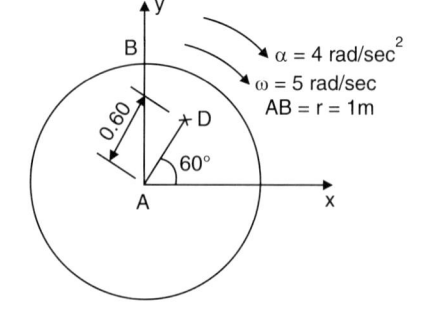

Fig. 22.4(a)

$$a_n = \frac{v_{B/A}^2}{r} = \frac{5^2}{1} = 25 \text{ m/sec}^2$$

Hence, the acceleration of B is given by the three vectors shown in Fig. 22.4(b).

GENERAL PLANE MOTION OF RIGID BODIES

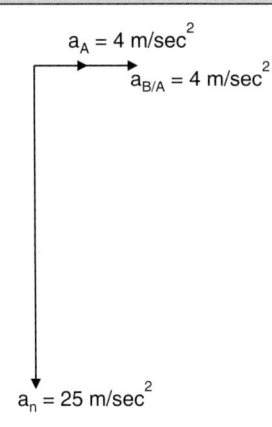

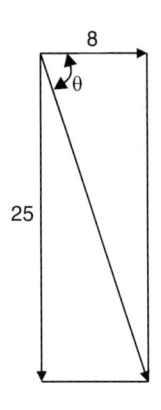

Fig. 22.4(b)

$$\mathbf{a_B} = \sqrt{8^2 + 25^2} = \mathbf{26.249 \text{ m/sec}^2} \quad \textbf{Ans.}$$

and its inclination to the horizontal is given by
$$\tan \theta = 25/8$$
$$\therefore \quad \theta = \mathbf{72.26°} \quad \textbf{Ans.}$$

Now, consider the rotation of point D.

Due to rotation about A, D has a linear velocity of $v_{D/A} = r\omega = 0.6 \times 5 = 3$ m/sec tangential to AD i.e., at 60° to vertical

$$v_{Dx} = v_A + v_{D/A} \sin 60°$$
$$= 5 + 3 \sin 60° = 7.598 \text{ m/sec.}$$
$$v_{Dy} = v_{D/A} \cos 60°$$
$$= 3 \cos 60° = 1.5 \text{ m/sec} \downarrow$$
$$v_D = \mathbf{7.745 \text{ m/sec}} \quad \textbf{Ans.}$$

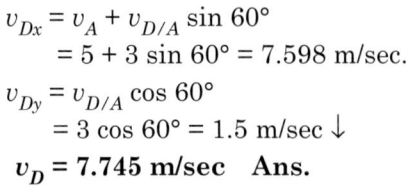

and its inclination to horizontal, θ, is given by $\tan \theta = \dfrac{1.5}{7.598}$

$$\therefore \quad \theta = \mathbf{11.17°}$$

Due to rotation about A, D has a tangential acceleration $= r\alpha = 0.6 \times 4 = 2.4$ m/sec² and radial inward acceleration

$$\frac{v_{D/A}^2}{r} = \frac{3^2}{0.6} = 15 \text{ m/sec}^2$$

Since $\quad v_{D/A} = 0.6 \times 5 = 3$ m/sec²

Hence total acceleration of D is vectorial sum of (i) acceleration due to translation $a_A = 4$ m/sec² (ii) tangential acceleration $r\alpha = 2.4$ m/sec² due to rotation and (iii) radial inward acceleration of 15 m/sec².

Let a_D be acceleration of D and its inclination to horizontal be θ as shown Fig. 22.4(f).

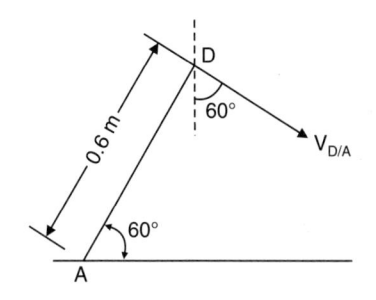

Fig. 22.4(c)

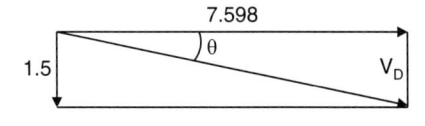

Fig. 22.4(d)

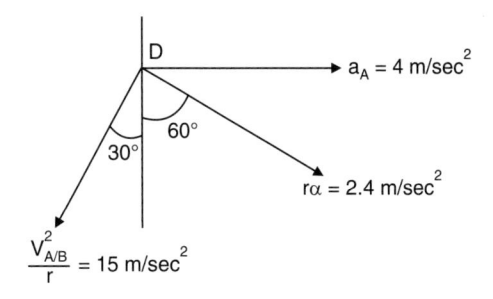

Fig 22.4(e)

Then $a_D \sin \theta = Y = 2.4 \cos 60° + 15.0 \cos 30°$
$= 14.190 \text{ m/sec}^2 \downarrow$

$a_D \cos \theta = 4 + 2.4 \sin 60° - 15 \cos 30°$

$= \overrightarrow{-1.422} \text{ m/sec}^2$

$= \overleftarrow{1.422} \text{ m/sec}^2$

$\boldsymbol{a_D} = \sqrt{14.190^2 + 1.422^2}$

$= \mathbf{14.261 \text{ m/sec}^2}$ **Ans.**

$\theta = \tan^{-1} \dfrac{14.190}{1.422}$

$= \mathbf{84.277}$ **Ans.**

Fig. 22.4(f)

Example 22.3. *A slender beam AB of length 3 m which remains always in the same vertical plane has its ends A and B constrained to remain in contact with a horizontal floor and a vertical wall respectively as shown in Fig. 22.5(a). Determine the velocity and acceleration of the end B at the position shown in the figure, if the point A has a velocity of 2 m/sec and an acceleration of 1.6 m/sec² towards left.*

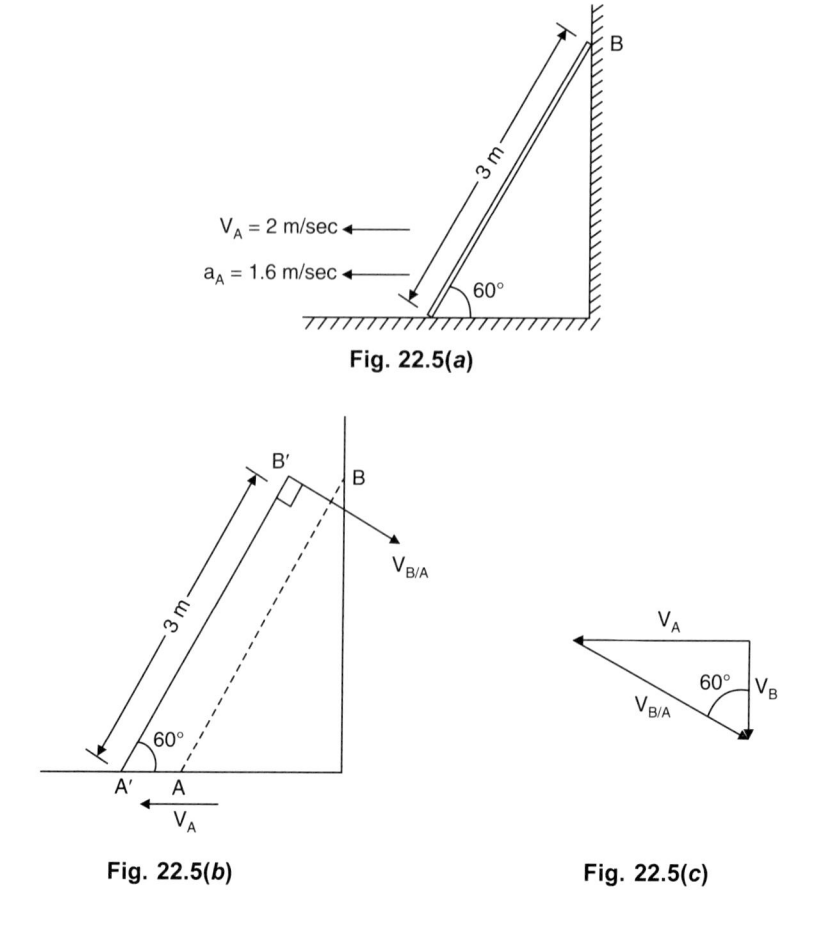

Fig. 22.5(a)

Fig. 22.5(b)

Fig. 22.5(c)

GENERAL PLANE MOTION OF RIGID BODIES

Solution. The plane motion of B can be split into translation of A and rotation about A as shown in Fig. 22.5(b). Due to rotation about A, B has the relative velocity $v_{B/A} = 3\omega$ where ω is angular velocity of B about A. Since the actual velocity of B is vertically downwards the vector diagram for the velocity of B is as shown in Fig. 22.5(c). From vector diagram

$$v_B = v_A \cot 60° = 2 \cot 60°$$
$$= 1.555 \text{ m/sec} \quad \text{Ans.}$$

$$v_{B/A} = \frac{v_A}{\sin 60°} = \frac{2}{\sin 60°} = 2.309 \text{ m/sec}$$

Acceleration of point B: There are three components of acceleration as shown in Fig. 22.5(d).

(i) Due to translation of $A = a_A = 1.6$ m/sec^2

(ii) Due to rotation about A,

tangential acceleration $= r\alpha = 3\alpha$ where α is angular acceleration, and

(iii) Radial inward acceleration $= \dfrac{v_{B/A}^2}{r} = \dfrac{2.3094^2}{3} = 1.778$ m/sec^2

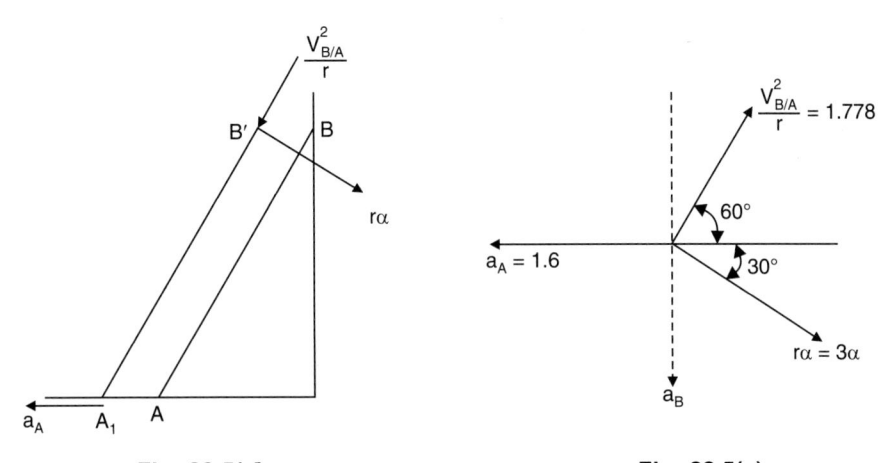

Fig. 22.5(d)   **Fig. 22.5(e)**

The net resultant acceleration of B, a_b is vertically downwards as shown in vector diagram 22.5(e) Since the resultant acceleration a_B is not having any component in the horizontal direction, considering horizontal forces, we get,

$$1.6 + 1.778 \cos 60° - 3\alpha \cos 30° = 0 \quad \text{or} \quad \alpha = 0.958 \text{ rad/sec}^2.$$

Considering vertical components, we get,

$$a_B = 1.778 \sin 60° + 3\alpha \sin 30°$$

i.e., $$a_B = 2.9766 \text{ m/sec}^2 \quad \text{Ans.}$$

Example 22.4. *Length of crank AB is 100 mm and that of connecting rod is 250 mm as shown in Fig. 22.6(a). If the crank is rotating in clockwise direction at 1500 rpm, for the position shown in Fig. 22.6(a), determine the angular velocity of the connecting rod and the velocity of the piston.*

478   FUNDAMENTALS OF ENGINEERING MECHANICS

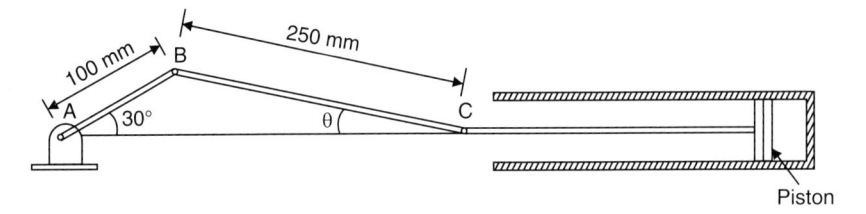

Fig. 22.6(a)

Solution. Angular velocity of the crank

$$\omega = 1500 \text{ rpm} = \frac{1500 \times 2\pi}{60} = 157.08 \text{ rad/sec}$$

∴ Tangential velocity of end B (Fig. 22.6(b)) is given by

$$v_B = r\omega = 0.100 \times 157.080$$
$$= 15.708 \text{ m/sec}$$

Now consider the motion of connecting rod BC. Inclination of BC to horizontal 'θ' is given by

$$\frac{100}{\sin \theta°} = \frac{250}{\sin 30°}$$

$$\sin \theta = \frac{100 \times \sin 30°}{250}$$

or $\theta = 11.537°.$

Fig. 22.6(b)

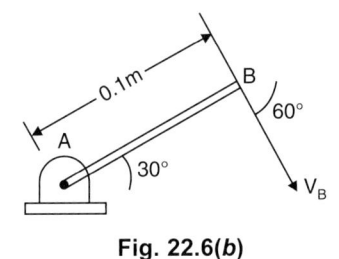

The motion of BC can be split into translation of B and rotation of BC about point B as shown in Fig. 22.6(c).

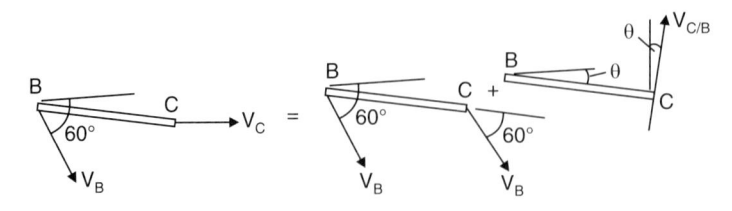

Fig. 22.6(c)

Let ω' be the angular velocity of BC. Tangential velocity of point C w.r.t. B, due to rotation, $v_{C/B}$ is given by

$$v_{C/B} = r\omega' = 0.25 \, \omega'$$

The resultant velocity of point C is v_C and is horizontal. Considering the vertical component of velocities, we get

$$0 = v_B \sin 60° + v_{C/B} \cos \theta = 15.708 \sin 60° + 0.25\omega' \cos 11.537°$$
$$= 13.6035 + 0.244\omega'$$

i.e., $\omega' = \mathbf{55.547 \text{ rad/sec}}$ **Ans.**

Considering the horizontal component of velocities we get,

$$v_C = v_B \cos 60° + v_{C/B} \sin \theta$$
$$= 15.7080 \cos 60° + 0.25 \times 55.547 \times \sin 11.5378°$$

i.e., $v_C = \mathbf{10.631 \text{ m/sec}}$ **Ans.**

GENERAL PLANE MOTION OF RIGID BODIES

22.3. INSTANTANEOUS AXIS OF ROTATION

In the previous article it is shown that plane motion of a point (v_B) can be split into translation of another point (v_A) and rotation of about that point ($r\omega = v_{B/A}$). At any instant it is possible to locate a point in the plane which has zero velocity and hence plane motion of other points may be looked upon as pure rotation about this point. Such point is called **Instantaneous Centre** and the axis passing through this point and at right angles to the plane of motion is called **Instantaneous Axis of Rotation**.

Consider the rigid body shown in Fig. 22.7 which has plane motion. Let B be a point having velocity v_B at the instant considered. Now locate a point C on perpendicular to the direction v_B at B at a distance r_B from the axis of rotation. Now plane motion of the B can be split into translation of C and rotation about C.

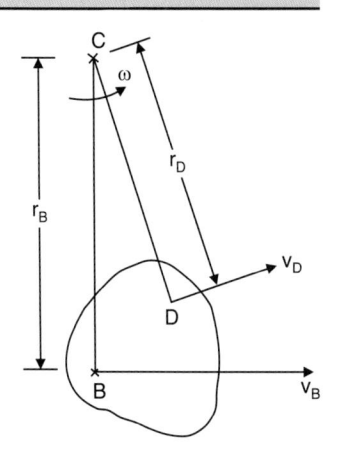

Fig. 22.7

$\therefore \quad v_B = v_C + r_B\,\omega$ and the direction of the velocity is at right angles to CB. If we write $r_B = v_B/\omega$, then from the above relation we get

$$v_B = v_C + \frac{v_B}{\omega}\cdot\omega \quad i.e., \quad v_C = 0.$$

Thus if point C is selected at a distance v_B/ω along the perpendicular to the direction of velocity at B, the plane motion of B reduces to pure rotation about C.

Hence C is instantaneous centre.

If D is any other point on the rigid body, its velocity will be given by

$$v_D = v_C + r_D\omega$$
$$= r_D\omega \quad \text{since } v_C = 0,$$

and   it will be at right angles to CD.

Thus if the instantaneous centre is located, motion of all other points at that instant can be found by pure rotation about C.

Methods of Locating Instantaneous Centre

Instantaneous centre can be located by any one of the following two methods:

(i) if the angular velocity ω and linear velocity v_B are known, instantaneous centre is located at a distance v_B/ω along the perpendicular to the direction of v_B at B.

(ii) If the linear velocities of two points of the rigid body are known, say v_B and v_D, drop perpendiculars to them at B and D. The intersection point is the instantaneous centre.

Use of Instantaneous Centre

If instantaneous centre is located, the velocities of all other points on the rigid body at that instant can be determined by treating the plane motion as pure rotation about the instantaneous centre. It may be noted that point C located as instantaneous centre, has zero velocity only at that instant. If it has zero velocity at time t, at time $t + dt$ it has some velocity. It means, the instantaneous centre is not having zero acceleration. Hence for the acceleration calculations we can not treat the plane motion as pure rotation about instantaneous centre. The plane motion can be treated as a case of pure rotation about instantaneous centre only for velocity calculations.

Example 22.5. *Determine the velocities of the points B and D given in example 22.2 by instantaneous centre method. Given*

Angular velocity $\omega = 5$ rad/sec.
Wheel radius = 1 m.

Solution.

∴ Velocity of geometric Centre A, $v_A = r\omega$ = 1 × 5 = 5 m/sec and it is horizontal in the rightward direction. Hence instantaneous centre is in the vertically downwards direction at a distance

$= v_A/\omega = 5/5 = 1$m, *i.e.,* at point C,

which is in contact with floor (see Fig. 22.8).

∴ $v_B = CB \times \omega = 2 \times 5$
$= 10$ m/sec.

$v_D = CD \times \omega$

Now $CP = CA + AP$ where P is foot of $\perp r$ of D on CB.

∴ $CP = 1 + 0.6 \times \sin 60° = 1.520$ m
$PD = 0.6 \times \cos 60° = 0.3$ m

∴ $CD = \sqrt{CP^2 + PD^2} = \sqrt{1.520^2 + 0.3^2} = 1.549$ m

∴ $v_D = CD \times \omega = 1.549 \times 5.0$

i.e., $v_D = 7.745$ **m/sec Ans.**

Its direction is at right angle to CD. Its inclination to horizontal, θ is given by

$$\theta = \angle PCD = \tan^{-1}\frac{0.3}{1.520}$$

i.e., $\theta = 11.17°$ **Ans.**

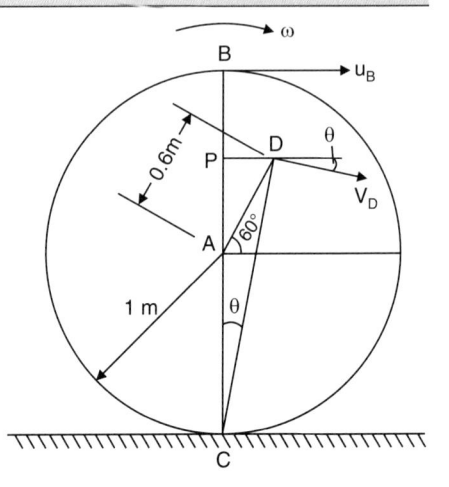

Fig. 22.8

Example 22.6. *Find the velocity of B in example 22.3 by instantaneous centre method.*

Solution. Since velocity at A (v_A) is horizontal and velocity at B (v_B) is vertical, instantaneous centre is point C, which is obtained by dropping perpendicular to the directions v_A and v_B at points A and B respectively. Now,

$v_A = AC \times \omega$

i.e., $2 = 3 \sin 60° \times \omega$
$\omega = 0.770$ rad/sec

and $v_B = BC \times \omega = 3 \cos 60° \times 0.770$

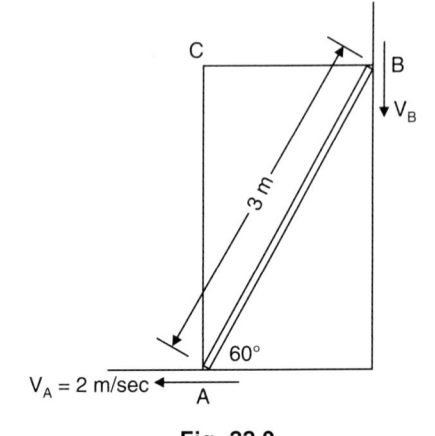

i.e., $v_B = 1.555$ **m/sec Ans.**

Fig. 22.9

22.4. KINETICS OF ROLLING BODIES

It is already known that the plane motion of a rolling body can be split into linear motion and rotation about the geometric axis. Hence the system of forces acting on the body are equal to an effective force $\dfrac{W}{g} a$ through the geometric centre in the direction of the resultant forces

GENERAL PLANE MOTION OF RIGID BODIES

(the direction of motion) and a rotational moment $I\alpha$ as shown in Fig. 22.10(a). The reverse effective forces may be applied on the body along with the system of forces acting on it (see Fig. 22.10(b) and then the system may be treated as in equilibrium. This is the *D'Alembert's* principle for rolling bodies.

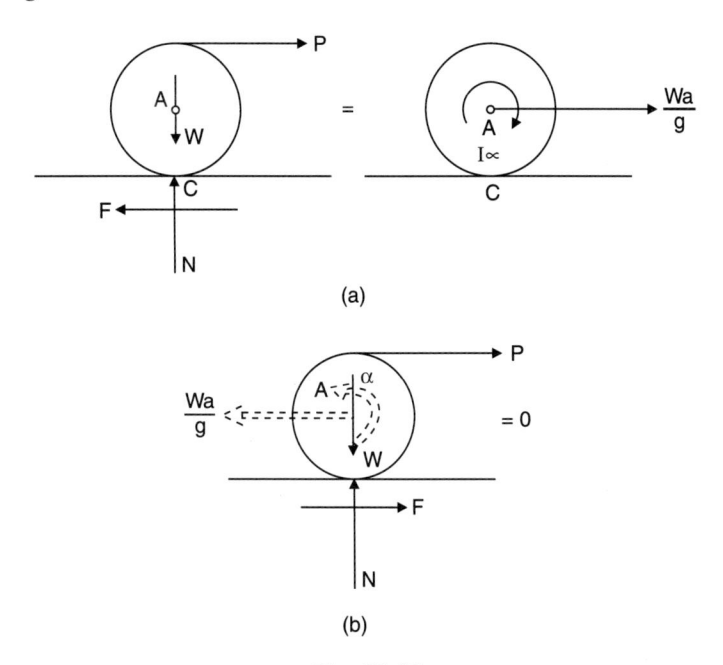

Fig. 22.10

Note: The relation $a_A = r\alpha$, holds good only for the bodies rolling without slipping. If slipping takes place, this relation is not applicable.

Example 22.7. *A solid cylinder weighing 1200 N is acted upon by a force P horizontally as shown in Fig. 22.11(a). Determine the maximum value of P for which there will be rolling without slipping. If P = 1000 N, determine the acceleration of the mass centre and the angular acceleration, given that the coefficient of static friction $\mu_S = 0.2$ and the coefficient of kinetic friction $\mu_K = 0.15$.*

Solution. Let the linear acceleration of geometric centre be a_A and the angular acceleration be α. The free body diagram of the cylinder along with reverse effective forces $\dfrac{W}{g} a$ and $I\alpha$ is shown in Fig. 22.11(b).

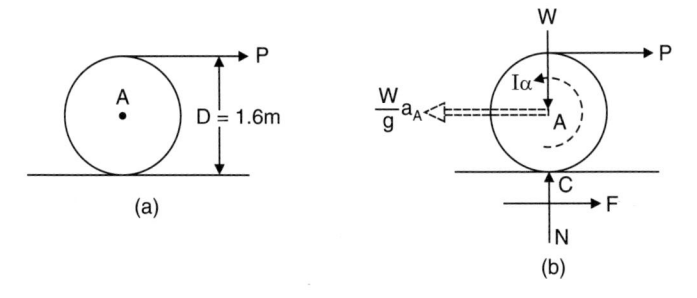

Fig. 22.11

(*i*) Maximum Value of P for Rolling Without Slipping:

Since there is no slipping.
$$a_A = r\alpha = 0.8\alpha \qquad \ldots(i)$$

$$I = \frac{1}{2}\frac{W}{g}r^2 = \frac{1}{2} \times \frac{1200}{9.81} \times 0.8^2 = 39.144$$

$$\Sigma F_v = 0 \quad \rightarrow \quad N = W = 1200 \text{ N}$$

From the law of friction,
$$F = \mu_S N = 0.2 \times 1200$$
$$= 240 \text{ N}$$

$$\Sigma F_H = 0 \quad \rightarrow$$

$$P - \frac{w}{g}a_A + F = 0$$

$$P - \frac{1200}{9.81} \times 0.8\alpha + 240 = 0$$

$$97.859\alpha - P = 240 \qquad \ldots(1)$$

Writing moment equilibrium equation about C, we get

$$P \times 1.6 - \frac{w}{g}a_A \times 0.8 - I\alpha = 0$$

$$P \times 1.6 - \frac{1200}{9.81} \times 0.8\alpha \times 0.8 - 39.144\alpha = 0$$

i.e. $\qquad 1.6 \times P = 117.431\alpha \quad \text{or} \quad P = 73.394\,\alpha \qquad \ldots(2)$

From (1) and (2), we get
$$97.859\alpha - 73.394\alpha = 240$$
$$\alpha = 9.81 \text{ rad/sec}$$
$$\therefore \qquad P = 73.394 \times 9.81$$
$$\boldsymbol{P = 720 \text{ N} \quad \textbf{Ans.}}$$

(*ii*) When $P = 1000$ N;

In this case as slippage occurs the relationship, $a_A = r\alpha$ does not hold good. In this case we have the relationship
$$F = \mu_K N = 0.15 \times 1200 = 180 \text{ N.}$$

Taking moment about geometric centre A, we get
$$P \times 0.8 - F \times r - I\alpha = 0$$
$$1000 \times 0.8 - 180 \times 0.8 - 39.144\,\alpha = 0$$
$$\boldsymbol{\alpha = 16.759 \text{ rad/sec}^2 \quad \textbf{Ans.}}$$

$\Sigma F_H = 0$ gives
$$\frac{1200}{9.81}a_A = 1000 + 180$$
$$\boldsymbol{a_A = 9.647 \text{ rad/sec}^2 \quad \textbf{Ans.}}$$

GENERAL PLANE MOTION OF RIGID BODIES

Example 22.8. *A solid cylinder of weight W and radius r rolls, down an inclined plane which makes $\theta°$ with the horizontal axis. Determine the minimum coefficient of friction and the acceleration of the mass centre for rolling, without slipping.*

Solution. The free body diagram of the cylinder along with reverse effective forces $\dfrac{W}{g} a_A$ and $I\alpha$ is shown in Fig. 22.12, where a_A is the acceleration of the mass centre and α is the angular acceleration.

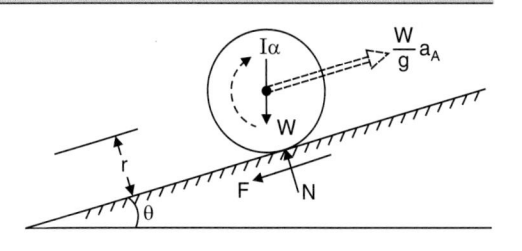

Fig. 22.12

Now,
$$I = \frac{1}{2} \frac{W}{g} r^2$$

$$a_A = r\alpha$$

Taking moment about the point of contact of cylinder with the plane, we get

$$W \sin \theta \times r - \frac{W}{g} a_A r - I\alpha = 0$$

$$W \sin \theta \times r = \frac{W}{g} \cdot r\alpha \, r + \frac{1}{2} \frac{W}{g} r^2 \alpha = \frac{3}{2} \frac{W}{g} r^2 \alpha$$

∴
$$\alpha = \frac{2}{3} \frac{g}{r} \sin \theta$$

∴
$$a_A = r\alpha = \frac{2}{3} g \sin \theta \quad \textbf{Ans.}$$

Now, Σ Forces normal to the plane = 0, gives

$$N = W \cos \theta$$

∴
$$F = \mu N = \mu W \cos \theta$$

Σ Forces parallel to the plane = 0, gives

$$F + W \sin \theta - \frac{W}{g} a_A = 0$$

$$F = -W \sin \theta + \frac{W}{g} \times \frac{3}{2} g \sin \theta = -\frac{1}{3} W \sin \theta$$

Hence, direction of F is to be reversed.

$$F = \frac{1}{3} W \sin \theta$$

i.e.,
$$\mu W \cos \theta = \frac{1}{3} W \sin \theta$$

$$\mu = \frac{1}{3} \tan \theta \quad \textbf{Ans.}$$

22.5. KINETICS OF GENERAL PLANE MOTION

Consider the body shown in Fig. 22.13 which has got plane motion. The plane motion of the body may be split into linear motion and rotation about its mass centre.

Then the effective force on the body is a force $\dfrac{W}{g} a$ at mass centre and a moment $I\alpha$ as shown in Fig. 22.13(b).

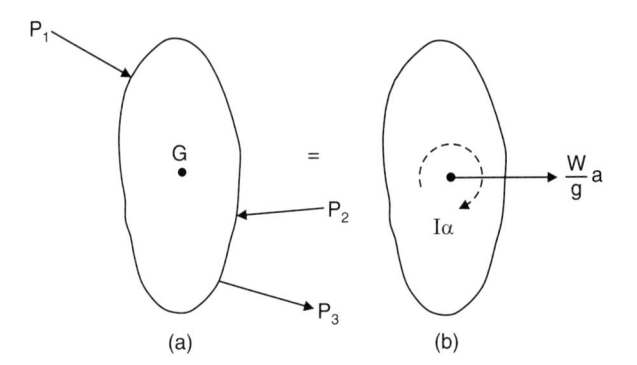

Fig. 22.13

Hence by equating the component of forces acting on the body to the respective components of effective force, $\dfrac{W}{g} a$, and equating moment of the force about any point to that of effective moment about the same point, kinetic equations may be formed and solved. Another method is one can use D' Alembert's principle, *i.e.* apply reverse effective force to a given force system and solve the equation of equilibrium. The system of force to be considered in this approach are shown in Fig. 22.14.

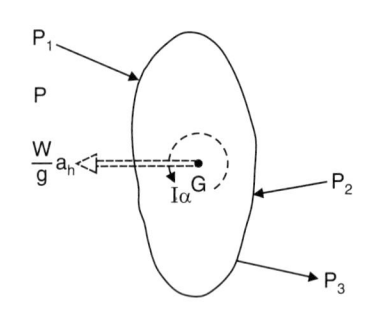

Fig. 22.14

Example 22.9. *A uniform bar 1.6 m long and weighing 300 N is held in position with its ends in contact with smooth surfaces as shown in Fig. 22.15(a). If the bar is released with no velocity, determine the angular acceleration of the rod and the reactions at ends, just after release.*

Solution. Let α and a be the angular acceleration and linear acceleration of centre of gravity of the rod respectively. The direction of accelerations a_A and a_B of points A and B are known (Fig. 22.15(b)). The relative velocity of B with respect to A is 1.6α and is directed at right angles to the position of rod. Hence, from vector diagram of the acceleration shown in Fig. 22.15(c), we get

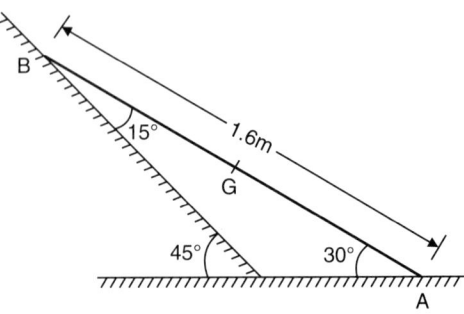

Fig. 22.15(a)

$$\frac{a_A}{\sin 75°} = \frac{a_B}{\sin 60°} = \frac{1.6\alpha}{\sin 45°}$$

and
$$a_A = 2.1856\alpha$$
$$a_B = 1.9596\alpha$$

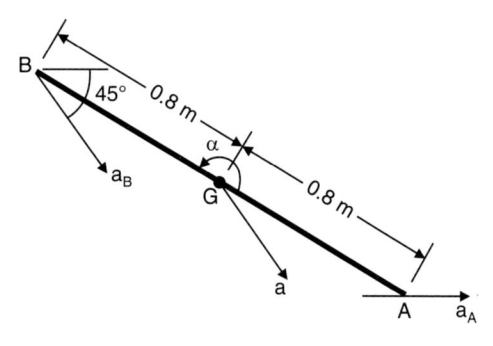

Fig. 22.15(b)

Fig. 22.15(c)

Now, the relative acceleration of G with respect to A is directed at right angles to the rod and is equal to
$$a_{G/A} = 0.8\,\alpha.$$
From the vector diagram shown in Fig. 22.15(d) for acceleration of G, we get
$$a_{Gx} = a_A - a_{G/A}\cos 60°$$
$$= 2.1856\alpha - 0.8\alpha \cos 60°$$
$$= 1.7856\alpha$$
$$a_{Gy} = a_{G/A}\sin 60°$$
$$= 0.8\alpha \sin 60° = 0.6928\alpha$$

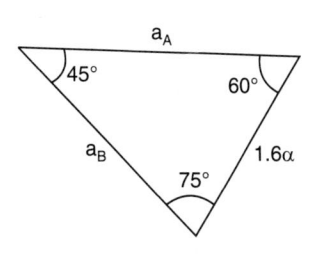

Fig. 22.15(d)

$$\therefore \quad \tan\beta = \frac{0.6928}{1.7856}$$
$$\beta = 21.21°$$

and
$$a_G = \sqrt{1.7856^2 + 0.6928^2} = 1.915\alpha$$

Now consider the kinetics of the rod.

Let E be the point of interaction of the reaction N_A and N_B. Then the required distances for taking moment about E can be easily worked out. Given system of forces (Fig. 22.15(e)) are the same as the effective force system shown in Fig. 22.15(f).

Now,
$$I = \frac{1}{12}\frac{W}{g}l^2 = \frac{1}{12} \times \frac{300}{9.81} \times 1.6^2$$
$$= 6.5240$$

Equating the moment of the two systems about E, we get,
$$300 \times 0.6928 = \frac{300}{9.81} \times 1.7856\alpha + \frac{300}{9.81} \times 0.6928 + 6.5240\alpha$$
$$= 82.3160\alpha$$

or
$$\alpha = \mathbf{2.5249\ rad/sec^2}\ \mathbf{Ans.}$$

Fig. 22.15(e)   Fig. 22.15(f)

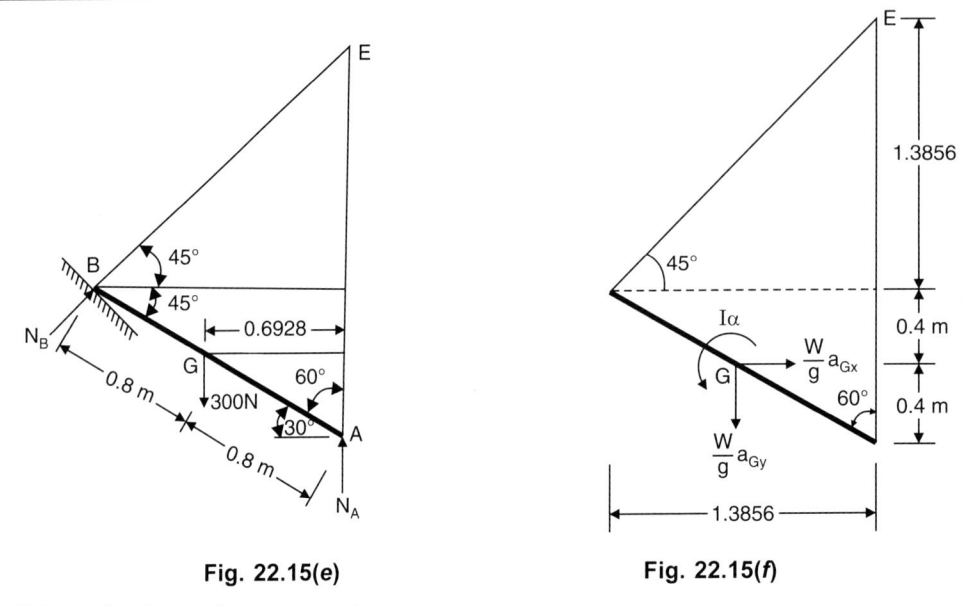

Taking horizontal components, we get,

$$N_B \cos 45° = \frac{w}{g} a_{Gx} = \frac{300}{9.81} \times 1.7856 \times 2.5249$$

$N_B = 194.98$ N   **Ans.**

Taking vertical components of the forces, we get

$$N_A + N_B \sin 45° - 300 = \frac{w}{g} a_{Gy}$$

$$N_A + 194.98 \sin 45° - 300 = \frac{300}{9.81} \times 0.6928 \times 2.5249$$

∴ $N_B = 215.62$ N   **Ans.**

Example 22.10. *A 5 N piston in connected to the 80 mm rotating crank by a 200 mm connecting rod. The connecting rod weighs 12 N. In this position shown in Fig. 22.16(a), if a force of 4000 N acts on the piston, and crank rotates at 1800 rpm, determine the reactions at the pins A and B.*

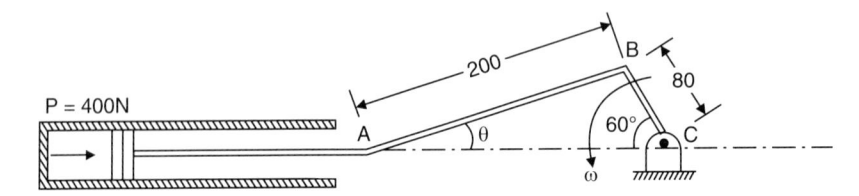

Fig. 22.16(a)

Solution. Let $\angle BAC = \theta°$. Then applying sine rule we get,

$$\frac{80}{\sin \theta} = \frac{200}{\sin 60°} \quad \text{or} \quad \theta = 22.2679°$$

GENERAL PLANE MOTION OF RIGID BODIES

Let ω be the angular velocity of the crank.

$$\omega = \frac{2\pi \times 1800}{60} = 188.4956 \text{ rad/sec}$$

∴ Velocity of B is $v_B = 0.08\omega$
$$= 15.0796 \text{ m/sec}$$

and is at right angles to BC.

Since BC is having pure rotation about C with uniform tangential velocity v_B, B has acceleration along BC equal to

$$a_B = \frac{v_B^2}{r} = \frac{15.0796^2}{0.8} = 2842.4292 \text{ m/sec}^2$$

Now consider connecting rod BA shown in Fig. 22.16(b).

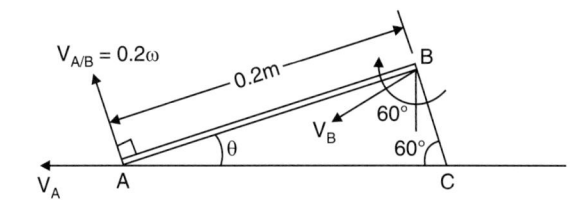

Fig. 22.16(b)

Fig. 22.16(c)

Let ω' be the angular velocity of BA about B. Then $v_{A/B} = 0.2\omega'$

$$v_A = v_B + v_{A/B}$$

From vector diagram shown in Fig. 22.16(c)

$$v_{A/B} \cos\theta = v_B \cos 60°$$
$$0.2\omega' \cos 22.2679° = 15.0796 \cos 60°$$
$$\omega' = 40.1873 \text{ rad/sec.}$$
$$v_{A/B} = 0.2\omega = 8.0375 \text{ m/sec}$$

and
$$v_A = v_B \sin 60° + v_{A/B} \sin\theta$$
$$= 15.8436 \text{ m/sec}$$

Consider the acceleration of AB. Let α be angular acceleration (Fig. 22.16(d)).

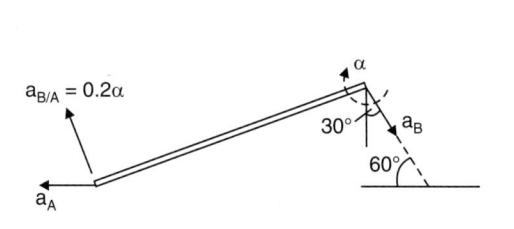

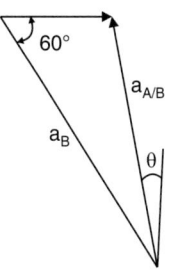

Fig. 22.16(d)

Fig. 22.16(e)

Now, $a_B = 2842.4292$ making 60° with horizontal.
$$a_{A/B} = 0.2\alpha$$

Now, $\vec{a_A} = \vec{a_B} + \vec{a_{A/B}}$

Looking at vector diagram shown in Fig. 22.16(e),
$$a_B \sin 60° = a_{A/B} \cos \theta$$
$$2842.4292 \sin 60° = 0.2\alpha \cos 20.2679°$$
$$\alpha = 13120.457 \text{ rad/sec}^2$$
∴ $a_{A/B} = 0.2\alpha = 2624.0913$
∴ $a_A = a_B \cos 60° - a_{A/B} \sin 20.2679° = 512.2027 \text{ m/sec}^2.$

Now the acceleration of the centre of gravity of bar AB can be obtained.

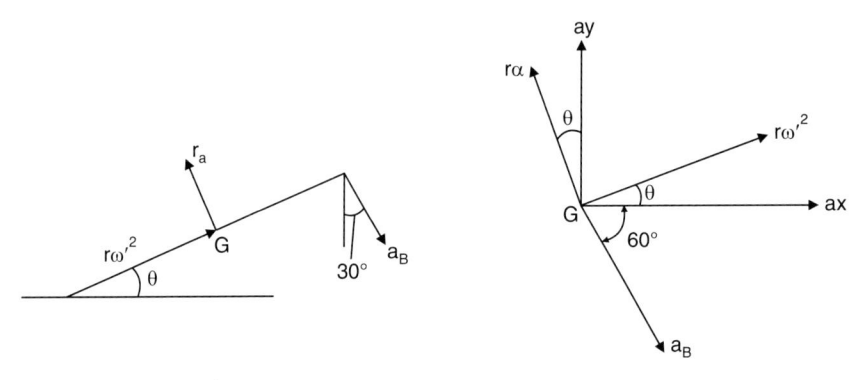

Fig. 22.16(f)   Fig. 22.16(g)

Various components of acceleration at G are shown in Figures 22.16(f) and 22.16(g). Let a_X be horizontal component and a_Y be vertical components of acceleration of C.G. of connecting rod AB.

Then, $a_X = a_B \cos 60° + r\omega'^2 \cos \theta - r\alpha \sin \theta$
$$= 1118.2109 \text{ m/sec}^2$$
$$a_Y = r\alpha \cos \theta + r\omega'^2 \sin \theta - a_B \sin 60°$$
$$= -1174.862 \text{ m/sec}^2$$

i.e., a_Y is downward 1174.862 m/sec².

Now consider the dynamic equilibrium of piston A (Fig. 22.16(h)).

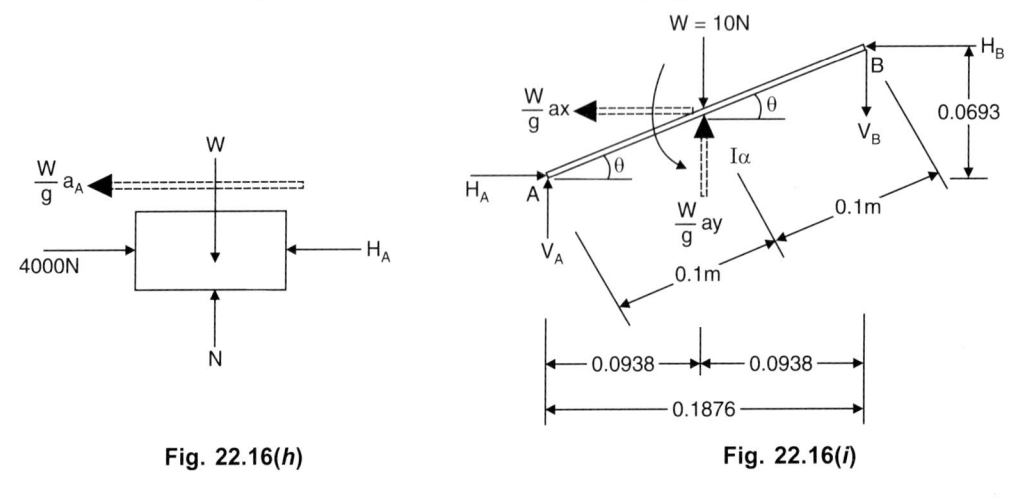

Fig. 22.16(h)   Fig. 22.16(i)

$\Sigma F_H = 0$, gives

$$H_A = 4000 - \frac{50}{9.81} \times 512.2027$$

i.e., $\quad H_A = 3738.38 \text{ kN}$ Ans.

Inertia forces are: $\dfrac{W}{g} a_x = \dfrac{10}{9.81} \times 1118.2109 = 1139.87\text{N}.$

and $\quad \dfrac{W}{g} a_Y = \dfrac{10}{9.81} \times 1174.862 = 1197.62 \text{ N}$

$$I\alpha = \frac{W}{g}\frac{l^2}{12}\alpha = \frac{10}{9.81} \times \frac{0.2^2}{12} \times 13120.457 = 44.8 \text{ N-m}$$

Taking moment about B, we get

$$v_A \times 0.1876 - H_A \times 0.0693 - 10 \times 0.0938 + \frac{W}{g} a_Y \times 0.0938$$
$$+ \frac{W}{g} a_x \times 0.03478 - I\alpha = 0$$

or $\quad v_A = 813.95 \text{ N}$ Ans.

$\Sigma F_V = 0$, gives

$$V_B = V_A + \frac{W}{g} a_Y - W$$

i.e., $\quad V_B = 2001.57 \text{ N} \downarrow$ Ans.

$$H_B = H_A - \frac{W}{g} a_x$$

i.e., $\quad H_B = 2598.51 \text{ N}$ Ans.

Important Formulae

When a wheel rotates without slip, the relationship between motion of its geometric centre and its angular motion are

$$s_A = r\theta$$
$$v_A = r\omega$$
$$a_A = r\alpha$$

PROBLEMS FOR EXERCISE

22.1 The wheel of a tractor is 1 m in diameter and the tractor is travelling at 9 kmph. What are the velocities at the top and bottom of the wheel relative to (i) a person seated in tractor; (ii) a person standing on the ground.

[**Ans.** (i) $v_{\text{Top}} = 2.5$ m/sec; $v_{\text{bottom}} = 2.5$ m/sec; (ii) $v_{\text{Top}} = 5.0$ m/sec; $v_{\text{bottom}} = 0$]

22.2 Rod *AB* shown in Fig. 22.17 is 2 m long and slides with its ends in contact with the floor and the inclined plane. At the instant when *AB* makes 40° with the horizontal, point *A* has left ward velocity of 5 m/sec. Determine the angular velocity of the rod and the velocity of end *B*.

[**Ans.** $\omega = 2.3040$ rad/sec; $v_B = 4.0760$ m/sec.]

22.3 In problem 19.2, if *A* has acceleration of 3m/sec² to the left at the instant as shown in Fig. 22.17, determine angular acceleration of the rod and resultant acceleration of end *B*.

[**Ans.** $\alpha = 3.3145$ rad/sec²; $a_B = 13.7438$ m/sec²]

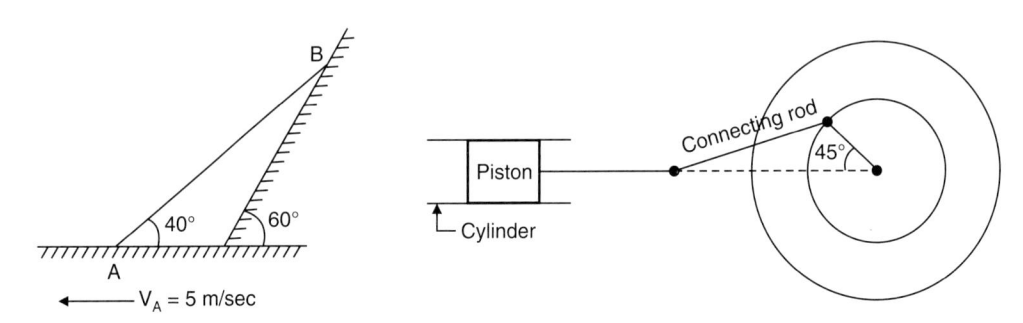

Fig. 22.17

Fig. 22.18

22.4 Figure 22.18 shows a reciprocating pump driven by a driving wheel. If crank is 80 mm long and connecting rod 200 mm, determine the velocity of the piston in the position shown. The driving wheel rotates at 2000 rpm in anticlock direction. [**Ans.** 15.34 m/sec]

22.5 Solve problem 22.2 using instantaneous centre method.

22.6 A solid sphere of weight *W* and radius *r* rolls down an inclined plane, which makes angle θ to horizontal. Determine the minimum frictional resistance required to prevent slippage and also the corresponding linear acceleration of mass centre. [**Ans.** μ = 1/3 tan θ]

22.7 A solid cylinder of weight 1000 N and radius 1.2 m is to be rolled up an inclined plane which makes 30° with horizontal. Determine the force *P* to be applied on the periphery in the direction parallel to the plane (see Fig. 22.19), such that it has maximum acceleration but no slippage. Take coefficient of friction of 0.2. [**Ans.** 519.61 N]

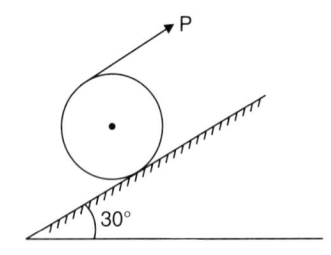

Fig. 22.19

Index

A

Acceleration 3, 296
Acceleration due to gravity 304
Acceleration-time curve 301
Angle of friction 118
Angle of projection 323
Angle of repose 119
Angular motion 447
Applied forces 25

B

Banking of roads 437
Both ends hinged 66

C

Cantilever 65
Central impact 417
Centre of gravity 226, 266
Centre of gravity from first principles 267
Centre of gravity of composite bodies 270
Centroid 227
Centroid of a circle 230
Centroid of a parabolic spandrel 231
Centroid of a semicircle 230
Centroid of a triangle 229
Centroid of composite sections 233
Characteristics of a force 10
Circular motion 434
Classical mechanics 1
Coefficient of friction 117
Coefficient of restitution 417
Composition of two force system 17
Concentrated loads 68
Cone of friction 119
Conservation of momentum 406
Continuous beam 67
Continuum 3
Couple 47

D

D' Alembert's principle 360
Designed speed 437
Differential screw jack 166
Dimensional homogenity 12
Direct central impact 420
Direct impact 417
Displacement 3, 295
Displacement-time curve 300
Distance 295
Double purchase winch crab 168
Dynamic friction 117
Dynamics 2

E

Eccentric impact 417
Efficiency 144
Effort 144
Energy 377
Engineering mechanics 1
Equilibrium of bodies 25
Equilibrium of connected bodies 33
Equilibrium of non-concurrent system of forces 56
External moment 69

F

First order pulley system 153
Fixed support 65
Fluids 2
Force of jet on a vane 404
Free body diagram 26
Frictional force 117

G

General loading 68
General method of composition of forces 21
General plane motion 299
General plane motion of rigid bodies 473

H

Hinged or pinned support 65
Horizontal range 323

I

Ideal effort 145
Ideal load 145
Ideal machine 145
Idealisations in mechanics 14
Impact of elastic bodies 417
Impulse momentum 395
Impulse momentum method 395
Inclined plane 158
Inclined projection on level ground 326
Input 144
Instantaneous axis of rotation 479
Instantaneous centre 479

K

Kinematics 2
Kinematics of general plane motion 474
Kinetic energy of rotating bodies 453
Kinetics 2
Kinetics of general plane motion 484
Kinetics of rolling bodies 480

L

Lami's theorem 27
Law of machine 147
Law of transmissibility of force 6
Laws of friction 118
Laws of mechanics 4
Length 3
Lifting machine 144
Limiting friction 117
Line of impact 417
Load 144

M

Mass 2
Mass moment of inertia 276
Mechanical advantage 144
Mechanics 1
Method of joint 86
Method of section 97

Moment of a force 44
Moment of inertia 240
Moment of inertia from first principles 243
Moment of inertia of standard sections 245
Moment of inertia of composite sections 249
Momentum 3
Motion 295
Motion curves 300
Motion of body projected horizontally 323
Motion with uniform acceleration 301
Motion with uniform velocity 301
Motion with varying acceleration 315

N

Newton's first law 297
Newton's second law 297
Newton's third law 297
Newton's law of gravitation 297
Newtonian mechanics 1
Non-applied forces 25

O

Oblique impact 417, 424
One end hinged and the other on roller 66
Output 144
Overhanging beam 66

P

Parallel axis theorem 242
Parallel axis theorem/transfer formula 285
Parallelogram law of forces 7
Particle 4
Perfect, deficient and redundant frames 83
Perfect frame 83
Perpendicular axis theorem 242
Pile and pile hammer 410
Pin jointed frame 83
Plane frame 83
Polar moment of inertia 241
Polygon of law of forces 8
Power 378
Projectiles 323
Projection on inclined plane 340
Propped cantilever 67
Pulleys 152

INDEX

Q
Quantum mechanics 1

R
Radius of gyration 241, 277
Reactions 25
Relative velocity 349, 347
Relativistic mechanics 1
Resolution of forces 19
Resultant of non-concurrent force system 49
Resultant velocity 349
Retardation 296
Reversibility of a machine 151
Rigid bodies 1
Rigid body 3
Roller support 64
Rolling friction 117
Rope friction 135
Rotation 298
Rotation of rigid bodies 447

S
Screw jack 163
Second order pulley system 154
Self weight 25
Simple support 64
Simply supported beam 65
Single purchase winch crab 168
Sliding friction 117
Space 3
Space frame 83
Speed 296
Static friction 117
Statics 1
Super elevation of rails 437
System of forces 11

T
Theorems of moment of inertia 242
Theorems of Pappus-Guldinus 273
Third order pulley system 154
Time 2
Time of flight 323
Triangle law of forces 7
Translation 298
Trajectory 323
Types of supports 64

U
Uniformly distributed load (UDL) 68
Uniformly varying load 68
Unit of constant of gravitation 9
Unit of forces 8
Units 8

V
Varignon's theorem 45
Vectors 12
Velocity 3, 263
Velocity of projection 323
Velocity ratio 144
Velocity-time curve 300
Virtual work method 199

W
Wedges 129
Weston differential pulley block 157
Wheel and axle 155
Wheel and differential axle 156
Work 376
Work done by a spring 388
Work energy equation 379
Work energy method 376

X
x and y intercepts of resultant 50